Advances in the Science and Engineering of

CASTING SOLIDIFICATION

An MPMD Symposium Honoring Doru Michael Stefanescu

TMS2015

144th Annual Meeting & Exhibition

March 15-19, 2015 • Walt Disney World • Orlando, Florida, USA

New proceedings volumes from the TMS2015 Annual Meeting, available from publisher John Wiley & Sons:

- 6th International Symposium on High-Temperature Metallurgical Processing
- Advanced Composites for Aerospace, Marine, and Land Applications II
- Advances in the Science and Engineering of Casting Solidification
- Characterization of Minerals, Metals, and Materials 2015
- Drying, Roasting, and Calcining of Minerals
- Energy Technology 2015
- EPD Congress 2015
- Friction Stir Welding and Processing VIII
- Light Metals 2015
- Magnesium Technology 2015
- Rare Metal Technology 2015
- TMS 2015 Supplemental Proceedings

Advances in the Science and Engineering of
CASTING SOLIDIFICATION
An MPMD Symposium Honoring Doru Michael Stefanescu

Proceedings of a symposium sponsored by
the Materials Processing & Manufacturing Division of
The Minerals, Metals & Materials Society (TMS)
held during

TMS2015
144th Annual Meeting & Exhibition
March 15-19, 2015
Walt Disney World • Orlando, Florida, USA

EDITORS
Laurentiu Nastac, Baicheng Liu, Hasse Fredriksson,
Jacques Lacaze, Chun-Pyo Hong, Adrian V. Catalina,
Andreas Buhrig-Polaczek, Charles Monroe, Adrian S. Sabau,
Roxana Elena Ligia Ruxanda, Alan Luo, Subhayu Sen, and
Attila Diószegi

Editors
Laurentiu Nastac
Baicheng Liu
Hasse Fredriksson
Jacques Lacaze
Chun-Pyo Hong
Adrian V. Catalina
Andreas Buhrig-Polaczek
Charles Monroe
Adrian S. Sabau
Roxana Elena Ligia Ruxanda
Alan Luo
Subhayu Sen
Attila Diószegi

ISBN 978-3-319-48605-5 ISBN 978-3-319-48117-3 (eBook)
DOI 10.1007/978-3-319-48117-3

Chemistry and Materials Science: Professional

Published by Springer International Publishers, Switzerland, 2016
Reprint of the original edition published by John Wiley & Sons, Inc., 2015, 978-1-119-08238-5

Printed on acid-free paper

This Springer imprint is published by Springer Nature
The registered company is Springer International Publishing AG
The registered company address is: Gewerbestrasse 11, 6330 Cham, Switzerland

TABLE OF CONTENTS
Advances in the Science and Engineering of Casting Solidification

Solidification Processing I

Solidification Processing II

Solidification Processing III

Solidification Processing IV

Microstructure Evolution I

Microstructure Evolution II

Cast Iron I

Cast Iron II

Novel Casting and Molding Processes

Solidification Processing V

PREFACE

This book contains the proceedings of the honorary symposium, "Advances in the Science and Engineering of Casting Solidification, held at the TMS 2015 Annual Meeting & Exhibition in Orlando, Florida, March 15-19, 2015. The symposium was held in honor of Professor Doru Michael Stefanescu, emeritus professor of The Ohio State University and The University of Alabama at Tuscaloosa, USA.

This book encompasses the following four areas:

(1) *Solidification Processing*: theoretical and experimental investigations of solidification processes including castings solidification, directional solidification of alloys, electromagnetic stirring, ultrasonic cavitation, mechanical vibration, active cooling and heating, powder bed-electron beam melting additive manufacturing, etc. for processing of metals, polymers and composite materials.
(2) *Microstructure Evolution*: theoretical and experimental studies related to microstructure evolution of materials including prediction of solidification-related defects and particle pushing/engulfment aspects.
(3) *Novel Casting and Molding Processes*: modeling and experimental aspects including high pressure die casting, permanent casting, centrifugal casting, low pressure casting, 3D silica sand mold printing, etc.
(4) *Cast Iron*: all aspects related to cast iron characterization, computational and analytical modeling, and processing.

The topics included in this volume were selected to reflect the broad research interests of Prof. Stefanescu. The participants in this symposium are leading scientists in the field of casting and solidification from around the world. A significant number of the participants were also either his students or colleagues.

This book advances the state-of-the art science and engineering of casting solidification and microstructure evolution of materials. Also, it introduces to the audience advanced characterization and multiscale modeling and simulation techniques as well as related novel processing methods.

Finally, this symposium has been organized to recognize the important contributions of Prof. Stefanescu to our community as a leading scholar in the casting and solidification field, a mentor, an advisor, and a great friend. We would like to thank Doru for being a role model for all of us.

On behalf of all symposium organizers and participants,

Prof. Laurentiu Nastac
The University of Alabama
Dept. of Metallurgical & Materials Engineering

ABOUT THE HONOREE

Professor Doru Michael Stefanescu's scientific career spans more than four decades. He graduated with a Dipl. Eng. degree in Metallurgical Engineering from the University Politehnica Bucharest, Romania in 1965 and obtained a Ph.D. in Physical Metallurgy from the same institution in 1973. In 1980 he served as a Visiting Professor at the University of Wisconsin, Madison and then joined The University of Alabama where he taught and did research until 2005 when he retired as Cudworth Professor of Engineering Emeritus. He continued his scholarly activity as Ashland Designated Research Professor at The Ohio State University until 2010. During his activity at these universities Professor Stefanescu has conducted 39 Master of Science theses and 17 Philosophy Doctor dissertations. He has authored or co-authored 393 publications.

The third edition of his book *Science and Engineering of Casting Solidification* will be published by Springer in 2015. He is a Doctor Honoris Causa of the Technical University of Cluj-Napoca, Romania (1998); the University Transylvania, Brasov, Romania (2001); and Jonkoping University, Sweden (2012).

Prof. Stefanescu's research interests include experimental and numerical aspects of solidification processing, influence of low-gravity on solidification, processing of metal-matrix composites, processing of ceramic superconductors, manufacturing technologies and physical metallurgy of cast iron, steel and nonferrous alloys, and cast metals technology. He is recognized as a world expert in cast iron.

Professor Stefanescu was the Key Foundry Educational Foundation faculty at The University of Alabama between 1982 and 2005, and then served in the same position at The Ohio State University between 2005 and 2010. On November 5, 1999 the Foundry Educational Foundation in cooperation with the American Foundry Society presented him with the AFS Director's Award to recognize his efforts as an educator.

The University of Alabama has recognized Dr. Stefanescu's achievements by honoring him with the three most prestigious awards that it confers to deserving faculty: the Burlington Northern Foundation Faculty Achievement Award (1988), the Burnum

Distinguished Faculty Award (1990), and the Blackmon-Moody Outstanding Professor Award (1997). He is also the recipient of the 2009 Lumley Research Award, College of Engineering, The Ohio State University.

Dr. Stefanescu honors and awards include Fellow of the ASM International (1997) *"For fundamental contributions to the science of solidification through the modeling of solidification processes and the microstructural evolution of castings"*, the Award for Scientific Merit of the American Foundry Society (2000) for *"research involving cast iron solidification modeling, inoculation and cooling curve analysis as well as work with aluminum alloys and cast metal matrix composites"*, the Joseph Seaman Gold Medal of the American Foundry Society (2011), Honorary Member of the Romanian Academy of Technical Sciences (2012), and the John Campbell Medal from the Institute of Cast Metals Engineers, United Kingdom (2012). He was also awarded four NASA Certificates of Recognition for the Creative Development of Technical Innovations.

EDITORS

Laurentiu Nastac

Laurentiu Nastac is currently an Associate Professor of Metallurgical and Materials Engineering at The University of Alabama, Metallurgical and Materials Engineering Department, Tuscaloosa, Alabama. He is also the Key Foundry Educational Foundation (FEF) Professor and the Director of the Solidification and Ultrasonic Laboratories and of the MTE foundry. Dr. Nastac received the Diploma Engineering degree in Metallurgy and Materials Science from the University Politehnica of Bucharest, Romania in 1985 and the M.S. and Ph.D. degrees in Metallurgical and Materials Engineering from The University of Alabama (Advisor, Prof. Doru M. Stefanescu), Tuscaloosa in 1993 and 1995, respectively. He has held various engineering, research, and academic positions in Romania (1985-1991) and the United States (1991-present). At Caterpillar (1994-1996), he conducted research in the area of macro transport and solidification-kinetics modeling and developed specialized casting simulation software tools. At Concurrent Technologies Corporation (CTC; 1996-2011) he conducted research primarily in the area of advanced metal-casting and solidification processes with emphasis on the modeling and simulation of casting phenomena. In 1999, in recognition of his work on solidification of Ti-based alloys and superalloy remelt ingots, he received the prestigious "Bunshah Best Paper Award" from the American Vacuum Society, Vacuum Metallurgy Division. More recently, he received the NMC (Navy Metalworking Center) achievement award and two CTC awards.

Dr. Nastac developed eight software tools, made over 150 presentations, co-authored three patents and over 195 publications in the materials science and manufacturing fields, and co-authored eight books, one of which is a monograph titled *Modeling and Simulation of Microstructure Evolution in Solidifying Alloys* published by Springer in 2004. He is a Key Reader for *Metallurgical and Materials Transactions*, a member of the Editorial Board of the *International Journal of Cast Metals Research* and of the ISRN Materials Science, and a member of the TMS Solidification Committee. He served in scientific committees and as an organizer for international conferences dedicated to casting and solidification processes and CFD modeling and simulation in materials processing.

Baicheng Liu
Baicheng Liu graduated with Gold Medal Award from Tsinghua University in 1955. As a visiting scholar, he studied at University of Wisconsin-Madison and Massachusetts Institute of Technology from 1978 to 1981. He is now a professor in the School of Materials Science and Engineering and School of Mechanical Engineering of Tsinghua University. He was elected as an academician of the Chinese Academy of Engineering in 1999. He won "Guanghua Engineering Science and Technology Prize" in 2002 and "Outstanding Contribution for Casting Industry" by the Chinese Mechanical Engineering Society in 2012.

His major research interests are multiscale modeling and simulation of solidification process of shape casting, physical metallurgy of cast alloys and strategy study of development of advanced manufacturing industry as well. He won a number of first and second class awards of scientific and technological achievements from Ministry of Education and Ministry of Machinery Industry and also two awards from the United States.
He has published more than 300 papers. He was invited to visit and give presentation to 30 more universities of different countries, and present more than 30 papers including several keynote speeches in different international conferences.

Hasse Fredriksson
Hasse Fredriksson has been a professor in Casting of Metals at KTH since 1975 and is responsible for the research and education in the field of casting and solidification of metals at KTH. He has been the supervisor to 40 doctoral theses and to more than 50 master's theses. Prof. Fredriksson is the author or co-author of more than 250 scientific publications in the field of solidification and casting and has organized and edited seven international conferences. He is coauthor of the following books: *Materials Processing during Casting, Physics of Functional Materials*, and *Solidification Processing of Metals*, all published by John Wiley & Sons.

Jacques Lacaze

After engineering studies in Nantes (west cost of France), Jacques Lacaze prepared a Ph.D. thesis (1979-1982) in Paris dealing with solidification of continuous cast aluminum alloys in relation with their hot tearing tendency. Dr. Lacaze then got a position as associate researcher at the Centre National de la Recherche Scientifique (CNRS) and moved to Nancy where he stayed until 1994. That year he moved to Toulouse where he matured as a senior scientist at CNRS.

His work in Nancy was mainly concerned with microscopic aspects of solidification of metallic alloys, including aluminum alloys, steels, and nickel-base superalloys. This research was carried out within a team, headed by Gérard Lesoult, and was concerned with multiscale studies of solidification, from microscopic to macroscopic aspects. Focus was mainly put on designing original experiments, but the possibility of resorting to modelling was always considered.

In 1985-1986, Dr. Lacaze took a one year post-doctoral position in the Royal Institute of Technology in Stockholm (KTH) where he started works on cast irons (with Prof. Hasse Fredriksson) and thermodynamic optimization and calculations (with Bo Sundman). Back to Nancy, he became involved in most of the research performed in the team, including several projects on solidification of cast irons. These latter projects were mainly dealing with nodular cast irons in relation with the tube manufacturer Pont-à-Mousson. In the early 1990s, he was asked by this manufacturer to extend his investigations to the solid-state eutectoid transformation of nodular cast irons.

Moving to Toulouse, he joined a so-called transversal team dealing with mechanics, microstructure, oxidation, and corrosion which now includes 14 permanent staff, 12 permanent researchers and professors, one technician, and one research engineer. Within this team, Dr. Lacaze has brought his knowledge about heterogeneous phase transformations, including solidification and solid-state transformation of metallic alloys. Most of the studies there deal with structural alloys that have to undergo severe service conditions, including mechanical loading, high temperatures, and aggressive environments (oxidation) for energy and transport.

Studying cast irons has remained his preferred research area, and half of the 50 papers he has co-authored in recent years are concerned with these alloys. These papers deal with solidification, solid-state transformation and mechanical properties of nodular irons, though most of the principles put forward would apply to other types of cast irons. His main present interest is on the role of impurities.

Chun-Pyo Hong
Chun-Pyo Hong graduated from Yonsei Univesity, Korea in 1972, received a M.S. degree in Materials Processing from Tohoku University, Japan in 1975, and received a Ph.D. degree in Materials Science and Engineering from University of Tokyo, Japan in 1985. From 1975 to 1985, he taught at Kyung Pook National University, Korea. Since 1986, he has taught at the Department of Materials Science and Metallurgy at Yonsei University, Seoul, Korea. During this period, he has published more than 150 journal papers, 50 conference papers, and published about 30 patents. He has also published several textbooks, including Co*mputer Modelling of Heat and Fluid Flow in Materials Processing* from IOP in 2004. One of his invented techniques, named "nano-cast" which is a new rheo-diecasting method, is being commercially used for the production automobile cast parts in Japan and Korea.

Adrian V. Catalina
Adrian V. Catalina received his bachelor's degree in Metallurgical Engineering from the University Politehnica of Bucharest (Romania) in 1985. He worked as a foundry engineer (ICM Resita), research engineer (ICSITPSCM Bucharest), and lecturer (University Politehnica of Bucharest) before joining the Graduate School at the The University of Alabama from which he obtained his Ph.D. degree in Metallurgical and Materials Engineering. In 1999, Dr. Catalina joined the Microgravity Materials Science program at USRA/NASA Marshall Space Flight Center in Huntsville, Alabama and since 2004 he is a Senior R&D Engineer with the Virtual Product Development Division at Caterpillar. Among Dr. Catalina's research interests are

modeling of microstructure evolution during solidification processes and solid-state transformations of metallic alloys, prediction of defects formation in cast alloys, processing–microstructure–properties relationships, and development of high strength lightweight alloys are the most prominent. Dr. Catalina has published more than 20 journal articles, 25 conference papers, and has four patent applications.

Andreas Buhrig-Polaczek

Anderas Buehrig-Polaczek is head of the Foundry-Institute at RWTH-Aachen University, Germany. He received his Diploma (1987) and Doctorate of Engineering (Dr.-Ing.; 1992) in metallurgical engineering from the RWTH Aachen University, Germany. From 1998 to 2002 he was full professor at the Foundry Institute, University of Leoben, Austria, and General Manager of the Austrian Foundry Research Institute, Leoben, Austria. Since 2002 he has also served as Chairman of the Supervisory Board, ACCESS e.V. Materials & Processes, Aachen, Germany.

Prof. Buehrig-Polaczek is working in the fields of metallurgy, alloy and process development, solidification, simulation of casting processes, bionic and biomaterials, MMC's and metallic foams. He has more than 150 publications, and he likes to work in interdisciplinary teams in order to bring different aspects together. In cooperation with universities, he builds a national network for the faculties in materials science and engineering. Knowledge transfer and a close collaboration with the industry is a further aspect of his research activities. Prof. Buehrig-Polaczek is also member in several national and international boards of foundry related organizations, companies and journals, such as *International Journal of Cast Metals Research*.

Charles Monroe

Charles Monroe is an Assistant Professor at the The University of Alabama at Birmingham (UAB). He received his B.S. in Mechanical Engineering from Pennsylvania State University in 2003 and his M.S. and Ph.D in Mechanical Engineering from University of Iowa in 2005 and 2008 respectively. Dr. Monroe grew up learning about steel castings from his father, Raymond, who actively supports research in this area through the trade association, Steel Founders' Society of America. After his education, Dr. Monroe spent several years working for Caterpillar running analysis of all types of metal castings and processes. Here he developed many interests in research including hot tearing, thin wall filling, microstructure property relationships, cost analysis of casting manufacture, and more. At UAB, Dr. Monroe is the Key Professor in the Foundry Education Foundation and teaches classes to undergraduates and graduate students focusing on metal casting processing.

Adrian S. Sabau

Adrian S. Sabau received his Diploma of Inginer in Mechanical Engineering/Materials Processing from the University of Craiova, Romania in 1992 and a Ph.D. degree in Mechanical Engineering from Southern Methodist University in 1996. In 1999, Dr. Sabau joined Oak Ridge National Laboratory as a Research Staff Member of the Materials Science and Technology, where he has been a Senior Research Staff Member since 2008. He is the recipient of two R&D 100 awards in process sciences. Dr. Sabau seeks to advance the materials processing, metal casting, photonic processing, and materials for energy applications through the development of solution algorithms, computational and experimental methodologies for the property measurement, process analysis, and materials behavior in response to conditions experienced in service, such as oxide exfoliation in steam boiler tubes and high-heat flux of tungsten-based materials for fusion applications. Dr. Sabau has published more than 50 journal papers, 73 conference papers, and has three patent applications.

Roxana Elena Ligia Ruxanda

Roxana Elena Ligia Ruxanda received her Diploma Engineer in Casting and Solidification and Ph.D. degree in Thermal and Thermochemical Treatment of Alloys from the University Politehnica of Bucharest, Romania. In 1998, Dr. Ruxanda joined Dr. Stefanescu's group at the The University of Alabama as a post-doc NSF-NATO grant recipient and later as Research Assistant. In 2003, Dr. Ruxanda joined Emerson Climate Technologies in Sidney, Ohio, where currently occupies the position of Senior Lead Research Engineer in the Research Department. Dr. Ruxanda seeks to advance the materials characterization of metallic alloys, composite materials, polymers, solid residues, as well as the practical aspects of failure analysis and fractography, through metallography, image analysis, optical and electron microscopy, X-Ray fluorescence and diffraction, NDT ultrasonic, resonance acoustic methods, and X-Ray imaging. Dr. Ruxanda co-authored five chapters in ASM Handbook Volume 9 – Metallography and Microstructures, co-edited the proceedings of the Intern*ational Conference on the Science of Casting and Solidification*, published more than 70 refereed publications (journals and conference proceedings) and has two patent applications.

Alan Luo

Alan Luo is Professor of Materials Science and Engineering and Professor of Integrated Systems Engineering (Manufacturing) at The Ohio State University (OSU) in Columbus, Ohio, USA. Dr. Luo is also Director of OSU Light Metals and Manufacturing Research Laboratory (LMMRL). Prior to joining OSU in July 2013, Dr. Luo was a GM Technical Fellow at General Motors Global Research and Development Center (Warren, Michigan, USA) with 20 years of industrial experience.

Dr. Luo won two John M. Campbell Awards for his fundamental research, and three Charles L. McCuen Awards for research applications at GM. He has 17 patents and more than 180 technical publications in advanced materials, manufacturing and applications. Dr. Luo is an elected Fellow of ASM International and the Society for Automotive Engineers International (SAE). He received the TMS (The Minerals, Metals & Materials Society)

Brimacombe Medalist Award, SAE Forest R. McFarland Award in 2013, USCAR (United States Council for Automotive Research) Special Recognition Award in 2009, and ASM Materials Science Research Silver Medal in 2008. Dr. Luo's research is also recognized by several Best Paper awards from TMS, SAE, and AFS (American Foundry Society). He is the vice chair of TMS Light Metals Division and SAE Materials Engineering Activity.

Subhayu Sen

Subhayu Sen is currently serving as a Principal Engineer and Subject Matter Expert for Geocent, LLC in support of Boeing's Space Launch System. Dr. Sen received his Bachelor of Technology in Metallurgical Engineering from the Indian Institute of Technology, Kharagpur, India and his M.S. and Ph.D. in Metallurgical and Materials Engineering from the The University of Alabama, Tuscaloosa (Advisor, Prof. Doru M. Stefanescu). He has held various engineering and research positions between 1995 to present. He was the lead materials engineer for NASA, Marshall Space Flight Center, Materials and Processes (2004–2012) directly supporting material and process selection, testing, and qualification for first stage Solid Rocket Boosters, Launch Abort System, and previous Space Shuttle components. In addition, he was the lead metallurgist for technology development program on net shaped aluminum component development for cryogenic tanks, working in close collaboration with NASA and European partners. Between 1995 and 2003 Dr. Sen served as the Principal Scientist and Flight Project Scientist for Space Shuttle and International Space Station Payloads. He was directly responsible for ensuring the scientific integrity of the flight projects and providing the science and engineering liaison between the NASA project team and the principal investigator. He was also the principal experimentalist for glovebox space flight experiment on metal matrix composite solidification performed on USMP-4 shuttle mission. Dr. Sen has been the recipient of several NASA and Space Flight awards in recognition of his achievements in the field of aerospace materials. He has presented numerous invited scientific papers at international conferences and has over 50 publications in referred journals.

Attila Diószegi

Attila Diószegi became an engineer in material science at Technical University of Cluj in Rumania, 1984. The work carrier started up with employments as foundry technician at Rumanian foundries: IOB Bals and IAIFO Zalau. The foundry carrier was continued in Sweden 1986 as a foundry worker at Traryds Metallgjuteri and as a foundry trainee at Scandinavian Foundry School in Jönköping. Between 1990 and 2005, Dr. Diószegi was employed at Volvo's Foundry in Skövde as pattern designer, casting simulation engineer, research engineer for development of cast materials and processes.

A Ph.D. degree was earned in 2004 from Linköping University on the subject of microstructure formation and mechanical properties in grey cast iron. The scientific carrier was started 2005 as assistant professor in component technology at Jönköping University (JU). Further scientific merits were obtained 2008 as associate professor in foundry technology at Jönköping University and 2010 as associate professor in applied foundry technology at the Royal Institute of Technology, Stockholm. Since 2011, Dr. Diószegi has been a visiting professor at the Institute of Material Science at University of Miskolc, Hungary. He was appointed professor of foundry technology at Jönköping University in March 2013.

The research area of foundry technology practiced by Dr. Diószegi is interdisciplinary. The research includes all cast alloys with focus on cast iron produced in both sand and permanent molds. Liquid metallurgy, mold filling, molding materials, nucleation, solidification, microstructure and defect formation, modelling and simulation are covered.

Advances in the Science and Engineering of

CASTING SOLIDIFICATION

An MPMD Symposium Honoring Doru Michael Stefanescu

Solidification Processing I

Session Chair:

Adrian S. Sabau

Advances in the Science and Engineering of Casting Solidification
Edited by: Laurentiu Nastac, Baicheng Liu, Hasse Fredriksson, Jacques Lacaze, Chun-Pyo Hong, Adrian V. Catalina, Andreas Buhrig-Polaczek, Charles Monroe, Adrian S. Sabau, Roxana Elena Ligia Ruxanda, Alan Luo, Subhayu Sen, and Attila Diószegi
TMS (The Minerals, Metals & Materials Society), 2015

SCIENCE OF CASTING AND SOLIDIFICATION: ASM HANDBOOK CONTRIBUTIONS – HONORING PROFESSOR DORU MICHAEL STEFANESCU

[1]Afina Lupulescu, [1]Scott Henry, [1]Karen Marken, [1]Steven Lampman

[1]ASM International
9639 Kinsman Road, Materials Park, Ohio, 44073-002 USA

Keywords: Casting, Solidification, Computer Modeling, Microstructures

Abstract

Many of the metal casting processes are still empirical in nature. Many others are deeply rooted in mathematics and therefore, suitable for modeling. Science of casting and solidification is a major technical asset for foundry operations and of extreme importance in understanding different length scales microstructural changes and evolution as well as developing new processes and materials. In his attempt to describe combinations of solidification theory, research results and industrial practice, Professor Doru Michael Stefanescu (*ASM Fellow, 1997*) has made tremendous contributions to the field. Many of his views on casting and solidification are valued as important impacts within professional environments such as TMS and ASM International. He has written many articles for the *ASM Handbook* series on subjects including basic metallurgy of cast iron, compacted graphite irons, solidification, thermodynamic properties of iron-base alloys, and computational modeling. He was also Volume Chair for Volume 15: *Casting,* of the 9th Edition *Metals Handbook.*

Introduction

One of our authors first met Professor Doru Stefanescu in 1996, at *NASA*, Huntsville, Alabama while taking mission specific training for the space shuttle experiments that were soon to fly onboard space shuttle *Columbia.* At that point, Professor Doru Stefanescu was the Principal Investigator for the "Particle Engulfment and Pushing by Solidifying Interfaces" project while at University of Alabama, in Tuscaloosa. Since then, we had many professional group-to-group interactions and sharing common subjects during different professional meetings. Professor Doru Stefanescu's vast knowledge and experience in casting and solidification has been a valuable touchstone in all of our interactions. Without being exhaustive, this non-conventional paper focuses on some of Professor Doru Stefanescu professional highlights with emphasis on his *ASM Handbook* contributions.

Professional Highlights

Professor Doru Stefanescu had an interesting and complex professional journey starting with Polytechnic Institute in Bucharest, Romania. He is known for major accomplishments in casting and solidification while at University of Alabama. When he got closer to retirement, he transitioned to Ohio State University where he continued his work. His prestigious industrial positions, numerous professional services and editorial review boards of international archival journals are completing his professional profile. Professor Doru Stefanescu's professional activity has been centered on the science of casting and solidification. *Science and Engineering of Casting Solidification* [1], as the title of his textbook highlights, covers the essentials of solidification science of metals and alloys at different length scales and cooling rates specific to commercial castings and rapid solidification processes. He and his group contributed to outlining mathematical fundamentals necessary to build a working knowledge in the field, specifically partial differential equations and numerical analysis. His professional achievements resulted in 380 publications including 35 invited papers, 32 books and chapters in books, 73 refereed journal publications, 63 AFS Transactions papers, 73 refereed conference publications, 24 other conference proceedings publications, 68 other technical publications, and 12 patents. By applying his pedagogical and mentoring talent, Professor Doru Stefanescu directed and produced 39 Master students, 17 Ph. D students, and 25 Postdoctoral fellows. Professor John Morral mentioned that under Professor Stefanescu's supervision, while at Ohio State, "the students interested in solidification and foundry had much enthusiasm and were active going to the FEF National meetings in Chicago and participated when AFS had National meetings in Columbus." Professor Morral also thinks that Professor Stefanescu "graduate program involved only a few students, although he won a number of awards during that time and was actively attending meetings and writing papers [2]." The numerous awards that Professor Stefanescu received add to his professional activity and international recognition. The *John Campbell Medal* - Institute of Cast Metals Engineers, United Kingdom, 2012, the *ASSOFOND plaque* - The Italian Foundry Federation, 2012, Honorary Member of the Romanian Academy of Technical Sciences, 2012, the *Joseph Seaman Gold Medal* of the American Foundry Society (2011), the 3 *Dr. Honoris Causa* titles (Technical University of Cluj-Napoca – Romania, 1998, University of Transylvania – Romania, 2001, Jonkoping University - Sweden, 2012), the *ASM Fellow*, 1997, and the four *NASA* certificates of recognition for creative development of a technical innovation, are just a few of the very prestigious awards and honors that he has received [Figure 1].

Figure 1. Professor Doru Stefanescu receiving the Dr. Honoris Causa award, at University of Transylvania, Romania, 2001.

Professor Stefanescu and our ASM group interacted during many professional events from which the *Science of Casting and Solidification* (SOCAS) meeting organized by him and his group, TMS annual events, and ASM contributions are the highlights. He has been invited to many conferences and meetings, and he has organized numerous Symposia.

ASM Contributions

Many of his views on casting and solidification are valued as important impacts within professional environments such as TMS and ASM International. The time and dedication to perform each job is tremendous and highly appreciated by any professional society. Following his desire to disseminate worldwide the science of casting and metallurgy, Professor Doru Stefanescu has written many articles for the *ASM Handbook* series including articles about basic metallurgy of cast iron, compacted graphite irons, solidification, thermodynamic properties of iron-base alloys, solidification microstructures of aluminum and titanium alloys, as well as computational modeling. He also served as the Volume Chair for *Metals Handbook*, 9th Edition, Volume 15: *Casting*.

Classification and Basic Metallurgy of Cast Iron

This topic is one of the subjects covered by Professor Doru Stefanescu in the *first Volume* of the *ASM Handbooks* [3, 4]. Castings are major manufacturing processes and are used in 90% of all manufactured goods. Therefore, the subject of castings is vast and understanding of its basic metallurgy and logistics is a top priority. Proper consideration of the end results of the Fe-C phase diagram, and establishing criteria for classifying the cast iron it is not a trivial job.

Cast Iron identifies a large family of multicomponent ferrous alloys involving an eutectic. Cast irons have a higher carbon and silicon contents than steel, and therefore exhibit a rich carbon phase (>2% C). Depending primarily on composition, cooling rate, and melt treatment, cast iron can solidify according to the thermodynamically metastable Fe-Fe_3C carbide system or the stable Fe-Graphite system. Usually, the alloying elements may considerably change the maximum solubility of carbon in austenite (γ). Therefore, in exceptional cases, alloys with less than 2% C can solidify with a eutectic structure and still belong to the family of cast iron. The formation of stable or metastable eutectic is a function of many factors including the nucleation potential of the liquid, chemical composition, and cooling rate. The first two factors determine the graphitization potential of the iron. A high graphitization potential will result in irons with graphite, while a low graphitization potential will result in irons with iron carbide. Classification criteria for cast irons are established by fracture (white and gray iron), graphite shape, microstructure of the matrix (ferritic, pearlitic, austenitic, martensitic, bainitic or austempered, or any combination of these phases), commercial designation (common and special cast iron), and mechanical properties (as listed by various national and international specifications). Professor Stefanescu discusses 5 classes of cast irons (gray, ductile, compacted, malleable, and special) outlining carefully the differences among them in terms of microstructure, chemical compositions as well as cooling rates, liquid, and heat treatment effects.

Fundamentals of Solidification, Metallography and Microstructures for Steels and Cast Irons, Aluminum Alloys, Titanium Alloys; Computer Modeling of Solidification Microstructures

Solidification is the transformation of liquid matter into solid matter. Solidification of pure metals or alloys is one of the oldest manufacturing processes, for either cast or wrought as well as it is prone to modelling. This subject is presented by Professor Doru Stefanescu in *Volume 9* of the *ASM Handbooks* [5, 6, 7, 8, 9]. Solidification science initially evolved from the need of better understanding and developing casting processes. It further advanced to the science of many new developments that depart from traditional metal casting. The microstructure that results from solidification may be the final one, in which case it directly affects the mechanical properties of the product. In many cases, heat treatment or other processes may be used after solidification to further modify the initial microstructure. Usually, it results into a specific microstructure and needed properties applicable to a specific part. The effect of solidification on the morphology of the matrix can be deciphered at different length scales: a) *the macroscale* (10^{-3}-1m*)* describes shrinkage cavity, macrosegregation, cracks, surface roughness and casting dimensions; b) *the mesoscale* (10^{-4}m) shows mechanical properties that are affected by the solidification structure, and dependent upon grain size and type (columnar or equiaxed), chemical microsegregation, microshrinkage, porosity, and inclusions; c) *the microscale* *(*10^{-6}-10^{-5}m*)* describes the complex morphology of the grains during solidification. To evaluate the influence of solidification on the properties of the castings, it is necessary to know the as-cast grain morphology (i.e., size and type, columnar, or equiaxed) and the length scale of the microstructure (interphase spacing, e.g., dendrite arm spacing and eutectic lamellar spacing); d) *the nanoscale* (10^{-9} m*)* describes the morphology at the atomic level. At this scale, nucleation and growth kinetics of solidification are discussed in terms of the transfer of individual atoms from the liquid to the solid state. Features such as dislocations involving changes at atomic level are observed with different techniques (SEM, TEM, HRTEM, FIB, etc). Modeling (deterministic or/and probabilistic) supports all solidification scales. The dialogue between experiments and models has been extremely fruitful toward successful modeling of solidification microstructures. Professor Stefanescu belongs to a golden generation of solidification theoreticians and experimentalists (Dr. W.J. Boettinger, Dr. J.W. Cahn, Dr. S.R. Coriell, Dr. G.B. McFadden, Prof. A. Karma, Prof. S. Kobayashi, Prof. W. Kurtz, Prof. M. Rappaz, Prof. R. Trivedi, etc.) and stands out like a prominent figure with major advances in both theoretical and experimental field.

Thermodynamic Properties of Iron, Al and Cu-base Alloys

Not only was Professor Doru Stefanescu Volume Chair for Volume 15: *Casting,* of the 9th Edition *Metals Handbook* [10], he also contributed articles on thermodynamic properties of Fe-base as well as Al and Cu-base alloys. The structure and properties of cast metals are sensitive to numerous impurities. For example, purification of melts generally adds considerable cost to castings. Besides careful selection of metallic scrap, S and O removal from cast iron and steel, or alkali and alkaline earth elements removal from aluminum are highly desired. The structures and properties of steel and cast iron also depend on the control of the solidification structure in castings. It is then absolutely necessary to understand and control the thermodynamics of the liquid and solid phases as well as the kinetics of solidification (nucleation and growth of various phases). Knowledge of activities and activity coefficients is necessary in describing solution behavior and in solving problems that involve chemical equilibria. The thermal properties are useful in understanding the liquid state and solution behavior. The activity coefficients provide simple means of calculating the inter-diffusion coefficients and are used to correlate experimental data on dilute solutions. Calculation of solubility lines which are relevant to the construction of phase diagrams, the calculation of the activity of various components which in turn determines knowledge of probability of formation and relative stability of various phases, are factors that influence casting and solidification processes. Alloying elements are also considered in terms of their influence on the activity of carbon, which provides information on the stability of the main carbon-rich phases of iron-carbon alloys (graphite and cementite). Correlating all these elements becomes of tremendous importance in guiding the processes of casting and solidification.

Summary

Instead of a regular summary, we would rather quote one of Professor Doru Stefanescu close collaborators, Dr. Roxana Ruxanda [11]: "In the years passed, I witnessed first-hand some of his endeavors and I learned how much effort and dedication were invested in his undertakings. For as far as I know, I am certain of one thing: I should be better prepared to be excited by many of his future accomplishments." And more: "I know that he is very happy when his former students are successful; I know that he takes great pride in being one of the first to really understand the importance of the cooling curves and their interpretation. His research projects with *NASA* were very unconventional. Although his research made him well known in the field, he seemed to have enjoyed the most his teaching and interactions with his students." Professor Doru Stefanescu's contributions to metallurgy are very well outlined by his nomination for the *ASM Fellow* award, 1997. He was cited "for fundamental contributions to the science of solidification through the modeling of solidification processes and the microstructural evolution of castings." Congratulations Professor Doru Stefanescu for all your achievements and great volunteering work with the ASM International.

References

1. D.M. Stefanescu, *Science and Engineering of Casting Solidification*, (New York, NY: Second Edition, Springer, 2009), 402 pages.
2. John Morral: Personal Communication.
3. D.M. Stefanescu, "Classification and Basic Metallurgy of Cast Iron, Properties and Selection: Irons, Steels, and High Performance Alloys", *ASM Handbook, vol. 1*, (ASM International, OH: American Society for Metals, 1990), 3-11. Updated and revised by D.M. Stefanescu, under the direction of the ASM Handbook Committee, 2014.
4. D.M. Stefanescu, "Compacted Graphite Iron, Properties and Selection: Irons, Steels, and High-Performance Alloys", *ASM Handbook, vol. 1*, (ASM International, OH: American Society for Metals, 1990), 56–70.
5. D.M. Stefanescu and R. Ruxanda, "Fundamentals of Solidification, Metallography and Microstructures", *ASM Handbook, vol. 9*, (ASM International, OH: American Society for Metals, 2004), 71–92.
6. D.M. Stefanescu and R. Ruxanda, "Solidification Structures of Steels and Cast Irons, Metallography and Microstructures", *ASM Handbook, vol. 9*, (ASM International, OH: American Society for Metals, 2004), 97–106.
7. D.M. Stefanescu and R. Ruxanda, "Solidification Structures of Aluminum Alloys, Metallography and Microstructures", *ASM Handbook, vol. 9*, (ASM International, OH: American Society for Metals, 2004), 107–111.

8. D.M. Stefanescu and R. Ruxanda, “Solidification Structures of Titanium Alloys, Metallography and Microstructures”, *ASM Handbook, vol. 9*, (ASM International, OH: American Society for Metals, 2004), 116–126.
9. D.M. Stefanescu and R. Ruxanda, “Computer Modeling of Solidification Microstructures, Metallography and Microstructures”, *ASM Handbook, vol. 9*, (ASM International, OH: American Society for Metals, 2004), 127–131.
10. D.M. Stefanescu and S. Katz, “Thermodynamic Properties of Iron-Base Alloys, Casting, ASM Handbook”, *ASM Handbook, vol. 15*, (ASM International, OH: American Society for Metals, 2008), 41–55.
11. Roxana Ruxanda: Personal Communication.

Advances in the Science and Engineering of Casting Solidification
Edited by: Laurentiu Nastac, Baicheng Liu, Hasse Fredriksson, Jacques Lacaze, Chun-Pyo Hong, Adrian V. Catalina, Andreas Buhrig-Polaczek, Charles Monroe, Adrian S. Sabau, Roxana Elena Ligia Ruxanda, Alan Luo, Subhayu Sen, and Attila Diószegi
TMS (The Minerals, Metals & Materials Society), 2015

On the Solidification of Metal Alloys during Microgravity Conditions

Hasse Fredriksson[1]

[1]KTH (Royal Institute of Technology); Brinellvägen 23; Stockholm; SE-100 44; Sweden

Keywords: Solidification, Metals, Dendrites, Microgravity

Abstract

Thermal analysis and unidirectional solidification experiments of mainly Al-, Sn- and Pb-base alloys has been performed during Rocket Flights, Parabolic aircraft flights, Space Lab, and GAS- Shuttle flights over three decades. In this presentation the special equipments developed and the experimental
results from the different flight opportunities are to be presented. The cooling rates in the theraml analysis experiments have varied from around 0.01 to 10 C/Sec.The coarsnes fo the structure has been evaluated and the main effect is an increase in coarsness with decreasing gravity. The solidification undercooling was measured and it was mainly lower during microgravity conditions. The microgravity conditions gave mostly a larger effective partition coeffecient.
The result will be presented and shortly discussed from existing theoretical standpoints and the difference will be analysed.

Introduction

In the late sixties NASA sent out an information to different researchers in the world for an opportunity to perform material experiments in space during the Apollo misions. I was a PhD strudent at KTH in Stockholm at that time and I was interested in the effect ofnatural convection on the structure during the solidification of metall alloys. I sent in a proposal with that thema. As a result of that proposal I was later given the opportunity to do experiments in Space lab, Texus rocket flights, Parabolic air crafts fligths and developing a GAS container in a Shuttle flight and also to do some drop tower experiments. The experiemental results are in some cases confusing and some of those results will be presented here. The results will be discussed in relation to known theories about solidification of metals. Different types of experimental set up have been developed and used in the different opportunities. The experimental set up is in many cases uniq and will be presented together with the experimental results. The paper is a summury of the lecture given at the conference. The paper will mostly present results form alloys solidifiing by formation of dendrite crystals.

Experiments and Results

Introduction: A solidifiction process are often studied by thermal analysis or by unidirectional solidification experiments. During the thermal analysis the heat of solidification as well as the solidification temperaure can be found. Mostly it is interesting to measure how far below the equilibrium temperature the solidification starts and occure. It is also intereting to measure the concentration of the elements in the cenrtral part of a dendrite arms as function of the solidification temperature. The structure can be characterised by its coarsnes. Mostly one select a specific length in the structure to charaterise the coarsenes of the structure. For dendrites such a length are the distance between the primary or secondary dendrite arms. A large number of experiments have been done and four main differences have been observed between experiments performed under microgravity conditions and reference experiements on earth. In order to compare the result between mocrogravity experiments and reference experiments we will divide the measured result with each other and get a fraction between the differnce. Absolute values are presented elsewhere (1-5) . We will first present the result and later discuss them.

Texus rocket flight: In the late 70.ties the German Texus program started. The Swedish Space Cooperation together with KTH was invited to participate and developed a rocket module. This was one of seven modules in the rocket flight. The Swedish module contained ten mirror furnaces for thermal analysis and two gradient furnaces (1). In one of the gradient furnaces an experiment with one Al-6%Cu alloy. The growth rate and the primary and secondary dendrite arm space was measured. It is wellknown that most theories desribing dendritic growth shows that the growth rate times the dendrite arm space squared is equeal to a constant. The constant was evaluated and the following values were found:

$C_{Space}=3.8*10^9$
$C_{Earth}=2.5*10^9$
$C_{Space}/\ C_{Earth} = 1.5$

It is intereting to note that the constant are much larger for the sample processed under microgravity conditions then for the one processed on earth.

In one of the mirror furnaces thermal analysis of an eutectic AlSi alloy was performed and surprisinly it was found that the solidfication time was 20% longer in the samples processed under microgravity conditions in spite of a similar cooling rate . This indicated that latent heat should be larger when samples are processed in space. These two observations was not expected and could of course depend on misstakes in the evaluation of the experiemnts or other types of experimental misstakes, and it had to be investigated more carefully.

Experiment in Space Lab: One experiment with an Al-40%Sn alloy was performed in the isothermal heating facilitity (7). In the furnace the alloy was heated and melted and then solidifed under controled cooling conditions. The cooling rate of the liquid, prior solidification, was 0.045 C/sec. Figure 1 a and b shows a micrograph of the samples processed under earth conditions(Figure 1) and under microgravity conditions (Figure 1). The figures show that the dendrite structure are much coarser in the samples processed under microgravity conditions. The dendrite arm space was evaluted and the result are presented as cooling rate multiplied by dendrite arm space squared.This type of relation are comparable with the one were the growth rate are used intead of cooling rate. Normaly colling rate are related to growth rate by a constant. The result of such an evaluation gave:

$C_{Space}= 9*10^{-7}m^2*C/sec$
$C_{earth} = 6.5*10^{-7}\ m^2*C/sec$
$C_{Space}/\ C_{Earth} = 1.4$

Again it was found that the structure formed under microgravity conditions was coarser than the one found in the reference experiments on earth.

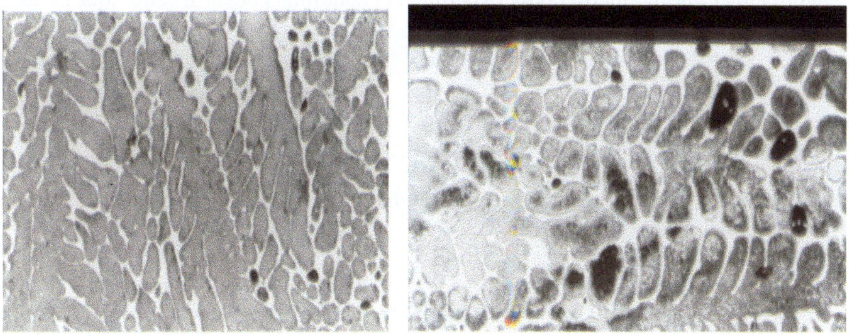

Figure 1. Microstructure from earth (left) and space (right) experiment

Shuttle flight: Four samples of lead–tin alloys was processed in a special grahite mould funace for slowly directional solidification (5). The aim was to investigate the effect or the natural convection on the macrosegregation (7)at the same time the structure coarsness was investigated The Cooling rate at the start of solidification was evaluated as well as the dendrite arm space. The characteristic constant was calculated. At this time the constant was found as follows:

For the Sn-10%Pb alloy:
$C_{Space} = 2.4*10^{-8}$ m^2 *C/sec; C_{earth}= $1.6*10^{-8}$ m^2*C/sec
$C_{Space}/ C_{Earth} = 1.5$

For the Pb-15%Sn alloy:
C_{Space}= $2.4*10^{-8}$ m^2*C/sec; C_{Earth}= $1.9*10^{-8}$ m^2*C/sec
$C_{Space}/ C_{Earth} = 1.26$

Also in this case the structure were coarser in the sample processed under microgravity conditions. The macrosegregations was completely eliminated.

Parabolic flights: By using the Shuttle or Rocket flights to do experiments the access time is normaly very long. The observations presented above needs a number of experiments to be verified. This will be possible to be performed during parabolic aircrafts flights. ESA has given those opportunities. In this case the mirror furnace was reconstructed (8) and used for thermal analysis experiments and several hundred experiments were performed.
Pure Metals. First a number of expereimtns with pure Al, Sn and Pb were performed. In this case the coarsness of the structure could not be evaluated. By analysing the solidification time the indication of changing in latent heat could be evaluated and an average value could be investigated. The average cooling rate was 50 C/Sec.

Tinn: L_{Space}/L_{Earth}=1.07
Lead: L_{Space}/L_{Earth}= 1.12
Aluminum: L_{Space}/L_{Earth}=1.02

The undercooling during the solidification process was also measured and the result is here presented as the equilibrium solidification temperature minus the measured solidification temperature either during microgravity experimental conditions or earth conditions divided by each other. The result is as follow:

Tinn: $\Delta T_{Space}/\Delta T_{Earth} = 0.5$
Lead: $\Delta T_{Space}/\Delta T_{Earth} = 0.45$
Aluminium: $\Delta T_{Space}/\Delta T_{Earth}=0.4$

The experimental results shows that the undercooling is lower in the samples solidified under microgravity conditions.

Alloys: A large number of Al-Cu, Al-Sn and Al-Si base alloys were investigated at cooling rates around 50 C/Sec. In all samples the growth undercooling was lower in the samples processed under microgravity condtions than in the samples processed on earth. Additional to those measurements the concentration in the central part of a dendrite arm was measured by microprobe measurements. It was found that the concentration in the dendrite arm was higher in the samples processed under microgravity conditions than in the samples processed on earth. One normaly present this type of mesurements as an effeciency partition coefficient, which is defined as the measured value divided by the original composition. It was found that the value was 25 % higher in alloys with 2 w-% Cu, 15 % higher in alloys with 5w-% Cu and 10% higher in alloys with 2 w-% Si.
The coarsnes of the structure followed the same observations(5) as is presented for the other alloys.

Discussion

Very carefully mesurements about the diffusion rate in metal melts during micrigravity conditions have been peformed by Nasa and presented by articals of Schmidt et.al. (9). The result shows that the diffusion rate are much lower under microgravity conditions. The diffusion rate influences on the solidifiaction process in many different ways. For the first it will be more difficult for crystal to grow and one should then expect a higher growth undercooling. Our experiments processed under microgravity conditions shows a lower undercooling than the one processed on earth.
The theories presented in the literature (10,11) about dendritic growth gives a relation between growth rate and the growing dendrite tip radius squared. Secondary dendrite arms are formed behind the tip and the distance between the secondary dendrite arms are direcectly related to the tip radius. In the expression for the tip radius the diffusion constant is included and the smaler the diffusion rate is the finer should the structure be. One could then expect a finer structure in the samples processed under microgravity conditions where the diffusion rate are smaler. However the opposite is observed. During the solidification process a coarsening of the structure occurs. This is often desribed by an Ostwald ripening process. Also in this case the process is diffusion controled and the coarsening should be slower under microgravity conditions and the structure finer.
Concentration and temperature gradients create convection in samples processed under earth conditions and the convection is close to eliminated during microgravity conditions. Convection corses a finer structure and one could expect a coarser structure in the samples processed under microgravity conditions due to an elimination of the natural convection. However the large difference in structure coarsnes between the samples processed under microgravity conditions compared to those one processed on earth cannot be explained by natural convection. This large changings can only be performed by strong forced convection (12) which is not the case.

The growth undercooling is related to the composition in the primary dendrite arm by the phase diagram. With the type of phase diagram relevant for the alloys investigated one could expect a lower concentration in the dendrite arms with lower growth undercooling (11). The experimental results showed that the growth undercooling was smaler in the space samples than in the samples processed on earth. One could thus expect a lower content of alloying element in the dendrite arms in the samples processed under microgravity conditions. However the opposite is observed. These observations cannot be explained with the known theories.
Finally one has observed that the latent heat evolved in the samples under microgravity conditions are larger than in the space samples.
As a conclusion there are mainly four differences observed between samples solidified under microgravity conditions compared with samples solidified on earth. These differences cannot be explained by the existing theories for dendritic growth.

Concluding Remarks

Aluminum and lead based alloys solidified under microgravity conditions show a lower growth undercooling, a coarser dendrite structure, a higher alloying content in the dendrite arms and a higher value on the latent heat. The existing theories for dendritic growth with the physical or thermodynamic models existing today cannot explain those observations.

References

1. T. Carlberg, H. Fredriksson, P. Sunnerkranz, S.Grahn, L.Stenmark. The Swedish Texus Experiment, a Technical Description and some Prelimnary Results. ESA Publication SP-135, 1978, 325-338.
2. Solidification of some Eutectic Alloys under Microgravity Conditions and at 1g. ESA Publication SP-142, 1979, 221-231.
3. Johan Dahlström,. On the solidification of Al alloys during Microgravity. PhD thesis 1999, KTH, Stockholm.
4. H. Fredriksson, J. Dahlström, J. Fjellstedt. On the solidification af Aluminium-base alloys unde different gravity conditions. Internatioanl confernce on Solidification Scince and Processing.
5. H. Fredriksson, S. Wallin. The effect of gravity in the solidification process inPb-Sn and Sn-Pb alloys. AAIIA Conference, Reno, USA,1993
6. D. Shapland, M. Rycroft. Spacelab, Cambridge University Press, 1984.
7. H. Shahani, G. Amberg, H. Fredriksson. Metallurgical Transaction. 23A, 1992, 2301-2311.
8. C. Lockowandt, K. Löth, L.Ekbom, A. Eliasson, A. Jarfors. The seventh European Symposium on Materials and fluid Science in Microgravity, April 1992, ESA SP-333,383-385.
9. R. W. Smith. The influence on G-Jitter on the measurement of solute diffusion in dilute liquid metals in a low Earth orbiting laboratory. Proc. Int Confrence Spacebound 2000. Canadian Space Agency.
10. W.Kurz. Fundamentals ofsolidification. Trans. Tech. Publ. 1989, Switzerland.
11. H Fredriksson, U Åkerlind. Solidifcation and Crystaliization Processing in Metals and Alloys. J. Wiley and sons, "012 Chichester UK.
12. H. Fredriksson, N. El-Mahallawy, M.Taha, Xiang Liu, G. Wänglöf. The effect of stirring on the Solidification Process in Metals. Scand. J. of Metallurgy 15, (1986), 127-137.

Advances in the Science and Engineering of Casting Solidification
Edited by: Laurentiu Nastac, Baicheng Liu, Hasse Fredriksson, Jacques Lacaze, Chun-Pyo Hong, Adrian V. Catalina, Andreas Buhrig-Polaczek, Charles Monroe, Adrian S. Sabau, Roxana Elena Ligia Ruxanda, Alan Luo, Subhayu Sen, and Attila Diószegi
TMS (The Minerals, Metals & Materials Society), 2015

FORMATION OF THE TIN RICH LAYER AND INVERSE-SEGREGATION IN PHOSPHOR BRONZES DURING CONTINUOUS CASTING

S. Saleem, M. Vynnycky, H. Fredriksson

Division of Casting of Metals, Royal Institute of Technology, Brinellvägen 23, SE-10044 Stockholm, Sweden

Keywords: Inverse segregation, Exudation, Thermal Stress, Continuous Casting

Abstract

Continuous casting of the phosphor bronzes has been investigated experimentally and analyzed with the help of a thermo-mechanical model. The microscopic investigation shows the spread of the tin rich liquid at the chill surface cause by the formation of flow channels underneath the chill surface. Precipitation of the secondary phases has also been observed under some casting conditions. The macro segregation profile along the solidification thickness predicts a strong casting parameter sensitive inverse segregation. The simulation results show high compressive stresses at the surface of the cast during solidification. The flow channels depth and thermal stress coupled with microsegregation calculations shows the possibility of the pressure driven flow of tin rich liquid towards the chill surface during solidification. The experimental observation and calculated results show that the inverse segregation can be homogenized and decreased by controlling the casting parameter that defines the liquid pool depth into the mould.

Introduction

The combination of solidification shrinkage and solubility difference of alloying elements in different phases makes the composition of the cast as the function of the casting dimensions. The macro-segregation caused by shrinkage driven flow and solute solubility difference during the solidification has been discussed in detail for both time dependent component casting and steady sate continuous casting problems [1-5].

The density difference of newly formed solid gives rise to a driving force that sucks the inter-dendritic liquid and the formation of a solute-rich region close to the chill surface normally referred as *inverse segregation*. Another consequence of the solidification is the spread of a solute rich layer at the surface of the cast. This layer is normally observed in DC castings. The phenomena is referred as *exudation* and is discussed as the result of the metallostatic pressure and that the liquid head forces the interdendritic liquid to flow towards the chill surface. At the same time thermal stress becomes strong enough to equalize the liquid head and start to separate the shell (solid/semi-solid) from the mould wall leaving an air gap between chill surface and the mould. The gap ease the liquid flow through the porous mushy zone as the surface contact is eliminated leaving the atmospheric pressure condition at the chill surface. The thermal stresses can also press the liquid towards the surface and increase the amount of liquid at the surface. In case of phosphorus bronzes the observed exudation layer thickness varies between 60-100 μm.

For the quality point of view, the exudate layer can be milled after casting, as it is inevitable during DC casting. Unfortunately, for phosphor bronzes, adverse effects appear during solidification. The exudate is normally higher in Sn contents with a considerably lower melting point and thermal conductivity. The heat transfer over the tin rich exudate layer and strand is

relatively poor. The thin tin rich layer at the chill surface acts as a solder between mould wall and cast and causes the strand to crack.
In addition to separate investigations of the mentioned phenomenon, the combined effect of solidification shrinkage and exudation on macrosegregation has been modeled [6-7] and studied experimentally [8]. However, the thermal stresses give rise to additional volume changes during solidification. The volume changes of the liquid filled mush zone can shift the net solute transfer depending upon whether the casting is under compression or tension and will be discussed here.

Experimental Methods

Casting strands were produced in the continuous caster. Phosphor bronzes with tin composition 6 - 8 wt. % at Sn were cast at different casting speeds. Samples were collected from relatively stable casting conditions.
The temperature at the center line of the casting was measured by inserting a K-type Inconel coated thermocouple. The melt pouring temperature for the casting is in the range of 1130-1140 K. At this temperature range the thermocouple can stay almost 1 min before dissolving into the melt. Due to the high cooling rates and small solidification thickness, thermocouple expose to high temperature for less than 30 sec. The thermocouple is certified according to the EN10204 3.1b and has an accuracy of ± 1.5 %.

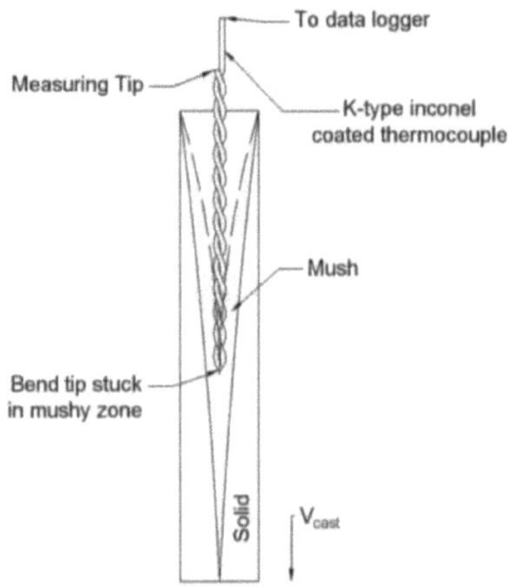

Figure 1 Continuous casting center line temperature measurement.

In order to make sure that the measuring tip of the thermocouple travels at the same speed as the casting and start the measurement at the entrance into the mould, thermocouple was bent at 0.5 m from the tip. The bent tip was stuck into the mushy zone and casting start pulling the thermocouple into the sump as shown in Figure 1. The thermocouple's bent parts were twisted around each other in order to keep the tip from moving with the turbulence of the melt and to get a stable temperature profile at the center of the casting strand. The measurements were taken at the middle and corner along the width of the casting for different casting speeds.
Samples were studied with the aid of LOM and SEM. XRD was performed over an area of 40 x 40 mm by removing the layers from the cast surface till the center at an interval of 0.5 mm.

Mathematical Model

The problem impose here is a steady-state two dimensional. The cast is assumed to be completely solid and leave the mould at the constant speed V_{cast}. The energy balance is given by

$$\rho C \mathbf{V}\ \nabla\cdot\mathbf{T} = \nabla\cdot(k\ \nabla\cdot\mathbf{T}) + \nabla\cdot[\rho(h_s + h_l)\mathbf{V}]$$

Where, $C = f_l C_l + f_s C_s$, $k = f_l k_l + f_s k_s$, $\rho = f_l \rho_l + f_s \rho_s$,

The subscript s and l denotes liquid and solid phases respectively. **V** is the velocity vector. The casting geometry is normally slender. The higher order derivatives in the length direction are small enough to be neglected as shown in a recent studies [14]. h_s and h_l are enthalpies. f_l and f_s are the volume fractions of solid and liquid. The following simplification and assumptions are imposed,

1. The shrinkage driven inter-dendritic liquid velocity does not contribute at the leading order for the energy balance.
2. There is no pore formation during the solidification i.e $f_l + f_s = 1$.
3. The physical properties are independent of the temperature but depends upon the phases.
4. The composition of the interdendritic liquid is uniform.
5. There is no diffusive flux into the interdendritic liquid.
6. There is a thermodynamic equilibrium throughout the solidification. The isotherms essentially represent the same volume fractions of liquid.
7. The solid formed above the coherence temperature is thermoelastic and not strained.

The liquid fraction is defined by

$$f_l = 1 - \frac{T_l - T + \frac{2}{\pi}(T_s - T_l)\left(1 - \cos\left(\frac{\pi}{2}\left(\frac{T - T_l}{T_s - T_l}\right)\right)\right)}{(T_l - T_s)\left(1 - \frac{2}{\pi}\right)}$$

The value of shrinkage coefficient is normally small (2 to 12%). This interdendritic shrinkage, although the main reason to give rise an additional component of the velocity field, does not contribute to the heat transfer at the leading order. However, for the calculations of the microsegregation one should look into the higher order contribution from the solute, mass, and momentum balance [9-11].

The uniform composition of interdendritic liquid implies a uniform liquid enthalpy. However, the enthalpy of the solid depends upon the concentration profile within the formed solid. While for the case of no diffusion in the solid, the instantaneous enthalpy will be an average value obtained by integration over the solid volume. For the same reference state and neglecting the effect of microsegregation on the already formed solid the enthalpies can be reduced to the effective latent heat ΔH. The values of the constant used are given in Table 1.

For the studied geometry the structural mechanics uncoupled from the heat flow as shown in the analysis before [13]. The thermal stresses can then be calculated with the help of the temperature field and positions of the solidus, liquidus and coherence boundaries. The detailed description of the model which calculated the mentioned boundaries explicitly is mentioned elsewhere [14].

The thermal boundary conditions can be defined from the heat extraction by the cooling water. However, to define the spread of the heat extraction as function of mould length need some considerations and sophisticated experimental approaches. For the present case the heat transfer coefficient is defined as function of mould length while keeping the fact that the total heat extraction should not exceed the heat taken by the cooling water. The extracted heat can be calculated by $Q = m\, C_W\, \Delta T$, where C_W is water specific heat and ΔT is the difference between inlet and outlet water temperatures.

Table 1. Sn-8 Pct Cu alloy physical properties

Cast superheat, K	$\Delta T_{superheat}$	20
Liquid density, Kg/m^3	ρ_l	8360
Solid density, Kg/m^3	ρ_s	8800
Solid/liquid conductivity, W/m-K	$k_{l,s}$	67
Latent heat of fusion, J/kg	ΔH	73000
Solid/liquid specific heat, J/kg-K	$C_{l,s}$	350
Liquidus temperature, K	T_l	1303
Solidus temperature, K	T_s	1173

The solubility difference of solute in the corresponding phases causes the enrichment of the remaining melt for $k_0 < 1$. Depending upon the diffusion of the solute elements into the solid, the composition of the solid could have a spatial dependent into the already formed solid or have a uniform composition. In figure 2 the solidification front is moving over the time from t_1 to t_2. However, the increment in the position of the front depends upon the volume changes cause by solidification shrinkage (β) and thermal strain (ξ). For $\beta > 0$, the volumetric shrinkage will be compensated by the bulk liquid. The additional flow of the liquid will be provided by the thermal shrinkage of the solid.

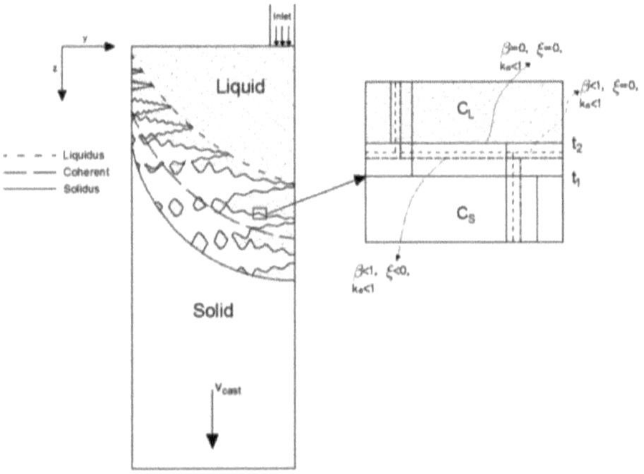

Figure 2 Effect of solidification shrinkage and thermal strain on the composition of mushy zone

In order to consider the effect of thermal strains and solidification shrinkage simultaneous on the net solute transfer, one should consider the point where the phenomenon takes effect. The solute transfer by shrinkage appears as soon as the melt loses the superheat and first solid appears. However, solid required a certain fraction in order to be able to resist the applied thermal stresses and act as a coherent solid, which corresponding to coherent temperature (T_c). The thermal stress appears at the later stage when, T= T_c at y =0

For the present case T_c is assumed at 66% solid fraction. The coherence temperature (T_C) can be measured experimentally. The direction of the flow caused by thermal stresses will be defined by the sign of the thermal volume changes. For a positive volume change, the shrinkage and thermal stress will act similarly. While for a negative volume change, the interdendritic liquid has to be expelled out of the mushy zone.

Results and Discussion

The measurement of the composition profile for the exudate layer shows an abrupt change at the visible interface between exudate layer and the strand (Fig 3). The cast thickness then follows the normally understood inverse segregation pattern. The spread of the tin over the surface of the strand is normally visible due to the whitish appearance of the tin. The strand surface looks bright and dark at different places. The bright areas are the consequence of the smooth tin rich surface while the surface under the dark areas is relatively rough.

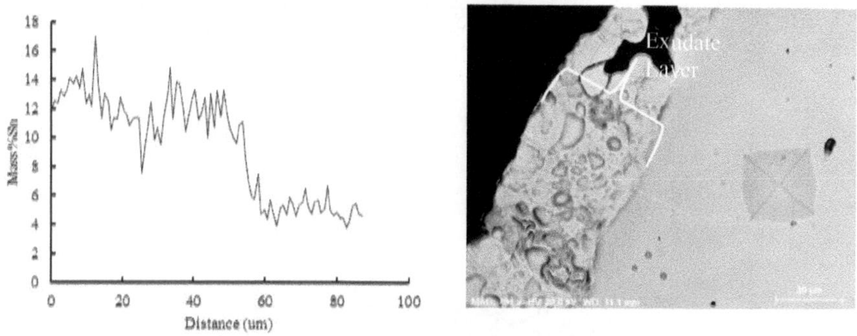

Figure 3 Exudate Sn rich layer, Cu6Sn

The tin rich surface is evident from the segregation profile. The depth of the liquid pool increases with the increase in casting speed as shown in figure 4. The casting leaves the mould at relatively higher temperature at increased speed. There is a risk for the casting to leave with certain fraction of liquid. The temperature at the exit (~ 800 °C) of the mould is close to the peritectic temperature for the Cu-Sn alloy. The centerline of the casting has a large fraction of the porosity. So the above mentioned model will not hold as soon as the two fronts will meet at a certain depth into the mould

The segregation measurements were performed without the removal of the chilled surface (Fig 5a). The first dip in the profile might be related to the formation of the air gap. The decrease in concentration close to the center of the casting is related to the center line feeding resistance. Figure 5 shows that the segregation profile close to the end of casting shifts almost 30% with the increase in casting speed by 15%. However, the last part solidifies almost at the same composition. The higher value of the Sn concentration at y = 0 is the result of the exudate layer. The lower casting speed causes the higher cooling rates and thermal gradients, while the increased casting speed spread out the leakage of the tin to the surface over the larger mould length. The spread out distributes the tin rich layer over the surface of the strand and avoids the excessive gradients caused by the concentrated thin film patches of the Sn. The flow rate of the tin towards the surface will be defined by the pressure head and thermal shrinkage. The composition of the melt pressed out will also takes its effect from the shrinkage driven interdendritic flow. Due to the small thickness and large two-phase region of the alloy, the liquidus fronts at the center will meet much earlier before the casting is completely solidifies.

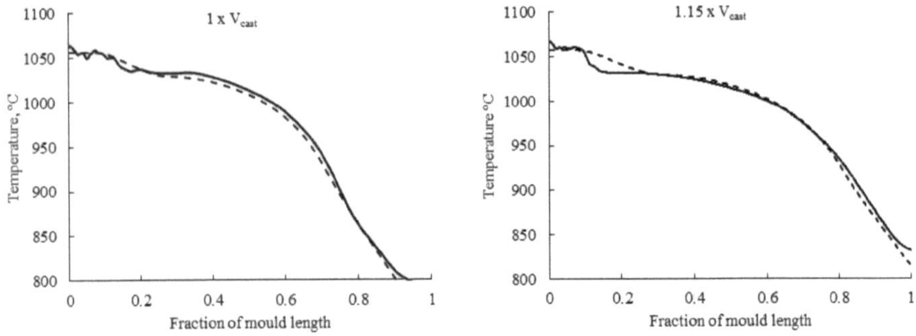

Figure 4 Centerline temperature of the strand during the casting. Experimental - Solid line, Simulated - Dot line. V_{cast} = Normal casting speed

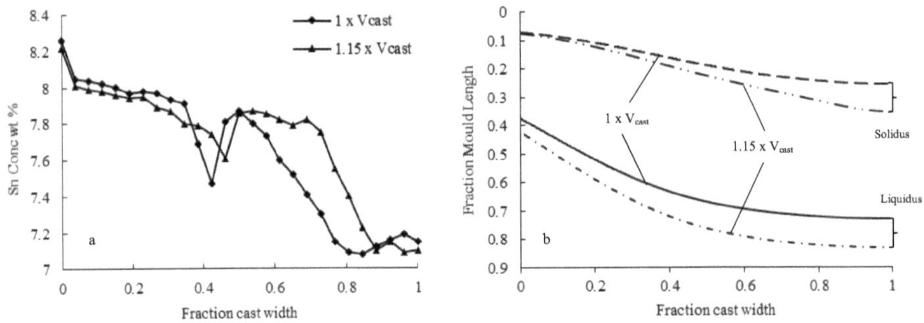

Figure 5 Cu8Sn a) Segregation profile across the cast thickness, b) Simulated solidus/liquidus isotherms, V_{cast} = Normal casting speed

The thermal stress at the surface of the strand is normally compressive as has been shown previously by [13]. The negative thermal stresses will give rise to a negative volume change for the interdendritic liquid. The inter-dendritic liquid which is incompressible will be forced towards the surface of the cast. However the condition for the formation of the air gap still holds. The liquid that flows under the influence of the thermal strain will get enriched due to the solidification and shrinkage driven flow. The contact between mould wall and cast will hold the pressure. Another parameter is if the thickness of the air gap will be of same magnitude as that of the air gap as described before [6]. However, the phenomena need a more careful modelling and validation with the experimental results.

The thermal stress becomes negative immediately after the start of the solidification. The maximum metallostatic pressure occurs at the maximum liquid pool depth. Even for the case in which casting solidifies at the bottom of the mould, the maximum pressure is comparatively small. From figure 6 it is evident that the thermal stress at the surface have much higher magnitude than the metallostatic head. For the initial thickness however, the magnitude for the applied pressure and stresses are comparable. The thermal stresses start to dominate before the casting is completely solid close to the mould wall.

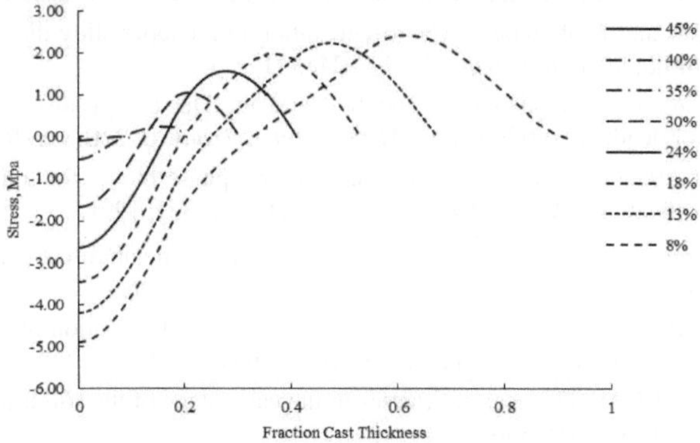

Figure 6 Thermal stress along the thickness of the strand at different mould levels

Conclusion

In the current study the possible role of thermal stress on the inverse-segregation and exudation has been investigated. The thermal stresses at the surface become negative for the initial stage of the solidification. As compare to thermal stresses, the metallostatic head is smaller for the case of continuous casting. In order to completely understand the flow of the solute rich liquid towards the cast surface, the effect of thermal stresses has to be incorporated in addition to the shrinkage driven flow.

The sharp exudate/strand interface could be the consequence of the trigger point defined by the thermal stresses. Nevertheless, the formation of the air gap make it possible for the solute rich liquid to come out at the surface, but the amount of the liquid suggest that the metallostatic head is not the driving force solely.

The increase in casting speed also increases the liquid pool depth and reduces the thermal gradients and cooling rates. The effects can be seen over the chilled surface as the exudate layer spread all over the surface rather producing a highly concentrated localized areas. The segregation profile also response to the cooling conditions and shifts toward the more homogenized distribution.

References

1. M. C. Flemings and G. E. Nereo. Macrosegregation. i. Trans. *TMS-AIME*, 239:1449–1461, 1967.
2. R. Mehrabian, M. Kaene and M.C Flemings. Interdendritic fluid flow and Macrosegregation; Influence of Gravity, *Met. Trans*, Vol 1, 1209-1220
3. M. C. Flemings 1974 Solidification Processing. McGraw-Hill, New York
4. M.C. Felmings, R. Mehrabian, and G.E. Nereo, *Trans AIME*, 242:41 (1968)
5. M. C. Felmings, and Nereo, G.E: *Trans AIME*, 242:50 (1968)
6. E. Haug, A. Mo, and H. J. Thevik. Macrosegregation near a cast surface caused by

exudation and solidification shrinkage. *Int. J. Heat Mass Transfer*, 38:1553–1563, 1995.

7. H. J. Thevik, A. Mo, and T. Rusten. A mathematical model for surface segregation in aluminum direct chill casting. *Metall. Mater. Trans.* B, 39B:135–142, 1999.
8. F. Kaempffer and F. Weinberg, Macrosegregation in a copper alloy directionally cast with exduation of liquid. *Metall. Trans 2*, 2477-2483 (1971)
9. D. R. Poirier, P. J. Nandapurkar, and S. Ganesan. The energy and solute conservation equations for dendritic solidification. *Metall. Mater. Trans.* B, 22B:889–900, 1991.
10. J. Ni and C. Beckermann. A volume-averaged two-phase model for transport phenomena during solidification. *Metall. Mater. Trans.* B, 22B:349–361, 1991.
11 M. J. M. Krane and F. P. Incropera. A scaling analysis of the unidirectional solidification of a binary alloy. Int. J. Heat Mass Transfer, 39:3567–3579, 1996.
12. M. Vynnycky. An asymptotic model for the formation and evolution of air gaps in vertical continuous casting. *Proc. Roy. Soc. A*, 465:1617–1644, 2009.
13. Kristiansson, J-O. 1982 Thermal Stresses in the early stage of the solidification of the steel. *J. Therm. Stresses 5*, 315-330
14. M. Vynnycky and S. Saleem. On the resolution of thermal gradients in a model for the continuous casting of binary alloys. Submitted to *Appl. Math. Comp.*, 2014.

Advances in the Science and Engineering of Casting Solidification
Edited by: Laurentiu Nastac, Baicheng Liu, Hasse Fredriksson, Jacques Lacaze, Chun-Pyo Hong, Adrian V. Catalina, Andreas Buhrig-Polaczek, Charles Monroe, Adrian S. Sabau, Roxana Elena Ligia Ruxanda, Alan Luo, Subhayu Sen, and Attila Diószegi
TMS (The Minerals, Metals & Materials Society), 2015

A MODEL OF CAVITATION FOR THE TREATMENT OF A MOVING LIQUID METAL VOLUME

G S Bruno Lebon[1], Koulis Pericleous[1], Iakovos Tzanakis[2], Dmitry Eskin[2]

[1]Centre for Numerical Modelling and Process Analysis;
The University of Greenwich; London, SE10 9LS, United Kingdom
[2]Brunel Centre for Advanced Solidification Technology (BCAST);
Brunel University; Uxbridge, Middlesex, UB8 3PH, United Kingdom

Keywords: Cavitation modeling, Ultrasonics, Light-metal alloys

Abstract

Ultrasonic cavitation is known to improve significantly the downstream properties and quality of metallic materials. The transfer of this technology to industry has however been hindered by difficulties in treating large volumes of liquid metal. To improve understanding of cavitation efficiency so that it can be applied to a moving melt volume, an improved cavitation model derived from the Keller-Miksis equation is developed, and applied to the two-phase problem of bubble breakup and propagation in the melt. Numerical simulations of the ultrasonic field are performed and the calculated acoustic pressure is applied to the source term of the bubble transport equation to predict the generation, propagation, and collapse of cavitation bubbles in the melt. The use of baffles to modify the flow pattern and amplify sound waves in a launder conduit is examined, to determine the optimum configuration that maximizes residence time of the liquid in high cavitation activity regions.

Introduction

Treating melt with ultrasound results in significant improvements in the quality and properties of metallic materials [1, 2, 3]. These improvements are primarily due to ultrasonic cavitation: the creation, growth, pulsation, and collapse of bubbles in the liquid [4, 5]. However, treating large volumes of liquid metal, as is required by processes such as continuous casting, remains difficult and this technology has not been successfully transferred to the industry as a result. To circumvent these difficulties, a fundamental study of the ultrasonic treatment of melt is required [6].

A plethora of transport equation-based cavitation models applied to the Navier-Stokes equations can be found in the literature, e.g. [7, 8]. Most of these have been developed to model cavitation in water, rather than metals, where evaporation and condensation are important parameters. Hence, in the model of Merkle *et al.* [9, 10], evaporation and condensation terms are both functions of pressure. For the Kunz *et al.* model [11], the evaporation term is a function of pressure; whilst condensation is a function of volume fraction. While both models produce satisfying predictions at different cavitation numbers [7], the full cavitation model of Athavale et al [12, 13] is more commonly encountered in the literature. This model provides the capability for multidimensional simulation of cavitating flows, the modelling of which is crucial to the design of many engineering devices. In their approach, the authors derived source terms for the bubble mass fraction transport equation from the Rayleigh-Plesset equation [14, 15], which governs the evolution of a spherical bubble [16, 17], to predict the formation and collapse of bubbles in cavitating flows. This model has been used in metals, for the modelling of solidification structure evolution in a crucible by Nastac [18].

In this study, the full cavitation model is modified to compute the bubble concentration in a launder. The Keller-Miksis equation, which is more appropriate than the Rayleigh-Plesset equation for large forcing amplitudes [19], is used to derived the source terms of the vapour transport equation. An additional turbulence source term is also derived from the Keller-Miksis equation to account for the increased turbulence created by bubbles as they cavitate. These changes are put forward to improve the understanding of the effect of ultrasonic treatment on flowing melt, paving the way to the ultrasonic treatment of liquid metal as a continuous process.

Theory

Governing Equations

The Keller-Miksis equation [19] governs the bubble evolution:

$$\left(1-\frac{\dot{R}}{c_l}\right)R\ddot{R}+\frac{3}{2}\left(1-\frac{1}{3}\frac{\dot{R}}{c_l}\right)\dot{R}^2=\left(1+\frac{\dot{R}}{c_l}\right)\frac{1}{\rho_l}[P_B-P_\infty-P_c]+\frac{R}{\rho_l c_l}\frac{dP_B}{dt} \quad (1)$$

where R is the bubble radius, c_l is the speed of sound in the surrounding liquid, ρ_l is the liquid density, P_B is the pressure inside the bubble, P_∞ is the pressure in the liquid far from the bubble center, and P_c is the partial pressure in the liquid at the center of the bubble if the bubble were to be absent. The dotted accents denote time derivatives.

Fluid flow is governed by the RANS equations:

$$\frac{\partial\rho}{\partial t}+\nabla\cdot(\rho\boldsymbol{u})=0 \quad (2)$$

$$\frac{\partial(\rho u_i)}{\partial t}+\nabla\cdot(\rho\boldsymbol{u}u_i)=\nabla\cdot[(\mu+\mu_t)\nabla u_i]+S_{u_i} \quad (3)$$

where ρ is the fluid density, $\boldsymbol{u}$ is the fluid velocity, μ is the dynamic viscosity, μ_t is the eddy viscosity, and S_{u_i} are the momentum sources, including buoyancy. The fluid density is related to the vapour and non-condensable gas mass fractions, f_v and f_g, according to:

$$\frac{1}{\rho}=\frac{f_v}{\rho_v}+\frac{f_g}{\rho_g}+\frac{1-f_v-f_g}{\rho_l} \quad (4)$$

where ρ_v is the bubble vapor density and ρ_g is the non-condensable gas density [12]. The standard $k-\varepsilon$ turbulence model is used for closure [20] with an additional source term to account for turbulence generated by bubble collapse. The velocity generated by bubble collapse is estimated from the Keller-Miksis equation (1), ignoring the second-order derivative of R and the time derivative of the bubble pressure, as:

$$v_k=\left[\frac{2}{3}\left(\frac{1+\frac{\sqrt{k}}{c}}{1-\frac{1}{3}\frac{\sqrt{k}}{c}}\right)\frac{p_v-p}{\rho_l}\right]^{\frac{1}{2}} \quad (5)$$

where the interfacial velocity of the bubble is approximated by $\sqrt{k}$. The additional turbulent source term in the k equation is:

$$S_k=f_g\rho\frac{v_k^2}{t_c} \quad (6)$$

where t_c is a characteristic time scale for the lifetime of the bubbles. The wave equation is:

$$\frac{\partial^2 p}{\partial t^2}-c^2\frac{\partial}{\partial x_i}\left(\frac{\partial p}{\partial x_i}\right)=c^2 S_p \quad (7)$$

where p is acoustic pressure, c is the speed of sound, and S_p are the wave source terms. The density and mass fraction transport equations are given by:

$$\frac{1}{\rho}=\frac{f_g}{\rho_g}+\frac{1-f_g}{\rho_l} \quad (8)$$

$$\frac{\partial(\rho_g f_g)}{\partial t}+\nabla\cdot(\rho_g\boldsymbol{u}f_g)=\nabla\cdot(\Gamma\nabla f_g)+S_G-S_C \quad (9)$$

where S_G and S_C are the growth and collapse source terms respectively. The source terms are derived from the wave equation (7) and the Keller-Miksis equation (1), ignoring the second-order derivative of R and the time derivative of the bubble pressure. The interfacial velocity of the bubble is approximated as $\sqrt{k}$, giving:
when $p < p_v$:

$$S_G = C_G \frac{\sqrt{k}}{\sigma} \rho_l \rho_g \left[\frac{2}{3} \left(\frac{1+\frac{\sqrt{k}}{c}}{1-\frac{1}{3}\frac{\sqrt{k}}{c}} \right) \frac{p_v - p}{\rho_l} \right]^{\frac{1}{2}} (1 - f_g) \tag{10}$$

when $p > p_v$:

$$S_C = C_c \frac{\sqrt{k}}{\sigma} \rho_l \rho_g \left[\frac{2}{3} \left(\frac{1+\frac{\sqrt{k}}{c}}{1-\frac{1}{3}\frac{\sqrt{k}}{c}} \right) \frac{p - p_v}{\rho_l} \right]^{\frac{1}{2}} (f_g) \tag{11}$$

where $C_G = 0.02$ and $C_C = 0.01$. σ is the surface tension between the liquid and the gas. p_v is the sum of the vapor pressure and an estimation of the local values of turbulent pressure fluctuations:

$$p_v = p_{sat} + 0.39 \frac{\rho k}{2} \tag{12}$$

where p_{sat} is vapor pressure [12].

Algorithm
The wave equation (7) is solved using the leapfrog scheme with a time step of 1 μs for a simulation time of 1 ms. The minimum and maximum acoustic pressures in the domain are stored at the end of the simulation, to be used in the source terms (10) and (11) respectively in the cavitation run.
The cavitation model is then run for a simulation time of 20 s, with a time step of 1 ms. The appeal of separating these two calculations lies in the reduced run time from weeks to hours, by enabling the use of large time steps for the cavitating flow simulations.

Problem Description

In order to model the continuous treatment of melt, a launder geometry as shown in Figure 1 is used as the computational domain. The sonotrode is located at the center of the domain between two baffles, and the tip is immersed 1 cm below the liquid surface. The frequency of the sonotrode is 20 kHz and the amplitude of the acoustic pressure below the sonotrode is 5.0 MPa. The inlet and outlet are at the low and high y boundaries respectively. The liquid enters the domain at a velocity of 0.01 m s^{-1} in the y direction. The launder problem is solved for both water at 20 °C and aluminium at 700 °C; both material properties are given in Table 1. The launder boundaries are assumed to be fully reflective to sound waves.
The distance between the baffles is a parameter, varied as a function of the sound wavelength. In water, the wavelength λ of ultrasound at 20 kHz is 7.4 cm, and in aluminium, the wavelength is 23 cm. The simulation is run for each liquid for the following distances L between baffles: 0.5 λ, 1 λ, and 1.5 λ using the procedure described in the theory section.

Results

Cavitation of a 100 μm bubble
Cavitating bubbles with radii of the order of 100 μm have been observed in X-ray imaging of Al-10Cu samples [21]. Solving the Keller-Miksis equation using the standard Runge-Kutta method for bubbles of this size in water and aluminium for a sinusoidal signal of amplitude 5.0 MPa

gives the radius evolution profiles depicted in Figure 2 and Figure 3 respectively. For both materials, bubble collapses occurs at 0.1 ms intervals: the characteristic time-scale t_c used in term (6) is therefore assumed to be 0.1 ms.

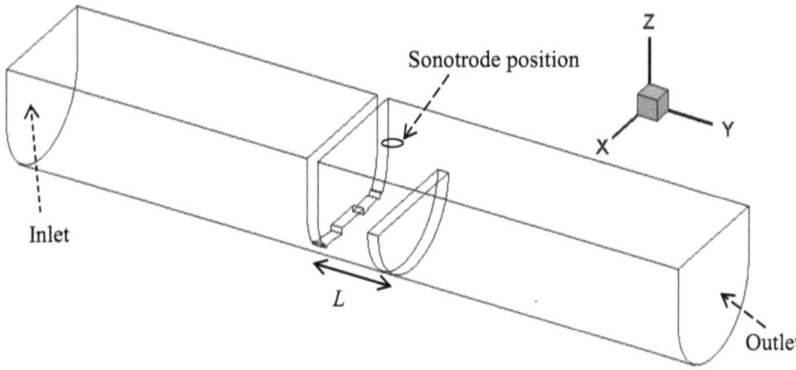

Figure 1. 50 cm x 9 cm x 8 cm launder. L denotes the length between baffles, of thickness 8 mm each. The sonotrode is immersed 1 cm into the free surface at the center of the domain.

Table 1. Material properties of liquid aluminium and water.

Material Property	Aluminium (700 °C)	Water (20 °C)
Sound speed c (m s^{-1})	4600	1481
Density ρ_l (kg m^{-3})	2350	1000
Dynamic viscosity μ (mPa s)	1.3	0.798
Surface tension (hydrogen interface) S (N m^{-1})	0.87	0.072
Vapour pressure p_{sat} (kPa)	0	4.24

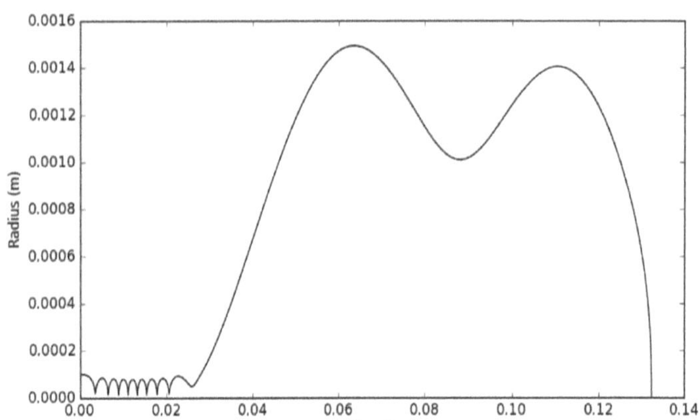

Figure 2. Radius evolution for a bubble of 100 μm in water.

Acoustic pressure solution

Solving the wave equation (7) yields minimum and maximum acoustic pressure contours. The minimum pressure contour is shown in Figure 4 and Figure 5. As expected, the extreme pressure values are found below the sonotrode, where the cavitation activity is expected to be most intense. The cavitation threshold of -100.0 kPa is achieved everywhere in the domain.

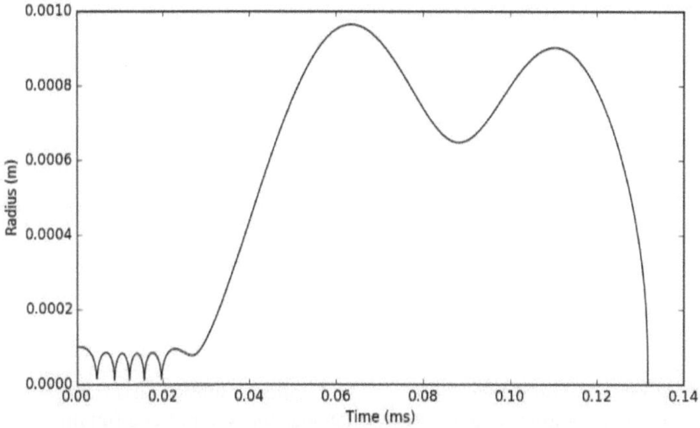

Figure 3. Radius evolution for a bubble of 100 μm in aluminium.

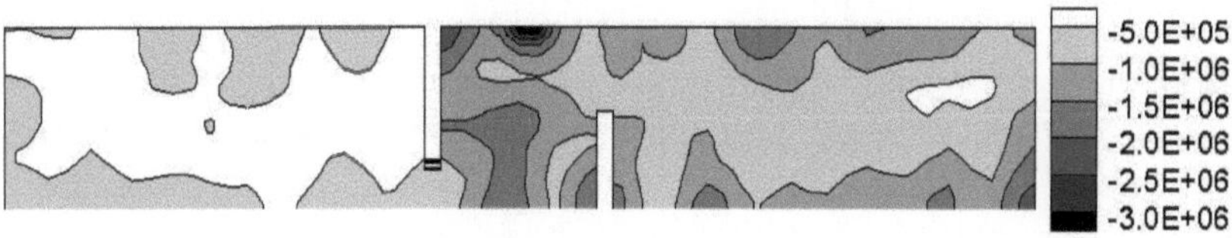

Figure 4. Predicted minimum instantaneous acoustic pressure (Pa) in domain for a configuration with baffles separated at a distance of 1.0 λ for water along the central x plane.

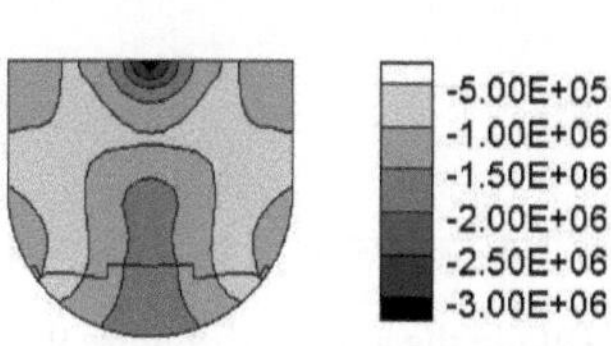

Figure 5. Predicted minimum instantaneous acoustic pressure (Pa) in domain for a configuration with baffles separated at a distance of 1.0 λ for water along the central y plane.

Cavitation run solution

The minimum and maximum pressure fields from the wave equation solution are used in the source terms of the bubble transport equation (9). The resulting bubble mass fraction plots along the axes of the domain are shown in Figure 6, Figure 7, Figure 8, and Figure 9. These results are obtained in a run time of the order of 1.5 days on a 3.60 GHz CPU, making this approach attractive for industrial applications.

For both water and aluminium, a baffle separation of 1.0 λ maximizes the bubble concentration in the domain. The flow generated with this baffle distance also convects the bubbles downstream, resulting in a larger bubble concentration at the outlet. The baffle distances at half integers of wavelength are equally poor at yielding a large bubble concentration downstream. A larger bubble concentration is desired, since higher cavitation activity – occurring as more bubbles collapse – is thought to promote better grain refinement.

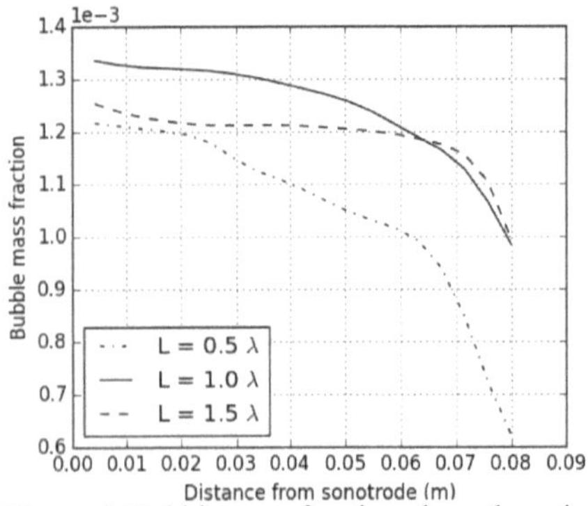

Figure 6. Bubble mass fraction along the axis of the sonotrode for water after a run time of 20 s. Mass fraction values are taken along the axis of the sonotrode.

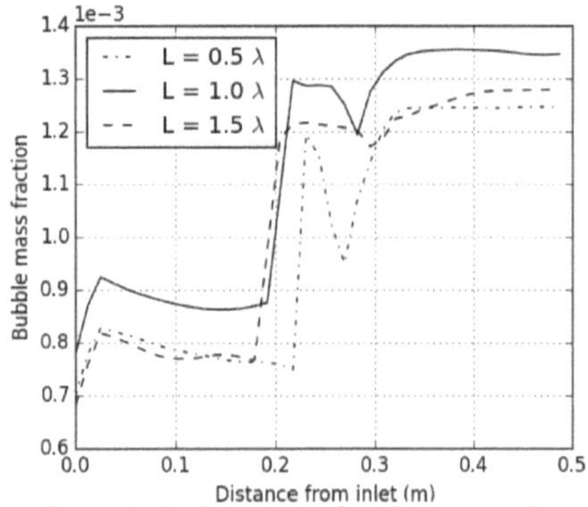

Figure 7. Bubble mass fraction across the launder for water after a run time of 20 s. Mass fraction values are taken along the axis of the launder.

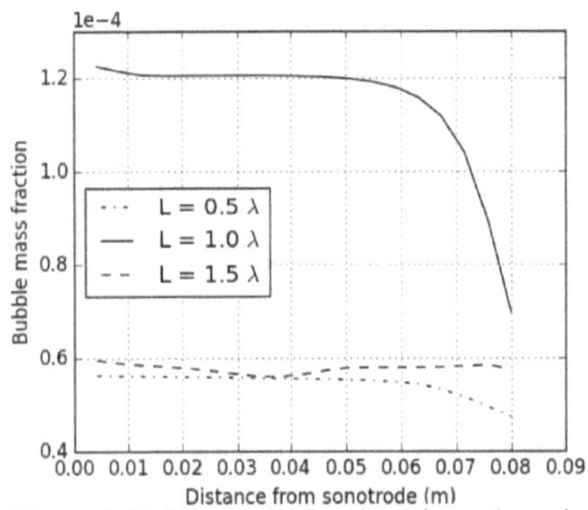

Figure 8. Bubble mass fraction along the axis of the sonotrode for aluminium after a run time of 20 s. Mass fraction values are taken along the axis of the sonotrode.

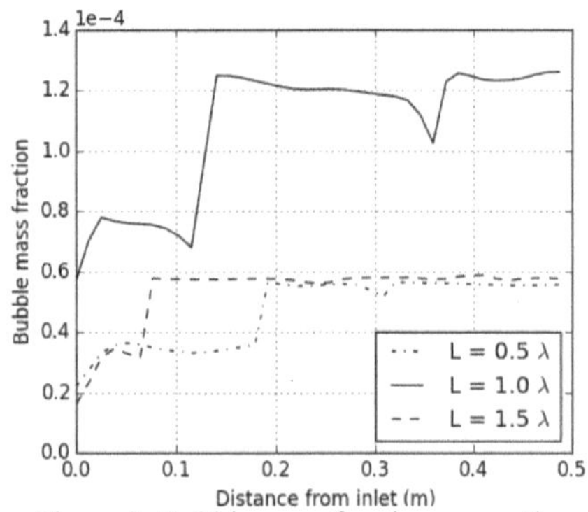

Figure 9. Bubble mass fraction across the launder for aluminium after a run time of 20 s. Mass fraction values are taken along the axis of the launder.

Observing the flow field, two recirculation patterns are seen in two regions: under the sonotrode and downstream of the second baffle. The recirculation pattern is depicted for the aluminium case with a baffle separation of 0.5 λ in Figure 10. The second recirculation pattern can be exploited by adding a second sonotrode above this region.

The minimum residence time, the time taken for the liquid to leave the domain along the shortest path, is given in Table 2. These values are dependent on the flow pattern that is altered due to the baffle separation and the change in local densities due to cavitation activity. The wavelength of aluminium is large compared with the size of the modeled launder, giving the unusually short time to leave the domain for a separation of 1.5 λ since the baffles are close to the inlet and outlets.

Figure 10. Recirculation in configuration with baffles separated at a distance of 0.5 λ for aluminium.

Table 2. Minimum residence time (s) for each case.

L (λ)	$L_{aluminium}$ (cm)	Aluminium (s)	L_{water} (cm)	Water (s)
0.5	11.5	7.0	3.7	2.6
1.0	23.0	7.3	7.4	1.2
1.5	34.5	0.2	11.1	2.4

Conclusions

A multi-scale model for the ultrasonic treatment of liquid metals is presented in this study. The bubble concentration for the flow in a launder with baffles to create recirculation zones in high cavitation activity regions is studied with an improved version of the full cavitation model derived from the Keller-Miksis equation.
The optimal baffle distance for both liquid water and aluminium has been found to be at one wavelength of the forcing frequency, implying that resonance with the driving frequency is a desired trait for ultrasonic cavitation treatment of liquid metal.

Acknowledgement

The authors are grateful to the UK Engineering and Physical Sciences Research Council (EPSRC) for financial assistance for this research in contract numbers: EP/K00588X/1 and EP/K005804/1.

References

[1] G. Eskin, Ultrasonic Treatment of Light Alloy Melts, Amsterdam: Gordon and Breach, 1998.

[2] J. Campbell, "Effects of vibration during solidification," *Int. Met. Rev.,* vol. 26, p. 71, 1981.

[3] N. Alba-Baena and D. Eskin, "Kinetics of ultrasonic degassing of aluminum alloys," in *Light Metals 2013*, John Wiley & Sons, Inc., 2013, pp. 957--962.

[4] G. I. Eskin, "Cavitation mechanism of ultrasonic melt degassing," *Ultrasonics Sonochemistry,* vol. 2, no. 2, pp. S137--S141, 1995.

[5] Y.-C. Wang and C. Brennen, "The noise generated by the collapse of a cloud of cavitation bubbles," in *American Society of Mechanical Engineers*, 1995.

[6] S. Komarov, K. Oda, Y. Ishiwata and N. Dezhkunov, "Characterization of acoustic cavitation in water and molten aluminum alloy," *Ultrasonics Sonochemistry,* vol. 20, no. 2, pp. 754--761, 2013.

[7] I. Senocak and W. Shyy, "Evaluation of cavitation models for Navier-Stokes computations," in *ASME 2002 Joint US-European Fluids Engineering Division Conference*, Montreal, 2002.

[8] I. Senocak and W. Shyy, "A pressure-based method for turbulent cavitating flow computations," *Journal of Computational Physics,* vol. 176, no. 2, pp. 363--383, 2002.

[9] C. L. Merkle, J. Feng and P. Buelow, "Computational modeling of the dynamics of sheet cavitation," in *3rd International Symposium on Cavitation*, Grenoble, 1998.

[10] V. Ahuja, A. Hosangadi and S. Arunajatesan, "Simulations of cavitating flows using hybrid unstructured meshes," *Journal of Fluids Engineering,* vol. 123, no. 2, pp. 331--340, 2001.

[11] R. Kunz, D. Boger, D. Stinebring, T. Chyczewski, J. Lindau, H. Gibeling, S. Venkateswaran and T. Govindan, "A preconditioned Navier-Stokes method for two-phase flows with application to cavitation prediction," *Computers & Fluids,* vol. 29, no. 8, pp. 849--875, 2000.

[12] A. Singhal, M. Athavale, H. Li and Y. Jiang, "Mathematical basis and validation of the full cavitation model," *Journal of Fluids Engineering,* vol. 124, no. 3, pp. 617--624, 2002.

[13] M. Athavale, Y. Jiang, H. Li and A. Singhal, "Application of the full cavitation model to pumps and inducers," *Int. J. Rot. Mach.,* vol. 8, no. 1, pp. 45--56, 2002.

[14] M. S. Plesset, "The dynamics of cavitation bubbles," *J. Appl. Mech.,* vol. 16, p. 227, 1949.

[15] M. S. Plesset and A. Prosperetti, "Bubble dynamics and cavitation," *Annual Review of Fluid Mechanics,* vol. 9, pp. 145--185, 1977.

[16] C. Brennen, Cavitation and Bubble Dynamics, Oxford: Oxford University Press, 1995.

[17] J.-P. Franc and J.-M. Michel, Fundamentals Of Cavitation, Kluwer Academic Publishers, 2004.

[18] L. Nastac, "Mathematical modelling of the solidification structure evolution in the pressence of ultrasonic stirring," *Metall. Mater. Trans. B,* vol. 42, no. 6, pp. 1297--1305, 2011.

[19] J. B. Keller and M. Miksis, "Bubble oscillations of large amplitude," *The Journal of the Acoustical Society of America,* vol. 68, no. 2, pp. 628--633, 1980.

[20] B. E. Launder and D. B. Spalding, "The numerical computation of turbulent flows," *Computer methods in applied mechanics and engineering,* vol. 3, no. 2, pp. 269--289, 1974.

[21] W. Wu, I. Tzanakis, P. Srirangam, S. Terzi, W. Mirihanage, D. G. Eskin and P. Lee, "In situ synchrotron radiography of ultrasound cavitation in a molten Al-10Cu alloy," in *TMS Annual Meeting*, Orlando (submitted), 2015.

Advances in the Science and Engineering of Casting Solidification
Edited by: Laurentiu Nastac, Baicheng Liu, Hasse Fredriksson, Jacques Lacaze, Chun-Pyo Hong, Adrian V. Catalina, Andreas Buhrig-Polaczek, Charles Monroe, Adrian S. Sabau, Roxana Elena Ligia Ruxanda, Alan Luo, Subhayu Sen, and Attila Diószegi
TMS (The Minerals, Metals & Materials Society), 2015

ULTRASONIC PROCESSING OF 6061-BASED NANOCOMPOSITES FOR HIGH PERFORMANCE APPLICATIONS

S. Jia[1], P.G. Allison[1], T.W. Rushing[2], L. Nastac[1]

[1]The University of Alabama, Tuscaloosa, AL, 35487, USA, email: lnastac@eng.ua.edu
[2]US ARMY ERDC, Vicksburg, MS 39190, USA

Keywords: Ultrasonic Cavitation Processing; As-cast 6061 Alloy; Microstructure Evaluation; Tensile Properties Evaluation

Abstract

Previously studies show that microstructure and mechanical properties of a cast component can be considerably improved when ceramic nanoparticles are used as a reinforcement to form a metal-matrix-nano-composite material.
In the present study, 6061 nanocomposite castings were fabricated by using the ultrasonic stirring technology (UST). The 6061 alloy and Al_2O_3/SiC nanoparticles were used as the matrix alloy and the reinforcement materials, respectively. Nanoparticles were inserted into the molten metal and dispersed by ultrasonic cavitation and acoustic streaming. The microstructure, mechanical behavior and mechanical properties of the cast nanocomposites have been investigated in detail in this work. The current experimental results showed that the tensile strength, yield strength of the cast reinforced 6061 alloy with Al_2O_3 or SiC nanoparticles increased slightly while the elongation increased significantly.

Introduction

Aluminum matrix nanocomposites have the potential to offer outstanding properties, including low density, high specific strength, high specific stiffness, excellent wear resistance and controllable expansion coefficient, which make them attractive for numerous applications in aerospace, automobile, and military industries field [1-4].
It is a promising approach to use nano-sized ceramic particles to fabricate metal matrix nano-composite (MMNC), while maintaining good ductility [5, 6]. Currently, there are several fabrication methods of MMNCs, including mechanical alloying with high energy milling [7], ball milling [8], nano-sintering [9], spray deposition, electrical plating, sol-gel synthesis, laser deposition, etc. The mixing of nano-sized ceramic particles is normally lengthy, expensive, and energy consuming.
Several recent studies revealed that ultrasonic cavitation processing is highly efficient in dispersing nanoparticles into the melt [10-11]. Ultrasonic vibration has been extensively used in purifying, degassing, and refinement of metallic melt [12-14], mainly because introducing the ultrasonic energy into a liquid will induce nonlinear effects such as cavitation and acoustic streaming. Ultrasonic cavitation can create small-size transient domains that could reach very high temperatures and pressures as well as extremely high heating and cooling rates. The shock force that takes place during ultrasonic cavitation processing coupled with local high temperatures can break the nanoparticle clusters and clean the surface of the particles [15-17]. Furthermore, ultrasonic vibration can improve the wettability between the reinforced nanoparticles and the metal matrix, which will assist to distribute the nanoparticles more uniformly into the metal matrix.

Al_2O_3 and SiC are widely used as reinforcement particles due to their relatively good thermal and chemical stability. In this article, the effects of the ultrasonically dispersed Al_2O_3 and SiC nanoparticles on the microstructure and mechanical properties of cast 6061 nano-composites are studied in detail.

Experimental Approach

Aluminum alloy 6061 was selected as the metallic matrix because it is readily castable and widely used. The ceramic nanoparticles used in this study were β-SiC (spherical shape, average diameter of about 50 nm) and Al_2O_3 (spherical shape, average diameter of 20 nm).

The ultrasonic processing system used in this study is illustrated in Figure 1. The main parameters of the ultrasonic equipment are: maximum effective power, P = 2.4kW and frequency, f = 18 kHz. An induction furnace with a capacity of 2.7 kg was used to melt the 6061 alloy. After the alloy is melted, the Nb ultrasonic probe was inserted to about 50 mm beneath the melt surface to perform ultrasonic stirring at 1.75 kW and 18 kHz frequency. 1 wt% nanoparticles (Al_2O_3 or SiC) were injected into the cavitation area (just beneath the ultrasonic probe) during a 15 min time-frame. The 6061 molten pool was protected by Argon gas atmosphere. A thermocouple was used to monitor the melt temperature to control the superheat. A higher pouring temperature of 750°C was used to minimize the formation of metal-mold filling defects including cold-shuts. The metal mold was preheated to 400°C. The specimen was extracted from the metal mold after 30 min and tested on a tensile test machine. The dimensions of the specimen are 50.8 mm length and 12.7 mm diameter. The experiments were repeated several times for statistical interpretation of the results.

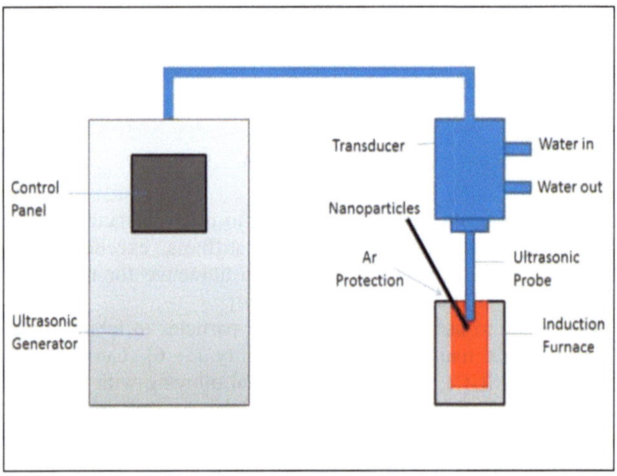

Figure 1 A sketch of the ultrasonic cavitation processing system.

Results And Discussion

<u>SEM Analysis Of 6061/1wt%SiC Sample</u>

Figure 2 shows the distribution of SiC nanoparticles in the 6061/1wt% SiC MMNC sample. It can be seen that the SiC nanoparticles were dispersed reasonably well into the 6061 matrix. It can also be seen from Figure 2(c) that some particle agglomeration occurred during the MMNC processing.

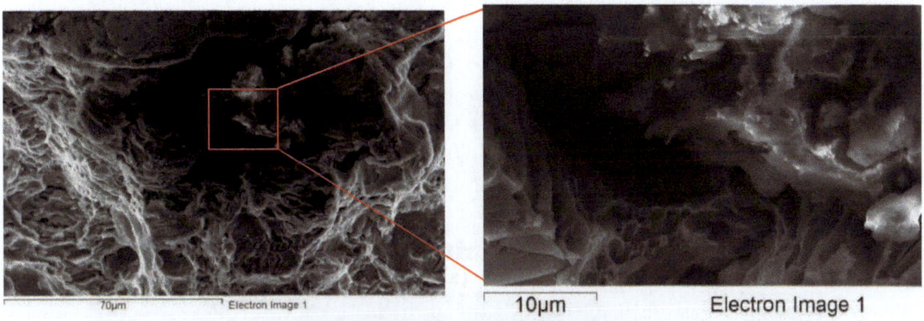

(a)

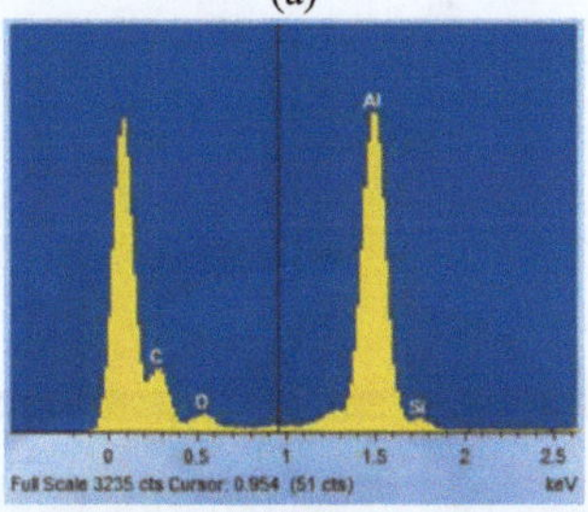

(b)

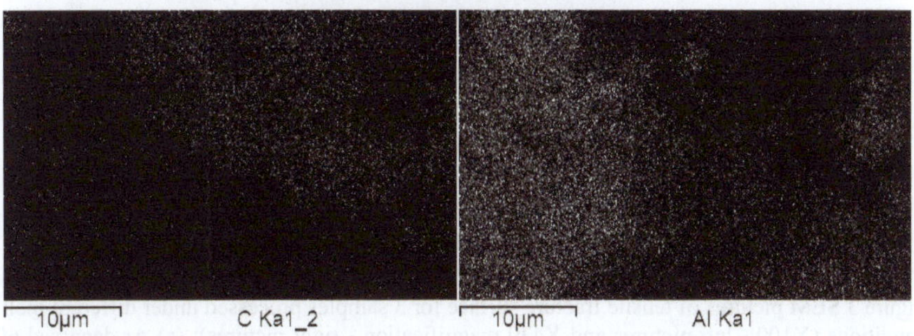

(c) (d)

Figure 2: (a) SEM tensile fracture surface of 6061/1% SiC nanocomposite; (b) EDS for the whole area on the tensile fracture surface; (c) EDS mapping for carbon; (d) EDS mapping for aluminum.

To double check the SiC distribution in the nanocomposites, EDS (Energy Dispersive X-ray Spectroscopy) analysis was also performed. Figure 2(b) shows the EDS spectrum of the carbon distribution in the 6061 nanocomposites, which means SiC nanoparticles are indeed reasonably well dispersed into the 6061 alloy.

From Figure 3, we can see that the samples failed in more of a brittle transgranular manner. Fracture path goes through the grains, and the brittle facets were caused by a fast propagation of the crack transgranularly across the grains.

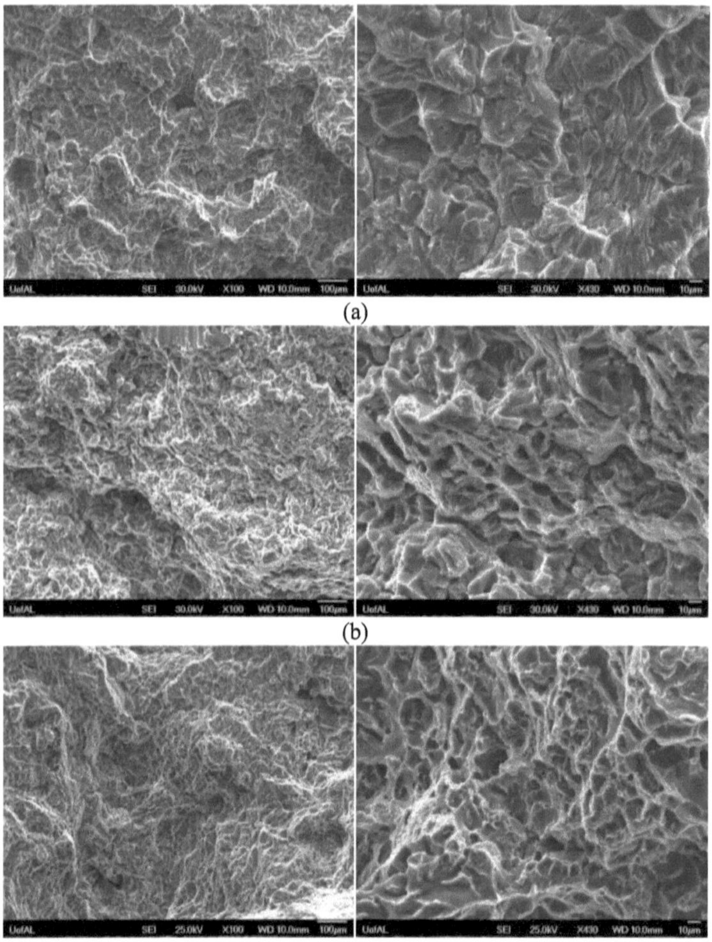

(a)

(b)

(c)

Figure 3 SEM pictures of tensile fracture surface for 3 samples processed under different melting conditions (X100 - left pictures and X430 magnification - right pictures): (a) Ar degassed 6061 alloy casting; (b) UST+1% SiC nanocomposite casting and (c) UST+1% Al_2O_3 nanocomposite casting.

One of the main mechanisms of ductile fracture is known as microvoid coalescence. In this type of failure, microscopic contaminants in the metal act as stress concentrators. When a triaxial stress state is concentrated at these points, the surrounding material pulls away from the contaminants forming voids in the metal. These voids grow until adjacent voids connect, coalescing into larger voids. Final failure of the material occurs when the voids grow larger and connect together. Also, the dimpled surface of nanocomposite samples indicates the fracture mechanism is dominated by microvoid coalescence, which is ductile fracture. The addition of nanoparticles really changed the fracture mechanism from brittle dominated shown in Figure 4(a) to ductile dominated shown in Figure 3(b) and 3(c).

Analysis of Mechanical Properties

A mechanical testing was performed to evaluate tensile strength and elongation (see Table I) of the samples obtanied under 3 different process conditions. As shown in Table I, by using the UST processing and the addition of the nanoparticles, the tensile strength almost remains the same. On the other hand, the elongation almost tripled after UST processing and the addition of nanoparticles.

Table I. Tensile Testing Results

Samples	Tensile strength MPa	Elongation %
Ar Degassed	163.3± 9.2	4.7±0.9
UST + 1% wt. SiC	158.6±10.6	13.1±2.4
UST + 1% wt. Al_2O_3	163.5±12.9	13.2±1.9

Conclusions

By using SEM and EDS analyses, it was shown in this study that the Al_2O_3/SiC ceramic nanoparticles were dispersed reasonably well into the 6061 matrix; but some insignificant particle agglomeration still occurred. This agglomeration could happen during the UST processing as well as during casting and solidification. A significant improvement in the ductility of the as-cast 6061 nanocomposite was achieved after the addition of the Al_2O_3/SiC ceramic nanoparticles via ultrasonic processing.

Acknowledgements

The authors would like to acknowledge Lyan Garcia and Jeb S. Tingle for their encouragement of this study. Permission to publish was granted by the Director, Geotechnical and Structures Laboratory.

References

1. B. Su, H.G. Yan, G. Chen, J.L. Shi, J.H. Chen and P.L. Zeng, "Study on the Preparation of the SiCp/Al-20Si-3Cu Functionally Graded Material Using Spray Deposition," *Mater Sci Eng A*, 527 (24) (2010), 6660-6665
2. S.M. Seyed Reihani, "Processing of Squeeze Cast Al 6061-30 vol% SiC Composites and Their Characterization," *J. Materials and Design*, 27 (3) (2004), 216-222

3. D. Bozic, B. Dimcic, O. Dimcic, J. Stasic and V. Rajkovic, "Influence of SiC Particles Distribution on Mechanical Properties and Fracture of DRA Alloys," *Materials & Design*, 31 (1) (2010), 134-141
4. P. Michael , De Cicco, X.C. Li and L.S Turng, "Semi-solid Casting (SSC) of Zinc Alloy Nanocomposites," *J. Mater. Process. Technol.*, 209 (18–19) (2009), 5881-5885
5. C.S. Goh, J. Wei, L.C. Lee and M. Gupta, "Simultaneous Enhancement in Strength and Ductility by Reinforcing Magnesium with Carbon Nanotubes," *Mater. Sci. Eng., A*, 423 (1–2) (2006), 153-156
6. A.B. Elshalakany, T. A. Osman, A. Khattab, B. Azzam, and M. Zaki, "Microstructure and Mechanical Properties of MWCNTs Reinforced A356 Aluminum Alloys Cast Nanocomposites Fabricated by Using a Combination of Rheocasting and Squeeze Casting Techniques," *J. Nanomater.*, 2014, 14
7. X.F. Chen, E.G. Baburai. F.H. Froes and A. Vassel, "Ti-6Al-4V/SiC Composites by Mechanical Alloying and Hot Isostatic Pressing," *Proc. Adv. Part. Mater. Processes*, 1997, 185-192
8. S.L Utriga Filho, R. Rodriguez, J.C. Eartman and E.J. Lavernia, "Powder Metallurgy-metal Matrix Composites-synthesis of Diamond Reinforced Al-Mg Nano Crystalline Composite Powder Using Ball Milling," *Mater Sci. Forum*, 416-418 (2003), 213-218
9. J.R. Groza, "Sintering of Nanocrystalline Powders", Int. J. of Powder Metall. 35 (1999), 59-66
10. Y. Yang, J. Lan and X.C. Li, "Study on Bulk Aluminum Matrix Nano-composite Fabricated by Ultrasonic Dispersion of Nano-sized SiC Particles in Molten Aluminum Alloy," *Mater Sci Eng A*, 380 (2004), 373-378
11. S. Mula, P. Padhi, S.C. Panigrahi, S.K. Pabi and S. Ghosh, "On Structure and Mechanical Properties of Ultrasonically Cast Al-2% Al2O3 Nanocomposite," *Mater Res Bull*, 44 (2009), 1154-1160
12. G.I. Eskin, "Effect of Ultrasonic Treatment of the Melt on the Microstructure Evolution During Solidification of Aluminum Alloy Ingots," *Zeitschrift fuer Metallkunde*, 93 (2002), 502-506
13. X. Jian, H. Xu, T. Meek and Q. Han, "Effect of Power Ultrasound on Solidification of Aluminum 6061 Alloy," *Mater. Lett.*, 59 (2 - 3) (2005), 190
14. G. I. Eskin, "Principles of Ultrasonic Treatment: Application for Light Alloys Melts," *Adv. Perform. Mater.*, 4 (1997), 223-232
15. J. Lan, Y. Yang and X.C. Li, "Microstructures and Microhardness of SiC Nanoparticles Reinforced Magnesium Composites Fabricated by Ultrasonic Method," *Mater Sci Eng A*, 386 (2004), 284-290
16. G. Cao, H. Choi, H. Konishi, S. Kou, R. Lakes and X. Li, "Mg-6Zn/1.5% SiC Nanocomposites Fabricated by Ultrasonic Cavitation-based Solidification Processing," *J. Mater. Sci.*, 43 (16) (2008), 5521
17. J.C. Yan,Z.W. Xu, L. Shi, X. Ma and S.Q. Yang, "Ultrasonic Assisted Fabrication of Particle Reinforced Bonds Joining Aluminum Metal Matrix Composites," *Mater. Des.*, 2011, 343-347

Advances in the Science and Engineering of Casting Solidification
Edited by: Laurentiu Nastac, Baicheng Liu, Hasse Fredriksson, Jacques Lacaze, Chun-Pyo Hong, Adrian V. Catalina, Andreas Buhrig-Polaczek, Charles Monroe, Adrian S. Sabau, Roxana Elena Ligia Ruxanda, Alan Luo, Subhayu Sen, and Attila Diószegi
TMS (The Minerals, Metals & Materials Society), 2015

Numerical Modeling of the Dispersion of Ceramic Nanoparticles during Ultrasonic Processing of A356-based Nanocomposites

Daojie Zhang and Laurentiu Nastac

The University of Alabama, Tuscaloosa, AL, 35487, USA, email: lnastac@eng.ua.edu

Abstract

The metal-matrix-nano-composite in this study consist of a A356 alloy matrix reinforced with 1.0 wt.% SiC-nanoparticles dispersed within the matrix via ultrasonic cavitation system, available in the Solidification Laboratory at The University of Alabama. The required ultrasonic parameters to achieve cavitation for adequate degassing and refining of the A356 alloy as well as the fluid flow and solidification characteristics for uniform dispersion of the nanoparticles into the aluminum alloy matrix are being investigated via CFD ultrasonic cavitation modeling. The multiphase CFD model for nanoparticle dispersion accounts for turbulent fluid flow, heat transfer and solidification as well as the complex interaction between the molten alloy and nanoparticles by using the Ansys's Fluent DDPM model. The modeling parametric study includes the effects of ultrasonic probe location, the fluid flow intensity, and the initial location where the nanoparticles are released into the molten alloy.

Keywords: Metal-matrix-nano-composites; Ultrasonic Processing; Modeling of Ceramic Nanoparticle Dispersion; Aluminum Alloys

1. Introduction

Aluminum-based metal matrix composites (MMCs) have been extensively studied and widely used in the aerospace, automotive and military industries due to their high strength-to-weight ratios and enhanced mechanical and thermal properties including specific modulus, superior strength, stiffness, good wear resistance, fatigue resistance and improved thermal stability [1-3]. However, the particles commonly used are micron-sized which has a counterpart that the ductility of the MMCs deteriorates with high ceramic particle concentration [4]. Consequently, more attention has been drawn to metal matrix nanocomposites (MMNCs), since the properties of metallic alloys reinforced by ceramic nanoparticles (with dimensions less than 100 nm) would be enhanced considerably while the ductility of the matrix is retained [5-11], because the incorporation of nano-sized reinforcements in metals and alloys will result in an Orowan-type strengthening mechanism where line defects are pinned by a uniform dispersion of particles within the grains, causing the dislocations to bow, and experimental evidence also shows that a significant degree of grain refinement (i.e., Hall-Petch strengthening) often occurs with additions of nanoparticles [12].

However, it is extremely difficult to obtain uniform dispersion of nano-sized ceramic particles in liquid metals due to high viscosity, poor wettability in the metal matrix, and a large surface-to-volume ratio, which results in agglomeration and clustering [4]. Currently, several fabrication technologies including high-energy ball milling [8, 11], in-situ synthesis [7], electroplating [13], and ultrasonic technology (UST) [4-5] are most commonly used, among which UST is supposed to be more reliable and cost effective.

During the solidification process, the particles will either be pushed or engulfed by the solidification front, among which particle pushing will always lead to segregation even clustering of the particulate reinforcement, which is undesirable as it results in non-homogeneous response and lower macroscopic mechanical properties. In general, it is considered that whether particles are pushed or engulfed during solidification depends on the velocity of the particle relative to the solidification front according to several previous models describing such particle engulfment and pushing phenomena [14-18]. However, these models only predict the behavior in the coarse (>> 1 μm) and fine particle (~ 1 μm) systems, and they don't accurately describe the ultrafine particle (<< 1 μm) system, presumably because the models rely on continuum mechanics. But these models cannot explain the evidence in MMNCs that nanoparticles can indeed be engulfed and distributed throughout the material and are not necessarily concentrated in grain boundary or interdendritic regions. As proposed by Ferguson [12], for sufficiently small particles, "Brownian Motion" can partially or completely counteract forces such as viscous drag, gravity and thermal/concentration gradients, thus leading to engulfment rather than pushing.
In this study, Ansys's Fluent Dense Discrete Phase Model (DDPM) [19] was adapted. The DDPM accounts for turbulent fluid flow, heat transfer, and the complex interaction between the molten alloy and nanoparticles during the melting and unidirectional solidification processes. Based on the theory proposed by Ferguson, all of the nanoparticles are assumed to be engulfed by the solidification front and no entrapment will occur. The dispersion of SiC nanoparticles with different injection positions, fluid flow intensities, and probe locations have been investigated in detail.

2. Model Description

The geometry of the model is shown in Figure 1. The ultrasonic probe has a diameter of 40 mm. The liquid aluminum is A356. It has a density of 2685 kg/m^3. The SiC nanoparticles with an average particle size of 55 nm and density of 3216 kg/m^3 are treated as inert-particles. The mass flow rate of the SiC nanoparticles is 0.014 kg/s. Thus, 1.0 wt.% of SiC nanoparticles can be injected at about 20 mm above the bottom of the furnace for 1 sec.

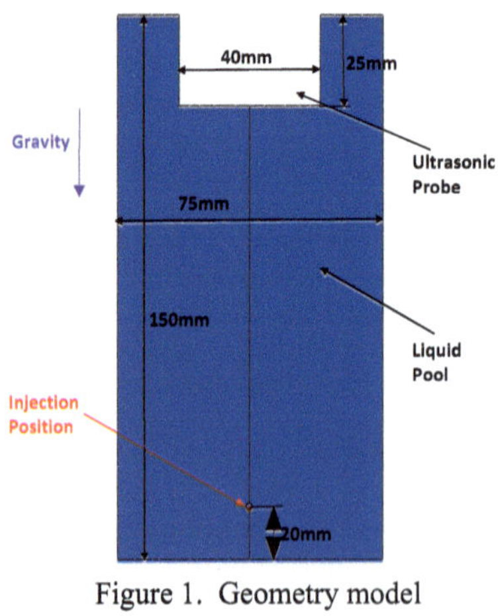

Figure 1. Geometry model

The multiphase computational fluid dynamics (CFD) model accounts for turbulent fluid flow, heat transfer, and the complex interaction between the molten alloy and nanoparticles by using the ANSYS Fluent DDPM and k-ω turbulence model [19]. The CFD model is described below.

2.1 Fluid Flow Model

In the Eulerian DDPM multiphase model an Eulerian treatment is used for each phase, and the discrete phase (nanoparticles) is designated as a granular phase. The volume fraction of the particulate phase is accounted for in the conservation equations.

The continuity equation for the phase q is

$$\frac{\partial}{\partial t}\left(a_q \rho_q\right)+\nabla \cdot\left(a_q \rho_q \mathbf{u}_q\right)=\dot{m}_{pq}-\dot{m}_{qp} \tag{1}$$

The momentum balance for the phase q yields

$$\frac{\partial}{\partial t}\left(a_q \rho_q \mathbf{u}_q\right)+\nabla \cdot\left(a_q \rho_q \mathbf{u}_q \mathbf{u}_q\right)=-a_q \nabla P+\nabla \cdot\left[a_q \mu_q\left(\nabla \mathbf{u}_q+\nabla \mathbf{u}_q{}^T\right)\right]+a_q \rho_q g+\mathbf{f}_{DPM}+\mathbf{f}_{other} \tag{2}$$

where a_q is the phasic volume fraction, ρ_q is the density, $\mathbf{u}_q$ is the velocity, μ_q is the molecular viscosity, and P is the pressure shared by all phases. $\dot{m}_{pq}$ characterizes the mass transfer from the pth to q^{th} phase, and $\dot{m}_{qp}$ characterizes the mass transfer from phase q to phase p. The momentum exchange term, $\mathbf{f}_{DPM}$, is considered only in the primary phase equations. The source term, $\mathbf{f}_{other}$, includes the virtual mass force, lift force, and turbulent dispersion force etc.

Equations (1) and (2) do not solve for the velocity field and volume fraction of the discrete phase. Their values are obtained from the Lagrangian tracking solution.

2.2 Solidification Model

The enthalpy method is used in the solidification model. The energy conservation equation of the enthalpy-formulation is

$$\rho \frac{\partial h}{\partial T}+\rho \nabla \cdot(\mathbf{u} h)=\nabla \cdot(k \nabla T)+Q_L \tag{3}$$

Where h is the sensitive enthalpy defined as

$$h=h_{ref}+\int_{T_{ref}}^{T} c_p dT \tag{4}$$

h_{ref} is the reference enthalpy at the reference temperature T_{ref} ; c_p is the specific heat; ρ and k are the density and thermal conductivity of the melting aluminum respectively and $\mathbf{u}$ is the velocity of the fluid.

The source term Q_L concerning the latent heat in a single phase solidification model can be written as

$$Q_L=\rho L \frac{\partial f_s}{\partial T} \tag{5}$$

Here, f_s is the solid fraction which is assumed to vary linearly in the mushy zone and it can be defined as

$$f_s = \begin{cases} 0 & T > T_L \\ (T_L - T)/(T_L - T_S) & T_S < T < T_L \\ 1 & T < T \end{cases} \qquad (6)$$

where L is the latent heat; T_L is the liquidus temperature; T_S is the solidus temperature.
The mushy region is treated as a porous medium, and the porosity in each cell is set equal to the liquid fraction f_l in that cell. The momentum sink due to the reduced porosity in the mushy zone takes the following form:

$$\mathbf{S} = \frac{f_s^2}{(f_l^3 + \varepsilon)} A_{mush}\mathbf{u} \qquad (7)$$

where ε is a small number (0.001) to prevent division by zero; A_{mush} is the mushy zone constant, which measures the amplitude of the damping.
The dynamic viscosity of the fluid also depends on the solid fraction which is related to temperature. Thus, the viscosity (unit: kg/(m·s)) is expressed as

$$\mu = \begin{cases} 0.5e-3 & T > T_L + 100 \\ 1.0e-3 & T_L < T < T_L + 100 \\ 3.0e-3 & T_S < T < T_L \\ 0.1 & 10 < T < T_S \end{cases} \qquad (8)$$

2.3 Particle Tracking Model

The trajectory of a discrete phase particle is predicted by integrating the force balance on the particle. The force balance equates the particle inertia with the forces acting on the particle, and can be written as

$$\frac{d\mathbf{u}_p}{dt} = \mathbf{F}_D + \mathbf{F}_G + \mathbf{F}_B + \mathbf{F}_{virtual-mass} + \mathbf{F}_{pressure-gradient} + \mathbf{F}_{lift} + \mathbf{F}_{interaction} \qquad (9)$$

where $\mathbf{u}_p$ is the particle velocity, and all the terms at the right-hand are with a unit of force/unit particle mass. The detailed description of the force terms shown in Equation (9) are presented in [19].
The chaotic effect of turbulence on the particle trajectories is accounted for using the stochastic tracking approach, i.e., the discrete random walk (DRW) model:

$$\mathbf{u} = \bar{\mathbf{u}} + \zeta\sqrt{2k/3} \qquad (10)$$

where $\bar{\mathbf{u}}$ is the mean fluid velocity in the trajectory equation (9), ζ is a normally distributed random number, and k is the local turbulent kinetic energy.
Equation (9) can be cast into the following general form:

$$\frac{d\mathbf{u}_p}{dt} = \frac{1}{\tau_p}(\mathbf{u} - \mathbf{u}_p) + \mathbf{a} \tag{11}$$

where the term $\mathbf{a}$ includes accelerations due to other forces except drag force.

Integrating the transport equation (11) for the path of each particle yields

$$\frac{d\mathbf{x}_p}{dt} = \mathbf{u} \tag{12}$$

where $\mathbf{x}_p$ is the particle position.

2.4 Boundary Conditions

The ultrasonic probe surface is set as velocity inlet during the first 2.0 sec, and adiabatic wall after that. The interface between liquid aluminum and air is pressure outlet and the other boundaries are set as wall all the time. The backflow temperature at the outlet is calculated in UDF. All the walls are considered to be adiabatic during the first 2.0 sec. After that, the heat transfer boundary condition for the bottom wall of the furnace is convection and the heat transfer coefficient is 1000 W/($m^2 \cdot s$), but all the other walls are still treated as adiabatic. All of the Discrete Phase BC Types are set as reflect. The velocity inlet profile imitating the vibration of the ultrasonic probe is defined in UDF, which is dependent on time as shown in Figure 2.

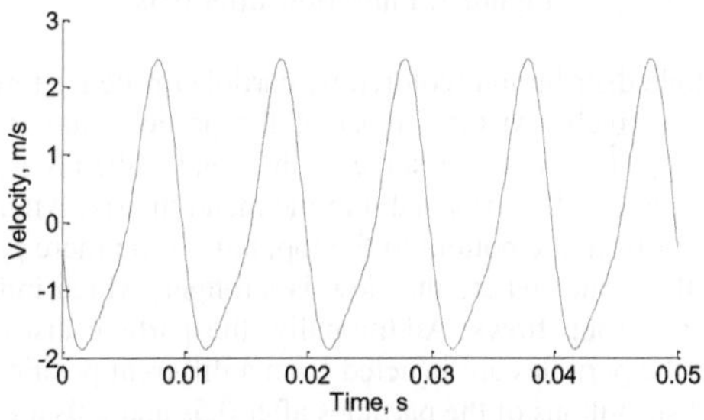

Figure 2. Velocity inlet profile

2.5 Solution Procedure

The SiC nanoparticles are injected at every fluid flow time step with a mass flow rate of 0.014 kg/s in the first sec. The distribution of the particle diameters varying from 45 nm to 65 nm follows the Rosin-Rammler expression. Particles are tracked at every time step after the fluid velocity field is solved. Because of the low volume fraction of the discrete phase, one-way coupling is employed, which neglects the effect of the discrete phase on the fluid turbulence.

3. Simulation Results and Discussion

Figure 3 shows the fluid flow (colored by velocity magnitude, similarly hereinafter) after 0.5s. It can be seen that the flow is much stronger at the center of the furnace. It is also assumed that the particles have little effect on the fluid flow (because of the one-way coupling), so the fluid flow after longer time (*e.g.*, 2s) is almost the same as that after 0.5s.

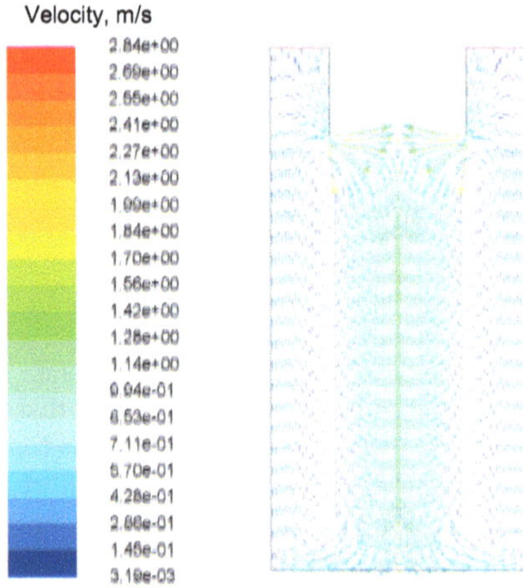

Figure 3. Fluid flow after 0.5s

Figure 4 shows the particle distribution (colored by particle residence time, similarly hereinafter) after 0.5s and 2.0s, respectively. At 0.5s, most of the particles are still at the bottom of the furnace where they are injected, however some of them have already gone to the top, but more time is needed for all of them to be dispersed into the metal matrix. After 2.0s, the particles have been dispersed pretty well from the bottom to the top, but a little more particles tend to stay near the wall, and fewer at the center where the flow is stronger, which indicates the nanoparticles couldn't disperse well in strong flows. Additionally, the particle distribution stays almost the same henceforth. When the particles are injected from a different position which is about 15 mm beneath the probe, the distributions of the particles after 0.5s and 2.0s are shown in Figure 5.

It is obvious that the particles are following the fluid flow. In the beginning, they are carried by the flow in the center to the bottom, and then back to the top near the wall. Nonetheless, after 2.0s when the distribution becomes stable, it has little difference with that when the particles are injected at the bottom, which demonstrates that the injection position will not affect the final distribution of the SiC nanoparticles.

To investigate the effect of the magnitude of the fluid flow on the dispersion of nanoparticles, the velocity magnitude is changed to be 1/10th of the original one. As the fluid flow is much weaker, the dispersion of the particles is much slower. After 20s, the distributions of the nanoparticles with different injection positions, which are shown in Figure 6, become almost stable. When the particles are injected at the bottom of the furnace, plenty of them aggregate at the bottom because the fluid flow near the wall is too weak to take them to the top. On the contrary, when they are injected at the top, they can be taken to the bottom by the relatively stronger flow at the middle of the furnace, which results in a much more uniform distribution of the particles.

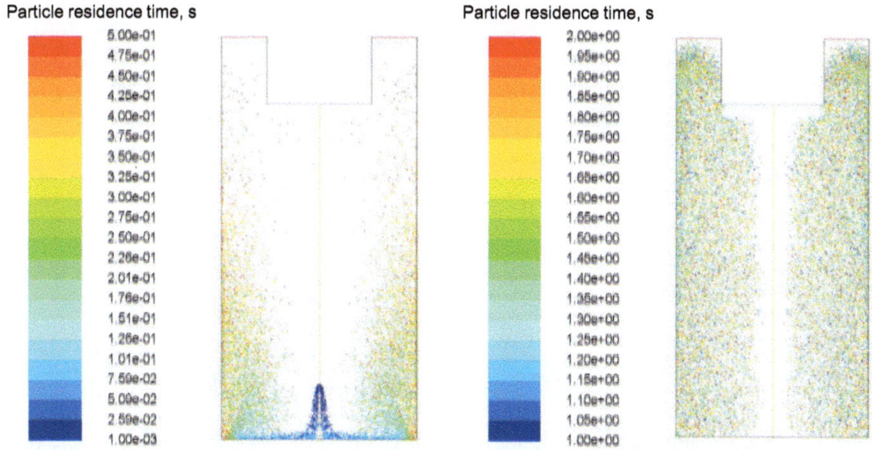

(a) Fluid flow (b) Particle distribution

Figure 4. Fluid flow and particle distribution after 0.5s ((a) and (b) and 2.0s ((c) and (d))

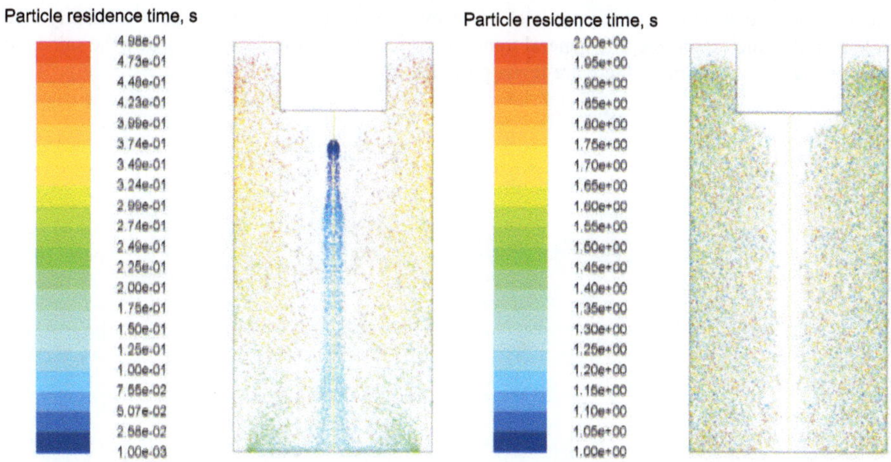

(a) After 0.5s (b) After 2.0s

Figure 5. Particle distributions after 0.5s (a) and 2.0s (b) with a different injection location

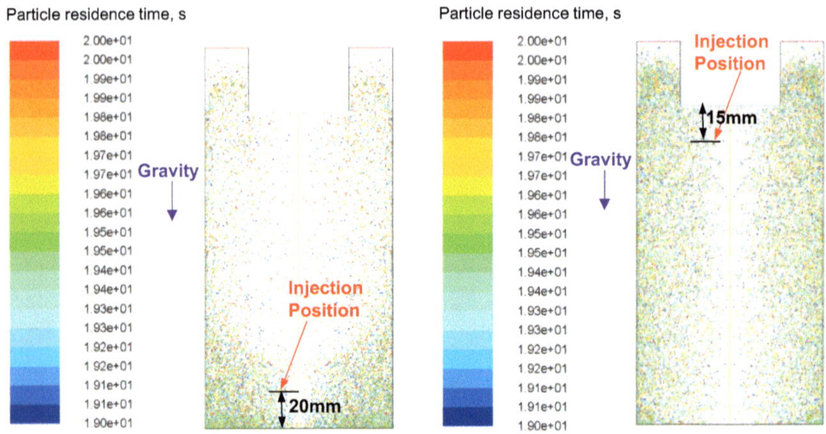

(a) At the bottom (b) At the top

Figure 6. Particle distributions after 20s with 2 different injection positions and a weaker flow

Figure 7 presents the fluid flow and particle distribution after 2.0s when the ultrasonic probe is placed at the bottom of the furnace with a stronger flow. Compared to Figure 3, the flow pattern is different because of the chance in the gravitational acceleration orientation, which had resulted in a different distribution of the particles. However, as the gravitational force is relatively weak compared to the drag force, the general trend of particle distribution stays basically the same, *i.e.*, where the flow is stronger, there are fewer particles, and vice versa.

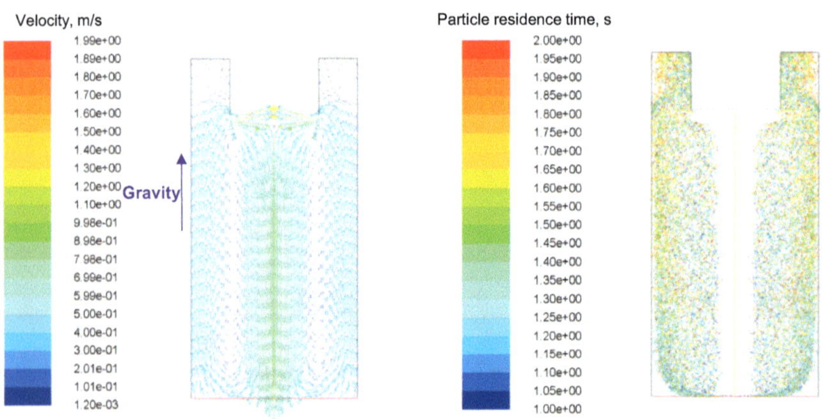

(a) Fluid flow (b) Particle distribution

Figure 7. Fluid flow and particle distribution after 2.0s with the ultrasonic probe placed at the bottom of the furnace (gravity vector up)

For the second case, the nanoparticles are injected beneath the ultrasonic probe, which is placed at the top of the furnace. In this case, the flow is strong enough to disperse the nanoparticles. After 2.0 sec all the particles are well dispersed and the ultrasonic probe is removed. In order to model the unidirectional solidification process, all the walls except the bottom one are treated as adiabatic. The particle distributions associated with the solidification after 2min and 4min are shown in Figures 8 and 9, respectively. As it can be seen from these figures, all the particles will be engulfed and uniformly distributed in the metal matrix based on the assumption that the Brownian motion will dominate during the particle engulfment/pushing process.

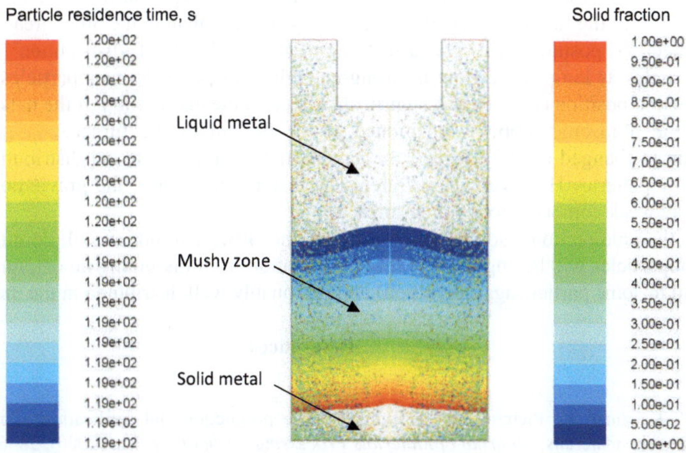

Figure 8. Particle distribution associated with the solidification after 2 min

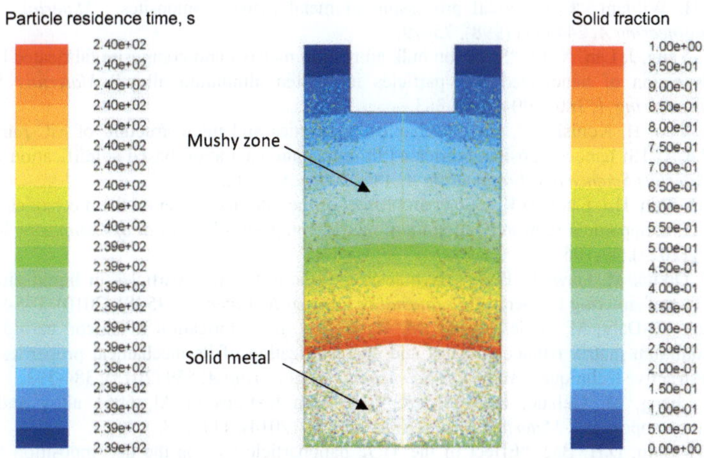

Figure 9. Particle distribution associated with the solidification after 4 min

4. Conclusions

The DDPM model coupled with the k-ω turbulence model is used to investigate the distribution of SiC nanoparticles with different injection positions and probe locations in the liquid A356 alloy under the ultrasonic stirring condition. Several forces acting on a particle including transverse drag force, buoyancy and gravitational force, virtual mass force, pressure gradient force, and lift force are accounted for to predict the trajectory of the particle. The simulation results reveal the following conclusions:

1. The particles are dispersed pretty well in the liquid pool except that there are fewer particles at the center of the furnace where the fluid flow is stronger than in the other areas.
2. The injection position will not affect significantly the final distribution of the SiC nanoparticles as long as the flow is strong enough to disperse the nanoparticles, otherwise, the injection position could have a significant effect on the distribution of the nanoparticles.
3. When the ultrasonic probe is positioned at the bottom of the furnace, *i.e.*, the gravity direction is changed to 180 degrees, the nanoparticles have a different distribution due to a new flow pattern. However, the difference is insignificant, since the gravitational force is relatively weak compared to the drag force.
4. During the unidirectional solidification process, the Brownian motion will dominate, and all the nanoparticles will be engulfed by the solidification front (assuming no entrapment, which can lead to some particle agglomeration) and reasonably well distributed in the metal matrix.

References

[1] J.W. Kaczmar, K. Pietrzak, W. Wlosinski. "The production and application of metal matrix composite materials," *Journal of Materials Processing Technology*, 106 (2000), 58–67.

[2] K. Durisinova, J. Durisin, M. Orolinova et al. "Effect of particle additions on microstructure evolution of aluminium matrix composite," *Journal of Alloys and Compounds*, 525 (2012), 137-142.

[3] C.H. William. "Commercial processing of metal matrix composites," *Materials Science and Engineering A*, 244 (1) (1998), 75–79.

[4] Y. Yang, J. Lan, X. Li. "Study on bulk aluminum matrix nano-composite fabricated by ultrasonic dispersion of nano-sized SiC particles in molten aluminum alloy," *Materials Science and Engineering A*, 380 (2004), 378-383.

[5] G. Cao, H. Konishi, X. Li. "Mechanical properties and microstructure of SiC-reinforced Mg-(2,4)Al-1Si nanocomposites fabricated by ultrasonic cavitation based solidification processing," *Materials Science and Engineering A*, 486 (2008), 357-362.

[6] J.H. Shin, H.J. Choi, M.K. Cho et al. "Effect of the inter face layer on the mechanical behavior of TiO_2 nanoparticle reinforced aluminum matrix composites," *Journal of Composite Materials*, 48 (1) (2014), 99-106.

[7] B. Dikici, M. Gavgali, Bedir. "Synthesis of in situ TiC nanoparticles in liquid aluminum: the effect of sintering temperature," *Journal of Composite Materials*, 45 (8) (2010), 895-900.

[8] A.A. El-Daly, M. Abdelhameed, M. Hashish et al. "Fabrication of silicon carbide reinforced aluminum matrix nanocomposites and characterization of its mechanical properties using non-destructive technique," *Materials Science and Engineering A*, 559 (2013), 384-393.

[9] X. Jiang, M. Galano, F. Audebert. "Extrusion textures in Al, 6061 alloy and 6061/SiC_p nanocomposites," *Materials Characterization*, 88 (2014), 111-118.

[10] J.H. Shin, D.H. Bae. "Effect of the TiO2 nanoparticle size on the decomposition behaviors in aluminum matrix composites," *Materials Chemistry and Physics*, 143 (2014), 1423-1430.

[11] C. Borgohain, K. Acharyya, S. Sarma et al. "A new aluminum-based metal matrix composite reinforced with cobalt ferrite magnetic nanoparticle," *Journal of Materials Science*, 48 (2013), 162-171.

[12] J.B. Ferguson, B.F. Schultz, P.K. Rohatgi et al. "Brownian Motion Effects on the Particle Settling and Its Application to Solidification Front in Metal Matrix Composites," Light Metals TMS, 2014, 1383-1388.

[13] F.K. Sautter. "Electrodeposition of dispersion-hardened Nickel-Al2O3 Alloys," *Journal of the Electrochemical Society*, 110 (1963), 557.

[14] J.K. Kim and P.K. Rohatgi, "An Analytical Solution of the Critical Interface Velocity for the Encapturing of Insoluble Particles by a Movinb Solid/Lquid Interface," *Metallurgical Materials Transaction A*, 29 (1998), 351-358.

[15] D.R. Uhlmann, B. Chalmers, and K.A. Jackson, "Interaction between Particles and a Solid-Liquid Interface," *Journal of Applied Physics*, 35 (1964), 2986-2993.

[16] D.M. Stefanescu, A. Moitra, A.S. Kacar, and B.K. Dhindaw, "The Influence of Buoyant Forces and Volume Fraction of Particles on th Particle Pushing Entrapment Transition During Directional Solidification of Al/Sic and Al/Graphite Composites," *Metallurgical Transactions A*, 21 (1990), 231-239.

[17] D. Shangguan, S. Ahuja, and D.M. Stefanescu, "An Analytical Model for the Interaction between an Insoluble Particle and an Advancing Solid Liquid Interface," *Metallurgical Transactions A*, 23 (1992), 669-680.

[18] G. Kaptay, "Interfacial Criterion of Spontaneous and Forced Engulfment of Reinforcing Particles by an Advancing Solid/Liquid Interface," *Metallurgical Materials Transaction A*, 32 (2001), 993-1005.

[19] Fluent 6.3: User's Guide Manual Fluent Inc. and Ansys's Fluent, (2006), http://ansys.com/.

Advances in the Science and Engineering of Casting Solidification
Edited by: Laurentiu Nastac, Baicheng Liu, Hasse Fredriksson, Jacques Lacaze, Chun-Pyo Hong, Adrian V. Catalina, Andreas Buhrig-Polaczek, Charles Monroe, Adrian S. Sabau, Roxana Elena Ligia Ruxanda, Alan Luo, Subhayu Sen, and Attila Diószegi
TMS (The Minerals, Metals & Materials Society), 2015

OPTICAL FLOATING-ZONE CRYSTAL GROWTH OF HEUSLER Ni-Mn-Sn ALLOY

Jinke Yu [1], Jian Ren [1], Hongwei Li [1], Hongxing Zheng [1*]

[1] Laboratory for Microstructures, Shanghai University, Shanghai 200072. China

Keywords: Heusler Ni-Mn-Sn, Optical floating zone method, Unidirectional growth, Martensitic transformation

Abstract

Heusler Ni-Mn-Sn unidirectional crystal was fabricated using optical floating zone furnace in the present work. Higher static argon pressure up to 7 bar was employed to suppress the volatilization, and meanwhile, a titanium bar was preheated to remove residual oxygen in the furnace chamber prior to the formal crystal growth. Experimental results show that the obtained $Ni_{50}Mn_{37}Sn_{13}$ unidirectional crystal was $\beta\{110\}$ preferentially oriented at the rate of 6 mm/h. Magnetization curves measured along different directions under a low magnetic field of 100 Oe demonstrates strong anisotropy.

1. Introduction

Heusler Ni–Mn based materials attracted much attention during the past decades due to their giant magnetocaloric effect [1,2]. Ni–Mn–Sn is highly potential for practical application with the advantages of excellent ductility and low-cost [3]. Present researches on the Ni-Mn-Sn have been concentrated on ribbons and bulk alloys [4-6]. On the other hand, it has confirmed that single crystals possessed larger magnetic-field-induced strain or magnetocaloric effect [7,8] and showed strong anisotropy [9,10], it is expected for enhanced magnetic properties in Ni–Mn–Sn single or unidirectional textured crystals. For the preparation of single crystal, Bridgman technique and Czochralski method have been reported to be used to Ni-Mn-Ga and Ni-Mn-In single crystal [11,12]. However, the preparation of Ni-Mn-Sn single crystal is challenging due to serious volatilization and the formation of copious green MnO [13]. In the present study, floating-zone technique was firstly applied to produce Ni–Mn–Sn single crystal and the orientation behavior was investigated.

2. Experimental Procedures

Polycrystalline $Ni_{50}Mn_{37}Sn_{13}$ (at.%) ingot about 25 g was arc-melted from Ni, Mn and Sn elements with purities of 99.99 wt.% under argon gas atmosphere. Additional 5 wt.% manganese was added to compensate for evaporation loss. The manganese chips were specially cleaned for ~20 s in 5 vol.% HNO_3 several times. The surface oxide of button ingot was mechanically removed prior to suction casting into a rod with diameter of 5 mm and ~60 mm in length. The

* Corresponding author at: Laboratory for Microstructures, Shanghai University, Shanghai 200072, China. Tel.: +86 21 56334045; e-mail: hxzheng@shu. edu. cn

first crystal was grown using traditional optical floating zone image furnace (Quantum and Design G2) under 1 bar 99.999% argon gas atmosphere. As shown in Figure 1a, heating the master Ni-Mn-Sn bar to about 1473 K formed a stable molten zone. The crystal growth started with a pulling rate of 6 mm/h and opposite rotating rate of 15 rpm for both the upper and lower master rods. Figure 1b is the real-time image of molten zone after 60 min, MnO around the molten zone can be seen clearly and the formation of MnO damaged the stability of the molten zone, eventually leading to the failure of the crystal growth. Figure 1c shows the miry quartz chamber after the crystal growth, a thick layer of black volatiles can be seen inside the chamber wall so that it prevented the light, and this was also an important factor for the failure of the crystal growth. The formation of MnO and volatiles has been reported by Laudise et al. [13].

Considering the formation of MnO may be associated with the residual oxygen in the chamber, the second experiment was performed by replacing the lower master Ni-Mn-Sn bar with a Ti bar. As shown in Figure 1d, preheating the titanium rod for 40 min to remove residual oxygen and then heating the upper master Ni-Mn-Sn bar above to about 1473 K and formed a stable molten zone above the surface of Ti bar .The pulling rate was 6 mm/h, and the same 15 rpm opposite rotating rate for the upper master rods and lower Ti bar. Figure 1e is the real-time image of molten zone after 240 min, there was no MnO around the molten zone. Figure 1f showed the clean quartz chamber after the second crystal growth and there was no clear compositional volatilization during the experiment. One high-quality Ni-Mn-Sn crystal was produced during the second experiment, as shown in Figure 2.

Specimen was cut along the crystal centerline and microstructural observations were performed on an optical microscope (Zeiss Axio Imager A2m). X-ray diffraction (XRD DLMAX-2200) was employed to detect the crystal structure and growth orientation. Magnetic measurements were performed using a multifunctional physical property measurement system (PPMS-9) under a magnetic field up to 100 Oe.

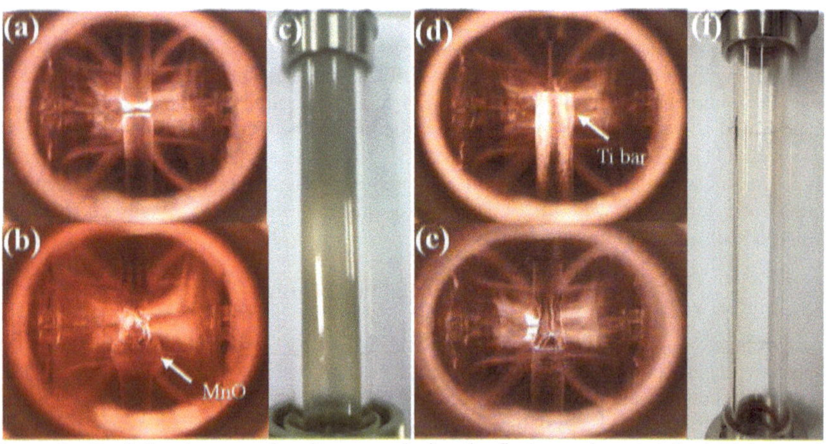

Figure 1: Real-time images of molten zone during the first experiment at the initial stage (a) and after 60 min (b), (c) quartz chamber with a thick black volatile layer. Real-time images of molten zone during the second experiment, (d) preheating Ti bar and (e) after 240min, (f) clean quartz chamber after finishing the second experiment.

3. Results and Discussion

Figure 2a presents the produced Ni–Mn–Sn unidirectional crystal with diameter of 5 mm and25mm in length. Figure 2b shows the transverse sectional macrostructure of the crystal. Parallel columnar crystals can be observed. Figure 2c is the enlarged image of the solid/liquid (S/L) interface where one can see the planar and straight interface feature.

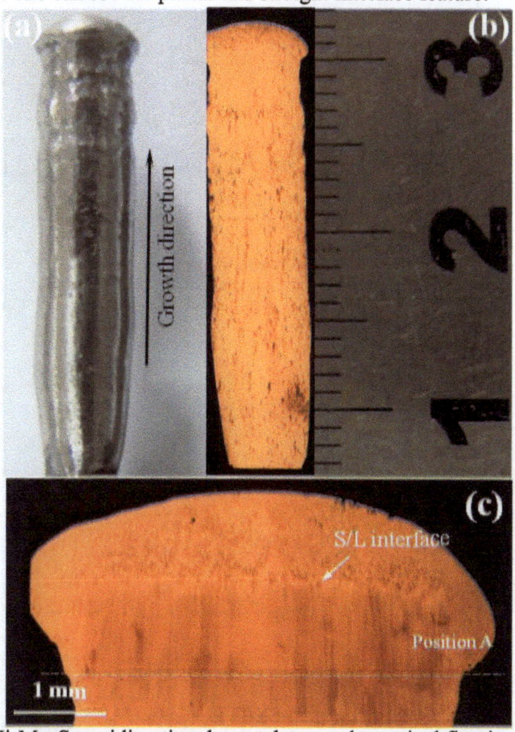

Figure 2: (a) The Ni-Mn-Sn unidirectional crystal grown by optical floating zone in the present study, (b) macrostructure of longitudinal section, (c) enlarged image of the final zone showing the planar solid/liquid interface and unidirectional feature of the grains.

Figure 3 gives the XRD patterns of the master Ni-Mn-Sn rod and the unidirectional crystal taken from the position A (marked by dashed line in Figure 2c). For the master Ni-Mn-Sn rod, most of the diffraction peaks are indexed well as modulated four-layered orthorhombic (4*O*) martensite. One weak peaks from *fcc* $L2_1$ austenite, $L2_1(333)$ is visible. The XRD pattern taken from the unidirectional crystal shows a strong $L2_1(220)$ peak from austenite and most of the martensitic peaks diminish, which is in accordance with our previous work [14]. Based on the XRD result, we can conclude that the martensitic transition temperature decreased slightly in the unidirectional crystal in contrast to the master alloy. During the crystal growth, it is inevitable that a little manganese would volatilized although no clear trace is visible inside the quartz wall. The volatization of manganese leads to the decrease of electron concentration (e/a) and thus lower transition temperatures[15].

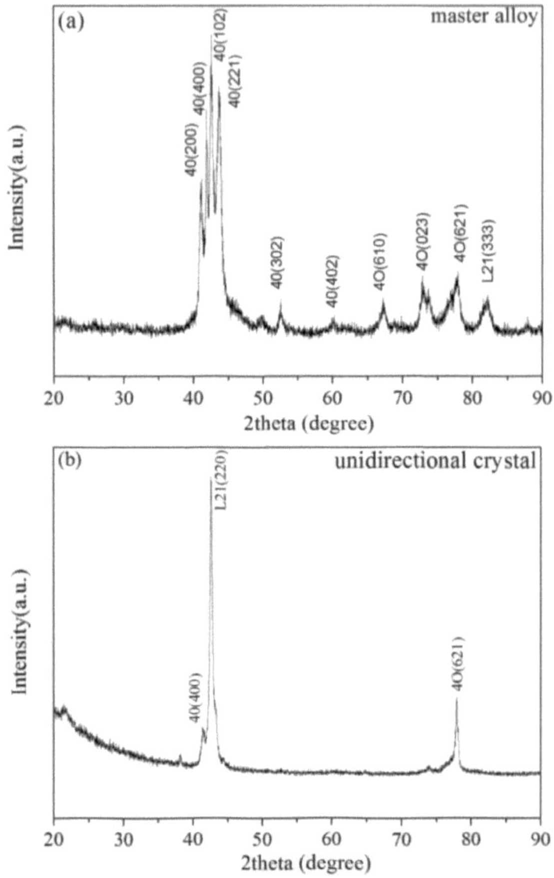

Figure 3: XRD patterns at room temperature, (a) master alloy and (b) unidirectional crystal taken from cross-section of the position A marked in Figure 2c.

Figure 4 shows the *M-T* curves under a low magnetic field of 100 Oe applied parallel and perpendicular to the grow direction of the Ni–Mn–Sn unidirectional crystal, respectively. The martensite start and finish temperatures (M_s and M_f), austenite start and finish temperatures (A_s and A_f) can be determined, as shown in Figure 4. The whole tendency of *M-T* curves measured with the magnetic field applied parallel and perpendicular to the grow direction are very similar. Upon cooling, magnetization initially keeps stable, and then increases sharply caused by the transition of austenite from paramagnetic to ferromagnetic. Further decreasing the temperature below M_s, the magnetization decreases dramatically to the minimum associated with the first-order martensitic transition from ferromagnetic austenite to weak magnetic martensite. In the last further cooling stage, the martensite transforms from weak magnetic to ferromagnetic. On the other hand, it can be seen that it is almost reversible for the magnetization change upon heating. It should be noted that the magnetization discrepancy (ΔM) between the austenite and martensite measured with the magnetic field applied perpendicular to the grow direction is 0.9 emu/g, it is about 1.3 times larger than that measured with the magnetic field applied parallel to the grow direction which is 0.7 emu/g. It demonstrates a strong magnetic anisotropy, which is potential for future engineering applications.

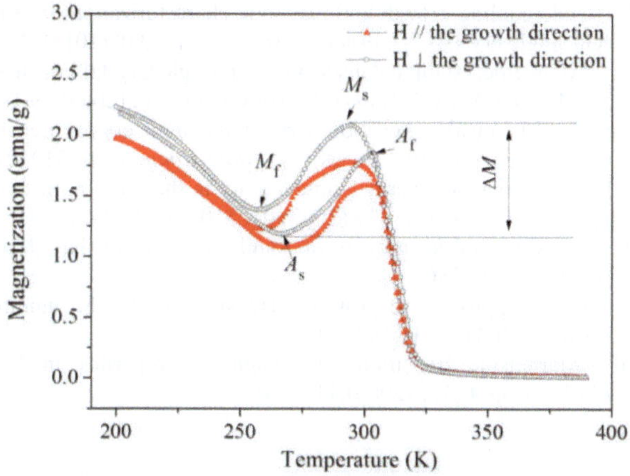

Figure 4: M-T curves under a low magnetic field of 100 Oe applied parallel and perpendicular to the grow direction of the Ni–Mn–Sn unidirectional crystal.

4. Conclusions

In summary, preheating the Ti bar is effective to reducing the formation of MnO and gray volatiles. The obtained Ni–Mn–Sn unidirectional crystal was $\beta\{110\}$ preferentially oriented and demonstrated a strong magnetic anisotropy.

Acknowledgement

The authors gratefully acknowledge the support from the National Natural Science Foundation of China (51474144, 51201096), the Specialized Research Fund for the Doctoral Program of Higher Education of China (20123108120019).

References

1. X. Moya et al., “Calorimetric study of the inverse magnetocaloric effect in ferromagnetic Ni–Mn–Sn,” *Journal of Magnetism and Magnetic Materials*, 316 (2007), e572 – e574.
2. D.Z. Wu et al., “Atomic ordering effect in $Ni_{50}Mn_{37}Sn_{13}$ magnetocaloric ribbons,”*Materials Science and Engineering A*, 534 (2012), 568–572.
3. S. Fähler et al., “Caloric Effects in Ferroic Materials: New Concepts for Cooling,” *Advanced Engineering Materials*, 14 (2012), 10-19.
4. H.X. Zheng et al., “Composition-dependent crystal structure and martensitic transformation in Heusler Ni–Mn–Sn alloys,” *Acta Materialia*, 61 (2013), 4648-4656.
5. X. Wang et al., “Origin of unusual properties in the ferromagnetic Heusler alloy Ni–Mn–Sn: A first-principles investigation,” *Scripta Materialia*, 89 (2014), 33–36.
6. Y. Zhang et al., “Large magnetic entropy change and enhanced mechanical properties of Ni–Mn–Sn–C alloys,” *Scripta Materialia*,75 (2014), 26-29.

7. S. J. Murray et al., "Large field induced strain in single crystalline Ni–Mn–Ga ferromagnetic shape memory alloy," *Journal of Applied Physics*, 87(2000),5774-5776.
8. F.X. Hu et al., "Large magnetic entropy change in a Heusler alloy $Ni_{52.6}Mn_{23.1}Ga_{24.3}$ single crystal," *Physical ReviewB*, 64(2001), 132412-1-132412-4.
9. J.F. Qian et al., "Undercooling growth and magnetic characterization of ferromagnetic shape memory alloy NiFeGa single crystals," *Journal Crystal Growth*, 388 (2014), 107-111.
10. H. Morito et al., "Magnetocrystalline Anisotropy in a Single Crystal Fe–Ni–Ga Ferromagnetic Shape Memory Alloy," *Materials Transactions*, 44 (2003), 661-664.
11. D.L. Schlagel et al., "Chemical segregation during bulk single crystal preparation of Ni–Mn–Ga ferromagnetic shape memory alloys," *Journal of Alloys Compounds*, 312 (2000), 77-85.
12. B. Zhang et al., "Giant magnetothermal conductivity in the Ni–Mn–In ferromagnetic shape memory alloys," *Applied Physics Letters*, 91 (2007) 012510-1-012510-3.
13. R.A. Laudise et al., "Czochralski growth of single crystals of $Ni_{3-x}Mn_xSn$," *Journal of Crystal Growth*, 118 (1992), 277-286.
14. J.K.Yu et al., "A new approach to grow the Heusler Ni–Mn– Sn unidirectional crystal," *Journal of Crystal Growth*, 402 (2014), 147-150.
15. X. Moya et al., "Martensitic transition and magnetic properties in Ni–Mn–X alloys," *Materials Science Engineering A* ,911 (2006),438-440.

Advances in the Science and Engineering of

CASTING SOLIDIFICATION

An MPMD Symposium Honoring Doru Michael Stefanescu

Solidification Processing II

Session Chair:
Charles Monroe

Advances in the Science and Engineering of Casting Solidification
Edited by: Laurentiu Nastac, Baicheng Liu, Hasse Fredriksson, Jacques Lacaze, Chun-Pyo Hong, Adrian V. Catalina, Andreas Buhrig-Polaczek, Charles Monroe, Adrian S. Sabau, Roxana Elena Ligia Ruxanda, Alan Luo, Subhayu Sen, and Attila Diószegi
TMS (The Minerals, Metals & Materials Society), 2015

Modeling of Macrosegregation during Solidification of Steel Ingot Casting

Wutao Tu, Houfa Shen, Baicheng Liu

Key Laboratory for Advanced Materials Processing Technology, Ministry of Education, School of Materials Science and Engineering, Tsinghua University, Beijing 100084, China

Keywords: multiphase modeling, macrosegregation, macrostructure

Abstract

Macrosegreation is an important transportation phenomenon during solidification of alloys, and hence deteriorates the homogeneity and the final properties of steel ingot casting. It still remains a great challenge to predict macrosegregation in practice with numerical models. Numerical simulation of macrosegregation in steel ingot casting, including mathematical model, model validations and engineering applications are introduced in this paper. The thermal-solutal convection, equiaxed grain sedimentation, and columnar to equiaxed transition (CET) are the main factors taken into considerations of current multiphase solidification model. Mathematical model is preliminarily verified to predict the macrosegregation of Sn-Pb alloy solidified in a rectangle cavity. Moreover, simulations with non-orthogonal grids are conducted for the macrosegregation prediction in a 3.3-ton steel ingot casting.

Introduction

Steel ingots are the fundamental parts for the monoblock forgings in manufacture industries. Although the vast majority of the world's steel is now continuously casted, the steel ingots are still indispensable for the production of some heavy industrial components considering their complex geometries, such as the pressure vessels for nuclear power generation. Macrosegregation or composition heterogeneity, one of the main defects encountered in the steel ingots, haunted the manufacturers over past decades. The relative motion of the solid grains and liquid melt is believed to be the main mechanism responsible for the macrosegregation [1]. The relative movement translates the solute partition effects on microscopic (microsegregation) to the difference of chemical composition on macroscopic (macrosegregation). The formation of relative movement is a multiphase phenomenon inherently, and many reasons, including the thermosolutal convection, the shrinkage flow, the grain sedimentation, contribute to it. Considering the huge cost and complexity of experiments, numerical simulations have been adopted as the main research tool for macrosegregation. The solute redistribution model [2], proposed by Flemings et al. in late 1960s, was the first pioneering work to investigate macrosegregation. Only the shrinkage flow in the mushy zone was considered in the solute redistribution model. Then various models have been developed [3-9]. Conservation equations were derived for different zones (liquid, solid and mushy) and coupled across the moving boundaries between zones in multi-domain model [3]. Eliminating the difficulties in tracking the moving boundaries of multi-domain model, the continuum model [4] and volume-averaging model [5] was proposed. A uniform set of conservation equations was derived by the mathematic theory for the system. Further considerations of interactions between solid and liquid phase on macrosegregation were taken into the two-phase solidification model by introducing the interfacial terms in conservation equations [6]. The two-phase model bridges the length scales between global transport phenomena and microscopic grain growth kinetics. Among the models, the most advanced and complicated ones are the multiphase models [7-9]. The first multiphase model was given by Wang and Beckermann [7] considering the non-uniform concentration in the interdendritic region. Ludwig and co-workers [8-9] developed a series of multiphase solidification models. The most sophisticated one is a five-phase model that accounts for columnar-to-equiaxed transition (CET), non-dendritic and dendritic crystal growth, and columnar primary dendritic tip tracking.

However, only a few research results have been reported to reveal the characteristic of macrosegregation in steel ingots using the two-phase or multiphase modeling owing to the difficulties brought by the huge computational resources required. A remarkable application was performed by Combeau and co-workers [10], in which macrosegregation of an industrial 3.3-ton steel ingot was measured and predicted. Li et al. [11] presented simulations for macrosegregation in an industrial 53-ton steel ingot using a two-phase multiscale solidification model. Comparisons were made between the measurements and predictions along the transverse sections of the ingot hot top. Besides, macrosegregation in a 2.45-ton steel ingot was investigated by a three-phase model for the mixed columnar-equiaxed solidification by Ludwig and co-workers [12]. In this paper, a multiphase model for predictions of macrosegregation is developed, which accounts for the CET in mixed columnar-equiaxed solidification along with the thermosolutal convection, grain sedimentation, and solidification shrinkage. Firstly, a benchmark recently proposed by Hachani et al. [13] is adopted to validate the basic performance of the mathematic model in predicting macrosegregation of Sn-Pb alloy solidified in a rectangle cavity. Then the multiphase model is applied to a 3.3-ton benchmark steel ingot along with the adoption of non-orthogonal grids for the boundary fitness of ingot geometry.

Model Descriptions

Table 1. Nomenclature

g_l, g_c, g_e	Volume fraction of liquid, columnar, and equiaxed phase
ρ_l, ρ_c, ρ_e	Density of liquid, columnar, and equiaxed phase
Γ_{cl}, Γ_{el}	Interfacial phase change rate of liquid-columnar interface, and liquid-equiaxed interface
$\mathbf{v}_l$, $\mathbf{v}_c$, $\mathbf{v}_e$	Velocity of liquid, columnar and equiaxed phase
p	Pressure
μ_l, μ_c, μ_e	Viscosity of liquid, columnar and equiaxed phase
$\boldsymbol{M}_c^d$, $\boldsymbol{M}_e^d$	Drag force between the columnar and liquid phase, and between the equiaxed and liquid phase
$\mathbf{g}$	Gravity force
C_l, C_c, C_e	Concentration of liquid, columnar and equiaxed phase
S_V	Interfacial area concentration
D_l, D_c, D_e	Solute diffusion coefficient of liquid, columnar and equiaxed phase
c_{pl}, c_{pc}, c_{pe}	Specific heat of liquid, columnar, and equiaxed phase
k_l, k_c, k_e	Thermal conduction coefficient of liquid, columnar and equiaxed phase
T	Temperature
L	Latent heat

The present model is a modification and continuation of previous two-phase solidification model [11, 14], which incorporates the macroscopic mass, momentum, heat and solute transfer with microscopic nucleation and growth of grains. According to the hydrodynamic characteristic, current model divides the solid phase into columnar phase and equiaxed phase. General assumptions are: (1) three phases are included: the liquid phase (l), the columnar solid phase (c), and the equiaxed solid phase (e). The corresponding volume fractions are defined as g_l, g_c, and g_e, with the relationship $g_l+g_c+g_e=1$; (2) the whole solidification region is divided into two subzones, a columnar one and an equiaxed one. Within the columnar zone, the solid phase is the columnar, and the morphology is cylinder. Within the equiaxed zone, the solid phase is the equiaxed, and the morphology is sphere; (3) the Lagrangian method is adopted to predict the final CET position using a front tracking method which tracks the moving curve of columnar dendrite tips, while the Eulerian method is used to solve the mass, momentum, energy and species conservation equations; (4) the interactions between the columnar phase and equiaxed phase are omitted. Nomenclature and conservation equations of the mathematic model are summarized in Table 1 and 2 respectively.

Table 2. Conservation equations

	Columnar zone	Equiaxed zone
Mass conservation	$\frac{\partial}{\partial t}(g_l\rho_l)+\nabla\cdot(g_l\rho_l\mathbf{v}_l)=-\Gamma_{cl}$	$\frac{\partial}{\partial t}(g_l\rho_l)+\nabla\cdot(g_l\rho_l\mathbf{v}_l)=-\Gamma_{el}$
	$\frac{\partial}{\partial t}(g_c\rho_c)=\Gamma_{cl}$	$\frac{\partial}{\partial t}(g_e\rho_e)+\nabla\cdot(g_e\rho_e\mathbf{v}_e)=\Gamma_{el}$
Momentum conservation	$\frac{\partial}{\partial t}(g_l\rho_l\mathbf{v}_l)+\nabla\cdot(g_l\rho_l\mathbf{v}_l\mathbf{v}_l)=-g_l\nabla p$ $+\nabla\cdot(g_l\mu_l\nabla\mathbf{v}_l)-\boldsymbol{M}_c^{d}+g_l\rho_l^{b}\mathbf{g}$	$\frac{\partial}{\partial t}(g_l\rho_l\mathbf{v}_l)+\nabla\cdot(g_l\rho_l\mathbf{v}_l\mathbf{v}_l)=-g_l\nabla p$ $+\nabla\cdot(g_l\mu_l\nabla\mathbf{v}_l)-\boldsymbol{M}_e^{d}+g_l\rho_l^{b}\mathbf{g}$
	$\mathbf{v}_c=0$	$\frac{\partial}{\partial t}(g_e\rho_e\mathbf{v}_e)+\nabla\cdot(g_e\rho_e\mathbf{v}_e\mathbf{v}_e)=-g_e\nabla p$ $+\nabla\cdot(g_e\mu_e\nabla\mathbf{v}_e)+\boldsymbol{M}_e^{d}+g_e\rho_e^{b}\mathbf{g}$
Species conservation	$\frac{\partial}{\partial t}(g_e\rho_e C_e)+\nabla\cdot(g_e\rho_e C_e\mathbf{v}_e)$ $=\frac{S_V\rho D_e}{\delta_e}(C_e^*-C_e)+C_e^*\Gamma_{el}$	$\frac{\partial}{\partial t}(g_e\rho_e C_e)+\nabla\cdot(g_e\rho_e C_e\mathbf{v}_e)$ $=\frac{S_V\rho D_e}{\delta_e}(C_e^*-C_e)+C_e^*\Gamma_{el}$
	$\frac{\partial}{\partial t}(g_l\rho_l C_l)+\nabla\cdot(g_l\rho_l C_l\mathbf{v}_l)$ $=\frac{S_V\rho D_l}{\delta_l}(C_l^*-C_l)-C_l^*\Gamma_{el}$	$\frac{\partial}{\partial t}(g_l\rho_l C_l)+\nabla\cdot(g_l\rho_l C_l\mathbf{v}_l)$ $=\frac{S_V\rho D_l}{\delta_l}(C_l^*-C_l)-C_l^*\Gamma_{el}$
Energy conservation	$\frac{\partial}{\partial t}[(g_c\rho_c c_{pc}+g_e\rho_e c_{pe}+g_l\rho_l c_{pl})T]+\nabla\cdot[(g_e\rho_e c_{pe}\mathbf{v}_e+g_l\rho_l c_{pl}\mathbf{v}_l)T]$ $=\nabla\cdot[(g_e k_e+g_l k_l)\nabla T]+(\Gamma_{cl}+\Gamma_{el})L$	

With respect to the discretization of conservation equations on non-orthogonal grids, cross-diffusions emerge due to the non-orthogonal grid lines [15]. Momentum interpolation method [16] is adopted to resolve the cell velocity to cure the pressure checkerboard problem on the collocated grids. Full details of the discretization on non-orthogonal grids are available in Reference 15. When it comes to the pressure-velocity coupling, the typical SIMPLE algorithm [17] is used. The program procedure is shown in Figure 1.

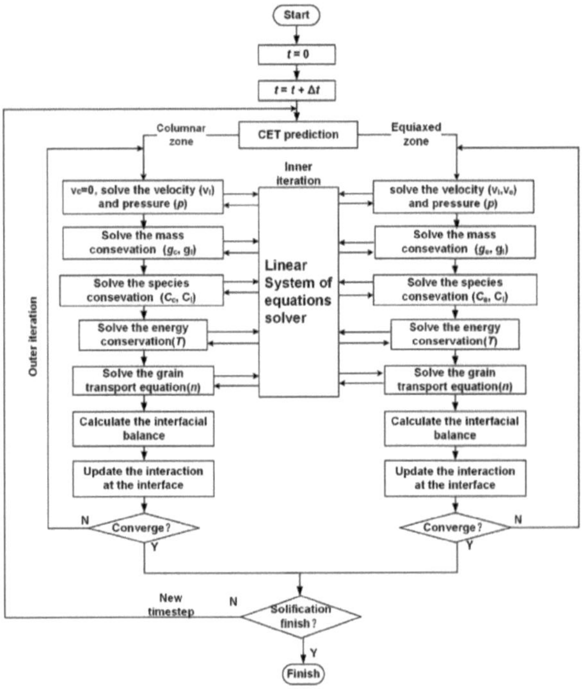

Figure 1. Flow chart of the calculation program.

Model Validation

Sn-3 wt.%Pb solidification benchmark

The Sn-3 wt.%Pb solidification benchmark was proposed by Hachani et al. in 2012 [13]. The schematic of the benchmark experiment is shown in Figure 2. In the benchmark experiment, Sn-3 wt.%Pb was solidified in a 100 mm×60 mm rectangle cavity. The cavity was differentially heated by side walls with heat exchangers. A 40K temperature gap was kept between the left heat exchanger and the right heat exchanger when solidification began, and temperature of left heat exchanger was higher than that of the right. During the solidification, both the heat exchangers cooled down at $-0.03\,K/s$ until complete solidification. The bottom and top surface was kept adiabatic. The final segregation map is quantitative analyzed by Inductively Coupled Plasma (ICP) spectrometry, and an array of 50 thermocouples was used to record the temperature field. Deduced temperature curves at side walls from temperature measurements in literature [18] are applied as Dirichlet boundary conditions in current simulations. The experimental results published by Hachani et al. [13], as well as the reference predictions by Carozzani et al. [18] are used in the following comparisons. All the thermo-physical parameters used are available in literature 18. The present solutions are calculated on a grids system of 50 CV×40 CV.

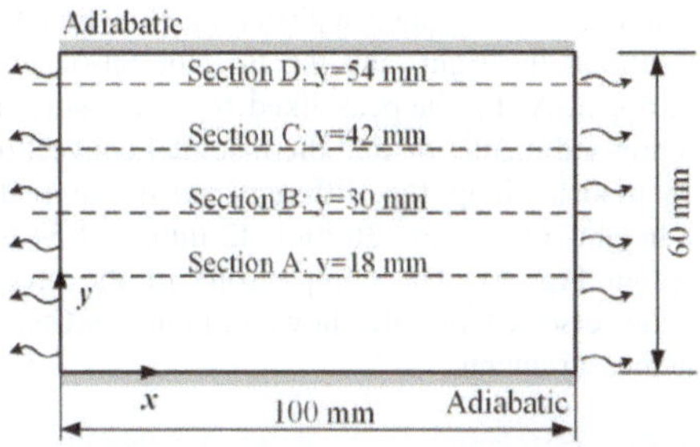

Figure 2. Schematic of the experiment benchmark.

Figure 3. Predicted CET position of the solidified Sn-3 wt.%Pb.

Figure 4. Macrostructure of the solidified Sn-3 wt.%Pb [13].

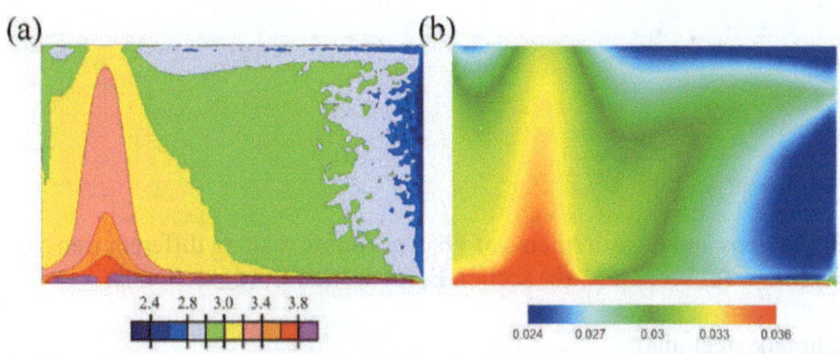

Figure 5. Predicted final macrosegregation of Sn-3 wt.%Pb in: (a) literature simulation [18], (b) present simulation.

As an important factor in the multiphase model, the predicted CET position is shown in Figure 3. From the prediction, it can be seen that the final full equiaxed region at the bottom goes wider than the region at the upper, which can also be noticed in the experiment macrostructure shown in Figure 4. This phenomenon can be explained by the sedimentation of equiaxed grains due to the gravity. Compared with experiment macrostructure, an overestimation of columnar width is predicted. This discrepancy may result from coarse grids used for the compromise between the accuracy and efficiency. Further discussions about the effects of larger equiaxed zone on the final macrosegregation are conducted in another research.

The predicted final macrosegregation maps both in reference simulation and in present simulation are presented in Figure 5. It can be seen that the main macrosegregation patterns predicted by present simulation are consistent with the ones in literature [18]. Early solidified melt forms the negative zone at the right, and the float of Sn-riched melt contributes to the negative segregation at the upper. As for the peak-liked zone in positive segregation, it forms due to the coupled effects of grain sedimentation and thermsolutal convection. Further investigations for the quantitative comparisons along the different transverse sections are depicted. The transverse sections are in heights of 18 mm, 30 mm, 42 mm, and 54 mm over the bottom face (shown by the dash lines in Fig. 2). The comparisons of the predicted and measured Pb concentrations along the transverse sections are shown in Figure 6. Generally, current predictions are in agreements with the measurements.

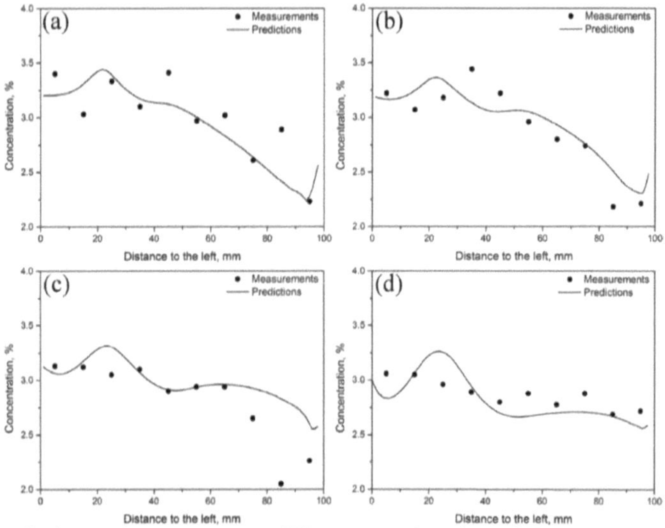

Figure 6. Predictions and measurements of Pb concentration along different transverse sections: (a) section A, (b) section B, (c) section C, (d) section D.

3.3-ton benchmark steel ingot

Furthermore, the model is used to predict the macrosegregation in a 3.3-ton steel ingot (about 2 m in height and 0.6 m in average diameter). The 3.3-ton steel ingot was reported by Combeau et al. [10] in 2009. Due to the detailed calculation parameters and concentration measurements along the ingot centerline, this 3.3-ton steel ingot is often taken as a benchmark steel ingot. In current simulation, non-orthogonal grids are generated according to the ingot geometry. Current predicted final CET position is shown in left half of Figure 7a. The average width of the columnar zone is about 40mm. This result is similar to the measured grain morphology map in Reference 10. The predicted final macrosegregation pattern is shown in right half of Figure 7a. The typical macrosegregation pattern can be observed: a negative zone is located at the bottom and a positive segregation is located at the top. Comparisons of measured and predicted carbon macrosegregation along the centerline of the ingot are illustrated in Figure 7b. Although some discrepancies could be found, the measured segregation tendency is reproduced well by current predictions.

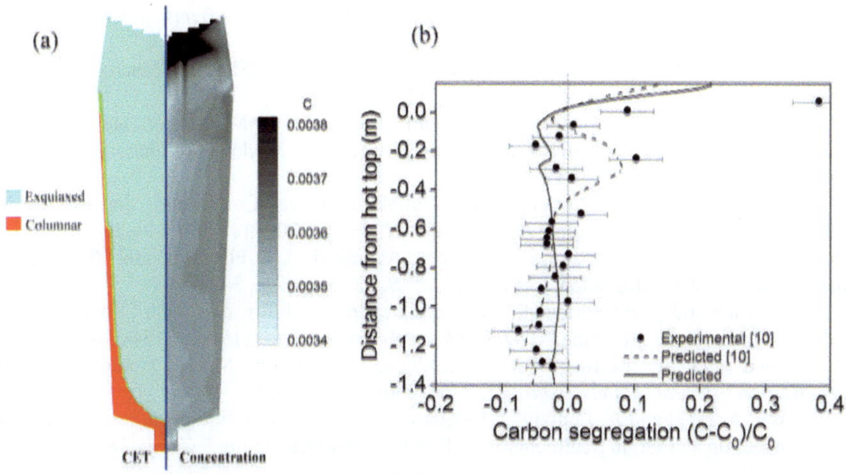

Figure 7. (a) Predicted final CET position (left) and macrosegregation map (right) in 3.3-ton benchmark steel ingot, (b) measured and predicted carbon segregation along ingot centerline.

Conclusions

A multiphase solidification model is presented in current research. The model incorporates the descriptions of heat transfer, melt convection, solute transport, columnar to equiaxed transition and solid movement on the process scale with microscopic relationship for gain nucleation and growth. Firstly, a Sn-3 wt.%Pb alloy solidification benchmark is adopted to validate the current multiphase model. Predicted CET position and concentration distributions along different transverse sections are compared with measurements. Then the model is applied to a 3.3-ton benchmark steel ingot. Typical macrosegregation patterns are reproduced including the negative segregation zone at the bottom and positive segregation zone at the top. Predictions are made for the CET position, which agrees with the measurements. Besides, the concentration distribution along the ingot centerline are taken into quantitative comparisons with measurements. It is shown a good tendency of agreement. Numerical efforts considering the multicomponent system and experiments investigating the macrostructure parameters remain to be done in the future.

Acknowledgements

This work was financially supported by the National Basic Research Program of China (No. 2011CB012900) and the National Science and Technology Major Project the Ministry of Science and Technology of China (No. 2012ZX04012011).

References

1. C. Beckermann, "Modelling of Macrosegregation: Applications and Future Needs," *Int. Mater. Rev.*, 47 (5) (2002), 243-261.
2. M. C. Flemings and G. E. Nereo, "Macrosegregation: Part. I," *Trans. Metall. Soc. AIME,* 239 (1967), 1449-1461.
3. S. D. Ridder, S. Kou and R. Mehrabian, "Effect of Fluid Flow on Macrosegregation in Axisymmetric Ingots," *Metall. Trans. B,* 12 (3) (1981), 435-447.
4. W. D. Bennon and F .P. Incropera, "A Continuum Model for Momentum, Heat and Species Transport in Binary Solid-Liquid Phase Change Systems: I. Model Formulation," *Int. J. Heat Mass Transfer,* 30 (1987), 2161-2170.
5. C. Beckermann and R. Viskanta, "Double-Diffusive Convection during Dendritic Solidification of a Binary System," *Physicochem. Hydrodyn*, 10 (2) (1995), 195-213.
6. J. Ni and C. Beckermann, "A Volume-Averaged Two-Phase Model for Transport Phenomena during Solidification," *Metall. Trans. B*, 22 (5) (1991), 349-361.
7. C. Y. Wang and C. Beckermann, "Equiaxed Dendritic Solidification with Convection: Part I. Multiscale/Multiphase Modeling," *Metall. Mater. Trans. A,* 27 (9) (1996), 2754-2764.
8. M. Wu and A. Ludwig, "A Three-Phase Model for Mixed Columnar-Equiaxed Solidification," *Metall. Mater. Trans. B*, 37 (5) (2006), 1613-1631.
9. M. Wu, A. Fjeld and A. Ludwig, "Modelling Mixed Columnar-Equiaxed Solidification with Melt Convection and Grain Sedimentation - Part I: Model Description," *Comp. Mater. Sci.*, 50 (1) (2010), 32-42.
10. H. Combeau et al., "Prediction of Macrosegregation in Steel Ingots: Influence of the Motion and the Morphology of Equiaxed Grains," *Metall. Mater. Trans. B*, 40 (3) (2009), 289-304.
11. W. Li et al., "Modeling of Species Transport and Macrosegregation in Heavy Steel Ingots,"*Matall. Mater. Trans. B.*, 45 (2) (2014), 464-471.
12. J. Li el al., "Modelling Macrosegregation in a 2.45 ton Steel Ingot," *IOP Conf. Ser: Mater. Sci. Eng.*, 33 (2012), p.012091.
13. L. Hachani et al., "Experimental Analysis of the Solidification of Sn-3 wt.%Pb Alloy under Natural Convection", *Int. J. Heat Mass Trans.*, 55 (2012), 1986-1996.
14. W. Tu, H. Shen, and B. Liu, "Two-phase Modeling of Macrosegregation in a 231t Steel Ingot," *ISIJ Int.*, 45 (2) (2014), 351-355.
15. W. Tu, H. Shen, and B. Liu, "Numerical Simulation of Macrosegregation: a Comparison between Orthogonal Grids and Non-orthogonal Grids," *Adv. Mater. Res.*, 968 (2012), 213-217.
16. C. M. Rhie, and W. L. Chow, "Numerical Study of the Turbulent Flow Past an Airfoil with Trailing Edge Separation," *AAA J.,* 21 (11) (1986), 1525-1532.
17. S. V. Patankar, *Numerical Heat Transfer and Fluid Flow* (New York, NY: Hemisphere Publishing Corporation, 1980), 113-134.
18. T. Carozzani et al., "Direct Simulation of a Solidification Benchmark Experiment," *Metall. Mater. Trans. B*, 44 (2) (2013), 873-887.

Advances in the Science and Engineering of Casting Solidification
Edited by: Laurentiu Nastac, Baicheng Liu, Hasse Fredriksson, Jacques Lacaze, Chun-Pyo Hong, Adrian V. Catalina, Andreas Buhrig-Polaczek, Charles Monroe, Adrian S. Sabau, Roxana Elena Ligia Ruxanda, Alan Luo, Subhayu Sen, and Attila Diószegi
TMS (The Minerals, Metals & Materials Society), 2015

SCALING ANALYSIS OF ALLOY SOLIDIFICATION AND FLUID FLOW IN A RECTANGULAR CAVITY

A. Plotkowski[1], K. Fezi[1], M. J. M. Krane[1]

[1]Purdue Center for Metal Casting Research
School of Materials Engineering, Purdue University
701 West Stadium Ave.; West Lafayette, IN 47906-2045, USA

Keywords: Alloy Solidification, Scaling Analysis, Natural Convection, Fluid Flow

Abstract

A scaling analysis was performed to predict trends in alloy solidification in a side-cooled rectangular cavity. The governing equations for energy and momentum were scaled in order to determine the dependence of various aspects of solidification on the process parameters for a uniform initial temperature and an isothermal boundary condition. This work improved on previous analyses by adding considerations for the cooling bulk fluid flow. The analysis predicted the time required to extinguish the superheat, the maximum local solidification time, and the total solidification time. The results were compared to a numerical simulation for a Al-4.5 wt.% Cu alloy with various initial and boundary conditions. Good agreement was found between the simulation results and the trends predicted by the scaling analysis.

Introduction

The properties of cast alloys are affected to a large extent by the local solidification conditions, which are in turn strongly coupled to the thermal and solutal transport during solidification and the resulting buoyancy driven convective flow in the bulk liquid. Therefore, it is desirable to understand the details of the transport phenomena that take place during solidification in order to predict the properties that result from a particular casting process. Due to the complex nature of the governing equations, numerical methods, such as the mixture formulation detailed by Bennon and Incropera [1,2], have generally been used to model these processes. However, these simulations are computationally expensive, each is specific to a particular material system and set of processes parameters, and general functional relationships are not easily identified.

Some previous analytical and semi-analytical methods have been presented, including similarity solutions [e.g., 3,4], the heat balance integral method [5-9], and scaling analyses [10-12]. These studies share a common shortcoming in neglecting the effects of the bulk fluid motion that occurs during the casting process. Amberg [12] briefly discussed the extinguishing of the superheat via conduction over a boundary layer at the liquidus interface. However, he did not consider the effect of the changing fluid temperature on the solidification conditions.

The purpose of this work is to combine the scaling analysis of the conduction heat transfer and solidification behavior, based on work by Krane and Incropera [10], with the changing temperature of the fluid resulting from the bulk fluid motion. The flow analysis was inspired by the work of Lin et al. [13] who considered the convection of a single phase fluid in a rectangular cavity cooled by an isothermal side-wall. This flow solution is applied at the solidification front at the liquidus temperature and is coupled to the solidification behavior.

Problem Statement

The system under consideration is shown in Figure 1. The domain is initially at a uniform temperature, T_0, above the liquidus temperature of the alloy, and solidification proceeds from the left isothermal wall, maintained at a temperature, T_w, below the solidus, while the remaining walls are assumed insulated. At short times, there are three spatial regions: (i) solid (ii) mush and (iii) bulk liquid. At some point, the liquid is consumed and only solid and mush remain. The mush is assumed impermeable and all properties are assumed uniform and constant. At the interface between the liquid and mush, a natural convection boundary layer forms, which carries cold fluid towards the bottom of the cavity at the expense of the hot fluid at the top of the cavity. Following Lin et al. [13], it is assumed that there is some discrete boundary that divides two fluid reservoirs, one at the initial temperature and the other at the liquidus temperature.

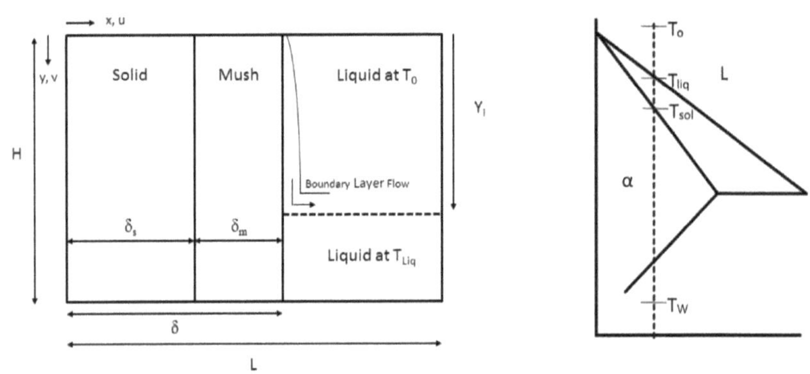

(a) *(b)*

Figure 1: (a) System schematic showing bulk fluid flow. (b) Simplified alloy phase diagram.

Scaling Analysis of the Solidification Behavior

The simplified one-dimensional governing equation for energy conservation, including the effect of the latent heat evolution and neglecting advection in the mushy zone, is:

$$\frac{\partial T}{\partial t} = \alpha \frac{\partial^2 T}{\partial x^2} + \frac{L_f}{c_p}\frac{df_s}{dt} \tag{1}$$

where α is the thermal diffusivity, L_f is the latent heat, and c_p is the specific heat. During the temporal regime in which all three spatial regions are present, Equation 1 is scaled over the combined mushy zone and thermal boundary layer using the following reference scaling variables:

$$\Delta T_{ref} \sim \Delta T_{sol} = T_{avg} - T_{sol}$$
$$\Delta x_{ref} \sim \delta_{m1} + \bar{\delta}_T \tag{2}$$

where T_{avg} is an average bulk fluid temperature, T_{sol} is the solidus temperature of the alloy, δ_{m1} is the mushy zone thickness during the first temporal regime, and $\bar{\delta}_T$ is the average natural convection boundary layer thickness. Here the boundary layer is assumed to be very thin, such that the temperature at the liquidus interface is equal to the bulk fluid temperature. Accordingly, the boundary layer thickness is dropped from the length scale. The change in the fraction solid over the mushy zone, Δf_s, is $O(1)$, and the latent heat term may be rewritten in terms of the freezing range, $\Delta T_m = T_{Liq} - T_{sol}$, and the Stefan number, St. The resulting scaled equation is therefore:

$$\frac{\Delta T_{sol}}{t} \sim \frac{\alpha \Delta T_{sol}}{\delta_{m1}^2}, \frac{\Delta T_m}{St\, t}, \qquad where\ St = \frac{c_p \Delta T_m}{L_f} \tag{3}$$

The thickness of the mushy zone can now be found:

$$\delta_{m1} \sim \left[(\alpha t) \frac{1}{1 + \frac{\Delta T_m}{\Delta T_{sol}} \frac{1}{St}} \right]^{1/2} \tag{4}$$

By writing a heat flux balance at the solid-mush interface, a relationship between the solid and mush thicknesses may be found.

$$\delta_{s1} \sim \delta_{m1} \left(\frac{T_{sol} - T_w}{T_{avg} - T_{sol}} \right) \tag{5}$$

The location of the solidification front is now simply the sum of the thicknesses of these two regions, $\delta \sim \delta_{s1} + \delta_{m1}$.

From here, the time at which the domain transitions to the second temporal regime which includes only solid and mush may be approximated by setting δ equal to the length of the cavity, L.

$$t_m(L) \sim \frac{L^2}{\alpha} \left[1 + \frac{\Delta T_m}{\Delta T_{sol}} \frac{1}{St} \right] \left[1 + \frac{T_{sol} - T_w}{\Delta T_{sol}} \right]^{-2} \tag{6}$$

where $t_m(x)$ is the time at which the mush-liquid front reaches a location x. A similar equation can be formulated for the solidus isotherm, denoted $t_s(x)$. The analysis must be repeated for the second temporal regime. The resulting equations for the solid and mush thicknesses are as follows:

$$\delta_{m2} \sim \delta_{m1}(t_m(L)) - \left[\frac{\alpha(t - t_m(L))}{1 + 1/St} \right]^{1/2} \tag{7}$$

and

$$\delta_{s2} \sim L - \delta_{m2}. \tag{8}$$

The local solidification time as a function of position can be found by finding from the difference in time between when the liquidus and solidus isotherms pass a particular point. The time at which the liquidus isotherm passes an arbitrary point, x, can be found from Equation 6 by replacing L with that position. The time at which the solidus isotherm passes that particular point can be found from Equations 5 and 8. These two regimes are split at the location of the solid front at time $t_m(L)$. This location is

$$\delta_{s1}(t_m(L)) \sim \left(\frac{T_{sol} - T_w}{\Delta T_{sol}}\right)\sqrt{B\tau_m(L)}$$
$$where \quad B = \alpha\left(1 + \frac{\Delta T_m}{\Delta T_{sol}}\frac{1}{St}\right)^{-1}. \tag{9}$$

The local solidification time can now be found for each regime by subtracting $t_m(x)$ from $t_s(x)$.

$$if\ x < \delta_{s1}(t_m(L)): t_{lst} \sim \frac{x^2}{B}\left[\left(\frac{\Delta T_{sol}}{T_0 - T_w}\right)^2 - \left(1 + \frac{T_{sol} - T_w}{\Delta T_{sol}}\right)^{-2}\right] \tag{10a}$$

$$if\ x > \delta_{s1}(t_m(L)): t_{lst} \sim \frac{1 + 1/St}{\alpha}\left(x - \delta_{s1}(t_m(L))\right)^2 + t_m(L) - \frac{x^2}{B}\left(1 + \frac{T_{sol} - T_w}{\Delta T_{sol}}\right)^{-2} \tag{10b}$$

Scaling Analysis of the Bulk Flow Behavior

The flow configuration is approximated by a quasi-static natural convection boundary layer for an isothermal flat plate at T_{liq}. The solution follows from Bejan [14] and is based upon the flow configuration proposed by Lin et al. [13] shown in Figure 1(a). The energy equation in this boundary layer is scaled using u_0 and v_0 for velocity, and the thermal boundary layer thickness, δ_T, and filling interface position, Y_i, as the x and y direction length scales, respectively.

$$u_0\frac{\Delta T}{\delta_T}, v_0\frac{\Delta T}{Y_i} \sim \alpha\frac{\Delta T}{\delta_T^2} \tag{11}$$

The incompressible continuity equation provides a relationship between the x and y direction velocities. When substituted back into Equation 11, the two advection terms are found to be equal and a relationship between the y-direction velocity and boundary layer thickness is found.

$$v_0 \sim \frac{\alpha Y_i}{\delta_T^2} \tag{12}$$

The vertical momentum equation will now be considered, in which the transient, pressure, and vertical viscous terms have been neglected. Like the energy equation, the two advection terms are equal by continuity. Equation 12 can be substituted into the viscous term to yield a relationship with the Prandtl number, Pr.

$$\frac{v_0^2}{Y_i}, Pr\frac{v_0^2}{Y_i} \sim g\beta\Delta T_{liq} \tag{13}$$

The first term represents inertia, the second viscous forces, and the third is the buoyancy driving force. Equation 13 suggests that the buoyancy term may balance with either the inertial or viscous term, the choice being determined by the Prandtl number. For liquid metals, *Pr* is small, and thus the viscous term is neglected.

$$v_0 \sim \left[Y_i g\beta\Delta T_{liq}\right]^{1/2} \tag{14}$$

Substitution into Equation 12 yields an estimate for the thermal boundary layer thickness.

$$\delta_T \sim \sqrt{\alpha}\left[\frac{Y_i}{g\beta\Delta T_{liq}}\right]^{1/4} \tag{15}$$

The fluid from the boundary layer fills the cavity with fluid at T_{liq}, and it is assumed that this fluid maintains a discrete interface with the fluid at the initial temperature at a level Y_i. Fluid from the boundary has no buoyancy driving force below the interface level, therefore, as the level of cold fluid rises, the fluid from the boundary layer will be ejected at increasingly higher locations. The interface position Y_i may be determined as a function of time by writing an equation for volume conservation of the liquid at the initial temperature that takes into account the fluid advected from the boundary layer and the fluid consumed by the advancing solidification front.

$$\frac{\partial}{\partial t}[Y_i(L-\delta)] \sim -H\frac{d\delta}{dt} - v_0\delta_T \tag{16}$$

This non-linear first order ODE for Y_i is solved numerically, coupled to the solidification solution through δ. Based upon Y_i, the average fluid temperature, T_{ave}, that affects the solidification behavior can be found using a weighted average of the initial and liquidus temperatures. The superheat is extinguished once Y_i is equal to zero. While the solidification scaling assumes that the average fluid temperature is constant with time, it can still be used to approximate the instantaneous velocity of the solidification front which is integrated in time numerically to find the interface position in Equation 16.

Comparison of Scaling Analysis to Simulation Results

The scaling analysis was compared to a full numerical simulation of the fluid flow and energy conservation during solidification of an Al-4.5 wt.% Cu alloy, the properties of which are shown in Table 1. The simulation is a mixture model based on the work of Bennon and Incropera [1,2] and the temperature formulation of the energy equation proposed by Voller and Swaminathan [15]. The simulations assume no macrosegregation takes place (the alloy solidifies at the nominal composition), a lever rule thermodynamic model, and no solutal buoyancy. Simulations were performed for a 0.25 m square cavity over a range of wall and initial temperatures.

The scaling analysis is an order-of-magnitude analysis which predicts functional relationships. Figure 2 compares the time at which the liquid is consumed ($t_m(L)$) as a function of wall temperature and initial temperature found by the scaling analysis and the full numerical simulation. The scaling analysis results are stretched by a constant multiplier so that the results for a case with a wall temperature of 500 K and an initial temperature of 1000 K can be easily compared to the simulation. Figure 2 shows good agreement in the trends of the results between the scaling analysis and simulations.

Table 1: Property values for Al-4.5 wt.% Cu Alloy

Density (kg/m^3)	2605	Latent Heat of Fusion (J/kg)	390,000
Specific Heat (J/kg-K)	1006	Eutectic Temperature (K)	821.2
Thermal Conductivity (W/m-K)	137.5	Eutectic Compositions (wt.% Cu)	0.33
Dynamic Viscosity (kg/m-s)	0.0014	Melting Temperature of Al (K)	933.5
Prandtl Number	0.015	Max Solubility of Cu in Al (wt.% Cu)	0.0565
Coef. of Thermal Expansion (K^{-1})	1.17×10^{-4}		

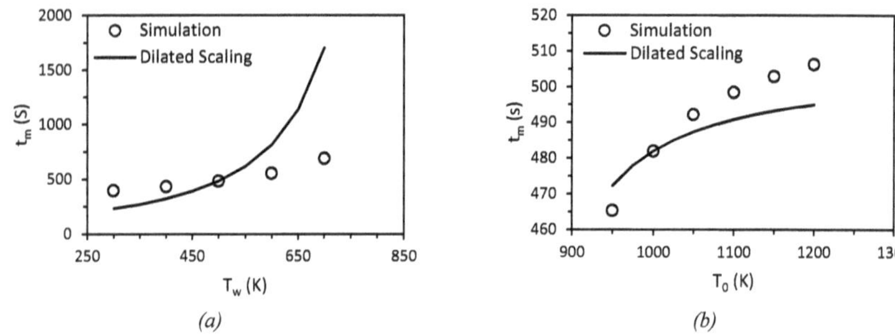

Figure 2: Comparison of scaling analysis trends to simulation results for an Al-4.5 wt.% Cu alloy in a 0.25 m sq. cavity for (a) various wall temperature and an initial temperature of 1000 K and (b) various initial temperatures with a wall temperature of 500 K.

Figure 3 shows a comparison of the predictions by scaling and simulation for the time required to extinguish the superheat in the bulk liquid. This time in the simulation was defined as the point at which the average liquid temperature fell 99% of the difference between the initial and the liquidus temperature.

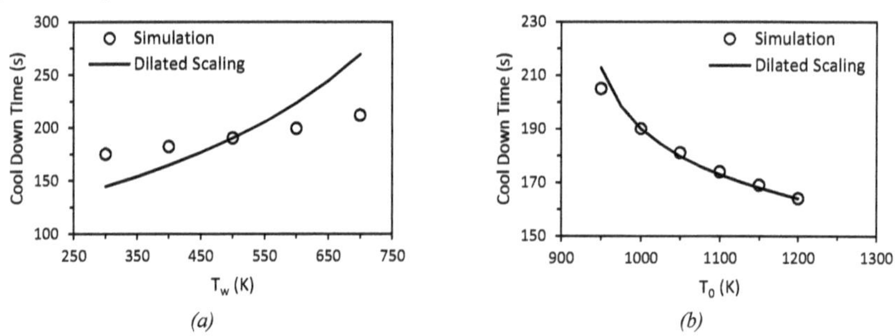

Figure 3: Comparison of scaling and simulation trends for the cool down time as a function of (a) wall temperature and (b) initial temperature.

A metallurgically interesting quantity is the local solidification time (LST) throughout the ingot. Figure 4a shows the simulation results for an initial temperature 1000 K and a wall temperature of 500 K, demonstrating that the maximum LST value does not occur at the opposite insulated sidewall. The existence of a local maximum in the LST is also predicted by the scaling analysis. The trends in maximum local solidification time as a function of the wall temperature are compared in Figure 4b.

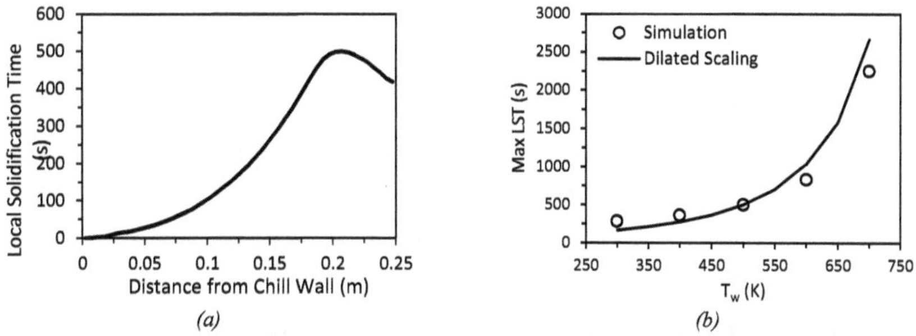

Figure 4: (a) Simulation results at $T_o = 1000$ K and $T_w = 500$ K showing a maximum LST and (b) a comparison of trends from the scaling analysis and simulation results for various wall temperature.

Conclusions

This work examines the solidification behavior of an alloy in a rectangular cavity cooled by an isothermal sidewall using a one dimensional scaling analysis of the freezing front motion and the natural convection boundary layer in the bulk liquid. The coupling of the bulk liquid motion and temperature to the front motion improves on earlier work. The analysis of the fluid flow is used to predict the change in temperature of the bulk fluid, which subsequently alters the solidification behavior. Comparison to a full numerical model for the solidification of an Al-4.5 wt.% Cu alloy shows good agreement with the relationships predicted by the scaling analysis for solidification time, bulk fluid cool down time, and the maximum local solidification time.

References

[1] W.D. Bennon, F.P. Incropera, A continuum model for momentum, heat and species transport in binary solid-liquid phase change systems - I. Model formulation, Int. J. Heat Mass Trans. 30 (1987) 2161–2170.

[2] W.D. Bennon, F.P. Incropera, A continuum model for momentum, heat and species transport in binary solid-liquid phase change systems- II. Application to solidification in a rectangular cavity, Int. J. Heat Mass Trans. 30 (1987) 2171–2187.

[3] V.R. Voller, A similarity solution for the solidification of a multicomponent alloy, Int. J. Heat Mass Trans. 40 (1997) 2869–2877.

[4] J.D. Chung, J.S. Lee, M. Choi, H. Yoo, A refined similarity solution for the multicomponent alloy solidification, Int. J. Heat Mass Trans. 44 (2001) 2483–2492.

[5] T.R. Goodman, The Heat-Balance Integral and Its Application to Problems Involving a Change of Phase, Trans. ASME. 80 (1958) 335–342.

[6] R.H. Tien, G.E. Geiger, A Heat-Transfer Analysis of the Solidification of a Binary Eutectic System, ASME J. Heat Transfer. 89 (1967).

[7] R.H. Tien, G.E. Geiger, The Unidimensional Solidification of a Binary Eutectic System With a Time-Dependent Surface Temperature, ASME J. Heat Transfer. 90 (1968).

[8] J.C. Muehlbauer, J.D. Hatcher, D.W. Lyons, J.E. Sunderland, Transient Heat Transfer Analysis of Alloy Solidification, ASME J. Heat Transfer. 95 (1973).

[9] S.H. Cho, J.E. Sunderland, Heat-Conduction Problems With Melting or Freezing, ASME J. Heat Transfer. 91 (1969).

[10] M.J.M. Krane, F.P. Incropera, A scaling analysis of the unidirectional solidification of a binary alloy, Int. J. Heat Mass Trans. 39 (1996) 3567–3579.

[11] G. Amberg, Parameter ranges in binary solidification from vertical boundaries, Int. J. Heat Mass Trans. 40 (1997) 2565–2578.

[12] S. Chakraborty, P. Dutta, Scaling analysis of momentum, heat, and mass transfer in binary alloy solidification problems, Mater. Sci. Tech. 18 (2002) 600–606.

[13] W. Lin, S.W. Armfield, J.C. Patterson, Cooling of a Pr<1 fluid in a rectangular container, J. Fluid Mech. 574 (2007) 85–108.

[14] A. Bejan, Convection Heat Transfer, 2nd ed., John Wiley & Sons, Inc., New York, 1995.

[15] V.R. Voller, C.R. Swaminathan, General Source-Based Method for Solidification Phase Change, Num. Heat Trans. B. 19 (1991) 175–189.

Advances in the Science and Engineering of Casting Solidification
Edited by: Laurentiu Nastac, Baicheng Liu, Hasse Fredriksson, Jacques Lacaze, Chun-Pyo Hong, Adrian V. Catalina, Andreas Buhrig-Polaczek, Charles Monroe, Adrian S. Sabau, Roxana Elena Ligia Ruxanda, Alan Luo, Subhayu Sen, and Attila Diószegi
TMS (The Minerals, Metals & Materials Society), 2015

STRUCTURE AND CASTING DEFECTS OF ALUMINUM BILLETS PRODUCED BY DIRECT-CHILL CASTING

D.G. Eskin[1,2]

[1]Brunel University, Brunel Centre for Advanced Solidification Technology, Uxbridge UB8 3PH, United Kingdom
[2]Tomsk State University, Tomsk, 634050 Russian Federation

Keywords: Aluminum, Cold crack, Direct-chill casting, Hot tear, Macrosegregation, Residual stress, Transition region, Thermal contraction

Abstract

Direct-chill (DC) casting is the main technology of round billets and flat ingots intended for further deformation. This casting technology has many advantages in control of the solidification and achieving high quality of the cast metal. On the other hand, ever increasing requirements to the soundness, mechanical and service properties of aluminium alloys as well as to increased output demand for deep understanding of the processes and mechanisms that control the formation of structure and defects during solidification. A combination of advanced computer simulation and dedicated experiments allows a modern scientist to achieve much better insight into the processing involved and suggest means of improving the quality of the castings. This paper presents an overview of research in the mechanisms of macrosegregation, hot and cold cracking in aluminium billets and the ways to control these defects by the control of process parameters and structure. The paper is illustrated with own results.

Introduction

Direct-chill casting of aluminum alloys is a well developed technology with a rather long history [1, 2, 3]. At the same time, despite 60 years of extensive research and practice, the formation of structure, defects and properties of billets and ingots produced by direct-chill (DC) casting is still not completely understood.

On the other hand, many phenomena have been discovered and interconnected, and a general pattern starts to emerge from different bits and pieces [4]. The development in the last 20 years of computer modeling and simulations, driven mainly by the industrial need to optimize and automate the casting process, gives a powerful tool to a researcher. Computer simulation allows a scientist to artificially uncouple the physical processes involved in the formation of a complex real-life pattern and to evaluate the individual effects. In addition, the quantities that are difficult or impossible to measure experimentally can be examined using computer simulations, for example temperature distribution and flow patterns in the sump of a billet.

Structure formation

The structure formed in billets and ingots during direct-chill casting is characterized by several distinct features that result from peculiar thermal and mass transfer during solidification and have consequences for the quality of the final as-cast product.

The grain size and dendrite arm spacing distribute unevenly across the thickness of the billet (ingot) as shown in Fig. 1. The main reason for the variation in size is the cooling regime with specific zones that are present in DC casting (see Fig. 1a): namely, primary (relatively slow) cooling in the mold where the shell of the billet is formed; air-gap in the lower part of the mold (very slow cooling) where a subsurface region of the billet is formed; secondary (very fast) cooling where the interior of the billet is formed; and downstream cooling where the solidification of the billet is finalized.

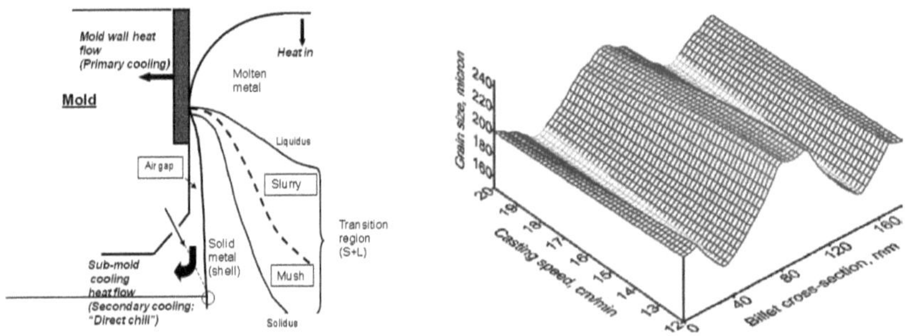

a b

Fig. 1 (a) Different zones in the billet sump and different regimes of cooling in DC casting [5]; (b) variation of grain size across the 200-mm billet diameter in dependence on casting speed for an Al–4% Cu alloy [6].

It is sometimes observed experimentally that the dendritic structure becomes finer in the center of the billet, though the cooling efficiency there should be less than in the immediate neighborhood. A thorough analysis of the available data, both experimental and numerical, shows that the accelerated solidification in the center of the billet has a reasonable physical explanation [4]. The formation of the solid phase produces latent heat of solidification and pumps additional thermal energy into the slurry zone, effectively slowing down the cooling. The central part of a casting is formed in the last stage of solidification when most of the latent heat has been already released and conducted out of the volume. Therefore, the cooling in the central-bottom part of the mushy zone is more efficient than in the slurry zone. In addition, the high solidification-front velocity (solidification rate) in the center of the billet may narrow the mushy zone as a result of a higher upward velocity of the solidus isotherm as compared to the liquidus velocity.

It is essential that there is no strict correlation between the cooling rate measured by thermocouples embedded into the billet and the particular structure found in the specific location. It is especially the case when so-called duplex structures are observed with a mixture of fine-cell and coarse-cell dendritic grains. The formation of duplex structures is important for macrosegregation.

Solidification under typical conditions encountered in most industrial processes, including DC casting, does not follow the equilibrium schedule with perfect partitioning and re-distribution of alloying elements. As a result, nonequilibrium phases form through solidification reactions that should not have occurred under equilibrium conditions. This phenomenon is called microsegregation and, along with the accompanying processes of back diffusion and dendrite coarsening, have a profound effect on the amount of eutectics. The eutectic is the last liquid that solidifies and its volume fraction plays an essential role in the occurrence of hot tears. Furthermore, the accumulation of solutes at the solid–liquid interface that also happens due to the limited diffusion influences the formation of dendritic grain structure (fragmentation of dendrites and columnar-to-equiaxed transition) and constitutes one of the mechanisms of grain refining by so-called growth restriction [7].

The amount of (nonequilibrium) eutectics is one of the main structure parameters that affect the susceptibility of an alloy to hot tearing. Hot cracking would not occur if there were no eutectics (pure metal), or the amount of eutectics were sufficient to assure good feeding conditions at the last stages of solidification. The presence of eutectics in the structure and the volume fraction of eutectics is a function of the chemical composition and solidification path (most of eutectics in wrought aluminum alloys is nonequilibrium) [8]. Our experimental and numerical studies show that the amount of eutectics is always lower in the center of the billet (ingot) where the hot tearing susceptibility is the highest [9]. The process parameters that facilitate hot tearing, e.g. a lower melt temperature and a lower alloying level, produce less eutectics. Hence, a higher melt temperature may hinder the development of hot cracks [10]. At the same time, it would result in general structure coarsening and affect the distribution of coarse-cell grains. The latter may have its effect on macrosegregation.

The formation of structure and defects during DC casting is tightly linked to the geometry of the billet sump and the flow patterns that exist in the slurry and mushy parts of the sump.

Transition region

It is important to distinguish the characteristic regions in the sump that are characterized by different behavior (see Fig. 1a). The liquid part of the sump or the liquid (molten) bath is located above the liquidus isotherm. Here, thermal convection is the main driving force for melt flow, although in the vicinity of the liquidus the solute-enriched and cooler liquid can exit the transition region and affect the flow pattern. The liquid bath has very little variation in temperature during steady-state casting and the liquidus isotherm is nearly flat. The transition region or solidification region is bordered by the liquidus and (nonequilibrium) solidus isotherms. This region distinctly widens towards the center of the billet, and its dimensions are critical for the occurrence of casting defects and structure formation [2, 3, 4].

The transition region can be further divided into two parts. The upper part is called slurry and is confined between the liquidus and so-called coherency isotherm (25–40% solid). Within the slurry zone the solid is suspended in the liquid and the entire mixture is fluid. As the solidification progresses and the partitioning of solutes between the solid and the liquid phases occurs, the concentration-driven (or solutal) convection joins the thermal convection as a driving force for the melt flow. The solid phase in the slurry zone can move with the melt flow but is also affected by gravity so it follows quite complicated trajectories. This phenomenon frequently results in the duplex grain structures that have been already mentioned. The difference in the velocities between the solid (diluted) and liquid (enriched) phases, as well as between the cooler (enriched) and hotter (diluted) liquid flows is the fundamental reason for macrosegregation caused by the thermo-solutal convection and solid-grain transport.

Further down the transition region, the so-called mushy zone lies between the coherency and the solidus isotherms. One of the main properties of the region close to the coherency isotherm and below it is the easiness of liquid penetration through it. This property is called permeability. High permeability enables the penetration of the thermo-solutal convective flow deeper into the slurry zone and even into the mushy zone and the transfer of the solute-rich, cool liquid into the slurry zone, frequently deciding the macrosegregation pattern. On the other hand, permeability also determines the extent of melt flow within the mushy zone. The main driving force for this flow is the solidification shrinkage and thermal contraction. The larger extent of the shrinkage deeper inside the mushy zone and the metallostatic head of the liquid bath and the slurry zone create the pressure drop over the mushy zone that forces the liquid in the direction normal to the coherency isotherm (also called the solidification front). The shrinkage-driven flow of the solute-rich liquid is one of the mechanisms of inverse macrosegregation. Within the mushy zone the solid grains interact with each other, first by short-term contact, then by entangling, and finally by forming the continuous network. The last stage is marked by another characteristic isotherm – rigidity (85–95 % solid). Starting from this moment the stresses, mostly caused by uneven thermal contraction in different sections of the billet, can be transferred by the rigid mushy zone. The rigidity

isotherm can be measured experimentally [11, 12]. A combination between the developing stresses (and corresponding strains) and limited permeability creates conditions for the formation of hot cracks.
Among casting parameters, the casting speed exerts the maximum and direct effect on the sump depth and the thickness of the transition (solidification) region [4]. The water-flow rate has a marginal effect, providing that the amount of water is adequate for the efficient cooling. The melt temperature does not affect much sump depth, but narrows the transition region by moving the liquidus isotherm down. In the transient stage of casting when the casting speed is changing the thermal inertia causes the delay in the sump development, which can explain the experimentally observed fact that the maximum of hot tearing susceptibility does not correlate necessarily to the highest casting speed in the transient casting regime, but rather to the deepest sump [4].

Macrosegregation

Several mechanisms of macrosegregation act simultaneously producing sometimes very peculiar segregation patterns [4]. It is known that the thermo-solutal convection tends to bring the solute to the center of the billet, whereas the shrinkage-induced flow acts in the opposite direction. Floating grains dilute the portion of the billet where they accumulate. Deformation triggered by thermal contraction can act in different directions, depending on the sign of strain and the presence of the open surface. Seemingly controversial experimental observations can be adequately explained, assuming that different mechanisms contribute to a different extent.
Among the casting parameters, the casting speed exerts the maximum effect on the extent of macrosegregation, Fig. 2a. This effect directly correlates with the changes in the geometry of the sump, in particular with the sump depth and the steepness of the solidification front, Fig. 2b. It is demonstrated both experimentally and numerically [2, 3, 4]. The centerline segregation increases with the casting speed. The shrinkage-induced flow is directly linked to the slope of the solidification front and, is therefore enhanced with the deepening of the sump, creating more inverse (negative) segregation. At the same time, a wider slurry zone and stronger penetration of the convective flow into the slurry region create more possibilities for floating grains to form and grow, further increasing the extent of negative macrosegregation. In this case, the contribution of the convective flow into the positive centerline segregation can be completely overtaken by other mechanisms.
A higher melt temperature results in a higher macrosegregation close to the surface of the billet and no significant changes in the rest of the billet [10].
The size of the billet is yet another major factor in the formation of macrosegregation pattern. It is possible to obtain normal (positive centerline) macrosegregation virtually for any size of a billet or ingot, providing that it is possible to cast at low speeds [3, 13]. It is demonstrated experimentally and numerically that the transition from positive to negative centerline segregation occurs for a certain billet diameter on increasing the casting speed (see Fig. 2c). Generally the obtained dependence can be explained well based on the thickness of the transition region and the slope of the solidification front; therefore from the ratio between the thermo-solutal convection and the shrinkage-induced flow. The actual casting speeds that may result in positive or negligible macrosegregation are, unfortunately, much lower than those that are currently used in industry. However, there exists a possibility to enhance the contribution of thermo-solutal convection by external action, e.g. mechanical or electromagnetic stirring [3, 4].
Grain refining shifts the transition from positive to negative macrosegregation to higher casting speeds through the interplay of a lower coherency temperature, decreased permeability of the mushy zone and wider spread of floating grains [4, 13] (see Fig. 2c).

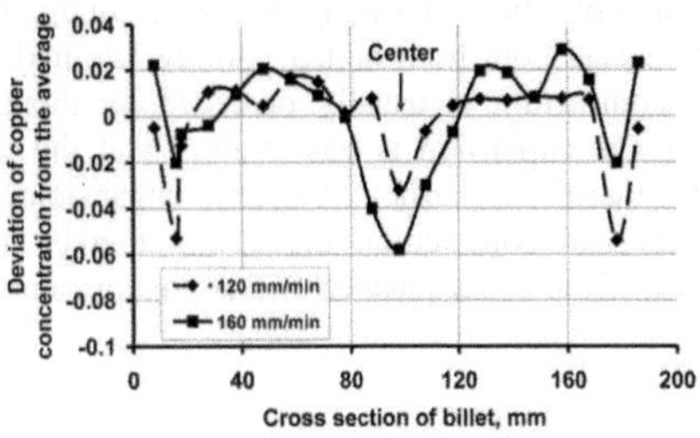

a

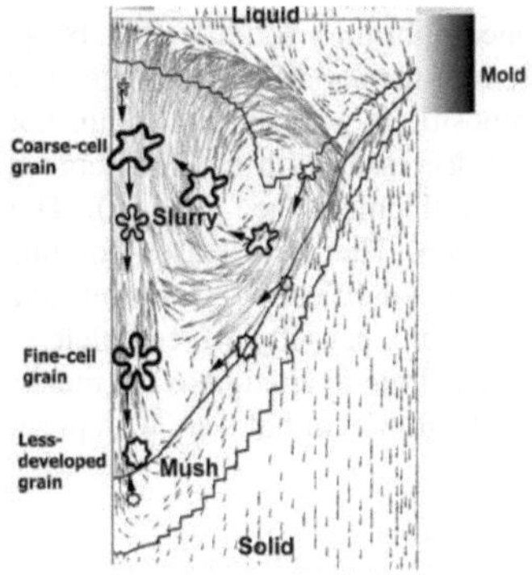

b

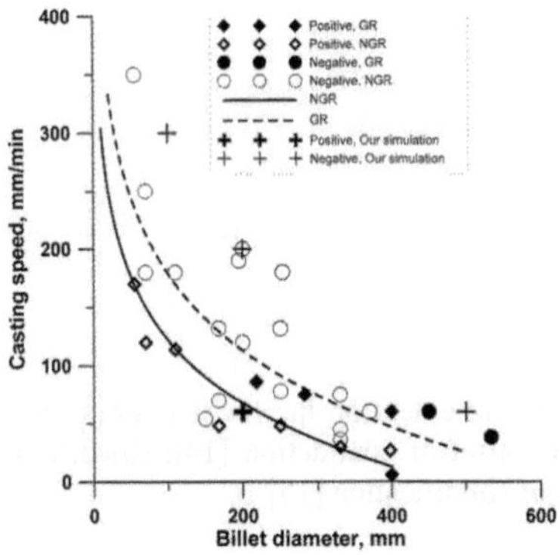

c

Fig. 2 Macrosegregation in DC casting: (a) effect of casting speed on macrosegregation of Cu in an Al–4% Cu alloy [6]; (b) flow patterns, characteristic zones and potential trajectories of floating crystals in the sump of a 200-mm billet; (c) effect of billet diameter and casting speed on the macrosegregation (positive centerline below the curves and negative – above the curves; dashed line – for grain refined alloys) [13].

Hot tearing

Further down the sump of the billet, closer to the (nonequilibrium) solidus the thermomechanical behavior of the semi-solid material becomes significant. Thermal contraction of the solid network saturated with liquid inside the billet bulk and the resistance to this contraction from the completely solid external part of the billet impose stress–strain conditions when the semi-solid material may fail and loose its integrity. This failure is called hot tearing or hot cracking. The fundamental reason for this defect is lack of liquid feeding that can compensate for the solidification shrinkage, and the formation of gaps (pores) that can develop into cracks under tension. Hence, the amount of liquid at the last stage of solidification (eutectics) and the permeability of the mushy zone play paramount role in the nucleation of hot cracks. On the other hand, mechanical properties of semi-solid alloys in the vicinity of the solidus and their dependence on temperature give an indication whether the material is able to withstand the thermal stress and can accommodate the thermal strain, or it will crack. The fracture can be prevented if the ductility or strength of semi-solid material is greater than the thermal strain or stress imposed on it [11]. And the fracture can be healed if the amount of liquid is sufficient to fill all the gaps created in the mush by deformation.

Thermal strain that is accumulated in the solidification range of an alloy and the temperature when the contraction starts (rigidity temperature) are studied in detail [4, 11, 12]. It is shown experimentally, using a specially developed technique, that the temperature of contraction onset corresponds to rather high fractions of solid as illustrated in Fig. 3a. In fact, in most aluminum alloys the thermal contraction starts at a temperature of equilibrium solidus, or between the equilibrium and nonequilibrium solidus. The total thermal strain in the (nonequilibrium) solidification range depends on the alloy system, and generally on the extent of the solidification range. The maximum of thermal contraction corresponds to alloy compositions with the maximum hot-tearing susceptibility. The temperature of contraction onset decreases with the structure refinement, caused both by a higher cooling rate and by grain refinement (see Fig. 3a). The thermal contraction is usually preceded by thermal expansion. The main reason for this pre-shrinkage expansion is gas evolution at the last stages of solidification. Grain refining facilitates thermal expansion. A greater expansion can compensate partially of completely for the thermal contraction, decreasing thereby the thermal strain exerted onto the mushy zone. One of the consequences of this phenomenon is the substitution of hot tears with porosity.

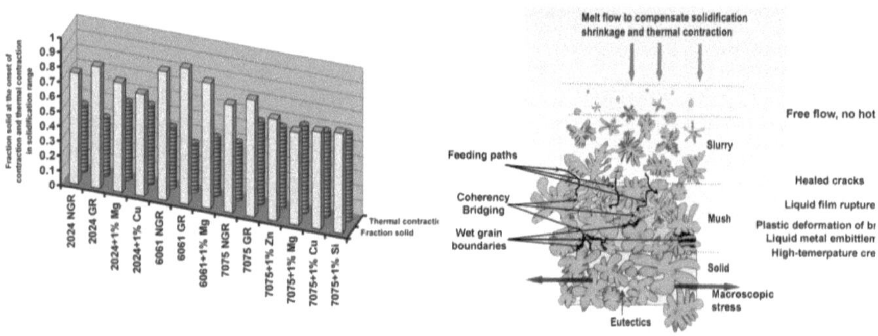

Fig. 3 Hot tearing: (a) effect if alloy type and grain refinement on the thermal strain in the solidification range and the fraction of solid at the onset of contraction [14]; (b) different mechanisms of hot tearing acting at different stages of solidification [15].

Grain refining decreases the probability of hot cracking upon DC casting of aluminum alloys through the delay in the rigidity development and improved tensile strength and ductility of the semi-solid material.

A number of theories and hypothesis has been suggested over the years for the mechanisms nucleation and propagation of hot cracks. Some of these mechanisms resulted in hot-tearing criteria, based on mechanical behavior of the alloy (critical strain, stress, or strain rate) or on its solidification characteristics (vulnerable solidification range, solidification time, feeding conditions, casting speed). Only those hot tearing criteria that take into account the dynamic nature of DC casting process, e.g. based on strain rate, or pressure variations, or casting speed; respond adequately to the change in hot tearing susceptibility with the change in process parameters. However, even these criteria can predict qualitatively only the probability of cracking, not the actual occurrence. A comprehensive model and a corresponding new hot tearing criterion should include nucleation and propagation of hot tears and connect these processes (i) to the microstructure evolution during solidification of the semi-solid material; (ii) to the macroscopic and microscopic thermo-mechanical situation in the mushy zone; and (iii) to the mechanical (or fracture-mechanical) properties of the mushy zone [15]. It is suggested that the mechanisms of hot crack nucleation and propagation can be different, depending on the composition of the alloy and the temperature range when the crack initiates and develops as demonstrated in Fig. 3b.

Cold cracks

Cold cracking phenomenon is the least studied, yet very important defect occurring during direct chill casting. The spontaneous nature of this defect makes its systematic study almost impossible. Computer simulation of thermomechanical behavior of an ingot during its cooling after the end of solidification requires constitutive parameters of high-strength aluminum alloys in the as-cast condition, which are not readily available.

The susceptibility to cracking is on one hand due to the segregation and precipitation of brittle intermetallics at grain boundaries and in interdendritic spaces, and on the other hand because of the development of high residual thermal stresses. Above the solidus, the low melting nonequilibrium eutectics provide potential sites for hot tearing. Further propagation of such cracks under tensile-stress fields that develop to some critical levels below the solidus may result in catastrophic failure of the ingot in the solid state; i.e. cold cracking. Other defects of certain size such as pores and inclusions can trigger cold cracking as well. What makes the material even more susceptible to cold cracking is the sever loss in ductility upon cooling after the end of solidification [3]. If the specific microstructure of the material in the as-cast condition provides potential crack sites for cold cracking, the stress development in the ingots during casting and further cooling is brought about by non-homogenous cooling conditions and certain thermo-physical properties of the material. For example, relatively higher coefficient of thermal expansion and lower thermal conductivity of some of 7xxx series alloys compared to other aluminum alloys result in steep temperature gradients and consequently high thermal stresses appear. In order to predict the cold cracking phenomenon in high-strength aluminum alloys, a true understanding of the residual thermal stress development during the casting is required.

Cold cracking susceptibility was studied using the simulation results and casting trials on several billets under various conditions as illustrated in Fig. 4 [16]. The simulation results showed that as the casting speed increases, the critical crack size (according to Griffiths' criterion) decreases leading to a higher failure probability of the billets. This was proven by casting trials. The critical crack sizes were validated upon experiments, where a 7 mm inclusion in the centre of the billet triggered the catastrophic failure (see Fig. 4b).

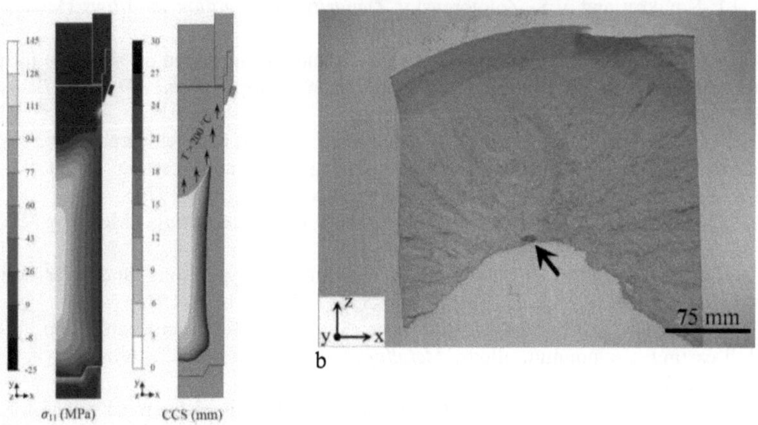

a

Fig. 4 Cold cracking: (a) computer simulation of principle stress and critical crack size (CCS) distribution in a 200-mm 7xxx-series billet; (b) cold crack in a 200-mm billet triggered by an oxide inclusion [16].

Concluding remark

This review on direct-chill casting of aluminum alloys demonstrates the complexity of mechanisms involved in the formation of structure, macrosegregation and hot tearing. It is a strong belief of the author that the control over structure formation and elimination of casting defects can be only achieved when the contributions of all the involved mechanisms and the ratio of these contributions are taken into account.

Acknowledgements

This paper presents an overview based on the results obtained under supervision of the author and Prof. L. Katgerman with participation of Drs. Suyitno, Q. Du, R. Nadella, M. Lalpoor, V.I. Savran and A. Stangeland. Their contribution is greatly appreciated.

References

1. W. Roth, "Ueber die Abkühlung des Stranges bein 'Wassergußʼ", *Aluminium*, 25 (1943) 283–291.
2. V.I. Dobatkin, *Continuous Casting and Casting Properties of Alloys* (Moscow: Oborongiz, 1948), 154 p.
3. V.A. Livanov, R.M. Gabidullin, and V.S. Shepilov, *Continuous Casting of Aluminum Alloys* (Moscow: Metallurgiya, 1977), 168 p.
4. D.G. Eskin, *Physical Metallurgy of Direct Chill Casting of Aluminum Alloys*. (Boca Raton, CA: CRC Press, 2008), 301 p.
5. J.F. Grandfield, D.G. Eskin, and I.F. Bainbridge. *Direct Chill Casting of Light Alloys: Science and Technology*. Hoboken: John Wiley & Sons, 2013, 412 p.
6. D.G. Eskin et al., Structure formation and macrosegregation under different process conditions during DC casting, *Materials Science and Engineering A*, 384 (2004) 232–244.
7. M.A. Easton and D.H. StJohn. A model of grain refinement incorporating alloy constitution and potency of heterogeneous nucleant particles, *Acta Materialia*, 49 (2001) 1867–1878.
8. I.I. Novikov and V.S. Zolotorevsky, *Dendritic Segregation in Alloys* (Moscow: Nauka, 1966), 44–109.
9. Suyitno et al., Effects of alloy composition and casting speed on structure formation and hot tearing during DC casting of Al–Cu alloys,*Metallurgical and Materials Transactions A*, 35A (2004) 3551–3561.
10. D.G. Eskin, V.I. Savran, and L. Katgerman, Effects of melt temperature and casting speed on the structure and defect formation during direct-chill casting of an Al–Cu alloy, *Metallurgical and Materials Transactions A*, 36A (2005) 1965–1976.
11. I.I. Novikov, *Hot Shortness of Non-Ferrous Metals and Alloys* (Moscow: Nauka, 1966), 188–196.
12. D.G. Eskin et al., Contraction of aluminum alloys during and after solidification, *Metallurgical and Materials Transactions A*, 35A (2004) 1325–1335.
13. D.G. Eskin, Q. Du, and L. Katgerman, Scale rules for macrosegregation during direct-chill casting of aluminum alloys, *Metallurgical and Materials Transactions A*, 39A (2008) 1206–1212.
14. A. Stangeland, A. Mo, and D.G. Eskin, Thermal strain in the mushy zone for aluminum alloys, *Metallurgical and Materials Transactions A*, 37A (2006) 2219–2229.
15. D.G. Eskin and L. Katgerman, A quest for a new hot tearing criterion, *Metallurgical and Materials Transactions A*, 38A (2007) 1511–1519.
16. M. Lalpoor et al., Cold cracking in DC-cast high strength aluminum alloy ingots: An intrinsic problem intensified by casting process parameters, *Materials Science and Engineering A*, 528 (2011) 2831–2842.

Advances in the Science and Engineering of Casting Solidification
Edited by: Laurentiu Nastac, Baicheng Liu, Hasse Fredriksson, Jacques Lacaze, Chun-Pyo Hong, Adrian V. Catalina, Andreas Buhrig-Polaczek, Charles Monroe, Adrian S. Sabau, Roxana Elena Ligia Ruxanda, Alan Luo, Subhayu Sen, and Attila Diószegi
TMS (The Minerals, Metals & Materials Society), 2015

THE FLUID FLOW AND SOLIDIFICATION PHENOMENON IN BILLET CONTINUOUS CASTING PROCESS WITH MOLD AND FINAL ELECTROMAGNETIC STIRRINGS

Dongbin Jiang and Miaoyong Zhu

School of Materials and Metallurgy, Northeastern University, Shenyang, 110819 China.

Keywords: continuous casting; electromagnetic stirring; fluid flow; solidification.

Abstract

Coupling electromagnetic field, flow field and solidification phenomena, a 3D numerical model was developed to simulate the SWRT82B steel billet by continuous casting with mold (M-EMS) and final electromagnetic stirrings (F-EMS). The columnar region was regarded as a porous media and a generalized approach was used to calculate permeability, while for the equiaxed zone, a variable apparent viscosity model was applied to simulate the fluid flow at the early stage of equiaxed dendrites. The model was validated by the measured data in plant. The simulation results show that the fluid flow pattern near the meniscus with considering solidification is quite different from that without considering solidification. And the maximum stirring velocity of molten steel in M-EMS and F-EMS zone increases with the increase of the current intensity. Moreover, the critical coherency solid fraction determined by the crystal morphology affects the fluid flow in F-EMS zone strongly.

1. Introduction

During the high carbon steel solidification, the negative solute elements reject from solid phase and transport under the influence of fluid flow, resulted in mascrosegregation of strand [1]. Even more serious is that the macrosegregation in billets or blooms cannot be eliminated in the subsequent heat treatment and affects the mechanical properties of final products. As the solute redistribution in the melt is strongly dependent on the melt convection flow [2], it is necessary to have an in-depth understanding of the melt flow during continuous casting process, especially with mold and final electromagnetic stirrings. In recent years, many researchers have devoted great efforts to the investigation of the fluid flow during the continuous casting progress.

Natarajan and El-Kaddah [3] used a fixed-grid method to calculate sub-mold electromagnetic field and fluid flow in continuous casting process. The results show that the secondary flow promoted mixing beyond the stirrer region. Liu *et al* [4] developed a simulation model coupled 3D electromagnetic field and fluid flow in a round-bloom mold with M-EMS. The electromagnetic force distribution and melt flow were analyzed. With the mould wall assumed as liquidus temperature, Yu and Zhu [5] studied fluid flow and temperature field in a round mould with EMS. The result shows the flow and temperature distribution were strongly influenced by EMS. Ayata *et al* [6] undertook a series of industry experiments of billet and bloom continuous casting to investigate the effect of EMS on solute segregation. It was found that the combined stirring was effective in improving the macrosegregation and there was an optimum condition in the stirring intensity and the solid fraction in the liquid pool of F-EMS. Tian [7] and Chen [8] respectively treated all the mushy zone as a porous media to simulated molten steel flow in mold region. The results both revealed that the temperature and thickness of solidified shell were affected by the fluid flow. Sun and Zhang [9] used a 3D and 2D hybrid method to simulate thermal solute transport during bloom casting. The solute elements redistribution and porosity were attributed to the gravity liquid flow.

In those studies, some authors [3-6] only simulated the fluid flow or confined to undertake plant trials to investigate the effect of EMS on the quality of strand. Although, both the flow and solidification inside the continuous casting strands have been simulated by several groups [7-9], the whole mushy zone was all treated as a porous media. In fact, it is not appropriate to treat all the mushy zone as porous medium, as the initial growing equiaxed dendrite crystal can move with the liquid [10]. Moreover, the permeability of columnar dendrite zone and equiaxed dendrite zone is strongly dependent on the dendrite structure and should be regarded separately. However, it is seldom reported in the published papers about continuous casting, especially with dual EMS. In the present work, a 3D simulation model coupling the electromagnetic field, fluid flow and solidification phenomena in the 160 mm × 160 mm billet continuous casting with dual EMS was developed, as shown in Figure 1. The columnar region was treated as a porous media, while a variable apparent viscosity model was developed to simulate the fluid flow at the initial growing stage of equiaxed crystals. The numerical model was validated by the measured strand surface temperature and the magnetic induction intensity of stirrer.

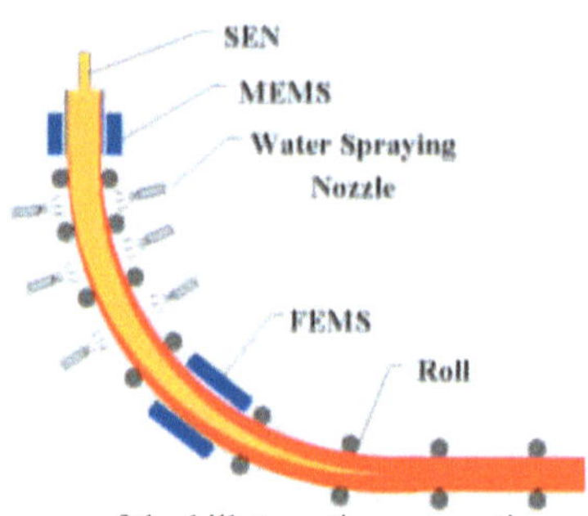

Figure 1. Schematic diagram of the billet continuous casting machine with dual EMSs

2. Mathematical Model

In the numerical model, the Maxwell's equations are solved to calculate electromagnetic forces which act as momentum sources in the Navier-Stokes equations. The electromagnetic field, fluid flow and solidification have to be coupled in order to describe reasonably the phenomenon occurring in the continuous casting process. The following assumptions are made in order to make it more efficient.

(1) The molten steel flow in the strand is assumed to be viscosity fluid flow and the increasing viscosity method was used to considering the initial crystal movement.

(2) Because of the small magnetic Reynolds number (about 0.01) [11] during the stirring process, the influence of flow on the electromagnetic field is ignored and the time-averaged electromagnetic force replaces of transient value.

(3) Compared with the latent heat of solidification, the heat source induced by the electromagnetic apparatus is ignored.

(4) In water spraying region, the same cooling intensity is applied to the strand surface, and the solidification shrinkage is not considered in simulation.

(5) As the strand is withdrawn from the meniscus to the secondary cooling zone and the influence of F-EMS on M-EMS flow is not considered due to the enough long distance between the M-EMS and F-EMS.

Thus, the governing equations of mathematical model coupling electromagnetic stirring, fluid flow and solidification phenomenon can be described as follows.

2.1 Electromagnetic Model

The distribution of magnetic induction intensity and the electromagnetic force is obtained by solving Maxwell's equations and the time averaged electromagnetic force replace the instantaneous value, calculated by Eq.(1) [5]. The M-EMS is placed in the lower region of the mold and the F-EMS is installed in the end of secondary cooling zone, which are 0.55 m and 8.2 m from meniscus respectively.

$$\boldsymbol{F}_{mag} = \frac{1}{2}\mathrm{Re}\left(\boldsymbol{j} \times \boldsymbol{B}\right) \tag{1}$$

2.2 Fluid Flow Model

The movement of steel in continuous casting is calculated by solving mass and momentum conservation equations [12], which are expressed as follows:

$$\frac{\partial \rho}{\partial t} + \nabla(\rho v) = 0 \tag{2}$$

$$\rho\frac{\partial v}{\partial t} + \nabla\cdot(\rho v v) = -\nabla P + \nabla\cdot[\mu_{eff}(\nabla\cdot v)] + \rho g \beta_T (T - T_{ref}) + F_{mag} + F_u S_m \tag{3}$$

$$\mu_{eff} = \mu_m + \mu_t \tag{4}$$

Where ρ is the steel density, $\boldsymbol{v}$ is velocity of steel, μ_m is the apparent viscosity calculated according to solid fraction, μ_t is turbulent viscosity which is calculated through the low-Reynolds number *k-ε* equations [9], F_u is a switch function, β_T is the thermal expansion coefficient, $\boldsymbol{g}$ is gravitational acceleration, S_m is the sink terms of momentum which is described as:

$$S_m = \frac{\mu_l}{K}(v - v_c) \tag{5}$$

$$f_l = \frac{T - T_s}{T_l - T_s} \quad (T > T_s, T < T_l) \tag{6}$$

Where f_l is the liquid fraction, v_c is the casting speed, K is the mushy zone permeability. It should be noted that the permeability is an important parameter for simulating the fluid flow during solidification and the determination of the permeability should be vary careful. As the the columnar dendrite grows from the billet surface and couldn't move with the liquid metal, as shown in Figure 2, it is reasonable to regard the columnar zone as a porous medium. As the solidification shrinkage is not considered in the calcualtion, the peremeability of fluid flow is determined by Eq.(7) [13], where λ_1 and λ_2 are primary and secondary arm spacing. The thicknesses of columnar and equixed dendrite zone are about 28 mm and 104 mm respectively. The measured primary and secondary arm spacing in the columnar zone are showed in Figure 3.

$$K_c = 0.0097\left[\frac{\lambda_2}{1+(\lambda_2/\lambda_1)+2(\lambda_2/\lambda_1)^2}\right]\frac{f_l^3}{f_s^2} \tag{7}$$

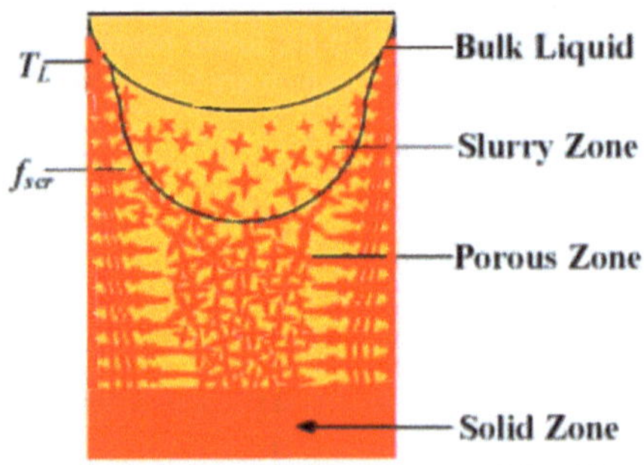

Figure 2. The schematic diagram of dendrite structure

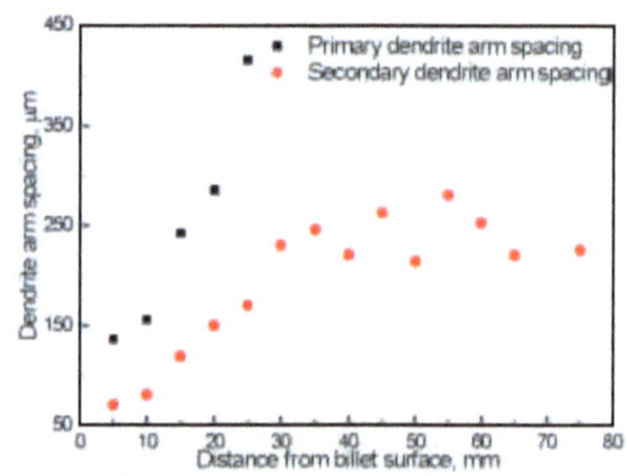

Figure 3. The measured primary and secondary arm spacing samples

For the equixed region, the initial formation of crystal is surrounded by liquid metal. The equiaxed dendrite move freely with liquid flow induced by solidification shrinkage, gravity or EMS forces. In this period, the mushy zone should be treated as slurry, not porous media. A varying apparent viscosity model shown as Eq.(8) [10], is used to simulate the fluid flow. With the decrease of temperature, the equixed dendrite grows continuously and sencondary dendrite arms gradually contact with one another, which is deemed to be the coherency point, where the solid fraction is defined as f_{scr} and it is determined by the dendrite morphology as Eq.(9) [14]. β is the sphericity [15]. After the impingement of equixed dendrite secondary arms, the solid network forms and the dendrite crystal couldn't move with the liquid mental, which should be considered as porous zone and the permeability is calculated by Eq.(10) [16].

$$\mu_m = \begin{cases} \mu_l(1+2.5f_s+10.05f_s^2+0.0071\exp(16.67f_s)) & f_s \le f_{scr} \\ \mu_l & f_s > f_{scr} \end{cases} \tag{8}$$

$$f_{scr} = \frac{\pi\beta}{6} \tag{9}$$

$$K_e = \frac{180}{\lambda_2^2}\frac{f_l^3}{f_s^2} \tag{10}$$

2.3 Solidification Model

In order to obtain the temperature field and solidification behavior in continuous casting process, the enthalpy equation is used, which can be rewritten as follows:

$$\frac{\partial}{\partial t}(\rho h)+\nabla\cdot(\boldsymbol{v}\rho h)=\nabla\cdot(k_{eff}\nabla T)+q_{EM} \tag{11}$$

$$h=\int_{T_{ref}}^{T} C_p dT + f_l L \tag{12}$$

where C_p is specific heat, L is latent heat of molten steel, h is the total enthalpy, k_{eff} is the thermal conductivity [10]. During the solidification of continuous casting, the whole heat transfer process is composed of the mold cooling, secondary cooling and air cooling [17]. The physical properties and geometrical parameters used in the numerical simulation are given in Table 1.

Table 1. The physical properties and process parameters for simulation

Item	Unit	Symbol	Value
Viscosity	$kg\cdot m^{-1}\cdot s^{-1}$	μ_l	0.006
Density	$kg\cdot m^{-3}$	ρ	7400
Solid thermal conductivity	$W\cdot m^{-1}\cdot K^{-1}$	k_s	30
Liquid thermal conductivity	$W\cdot m^{-1}\cdot K^{-1}$	k_l	20
Liquid specific heat	$J\cdot kg^{-1}\cdot K^{-1}$	$C_{p,l}$	825
Solid specific heat	$J\cdot kg^{-1}\cdot K^{-1}$	$C_{p,s}$	669
Latent heat	$J\cdot kg^{-1}$	L	275000
Solidus temperature	K	T_s	1633
Liquidus temperature	K	T_l	1738
Thermal expansion coefficient	K^{-1}	β_T	2.0×10^{-5}
Sphericity	1	β	0.28
SEN inner diameter	mm	Φ_{inner}	30
SEN outer diameter	mm	Φ_{outer}	70
SEN submergence depth	mm	h	100
Casting speed	m/min	v_c	1.8
Casting temperature	K	T_{in}	1769

2.4. Numerical Simulation Procedure

In the computational procedure, ANSYS is used to calculate the external electromagnetic field generated by the electromagnetic stirrer. The time averaged electromagnetic forces acted as added sources are interpolated into the fluid flow calculation. Then the SIMPLE algorithm based on controlled volume method is used to solve the momentum and energy equations.

3. Results and Discussion

3.1 The model validations

The numerical magnetic induction intensity and surface temperature of strand are compared with the measured as shown in Figure 4 and Figure 5 respectively. The magnetic induction intensities in the mold and final solidification stirrer are measured through CT-3 teslameter without casting steel. It can be seen from Figure 4 that the calculated magnetic induction intensity in the center of stirrers agrees well with the measured, as the M-EMS and F-EMS frequencies are 4Hz and 6Hz respectively. With increase current intensity, the calculated values are a bit larger than the measured. It is may be the efficiency of magnetic stirrer may decrease with the larger current intensity which has not been considered in the present calculation. Figure 5 shows the calculated and measured values are in a good agreement, although the predicted is a little larger than the measured, which may due to the effect of scale in the billet surface.

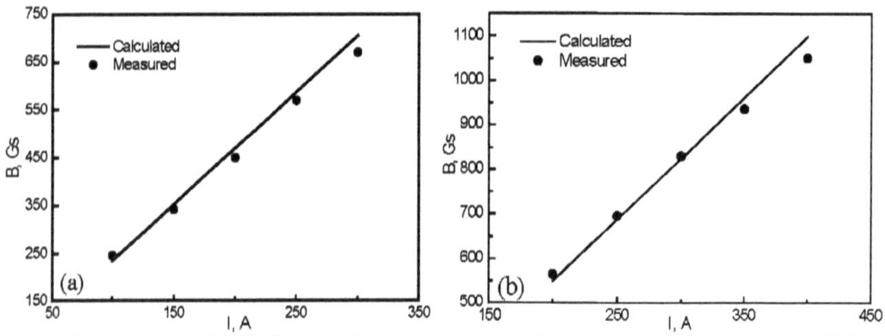

Figure 4. Comparison of calculated and measured magnetic induction intensity (a) M-EMS and (b) F-EMS

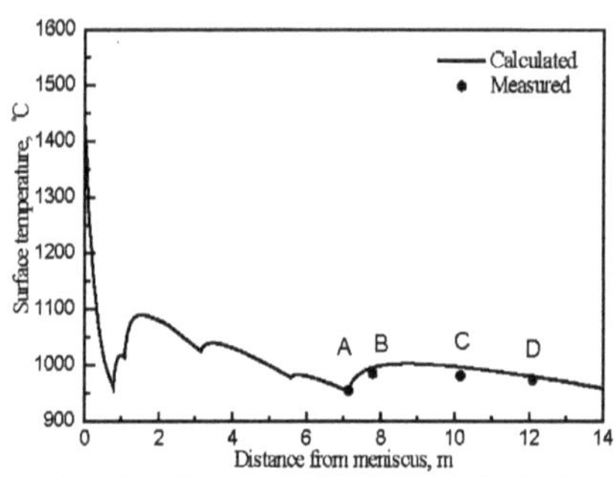

Figure 5. Comparison of calculated and measured values of the billet central surface temperature

3.2 The flow and temperature distribution in mold zone

Figure 6 shows the fluid flow pattern and liquid fraction in mold zone with M-EMS (200A 4Hz). It can be seen the molten steel flows into the mold with high velocity and penetrates deep down in the liquid pool in Figure 6(a). Some molten steel flows into the electromagnetic zone directly and the rotation flow forms due to the rotating electromagnetic forces, as shown in Figure 6(d). The other molten steel returning upside along solidification front and impinges the steel-slag interface. Figure 6(b) clearly shows that one part of the upward molten steel moves downside along the outside wall of SEN and flow into main stream while the other part turns back near the meniscus and moves downward. This is because of a certain amount of molten steel near the meniscus trans from liquid state to solid shell as a huge amount of heat is extracted and then withdrawn from mold zone, shown in Figure 6(c). Beside, the liquid steel near the solid shell moves downward at casting speed of strand which promotes the inverse circulating flow near the meniscus.

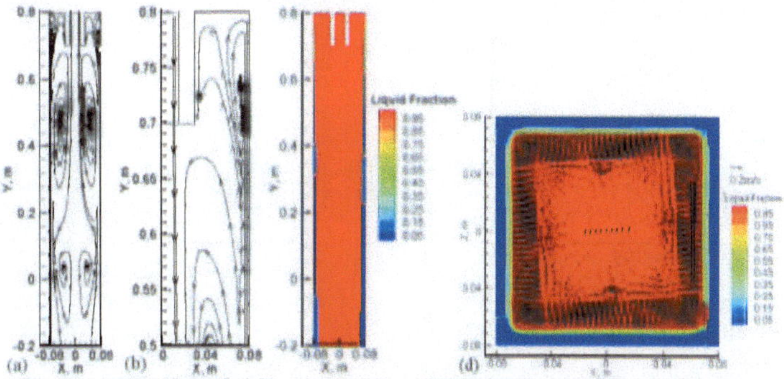

Figure 6. The flow field (a, b) and liquid fraction (c, d) in the mold zone

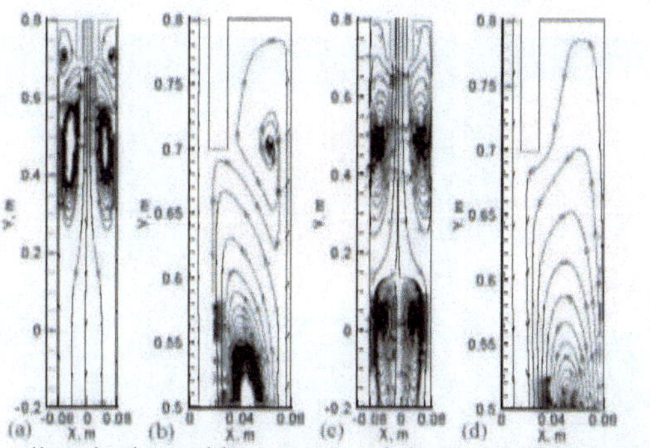

Figure 7. The streamlines in the mold zone: (a) and (b) No M-EMS but with considering solidification, (c) and (d) M-EMS(200A 4Hz) but without considering solidification

Figure 7(a) and (b) describes the fluid flow pattern at the longitudinal section of the mold zone without M-EMS. It clear shows that one part of the upward molten steel flows into main stream directly under the SEN and the other part moves upward along the outside wall of SEN and then moves back near the meniscus, which is seldom found in other papers. Figure 7(c) and (d) describes the fluid mechanism in the mold zone without considering solidification. The upward flow reaches the slag-steel interface directly and then penetrates into the main stream which is quite different from the results near the meniscus in Figure 6(a) and (b). So it is indispensable to consider the solidification phenomenon in the mold zone with EMS.

Figure 8 shows the stirring velocity and liquid fraction about 0.65m from meniscus with different electromagnetic parameters. The stirring velocity increases almost linearly with distance from the billet center and reaches its maximum near the solidification front. Then the stirring velocity decreases sharply in the porous zone. When the F-EMS current is fixed at 4 Hz, the maximum stirring velocity increases from 0.152 m/s to 0.286 m/s with the current intensity from 100 A to 250 A.

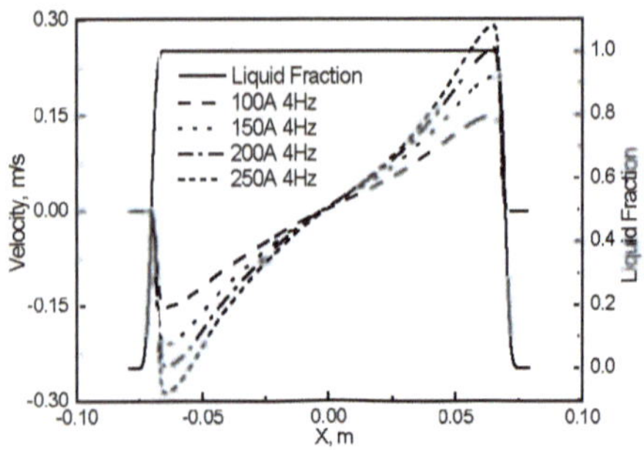

Figure 8. The stirring velocity and liquid fraction along the transverse direction

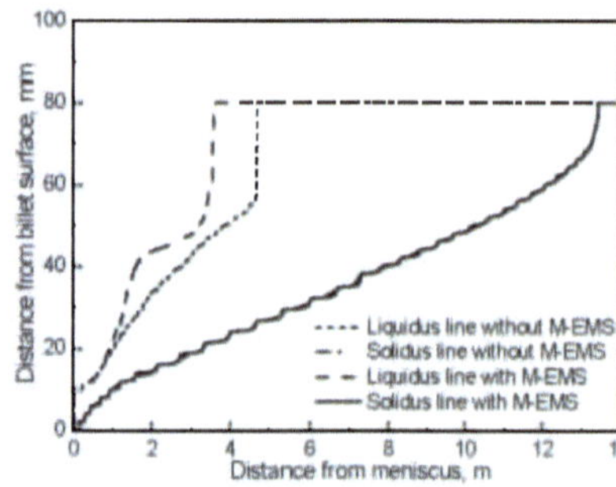

Figure 9. The liquidus and solidus line along the casting direction

3.3 The solidification and flow in the secondary cooling zone

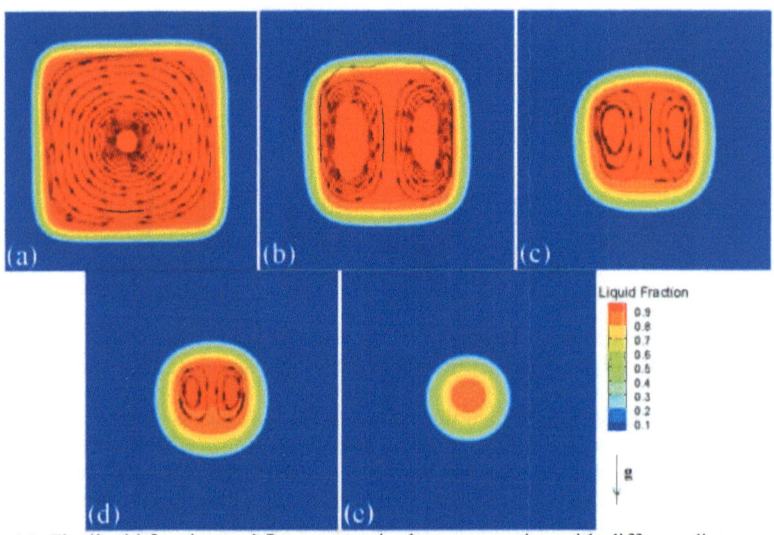

Figure 10. The liquid fraction and flow pattern in the cross section with different distances from meniscus: (a) Y=2m, (b) Y=4m, (c) Y=6m, (d) Y=8m, (e) Y=10m

Figure 9 shows the solidus and liqudus line along the casting direction with and without M-EMS. It can be seen that the liquidus and solidus line increase continuously with the distance from the meniscus. With the M-EMS applied(200A 4Hz), the molten steel washes the solidification front intensively and much superheat dissipates in the liquid pool. Therefore, the liquidus line increases quickly at the lower part of the mold and the end of liquidus line moves forward 1.14 m compared with case without M-EMS. It also shows that the molten steel flow in the mold has little effect on the movement of solidus line, which means the solid is insensitive to the flow pattern of molten steel [18].

Figure 10 describes the liquid fraction and flow pattern at the transverse section of billet with increase distances from the meniscus. As M-EMS is applied, the molten steel in the cross section with 2 m away from the meniscus still shows the rotational flow pattern, shown in Figure 10(a), which means the rotational flow is not restricted in the mold zone. With the further increasing the distance from the meniscus, a pair of opposite circulation flows are formed, as shown in Figure 10(b)~(d). This is because of low temperature molten steel is denser than that of high temperature and the secondary cooling zone is arc shaped. Under the effect of gravity, the cool liquid steel in the mushy zone moves downward and the elevated molten steel is pushed into the inner side. At the later stage of solidification, the liquid fraction decreases and flow resistance rises, resulted in the disappearance of fluid flow phenomenon shown in Figure 10(e).

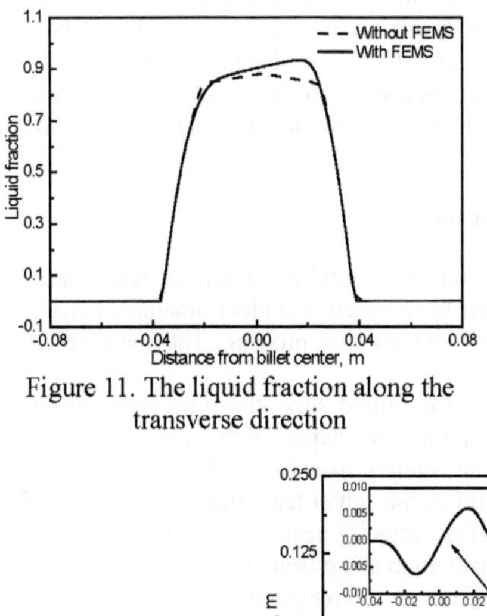

Figure 11. The liquid fraction along the transverse direction

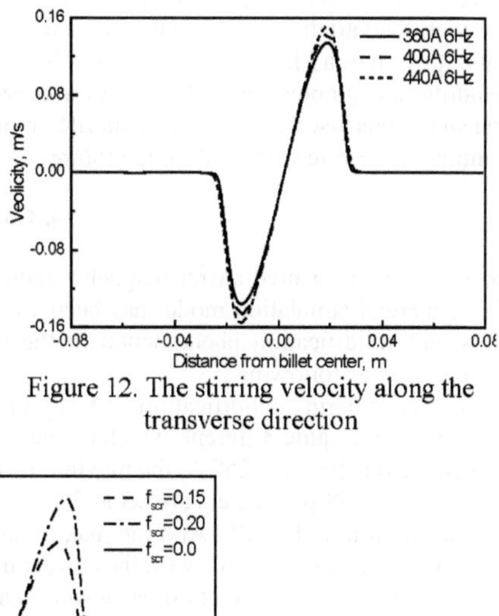

Figure 12. The stirring velocity along the transverse direction

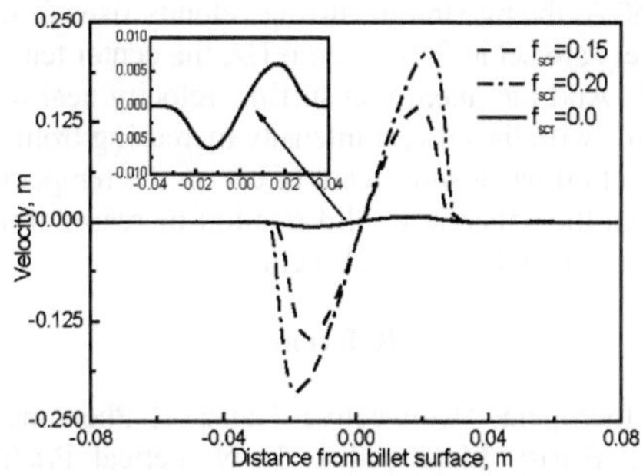

Figure 13. The stirring velocity along the transverse direction

Figure 11 shows the liquid fraction in the mushy zone is not uniform and the hot point shows a clear deviation from the billet center without F-EMS. As the F-EMS is applied, the molten steel in the mushy zone is forced to move around and washes the solidification front which destroys the nature convection mechanism and promotes the temperature to distribute evenly. It also can be seen that the maximum liquid fraction is lower than that without F-EMS. Figure 12 shows the distribution of stirring velocity in the mushy zone is cosine curve. With the current intensity increasing from 360 A to 440 A, the maximum stirring velocity near the solidification front rises from 0.135 m/s to 0.152m/s.

Figure 13 shows the stirring velocity along the transverse direction with different coherency solid fraction as the F-EMS current is set at 400 A and 6 Hz. When all the equixed dendrite zone is treated as a porous medium, the corresponding coherent solid fraction is set to be zero and the maximum stirring velocity is only 0.006 m/s which is in the same order of magnitude with Sun's results [9]. As the coherency solid fraction increases from 0.15 to 0.20, the maximum stirring velocity increases from 0.144 m /s to 0.215 m/s. So the coherency solid fraction which distinguishes the suspended crystal zone and rigid solid network zone has a significant influence on the fluid flow.

In the simulation, the influence of M-EMS on the movement of solidus line is small, which is different from the Oh's [19] measured results. It may be that the solute transport which strongly affected the solidus temperature during the continuous casting is not considered in the simulation. Besides, the variation of coherency solid fraction with stirring parameters is assumed. The forced fluid flow has a significant effect on the crystal growth [20] and can turn the crystal from dendrite into globular crystal. So it is necessary to couple the macro-scale fluid flow and heat transfer, micro-scale growing dendrite morphology and solute transportation during the continuous casting with EMS in the future.

4. Conclusions

Based on the measured stirrer magnetic induction intensities and billet surface temperature, the 3D numerical simulation model has been validated to represent the electromagnetic field, fluid flow and solidification phenomenon in the continuous casting process. The main results are summarized as following:

1. With considering solidification or EMS stirring in the mold zone, the fluid flow pattern near the meniscus is quite different, which is seldom found in other paper. With the increase of current intensity from 100 A to 250 A, the maximum stirring velocity rises from 0.152 m/s to 0.286 m/s.
2. As the F-EMS parameters are set at 360 A and 6 Hz, the center temperature drops about 5.5 K compared with no F-EMS. And the maximum stirring velocity near the solidification front rises from 0.135 m/s to 0.152 m/s with the current intensity increasing from 360 A to 440 A.
3. The coherency solid fraction has a significant effect on the temperature distribution and fluid flow in F-EMS zone. When the coherency solid fraction increases from 0.15 to 0.20, maximum stirring velocity increases from 14.4 cm/s to 21.5 cm/s.

References

1. P.J. Prescott and F.P. Incropera, "Convective Transport Phenomena and Macrosegregation during Solidification of a Binary Metal Alloy: I—Numerical Predictions," *Journal of Heat Transfer*, 116 (3) (1994), 735-741.

2. M. Wu et al., "Modeling Diffusion-Governed Solidification of Ternary Alloys–Part 2: Macroscopic Transport Phenomena and Macrosegregation," *Computational Materials Science*, 92 (2014), 267-285.

3. T.T. Natarajan and N. El-Kaddah, "Finite Element Analysis of Electromagnetic and Fluid Flow

Phenomena in Rotary Electromagnetic Stirring of Steel," *Applied Mathematical Modelling*, 28 (1) (2004), 47-61.
4. H. Liu et al., "Numerical Simulation of Fluid Flow in a Round Bloom Mold with In-Mold Rotary Electromagnetic Stirring," *Metallurgical and Materials Transactions B*, 43 (6) (2012), 1657-1675.
5. H.Q. Yu and M.Y. Zhu, "Three-Dimensional Magnetohydrodynamic Calculation for Coupling Multiphase Flow in Round Billet Continuous Casting Mold with Electromagnetic Stirring," *IEEE Transactions on Magnetics*, 46 (1) (2010), 82-86.
6. K. AYATA et al., "Improvement of Macrosegregation in Continuously Cast Bloom and Billet by Electromagnetic Stirring," *Transactions of the Iron and Steel Institute of Japan*, 24 (11) (1984), 931-939.
7. X.Y. Tian et al., "Numerical Analysis of Coupled Fluid Flow, Heat Transfer and Macroscopic Solidification in the Thin Slab Funnel Shape Mold with a New Type EMBr," *Metallurgical and Materials Transactions B*, 41 (1) (2010), 112-120.
8. R. Chen, H.F. Shen and B.C. Liu, "Numerical Simulation of Fluid Flow and Solidification in Continuous Slab Casting Mould Based on Inverse Heat Transfer Calculation," *Ironmaking & Steelmaking*, 38 (7) (2011), 546-551.
9. H. Sun and J. Zhang, "Study on the Macrosegregation Behavior for the Bloom Continuous Casting: Model Development and Validation," *Metallurgical and Materials Transactions B*, 45 (3) (2014), 1133-1149.
10. G.M. Poole and N. El-Kaddah, "An Improved Model for the Flow in an Electromagnetically Stirred Melt during Solidification" *Metallurgical and Materials Transactions B*, 44 (6) (2013), 1531-1540.
11. S.I. Chung, Y.H. Shin and J.K. Yoon, "Flow Characteristics by Induction and Gas Stirring in ASEA-SKF Ladle," ISIJ international, 32 (12) (1992), 1287-1296.
12. A. Vakhrushev et al., "Numerical Investigation of Shell Formation in Thin Slab Casting of Funnel-Type Mold," *Metallurgical and Materials Transactions B*, 45 (3) (2014), 1024-1037.
13. Y Natsume et al., "Quantitative Model to Determine Permeability for Columnar Dendritic Structures," *ISIJ International*, 53 (5) (2013), 838-847.
14. G. Poole and N. El-Kaddah, "Effect of Coil Design on the Temperature and Velocity Fields during Solidification in Electromagnetic Stirring Processes," *ISIJ International*, 54 (2) (2014), 321-327.
15. M. Wu, A. Ludwig and A. Fjeld, "Modelling Mixed Columnar-Equiaxed Solidification with Melt Convection and Grain Sedimentation–Part II: Illustrative Modelling Results and Parameter Studies," *Computational Materials Science*, 50 (1) (2010), 43-58.
16. I. Farup and A. Mo, "Two-Phase Modeling of Mushy Zone Parameters Associated with Hot Tearing," *Metallurgical and Materials Transactions A*, 31 (5) (2000), 1461-1472.
17. W.L. Wang et al., "Micro-Segregation Behavior of Solute Elements in the Mushy Zone of Continuous Casting Wide-Thick Slab," *Steel Research International*, 83 (12) (2012), 1152-1162.
18. J. Szekely and V Stanek, "On Heat Transfer and Liquid Mixing in the Continuous Casting of Steel," *Metallurgical Transactions*, 1 (1) (1970), 119-126.
19. S.K. Oh and Y.W. Chang, "Macrosegregation Behavior in Continuously Cast High Carbon Steel Blooms and Billets at the Final Stage of Solidification in Combination Stirring," *ISIJ International*, 35 (7) (1995), 866-875.
20. A.N. Turchin, D.G. Eskin and L. Katgerman, "Solidification under Forced-Flow Conditions in a Shallow Cavity," *Metallurgical and Materials Transactions A*, 38 (6) (2007), 1317-1329.

Advances in the Science and Engineering of Casting Solidification
Edited by: Laurentiu Nastac, Baicheng Liu, Hasse Fredriksson, Jacques Lacaze, Chun-Pyo Hong, Adrian V. Catalina, Andreas Buhrig-Polaczek, Charles Monroe, Adrian S. Sabau, Roxana Elena Ligia Ruxanda, Alan Luo, Subhayu Sen, and Attila Diószegi
TMS (The Minerals, Metals & Materials Society), 2015

COLUMNAR-TO-EQUIAXED TRANSITION IN Zn-27wt.%Al ALLOYS: A COMPARATION BETWEEN VERTICAL AND HORIZONTAL DIRECTIONAL SOLIDIFICATIONS

Alicia E. Ares[1,2], Carlos E. Schvezov[1,2]

[1] Materials Institute of Misiones, IMAM (CONICET-UNaM), University of Misiones; 1552 Azara Street, Posadas, Misiones, 3300 Argentina.

[2] Member of CIC of the National Research Council (CONICET) of Argentina.

Keywords: CET, Al-Zn Alloys, Vertical and Horizontal Solidification, Thermal Parameters.

Abstract

The columnar to equiaxed transition (CET) was studied in Zn-27wt.%Al alloys, which were directionally solidified from a chill face in both vertical and horizontal arrangement of experimental device. The main parameters analyzed include cooling rate of the melt, position of the solidification fronts for solidus and liquidus temperatures, velocity of solidification fronts and temperature gradients. In both type of solidifications the CET does not occur in abrupt form in the samples and the phenomenon is present when the gradient in the liquid ahead of the columnar dendrites reaches critical and minimum values, being negative in most of the cases. In addition, there is an increase in the velocity of the liquidus front faster than the solidus front, which increases the size of the mushy zone.

Introduction

Zinc-based foundry alloys have evoked the most rapidly growing interest among the research community. Cast Zn–Al (ZA) alloys have been popular for engineering components including those in the automotive industry (initially the ZA casting alloy family are ZA8, ZA12 and ZA27). Aluminum is one of the major alloying elements in Zn alloy systems where it imparts fluidity to the alloys [1, 2]. In practice, the amount of Al added to Zn-based alloys in order to attain good engineering properties varying over a wide range.

The mixed mode of solidification (columnar and equiaxed growth) occurs if equiaxed grains can nucleate and grow in the bulk liquid ahead of the advancing columnar front, resulting in a transition from a columnar zone to a central equiaxed zone in some as-cast structures, that is, the columnar-to-equiaxed transition (CET).

As was suggested by many previous studies, the CET, caused by the competition between columnar and equiaxed growth, is primarily governed by such casting parameters as the alloy composition, pouring superheat, nuclei density present in the melt, cooling capacity at the metal/mold interface, and melt convection. Qualitatively, it can be anticipated than the CET occurs earlier when an alloy has a higher solute level, lower pouring temperature, smaller thermal gradient, higher nuclei density present in the melt, and more vigorous melt convection. However, quantitative predictions of the CET require a more thorough understanding and a full account of all physical mechanisms involved. For small specimens of almost uniform temperature, experimental observations indicate that nucleation and growth of the equiaxed grains ahead of the columnar front are the most important mechanisms for causing the CET [3-7].

Thus, the purpose of the present work is to experimentally investigate the effect of direction (vertical and horizontal) during directional solidification processing on the thermal and structural parameters in the Zn-Al alloy (particularly Zn-27wt.%Al) and to compare the results with similar previous experimental results.

Experimental Procedure

Vertical and horizontal directional solidification

A vertical one-directional solidification device (Figure 1 (a)) was used to perform the experiments with Zn-27wt.%Al alloys. In the case of the vertical setup, a set of three samples were directionally solidified, using glass cylindrical molds of 22 mm of internal diameter, and cooled predominantly from the bottom, using a tubular heat exchanger device.

In the vertical solidification, the melt was heated to the required temperature; the heat unit was turned off while cooling water was forced to circulate into the heat extraction system. Temperature measurements were made during solidification using K-Type chromel/alumel thermocouples (1 mm and 1.5 mm stainless steel diameter and 300 mm length) introduced into Pyrex glass rods of 2.5mm internal diameter. The vertical distance between adjacent thermocouples was chosen in approximately 20 mm.

We used pure elements, electrolytic Zn and Al 99.99%. After preparation, the alloy was cast in a ceramic mold with suitable cylindrical dimensions to be used in the vertical furnace. The length of the samples was around 80 mm. It is noteworthy to point that the melting point of the alloys did not exceed the limiting operating temperature of the Pyrex® mold (Figure 1(c)).

Thermocouples were previously calibrated using Zinc and Aluminum at their melting points. During solidification experiments of Zn-Al alloys, the temperature measured by each thermocouple was recorded at regular intervals of about 1min.

In the case of horizontal solidification, ceramic molds of 50 mm in diameter were used for experiments cooled from both ends. Eight made K-type thermocouples were used in this experimental setup. For the horizontal arrangement, thermocouples were fabricated with thin Chromel-Alumel wires of 0.5 mm diameter that were inserted into bifilar ceramics of ~ 4.0 mm external diameter and ~ 1.0 mm hollow diameter and introduced inside Pyrex® glass rods of 7.0 mm external diameter and ~ 5.3 mm internal diameter. Adjacent thermocouples were located at a distance of ~ 20 mm. Temperatures were measured at regular intervals of 10 s.

A schematic drawing of the horizontal experimental device is shown in Figure 1 (b). Small 140 mm long hemicylindrical probes of Zn-Al alloy were solidified in the horizontal setup. The heat flux toward the ends of the sample was obtained by two cooling systems located at the ends of the ceramic crucible. In this setup, temperatures at eight different positions were measured using a TC 7003C acquisition system and recorded every 1min using SensorWatch® software in a compatible PC from the early beginning until the end of the solidification. Alloys were prepared with high purity metals (electrolytic Zinc and commercial grade Aluminum). For the horizontal setup, a set of the same alloy composition samples were prepared. The alloy was first melted and mixed in a graphite crucible using a conventional furnace and then poured into a previously heated ceramic crucible. The crucible with the alloy was located into the horizontal furnace and heated up above the melting point of the alloy (Figure 1(d)). The solidification of the sample was obtained by cooling down the alloy using the cooling system which extracts the heat toward both ends.

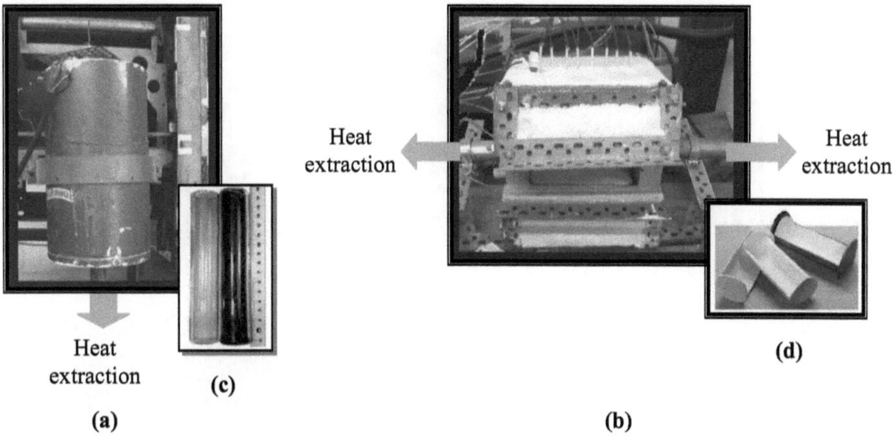

Figure 1. Details of experimental setup. (a) Vertical. (b) Horizontal. (c) Glass moulds. (d) Ceramic moulds.

Metallography

After solidification, both the vertical and horizontal directionally solidified Zn-based alloy samples were cut in the axial direction. The samples were polished and etched using concentrated Hydrochloric acid for ~ 3 s at room temperature followed by rinsing and wiping off the resulting black deposit. To develop the microstructures, the samples were etched with a mix containing chromic acid (50 g Cr_2O_3; 4g Na_2SO_4 in 100ml of water) for 10 s at room temperature [8].

The typical macrostructures of both the vertically and the horizontally solidified Zn-27wt%Al samples are shown in Figures 2 (a) to (e). The position of the transition in vertically solidified samples was located by visual observation and using an Arcano® optical microscope, and the distance from the chill zone of the sample was measured with a ruler. As reported before, it can be seen in Figure 3(b) that the CET is not sharp, showing a zone where some equiaxed grains coexist with columnar grains.

The size of the transition zone is in the order of up to 10mm. As previously reported no effects of the set of thermocouples in the transition, either acting as nucleating sites or changing the solidification structure were observed [9-11].

The position of the transitions in horizontally solidified samples was located by visual observation and the distance from the chill zone of the sample was measured with a ruler (two zones of transitions were observed (near each side of the samples). In this case, the thermocouples seem to act sometimes as nucleating sites in the proximities of the glass rods, probably due to the fact that they also behave as a small heat sink, since the glass rods are not immersed in the melt as it happens in the case of the vertical setup.

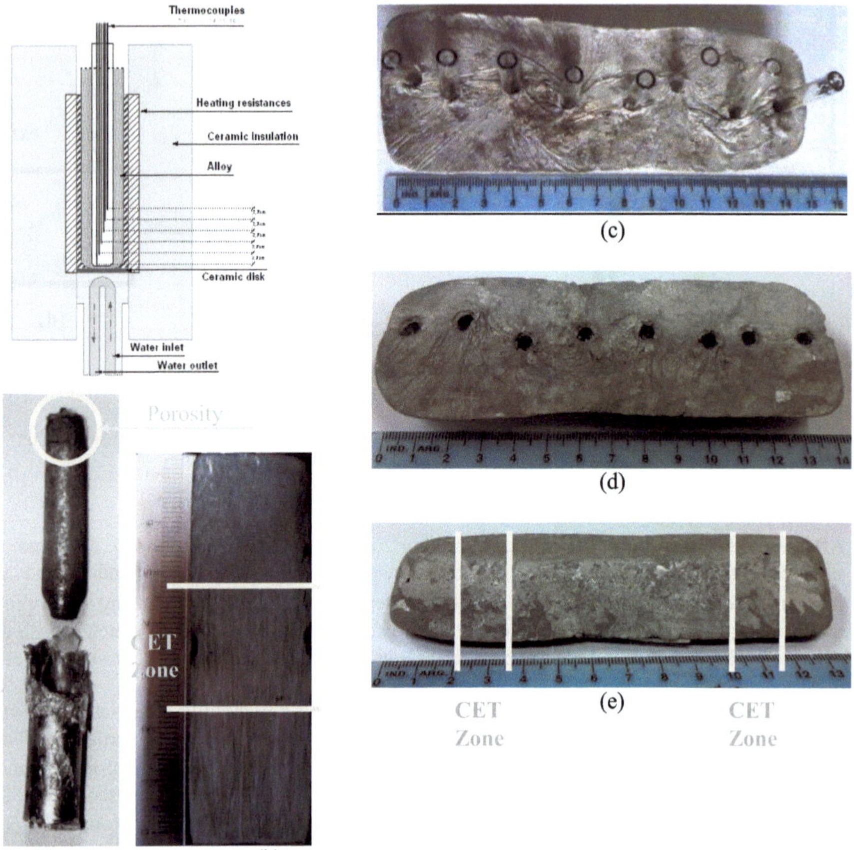

(a) (b)

Figure 2. Macrostructures of Zn-27wt.%Al alloys. (a) Vertical solidification. (b) Horizontal solidification.

Results and Discussion

In vertical solidification, the CET was obtained in all the experiments as can be seen as an example in the macrostructure of Figure 2 (b) for Zn-27wt.%Al alloy. As was reported before, no effects of the set of the thermocouples in the transition either acting as nucleating sites or changing the solidification structure were observed; Figure 2 (b) also shows that the transition in grain size is not sharp; there is a''zone'' of CET. The smallest diameter of the samples used in the present work produced less fluid flow than in other cases previously reported [9-11].
In the macrograph of Figure 2 (a) it is also possible to observe the pores and cavities resulting from the final contraction of the specimen. A typical set of cooling curves is shown in Figure 3 (a) to (d).

In the case of vertical solidification, the thermocouple T_1 is at the lowest position and the first to reach the solidification front whereas T_5 is at the highest position. In all the curves it is possible to identify a period corresponding to the cooling of the melt, a second period of solidification and the final period of cooling of the solid to ambient temperature. In addition, temperatures versus time curves were obtained for each running in the horizontal solidification (Figure 3 (b)). The beginning and the end of the solidification in each finite volume were estimated from the change in the slope of the temperature versus time curves, taking the time derivatives of each curve. Each thermocouple characterizes the temperature of the whole volume (approximately 20mm wide, which is the distance between thermocouples). In the case of horizontal solidification; T_1 and T_8 are in the extreme of the samples.

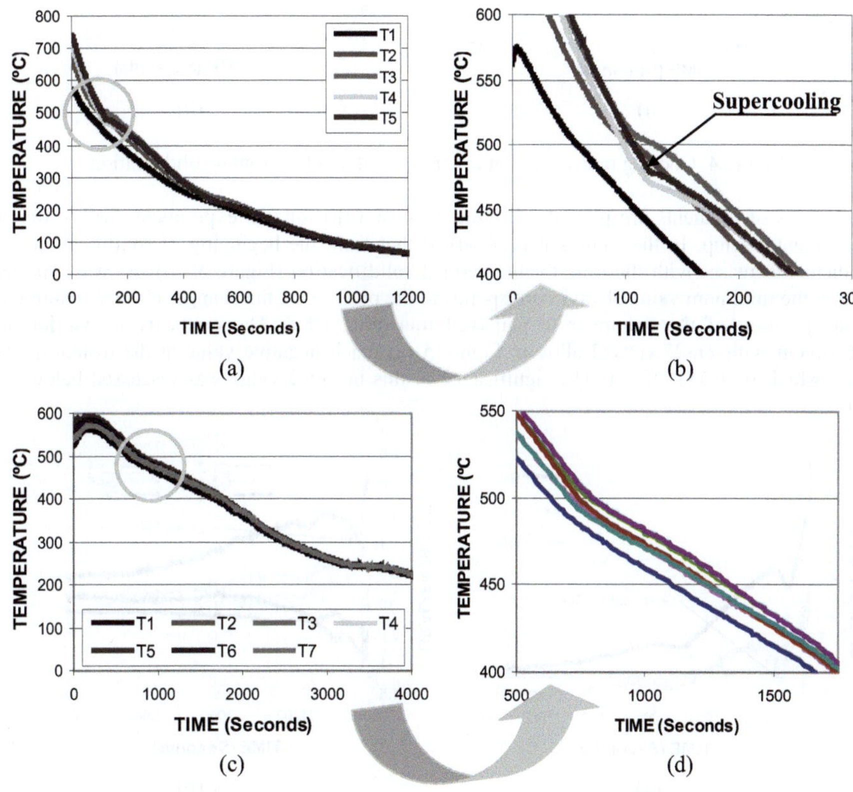

Figure 3. Temperatures versus time curves. (a) - (b) Vertical. (c) - (d) Horizontal solidification.

In Figure 3 (b) is feasible appreciate that in the case of vertical solidification was possible to observe and register the supercooling necessary for nucleation of equiaxed grains in front of columnar grains (for thermocouples located in the CET and equiaxed zones). In the case of horizontal solidification this phenomenon was not evident (Figure 3 (d)).
The cooling rate of the liquid alloy was determined from the temperature versus time curves at each thermocouple position (as can be observed in Figure 4 (a) and (b)) and taking the average slope. The values for all the experiment is presented in Table 1.

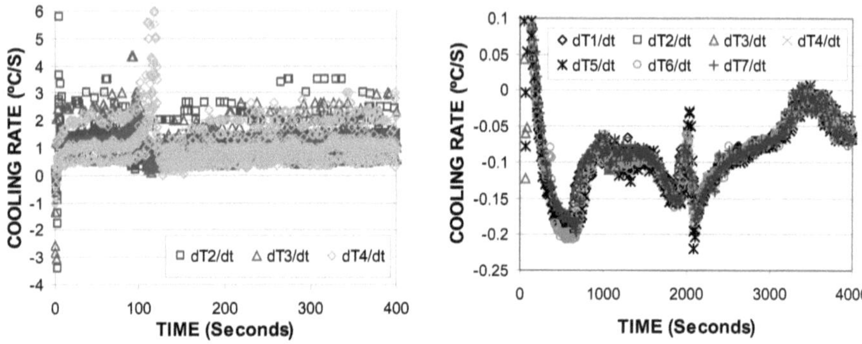

(a) (b)

Figure 4. Cooling rate versus time. (a) Vertical. (b) Horizontal solidification.

The values of gradients are plotted in Figure 5 (a) and (b) for two experiments with different experimental setup. In the figures it is observed that from the beginning of solidification, the gradients decreases with the time for the vertical solidification (Figure 5 (a)). As was reported before, the minimum value always corresponds to the position of the change of the structure that is the position of the columnar to equiaxed transition, CET. The gradients shown for the experiment with Zn-27wt.%Al alloy in Figure 5 (a) reach negative value at the instant of the CET, which is -0.381 °C/cm. The significance of this negative value was discussed before [9-11].

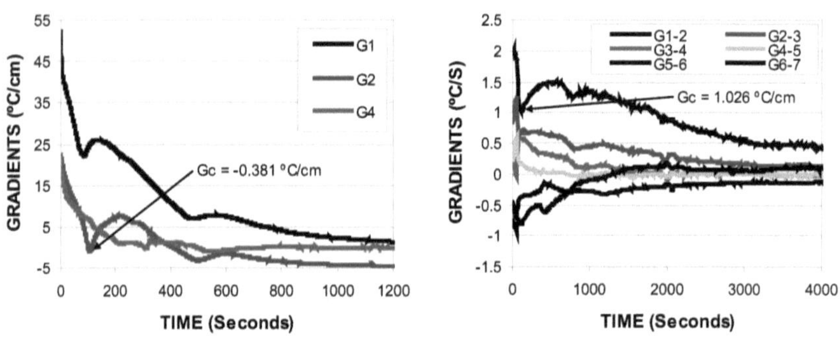

(a) (b)

Figure 5. Temperature gradient versus time. (a) Vertical. (b) Horizontal solidification.

For the horizontal solidification (Figure 5 (b)) the critical temperature gradient was minimum and positive in most of the cases.

The solidification velocities were determined by recording the start of the liquidus temperature and the end of the solidus temperature at each thermocouple position; with these data it was possible to represent the position of the solidification front from T_L to T_S at each thermocouple position as horizontal bars as was reported before [9-11]. From these types of figures the velocities of liquid and solid interphases can easily be calculated, see Figure 6 (a) and (b) for vertical and horizontal solidification, respectively.

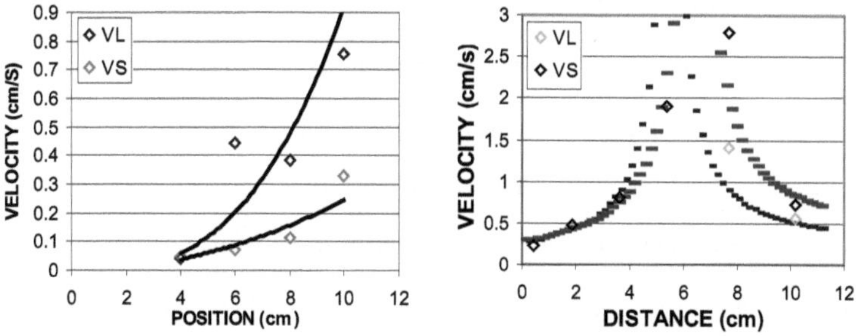

Figure 6. Velocities of liquidus and solidus interphases versus time. (a) Vertical. (b) Horizontal solidification.

Table 1. Values of thermal parameters during solidification of Zn-27wt%Al using two different experimental devices.

Vertical solidification					
CET_{Max} (cm)	CET_{Min} (cm)	Undercooling (°C)	Cooling rate (°C/S)	Temperature gradient (°C/cm)	Interphase velocity (cm/S)
2.3	4.7	3.2	1.25	-0.38	1.7
3.2	4.6	2.8	1.75	-0.10	1.9
4.6	6.8	2.4	2.15	-1.24	2.15

Horizontal solidification									
Left CET Position		Right CET Position		Cooling rate		Critical temperature gradient		Interphase velocity	
CET_{Max} (cm)	CET_{Min} (cm)	CET_{Max} (cm)	CET_{Min} (cm)	Left (°C/S)	Right (°C/S)	Left (°C/cm)	Right (°C/cm)	Left (cm/S)	Right (cm/S)
2.0	3.5	10.1	11.6	0.44	0.57	1.02	0	0.8	0.9
2.5	3.7	9.6	10.8	0.63	0.75	-0.15	0.66	1.1	1.2
3.1	3.9	8.3	9.6	0.82	1.02	1.25	-0.99	1.4	1.6

Conclusions

The columnar-to-equiaxed transition was obtained in both vertical and horizontal solidifications. In vertical solidification onet CET was obtained however samples horizontally solidified presented two CET. The main parameters analyzed include cooling rate of the melt, position of the solidification fronts for solidus and liquidus temperatures, velocity of solidification fronts and temperature gradients.
In both type of solidifications the CET does not occur in abrupt form in the samples and the phenomenon is present when the gradient in the liquid ahead of the columnar dendrites reaches critical and minimum values, being negative in most of the cases. In addition, there is an increase in the velocity of the liquidus front faster than the solidus front, which increases the size of the mushy zone. Could not be identified the supercooling in horizontal solidification.

Acknowledgements

This work was supported by PICT-2011-1378 of the National Agency for Promotion of Science and Technology.
Thanks are due to the Argentinean Research Council (CONICET) for the financial support.

References

[1] P. Choudhury, S. Das, B.K. Datta; Effect of Ni on the Wear Behavior of a Zinc-Aluminium Alloy, *J.Mater. Sci.*, 37 (2002) 2103–2107.

[2] B.K. Prasad, A.K. Patwardhan, A.H. Yegneswaran, Z. Metallkd.; Influence of aluminum content on the physical sliding wear properties of zinc-based alloys, *Z. Metallkd*, 88 (4) (1997) 333– 338.

[3] C. Degand, D.M. Stefanescu, G. Laslaz, Solidification Science and Processing, Edited by I. Ohnaka and D.M. Stefanescu, The Minerals, Metals and Materials Society (1996).

[4] D.M. Stefanesku, and C. Kanetkar, in H. Fredriksson (ed.): 'State of the art of computer simulation of casting and solidification processes', Paris, Les Editions de Physique, 255 (1986).

[5] D.M. Stefanesku, and C. Kanetkar, in 'Computer Simulation of Microstructural Evolution', (ed. D.J. Srolovitz), 171; Warrendale, PA, Metallurgical Society of AIME (1986).

[6] M. Rappaz, and D.M. Stefanescu, in ' Solidification Processing of Eutectic Alloys', (ed. Stefanescu, D.M. et al.), 133; Warrendale, PA, Metallurgical Society of AIME (1988).

[7] J.A. Spittle; Columnar to equiaxed grain transition in as solidified alloys, *International Materials Review*, 51 (2006) 247-269.

[8] ASM Handbook - Volumen 9 - Metallography and Microestructures 2004 – ASM International – The Materials Information Company – p. 2269, 2270.

[9] S. F. Gueijman, C. E. Schvezov, A. E. Ares, "Directional Solidification and Characterization of Zn-Al and Zn-Ag Diluted Alloys", *Materials Transactions*, 51 (2010) 1851-1870.

[10] A. E. Ares, C. E. Schvezov, Solidification parameters during the columnar-to-equiaxed transition in lead-tin alloys, *Metall. Mater. Trans.*, 31A (2000) 1611-1625.

[11] A. E. Ares, C. E. Schvezov, "Influence of solidification thermal parameters on the columnar-to-equiaxed transition of aluminum-zinc and zinc-aluminum alloys". *Metall. Mater. Trans.*, 38A (2007) 1485-1499.

Advances in the Science and Engineering of Casting Solidification
Edited by: Laurentiu Nastac, Baicheng Liu, Hasse Fredriksson, Jacques Lacaze, Chun-Pyo Hong, Adrian V. Catalina, Andreas Buhrig-Polaczek, Charles Monroe, Adrian S. Sabau, Roxana Elena Ligia Ruxanda, Alan Luo, Subhayu Sen, and Attila Diószegi
TMS (The Minerals, Metals & Materials Society), 2015

MICROSTRUCTURE INVESTIGATION OF ALUMINUM DIE-CAST PARTS WITH DIFFERENT GATING CONDITIONS TESTED IN FATIGUE

Roxana Ruxanda[1], Richard Obara[1]

[1]Emerson Climate Technologies
1675 West Campbell Road, Sidney, OH 45365, USA

Keywords: Microstructure, Fatigue, Die-casting, Gating

Abstract

Aluminum die-cast parts from a foundry being evaluated as an alternate source were subjected to fatigue testing. The results showed lower than expected fatigue life and a change in the failure location of the evaluated parts compared with the baseline parts, which prompted submission of both sets of samples for metallurgical evaluation. Optical microscopy, scanning electron microscopy and non-destructive X-Ray imaging tools were employed to characterize the material and fracture surface. Die-cast parts typically exhibit some level of microstructural inconsistencies, such as micropores or microshrinkage. However, in this case, the presence of atypical casting defects (such as cold shuts) in the evaluated samples pointed toward incorrect casting parameters and design as root cause of the lower fatigue life and change in failure mode. Further investigation determined that a different gating system, more prone to cold shut development - all other parameters being equal, was used to produce these parts.

Introduction

In the reality of industrial production it is rarely advisable to rely on a sole parts supplier, its product quality, availability or historical partnership notwithstanding. A battery of tests must be employed to assure the alternate product (*Alternate*) meets the stringent quality demands of the current production product (*Baseline*), which has excellent properties for service and performs very well. Among others, parts are often subjected to tensile and impact testing, chemical analysis, non-destructive testing and fatigue testing; however, at times a first test may be enough to pinpoint towards differences in the manufacturing process that could adversely affect the product quality. Such a case may be when the fatigue fracture changes location and the failure mode, although the material, mechanical properties or geometry appear to be similar.

Case Analysis

Parts from a foundry being evaluated as an *Alternate* source for A380 die castings were subjected to fatigue to failure testing and compared with the *Baseline* parts from current production, which perform exemplary and have a fatigue life exceeding the service demands. The results showed lower fatigue life of the *Alternate* parts compared with the fatigue life of the *Baseline* parts (Figure 1), but more concerning, a beta coefficient close to one was reported; this indicates that some fractures could occur early due to material inconsistencies such as porosity, cold shuts, etc. In-depth examination [1] of the *Alternate* fatigue data showed that these parts have two separate failure modes, with two different Weibull distributions indicating a strong probability of early cracking risk, followed by standard fatigue crack life (Figure 2).

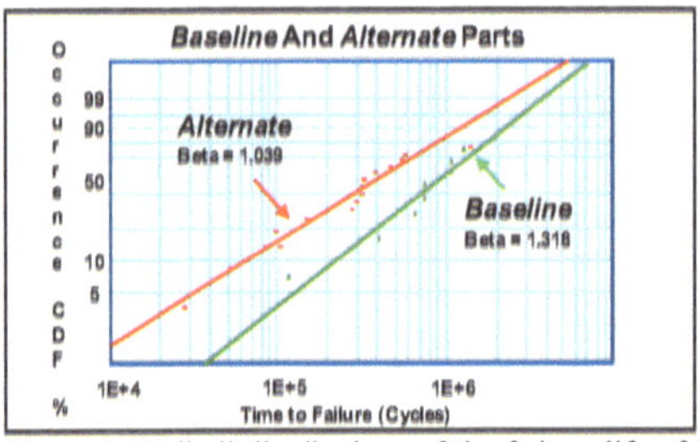

Figure 1. Weibull distributions of the fatigue life of the *Alternate* and *Baseline* parts (test to failure).

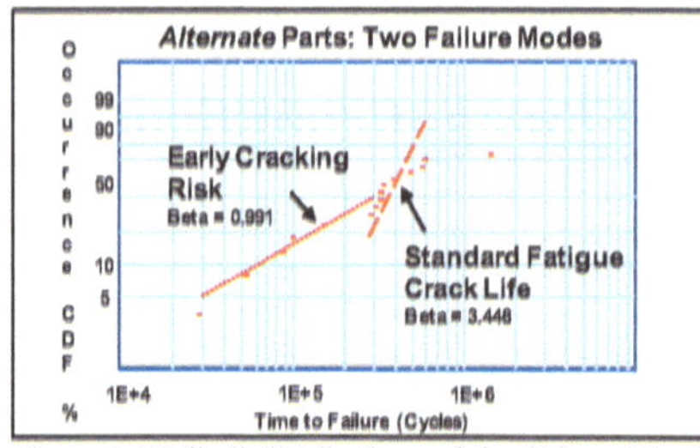

Figure 2. Weibull distributions of the fatigue life of the *Alternate* parts showing two separate failure modes.

Also it was noted a change in the failure location (Figure 3) and fracture morphology (Figure 4 and Figure 5). The fracture location of the *Alternate* part is at the thinnest area, while the *Baseline* fracture location is at the base of a large post, which is the fatigue crack location of a sound casting predicted by modeling [1]. The *Alternate* fatigue origin is a subsurface defect, whereas the *Baseline* origin is at the free surface. The fracture surfaces are different too: the *Alternate* fracture surface has numerous internal gas porosities and cold shuts, as opposed to the *Baseline* fracture, which has no apparent casting defects.

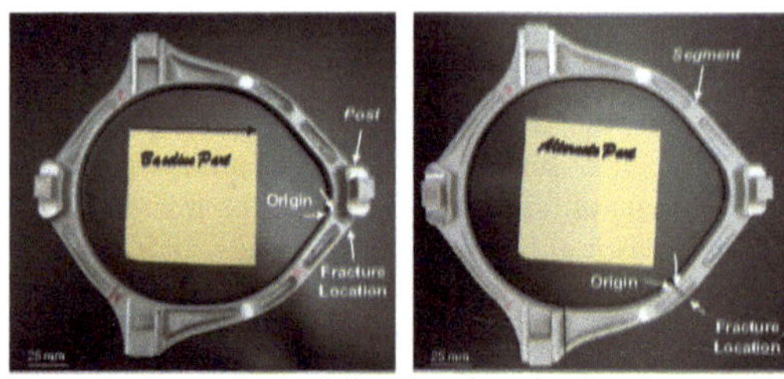

Figure 3. Change in the failure location: the *Baseline* part (left photo) has the fracture location at *Post*, while the *Alternate* part (right photo) has the fracture location at *Segment*.

Microstructure Analysis

The microstructure of the *Baseline* and *Alternate* parts (examined with an Axiovert 200MAT Zeiss microscope) indicates the presence of more internal porosity in the *Alternate* sample in the region of the fracture (Figure 6 and Figure 7). The internal porosity has a pattern that follows the direction of the melt movement during casting (flow-induced segregation [2]) and porosity. The constituents are similar for both the *Alternate* and *Baseline* samples.

Flow-Induced Segregation Melt movement during solidification leaves traces in microstructure in form of flow-induced segregation patterns. In *Alternate* parts (and at a much lesser extent in *Baseline* parts), the microstructure shows evidence of a rolling pattern, produced while pressing the material through the die (Figure 6, Figure 7, Figure 8). The fact that the material could not achieve structural uniformity indicates low casting temperature.

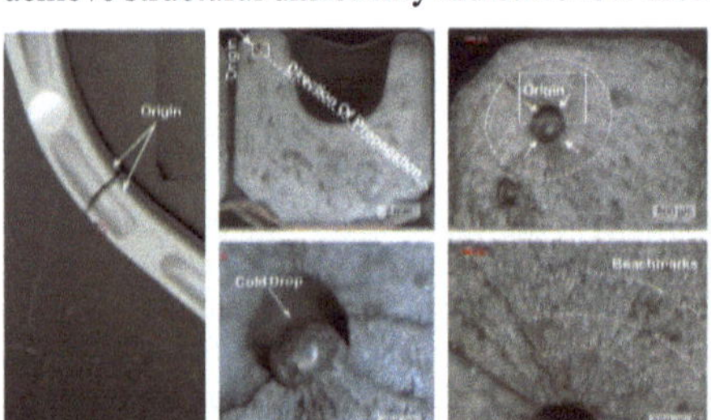

Figure 4. Fracture morphology of the *Alternate* part. Bottom images are details from top (see rectangles).

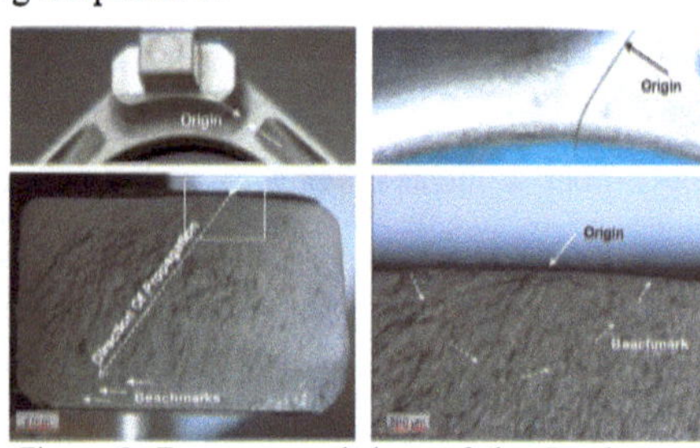

Figure 5. Fracture morphology of the *Baseline* part. Right image is a detail from left (see rectangle).

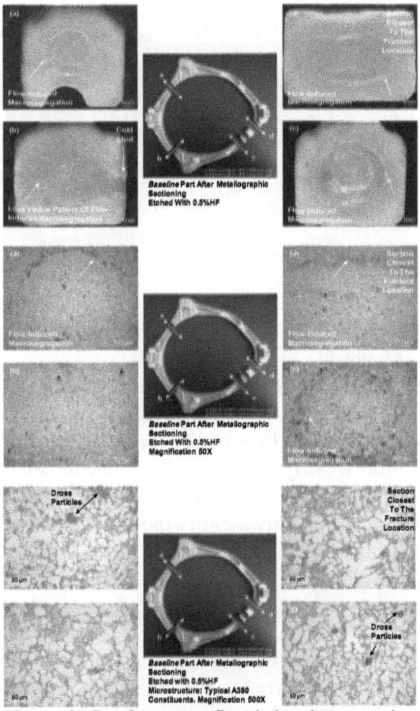

Figure 6. *Baseline* part: flow-induced segregation, typical A380 constituents and dross particles.

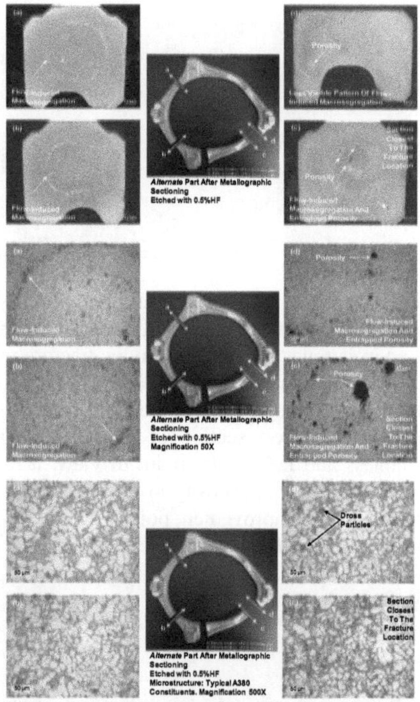

Figure 7. *Alternate* part: flow-induced segregation, typical A380 constituents and dross particles.

<u>Black Oxide Film And Dross</u> A black oxide film delineating the grains is present in regions with scattered and clustered porosity in the *Alternate* parts (Figure 9). Oxide films form easily at the free surface of the melt, due to the high affinity between aluminum and oxygen. If the flow is turbulent, these oxides become entrapped into the casting, being frequently associated with porosity, hot tearing and cracks [3-6]. Their presence decreases the fatigue life, strength, ductility and corrosion resistance of the material because they induce intergranular decohesion.
Dross is the layer consisting of primarily aluminum oxides (Al_2O_3), aluminum nitrides and metallic aluminum that forms on the surface of the melt [7] that later becomes entrapped into the casting. Dross appears mostly as distinct particles and is basically inevitable in die-castings, being rarely acknowledged, unless it exists in large amounts. Both the *Alternate* and *Baseline* castings have a reasonable small amount of dross in their microstructure (Figure 10).

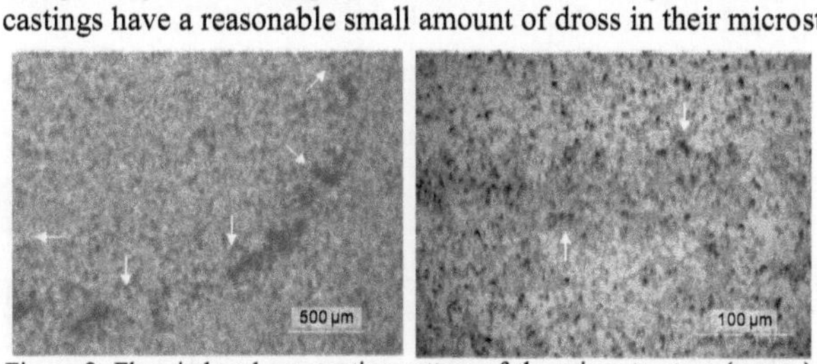

Figure 8. Flow-induced segregation pattern of the microstructure (arrows). Right: detail from left showing eutectic associated with small dross particles.

<u>Porosity: Gas Porosity And Microshrinkage</u> Due to the nature of the die-castings process, a certain low level of porosity is typically admissible in the final product, as long as it is not concentrated in sensitive regions. Gas porosity and microshrinkage,

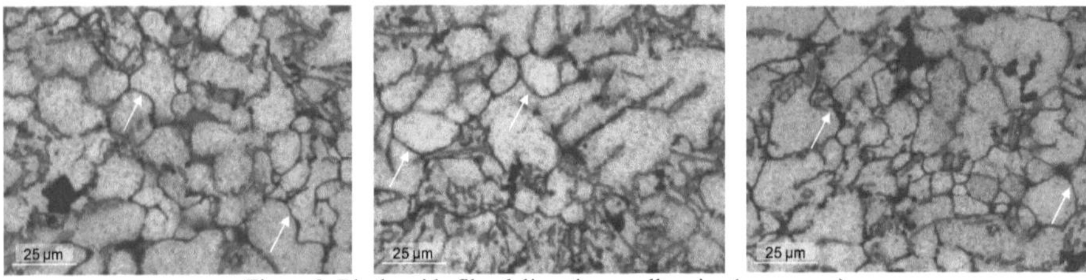

Figure 9. Black oxide film delineating small grains (see arrows).

present as either regions of clustered porosity or individual pores are present throughout the microstructure of both *Baseline* and *Alternate* parts; in the latter, more internal porosity was found in samples taken closest to the fracture region (Figure 11).

Scanning Electron Microscopy (SEM was performed using ZEISS EVO25MA) further revealed numerous shrinkage voids in the open fracture surface of the *Alternate* parts along the centerline, within the thickest section of the part, or in random places (Figure 12). From a material soundness and fatigue life point of view, it matters little what type of porosity is encountered, since is known that porosity in general decreases the fatigue life of the material [8]. However, from a process improvement point of view, it was critical to correctly identify the main source of porosity as solidification contraction, as opposed to gas porosity.

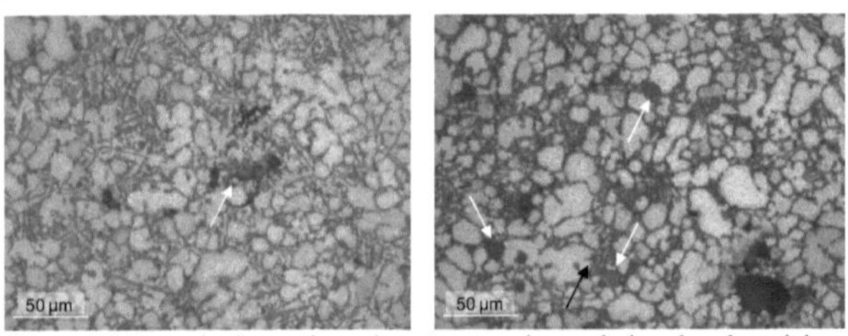

Figure 10. Dross in the casting microstructure: large, dark-colored particles.

Cold Drops And Cold Shuts

Cold drops are casting defects that nucleate internally in gas voids, due to delayed solidification of the undercooled eutectic, while a cold shut is a material discontinuity characterized by the presence of resolidification interfaces, ty-

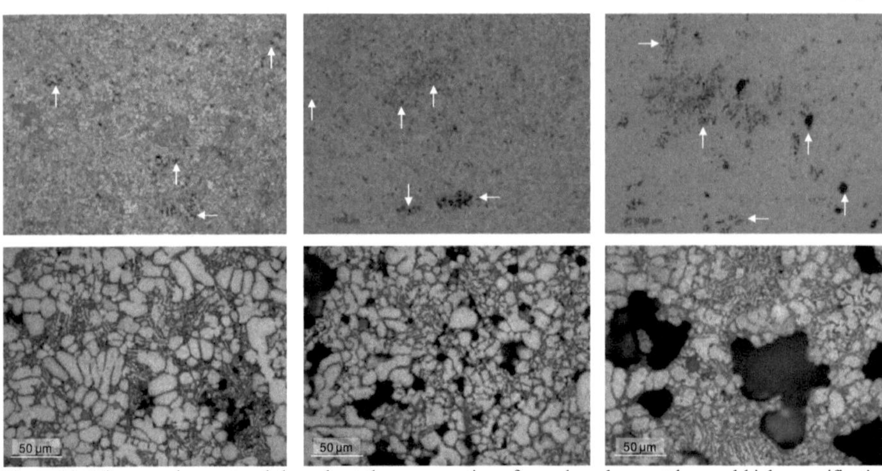

Figure 11. Microporosity scattered throughout the cross section of samples, shown at low and high magnification (top magnification 50X, bottom magnification 1000X).

Figure 12. Microshrinkage detected in fracture surface in various locations: along the centerline, the thickest section of the part, or in random places. Bottom images are details from the top images (see rectangles).

Figure 13. Partial cold shut observed on a metallographic sample showing segregated microstructure and open dendrites along the discontinuity line. The images at right are details from the images at left (see rectangles).

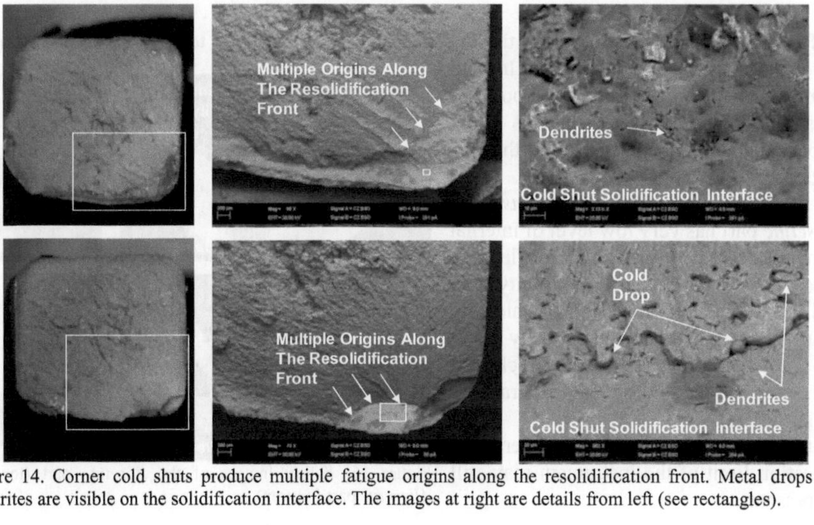

Figure 14. Corner cold shuts produce multiple fatigue origins along the resolidification front. Metal drops and dendrites are visible on the solidification interface. The images at right are details from left (see rectangles).

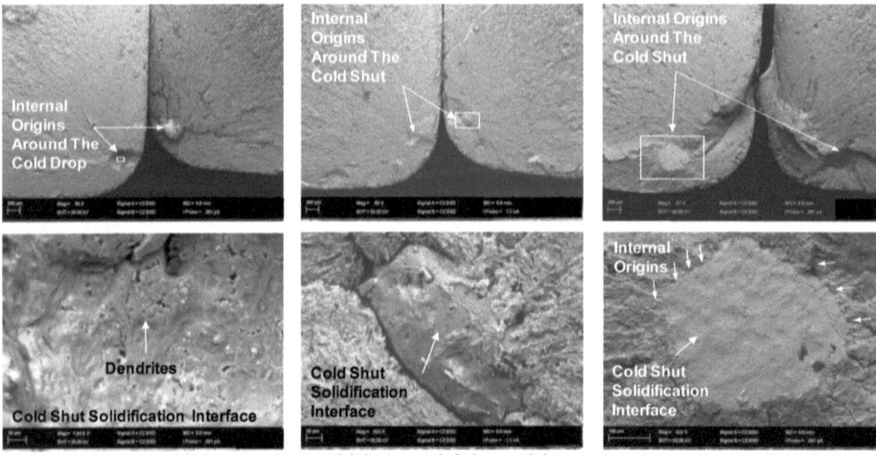

Figure 15. Internal cold shuts produce multiple internal fatigue origins around their perimeter. Cold drops and dendrites are visible on the solidification interface. Bottom images are details from the top images (see rectangles).

pically due to low casting temperatures, inadequate filling rates, cold dies, dirty metal, lack of venting, or too much lubricant These casting non-conformities found in the *Alternate* parts in the fracture surface can be partial cold shuts (Figure 13), where some level of connectivity between the two free surfaces is maintained, or regular cold shuts, where the free solidification surfaces are fully separated (Figure 14 and Figure 15).

The extent of the fatigue life is influenced by the presence of cold shuts in the sensitive area: samples with low fatigue life have a relatively larger cold shut, located at the edge of the part, while the parts with high fatigue life have a relatively small cold shut, located internally (without connection with the outer surface).

X-Ray Analysis

X-Ray analysis was conducted on the *Baseline* and *Alternate* parts to investigate and compare the overall soundness of the parts. In both samples, localized shrinkage-type porosity was found in the *Posts*, which is to be expected, since these are the thickest regions of the casting.

In the thinner areas (the *Segments*), the *Baseline* part has very low level of internal porosity or no porosity, while the *Alternate* sample has a discontinuity line, individual gas pores up to 0.7 mm and many regions of clustered porosity up to 2.4 mm throughout the sample, including the areas susceptible to fatigue cracking (Figure 16).

Note that microporosity under the detection limit of the instrument may still be present.

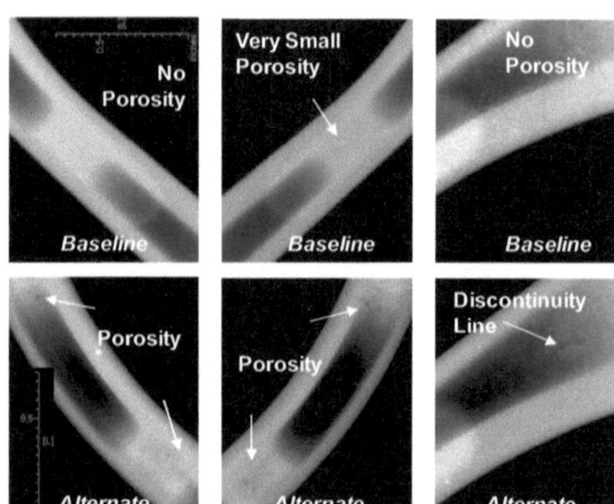

Figure 16. X-Ray images of the *Baseline* part (top) and *Alternate* part (bottom).

The Gating System Solution

The investigation conducted so far was categorical in assigning causes such as low casting temperature and inadequate filling rates (cold shut, porosity) and turbulent flow (flow-induced segregation and black film at grain boundary) to explain the decrease in fatigue life of the *Alternate* parts compared to the *Baseline* parts. Since the part geometry, material chemistry and hardness were similar (Table I, Table II), the investigation moved toward the solidification process and the gating and risering systems used for each casting.

As expected, it was found that the design for the *Baseline* castings [9] provided uniform flow and heat, while the design for the *Alternate* castings [10] showed propensity for cold shut and air entrapment (Figure 17). The *Alternate* castings fractured at the thinnest area opposite the gates, where the solidification fronts from the two unbalanced streams of fluid traveling from the gate met and the potential for cold shut is maximized if the local temperature is low (due to low casting temperature, cold die, etc.). The *Baseline* castings fractured at the base of the *Post*, which is in agreement with the fatigue crack location of a sound casting predicted by modeling [1].

Table I. Chemical Composition, wt%

Element	*Baseline*	*Alternate*
Si	8.7	9.2
Fe	0.74	0.75
Cu	3.6	3.44
Mn	0.18	0.21
Mg	0.01	0.04
Zn	1.7	1.84
Ni	0.07	0.09
Cr	0.03	0.03
Pb	0.05	0.05
Sn	0.02	0.02
Ti	0.05	0.13
Others	< 0.5	< 0.5
Al	Rem.	Rem.

Table II. Hardness Values

Method	*Baseline*	*Alternate*
HB_{500kg}	80	79
HV_{300g}	101	102

Note that the microhardness values may differ in trend from the hardness values, since these readings are taken at a constituent/ phase level.

As a result of the investigation, the foundry being evaluated as an alternate source proceeded with modifications to the gating system of the *Alternate* castings, including increasing the stroke speed to assure that the cast alloy traveled longer distances while still hot [10], avoiding therefore as much as possible the cold shut formation. The resultant *Modified* castings tested in fatigue showed that the *Modified* and *Baseline* Weibull distributions are statistically equivalent (Figure 18). The fracture location was moved closer to the *Post* and the origin stayed consistently closer to the edge, although in some cases a subsurface location was still found. Fracture surface of the *Modified* parts show characteristic fatigue features such as beachmarks and striations, much better defined than in the *Alternate* or even *Baseline* castings (Figure 19). The microstructure (Figure 20) is sound and homogeneous, with less visible pattern of flow-induced segregation and minimum porosity.

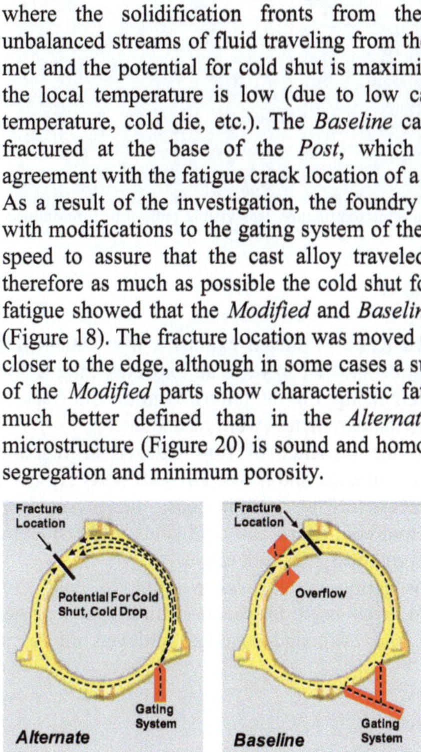

Figure 17. Gating systems showing propensity for cold shut, cold drop of the *Alternate* casting compared with the *Baseline* casting.

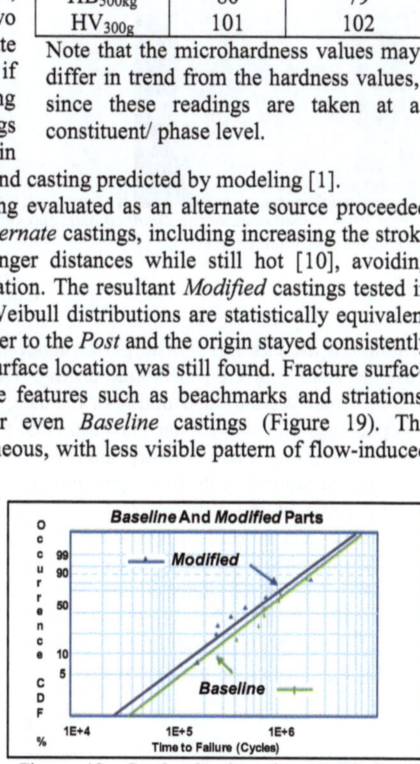

Figure 18. Graph showing the *Modified* and *Baseline* Weibull distributions being statistically equivalent.

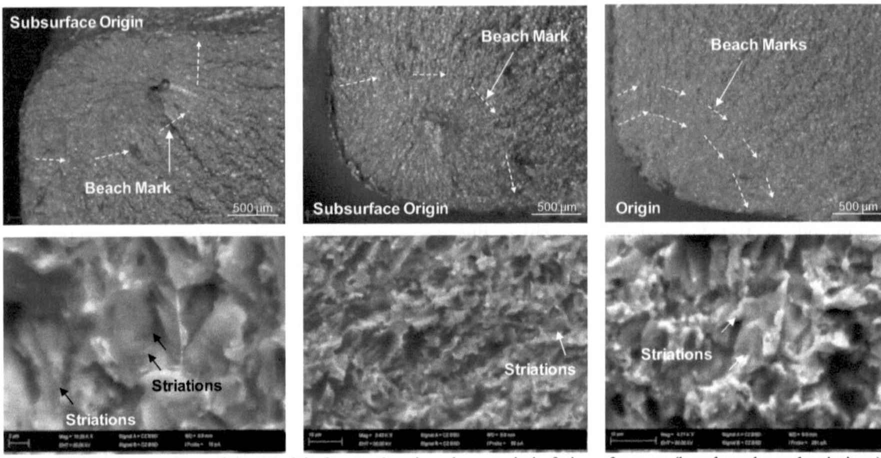

Figure 19. Fracture surface of the *Modified* parts showing characteristic fatigue features (beachmarks and striations).

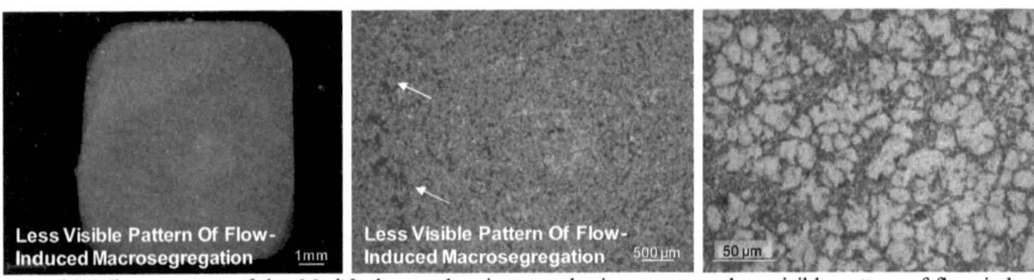

Figure 20. Microstructure of the *Modified* parts showing sound microstructure, less visible pattern of flow-induced segregation and minimum porosity.

Conclusions

Aluminum die-castings whose performance needed to be evaluated against a well established baseline were subjected to fatigue testing. The results showed lower than expected fatigue life and incorrect Weibull distribution of the data, as well changes in the failure location and fracture morphology. This was a clear indication that the casting parameters were inadequate.

In addition, the microstructure analysis revealed the presence of flow-induced segregation patterns, entrapped oxide films, gas porosity, microshrinkage and dross. Die-cast parts typically exhibit some of these microstructural inconsistencies; however, in this case, the presence of atypical casting defects such as the numerous internal cold drops and cold shuts in the fracture surface pointed toward incorrect casting parameters and design as root cause.

Further investigation determined that a different gating system, more prone to cold shut development - all other parameters being equal, was used to produce these parts. After improvements were made to the gating system, the evaluated castings achieved the same excellent quality as the baseline parts.

References

1. Internal Communication, Applied Mechanics Department, Emerson Climate Technologies, (2012).

2. S. Eckert et al., "Electromagnetic Melt Flow Control During Solidification of Metallic Alloys", *The European Physical Journal Special Topics*, 220 (1) (2013), 123-137.
3. Yun Wang, Hu-Tian Li, and Zhongyun Fan, "Oxidation of Aluminium Alloy Melts and Inoculation by Oxide Particles", *Trans Indian Inst Met*, DOI 10.1007/s12666-012-0194-x.
4. John Campbell, "An Overview of the Effects of Bifilms on the Structure and Properties of Cast Alloys", *Metallurgical and Materials Transactions B*, 37 (6) (2006), 857-863.
5. Q. G. Wang et al., "Oxide Films, Pores and the Fatigue Lives of Cast Aluminum Alloys", *Metallurgical and Materials Transactions B*, 37 (6) (2006), 887-895.
6. Q. G. Wang and P. E. Jones, "Prediction of Fatigue Performance in Aluminum Shape Castings Containing Defects", *Metallurgical and Materials Transactions B*, 38 (4) (2007), 615-621.
7. S. Wang et al., "Dross Recovery of Aluminum Alloy 380", *NADCA Cast Expo 2008*, Atlanta GA, May 17-20, (2008).
8. C.L. Burton et al., "Failure Analysis of a Cast A380 Aluminum Alloy Casting Using a Microstructurally Based Fatigue Model", *http://www.afsinc.org/files/images/CPedia.pdf*, CAVS, Mississippi State University.
9. Private communication with the foundry involved in producing the *Baseline* castings.
10. Private communication with the foundry involved in producing the *Alternate* and *Modified* castings.

Advances in the Science and Engineering of

CASTING SOLIDIFICATION

An MPMD Symposium Honoring Doru Michael Stefanescu

Solidification Processing III

Session Chair:
Amber Genau

Advances in the Science and Engineering of Casting Solidification
Edited by: Laurentiu Nastac, Baicheng Liu, Hasse Fredriksson, Jacques Lacaze, Chun-Pyo Hong, Adrian V. Catalina, Andreas Buhrig-Polaczek, Charles Monroe, Adrian S. Sabau, Roxana Elena Ligia Ruxanda, Alan Luo, Subhayu Sen, and Attila Diószegi
TMS (The Minerals, Metals & Materials Society), 2015

A COUPLED THERMO-MECHANICAL SIMULATION ON SQUEEZE CASTING SOLIDIFICATION PROCESS OF THREE-DIMENSIONAL GEOMETRICALLY COMPLEX COMPONENTS

Jie Tang[1], Zhiqiang Han[1], Feifan Wang[1], Jue Sun[2], Shanxin Xu[2]

[1]Key Laboratory for Advanced Materials Processing Technology, Ministry of Education, School of Materials Science and Engineering, Tsinghua University, Beijing 100084, China
[2]Suzhou Sanji Foundry Equipment Co., Ltd., Suzhou, Jiangsu Province 215106, China

Keywords: Squeeze Casting, Thermo-Mechanical Coupled Simulation, Three-Dimensional Component, Pressure Transmission

Abstract

A coupled thermo-mechanical simulation method for three-dimensional squeeze casting components has been developed. The simulation was achieved by using ANSYS Parametric Design Language (APDL). The effect of volume shrinkage due to cooling and solidification, the effect of pressure on the latent heat release, the mutual dependence of interfacial heat transfer and casting deformation, and materials behavior under elevated temperatures were taken into account in the simulation. A step-shaped trial casting was simulated, which demonstrates the ability of the method to simulate the pressure transmission and decline inside the casting as well as the distribution and evolution of the interfacial heat transfer coefficient. Finally, the method was applied to simulate the solidification of an automotive sub-frame component, based on which the squeeze casting process of the component was optimized.

Introduction

Squeeze casting (SC) is an advanced processing technology where high pressure is applied on the molten metal during cooling and solidification process, and the manufactured components normally have high integrity and superior mechanical properties [1]. In the SC process, the heat transfer across the casting/mold interface is enhanced due to the tight interface contact caused by the applied pressure, and the molten metal is squeezed from hot spots to feed the volume shrinkage caused by cooling and solidification [2]. However, as the solidified shell becomes thicker during solidification, its hindering effect on pressure transmission increases, and the pressure in the unsolidified area decreases. When the pressure in the mush and liquid area is too low, shrinkage defects will probably form. Therefore, it is of significant importance to understand quantitatively the thermo-mechanical process of SC by numerical simulation.

Some researches focused on numerical analysis of the SC process are reported in the literatures [3-5]. Hu and Yu [3] developed a two-dimensional model based on the control-volume finite-difference approach and enthalpy method to simulate the heat transfer and solidification phenomena of squeeze cast magnesium alloy (AZ91D), however, their model does not include stress and deformation simulation. Lee et. al. [4] reported a two-dimensional finite element method for fully coupled heat transfer and deformation analysis based on ABAQUS to simulate an indirect squeeze casting process. The author used elastic-plastic model to calculate the stress distribution of the casting. However, viscoplastic behavior cannot be neglected for materials at elevated temperatures, and the behavior of liquid phase needs to be treated specially. Zhu et. al. [5] reported a coupled thermo-mechanical two-dimensional finite element model to simulate the temperature, stress and shape development during the solidification process of squeeze casting of aluminum alloy. However, two-dimensional simulation cannot fulfill the requirement of dealing with a real squeeze casting component usually with complicated three-dimensional geometry. In this paper, a coupled thermo-mechanical method for three-dimensional casting simulation has been developed, in which the effect of volume shrinkage, materials constitutive model and interfacial heat transfer model are taken into account. Experiments and numerical simulations of a step-shaped casting and a sub-frame component are carried out and the results are compared and discussed.

Method Description

Computational Method for the Coupled Thermo-mechanical Simulation

A coupled thermo-mechanical simulation method based on ANSYS has been implemented to solve the thermal stress in the squeeze casting solidification process. As we know that squeeze casting is a process where the interactions exist between heat transfer and deformations. The heat transfer condition at interface will change due to the variation of contact pressure or gap, resulting in a significant influence on the distribution of temperature in the casting. On the other hand, the mechanical properties of alloys depend on temperature, and the change of temperature will have a great effect on the stress field. Therefore, the thermal and mechanical aspects should be coupled to reflect the inherent physics of the system. In each time step, the temperature field was calculated first, and material properties were updated based on the temperature calculation results. Then, stress analysis was carried out, and the boundary conditions for thermal analysis were updated. The computational process is shown in Figure 1.

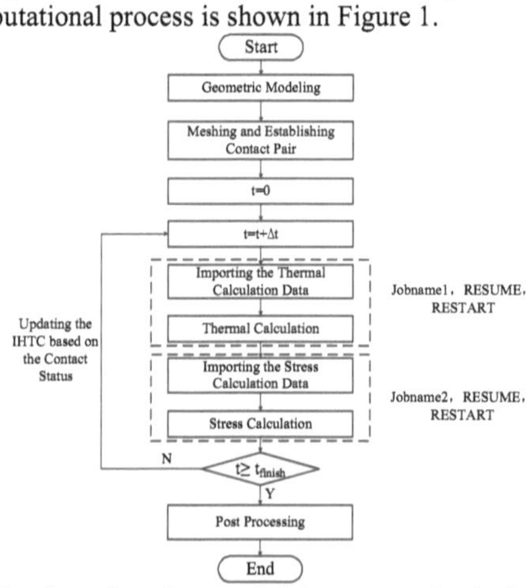

Fig. 1. The flow chart for coupled thermo-mechanical simulation

Materials Behavior under Elevated Temperatures

The mechanical behavior of alloy in the mold cavity is very complex during the solidification process. The deformations of the solidifying alloy include elastic deformation, plastic deformation and creep. Volume shrinkage happens during cooling and solidification.

Therefore, The incremental total strain ($\{\Delta\varepsilon\}$) is composed of elastic ($\{\Delta\varepsilon_{e}\}$), thermal ($\{\Delta\varepsilon_{th}\}$), and inelastic ($\{\Delta\varepsilon_{in}\}$) strain components, as given in Equation (1).

$$\{\Delta\varepsilon\}=\{\Delta\varepsilon_{e}\}+\{\Delta\varepsilon_{th}\}+\{\Delta\varepsilon_{in}\} \tag{1}$$

In this paper, the thermal strain increment is defined by volume changes caused by both temperature change and phase transformation. It is calculated by using the thermal linear-expansion (TLE [6]) function, as shown in follows.

$$\{\Delta\varepsilon_{th}\}^{t+\Delta t}=\left(TLE\left(T+\Delta T\right)-TLE\left(T\right)\right)\{1 \quad 1 \quad 1 \quad 0 \quad 0 \quad 0\} \tag{2}$$

Creep is significant at high temperatures and indistinguishable from plastic strain. For solid elements, a thermal elasto-viscoplastic model was used to describe the creep at high temperatures.

$$\dot{\varepsilon}_{cr}=A\left[sinh(\alpha\bar{\sigma})\right]^{n}exp(-\frac{Q}{RT}) \tag{3}$$

where A, α, and n are parameters depending on material properties, R is the gas constant, Q is the activation energy of the material. For the liquid phase, an elastic-perfectly plastic model was adopted, where the yield stress σ_Y is quite small. This treatment can not only effectively eliminate shear stress in liquid phase, but also avoid computational difficulties. The inelastic strain of liquid is calculated by the following equation [7].

$$\Delta\bar{\varepsilon}_{cr}=\begin{cases}0 & \hat{\bar{\sigma}}^{t+\Delta t}\le\sigma_{Y}\\ \dfrac{\hat{\bar{\sigma}}^{t+\Delta t}-\sigma_{Y}}{3G} & \hat{\bar{\sigma}}^{t+\Delta t}>\sigma_{Y}\end{cases} \tag{4}$$

Equations (3) and (4) are employed to express the constitutive relationship of materials, and the subroutine usermat was used to define constitutive relationship. In every Newton-Raphson iteration, usermat is called to update the stress vector, and the updated stress vector and the consistent tangent operator were supplied to the next calculation.

The Characteristics of Latent Heat Release

The solidification process is accompanied by the latent heat release. It is generally recognized that the relationship between the release of latent heat and the increase of solid fraction is linear. In SC process, the applied pressure has remarkable effects on the latent heat release. Computer-Aided Cooling Curve Analysis (CA-CCA) is a thermal analysis technique and has been widely used to investigate the solidification paths of ferrous or nonferrous alloys. In this study, the temperature variation inside the casting during solidification was measured and the pressurized solidification characteristics were investigated with CA-CCA where Fourier method was adopted to calculate the solid fraction curve as follows.

The temperature field of the casting can be described by the Fourier equation, which includes a heat source.

$$\frac{\partial T}{\partial t}=\alpha\nabla^{2}T+\frac{q_{S}}{C_{v}} \tag{5}$$

where $\partial T/\partial t$ is the cooling rate (K/s), α is the thermal diffusivity (m^2/s), C_v is the volumetric specific heat (J/m^3 K), and q_s is the volumetric heat generated during solidification (J/m^3 s). $\partial T/\partial t$ and $\nabla^2 T$ can be determined from the measured cooling curves. In the absence of heat generation, as is in the case of temperature higher than liquidus or lower than solidus, q_s=0, so the liquid thermal diffusivity α_L and solid thermal diffusivity α_S can be calculated through Equation (5). In the solidification process, thermal diffusivity can be assumed to be a function of the volume fraction of solid f_S and given by

$$\alpha = \alpha_L (1 - f_s) + \alpha_S f_s \tag{6}$$

The relationship between the release of latent heat and the increase of solid fraction is linear, which means

$$f_s(t) = \frac{\int_{fb}^{t} \left[\frac{\partial T}{\partial t} - \alpha \nabla^2 T \right] dt}{\int_{fb}^{te} \left[\frac{\partial T}{\partial t} - \alpha \nabla^2 T \right] dt} \tag{7}$$

An iterative procedure was adopted to solve Equations (5) and (7), and the solid fraction curve of A356 aluminum alloy is obtained as the solid line in Figure 2(a). It shows that the solid fraction increases very fast while the temperature drops a bit at the beginning of solidification process, and temperature drop is very obvious during the second stage of solidification process, which means less latent heat was released. At the end of solidification, because of eutectic reaction, the release of latent heat increases a bit. For the convenience of calculation, we use three linear segments to approximate the relationship between temperature and solid fraction as the dotted line shown in Figure 2(a). The relationship between enthalpy and temperature was employed to deal with latent heat release. In order to consider the effect of the applied pressure in squeeze casting, the relationship between the enthalpy and temperature based on the solid fraction curve we obtained through CA-CCA method was used in the simulation, which is shown as the dotted line in Figure 2(b). As a comparison, we provide the data retrieved from the ProCAST database, i.e. the solid line in Figure 2(b).

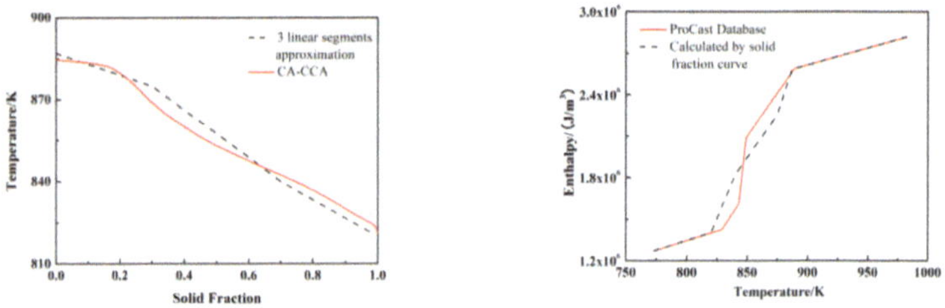

(a) solid fraction curve

(b) the relationship between enthalpy and temperature

Fig. 2. The characteristics of latent heat release

Interfacial Heat Transfer Model

The heat transfer between the casting and the mold has a significant influence on the solidification process. The method of updating the interfacial thermal resistance plays an important role in fulfilling the coupled thermo-mechanical simulation. Due to the shrinkage and deformation of the metal during solidification, the interfacial heat transfer coefficient (IHTC) is a function of time and position. In the present work, an interfacial heat transfer model considering gap or pressure is employed temporarily to calculate the IHTC [8].

$$h = 1 / \left(\frac{1}{h_{ini}} + \frac{\delta}{k_{air}} \right) + \frac{\sqrt{P}}{2} \tag{8}$$

where h_{ini} is the reference value of the IHTC, δ is the gap at the interface (m); P is the contact pressure (Pa).

Simulation Results and Discussions

Step-shaped experimental casting

A step-shaped trial casting was designed to test the effectiveness of the coupled simulation method. Squeeze casting experiments were carried out and the temperature variation inside the casting during the pressurized solidification process was measured. The applied pressure was set as 27 and 82MPa.

Figure 3 shows the IHTC at t=16s under different applied pressures. The IHTC is about 4000W/m^2·K at 16s under the pressure of 27MPa. The value is increased to about 6000 W/m^2·K under the pressure of 82MPa. Hence the increase of the applied pressure significantly enhances the heat transfer intensity. The value of the IHTC at the thinnest step of the casting is maximum. Apparently this region is solidified first during the solidification process, leading to the fact that the thinnest step is stronger to attain more pressure than other parts, which means higher values of IHTC according to Equation (8). Figure 4 shows the calculated and measured cooling curves at the centers of the three steps. Position 1 is located at the center of the thickest step and position 3 is located at the center of the thinnest step. It can be seen that the calculated temperature variations agree well with the measured results at position 1 and position 2. There is some deviation at position 3. The accuracy of the temperature field calculation largely depends on the heat exchange at the interface and the constitutive relations of materials, especially the reliability of the interfacial heat transfer model. However, the study on the relationship between heat transfer coefficient and the contact status at the interface is not adequate. More efforts on the research on IHTC for squeeze casting are still needed.

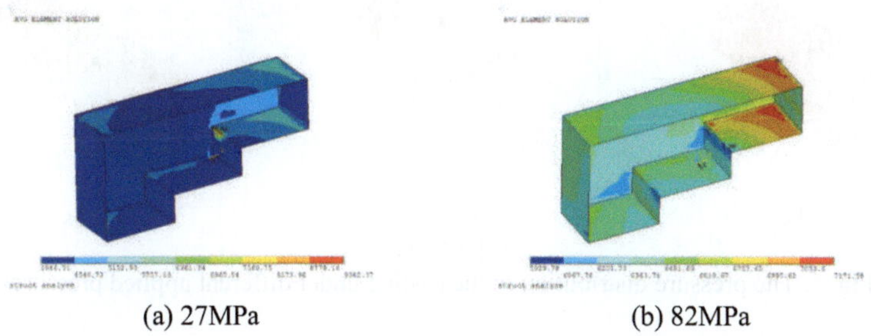

(a) 27MPa (b) 82MPa

Fig. 3. The interfacial heat transfer coefficient at t=16s under different applied pressures

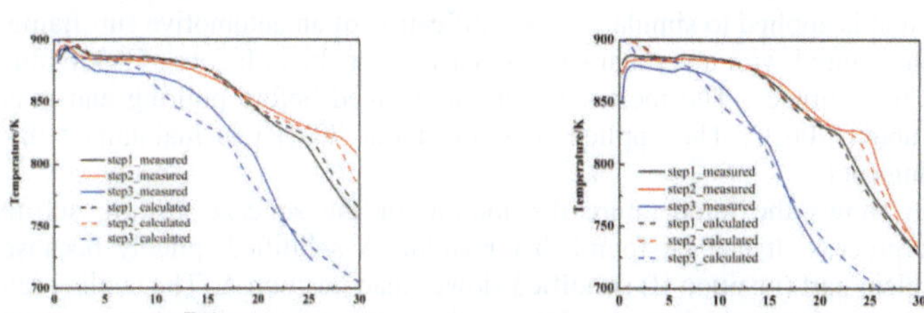

(a) 27MPa (b) 82MPa

Fig. 4. The simulated and experimental cooling curves under different applied pressures

Figure 5 shows the contours of pressure (here, pressure refers to static pressure) distribution in the casting under different applied pressures. The results show that during solidification process, the solidified shell withstands more pressure than alloys in the liquid and mushy zone, and the pressure in the area not completely solidified keeps decreasing with the increasing of the thickness of the solidified shell. When the applied pressure is 27MPa, the pressure decreases to about 15MPa at 20s, and then rapidly drops to zero at the end of the solidification. However, the pressure at the end of solidification could be 35MPa when the applied pressure is 82MPa as we can see in Figure 5. It can be found that under lower applied pressure (such as 27MPa), the pressure at the center decreased to zero before the melt are fully solidified, which indicates a great tendency to form shrinkage defects. For casting under higher applied pressure (such as 82MPa), the molten metal solidified under pressure during the whole solidification process, which indicates a good integrity in the casting.

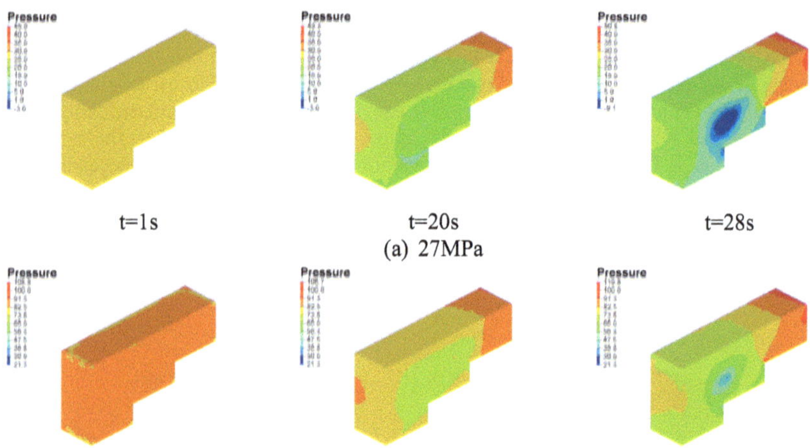

t=1s t=20s t=28s

(a) 27MPa

t=1s t=17s t=25s

(b) 82MPa

Fig. 5. The pressure distribution in the casting under different applied pressures

Automotive Sub-frame Component

The method is applied to simulate the solidification of an automotive sub-frame component, which is manufactured with a vertical squeeze casting machine. Ingots of A356 aluminum alloy were melted in a furnace. The molten metal was refined before pouring and maintained at a temperature about 700°C. The applied pressure about 70MPa is maintained throughout the squeeze casting cycle.

Figure 6 shows the temperature distribution of the squeeze casting sub-frame during solidification process. It can be found that position A solidified quickly because of its thin section. The plain part (position B) solidified slower than position A. The cooling rate at position C is lower due to its larger thickness, and the cooling rate of the gating system (position D) is the lowest. As shown in Fig.6(c), after 17.5s, most parts of the casting have been solidified, however, some thick wall areas were still not solidified. These regions have a great tendency to form shrinkage defects due to the lack of effective feeding.

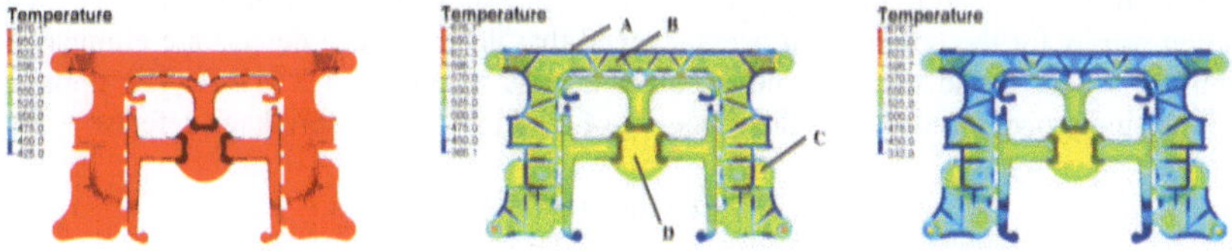

(a)t=0.01s (b)t=10s (c)t=17.5s

Fig. 6. The temperature distribution of squeeze casting sub-frame during solidification process

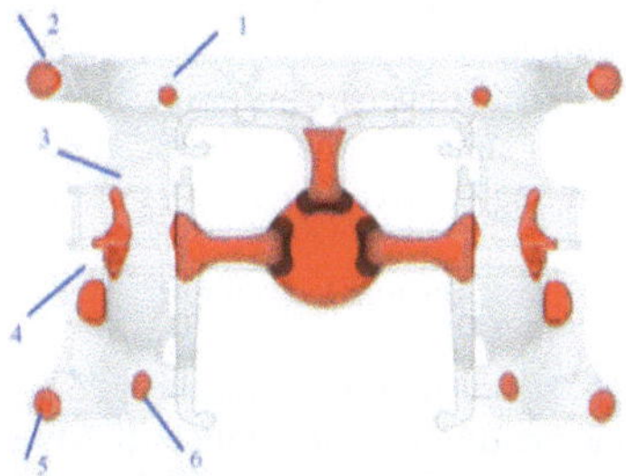

Fig. 7. The distribution of areas not completely solidified

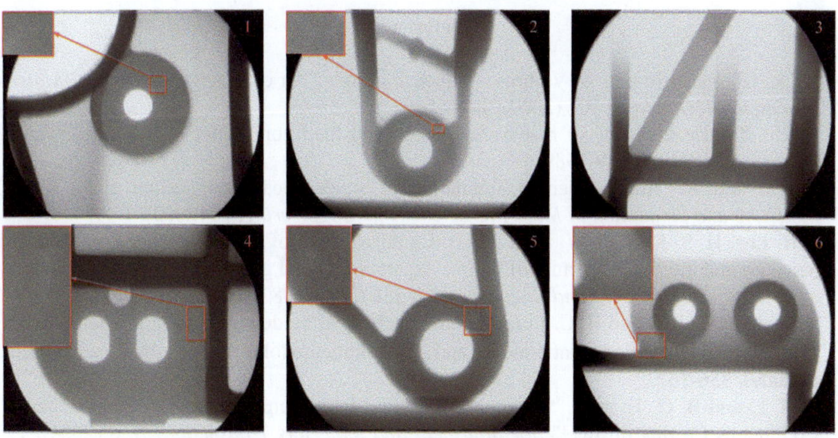

Fig. 8. The X-ray inspection results of squeeze cast sub-frame

The distribution of regions not completely solidified has been presented in Figure 7. The parts that are not solidified are the places with thick walls or the gating system. Those isolated molten pools without effective feeding have a great tendency to form shrinkage defects. X-ray inspection has been employed to detect the internal defects in castings, and the results are shown in Figure 8. The shrinkage defects are found in positions 1, 2, 4, 5 and 6. The simulation results are in good agreement with the X-ray inspection results. The squeeze casting process of the component is optimized based on the simulation work by enlarging the cross section of inner gates near positions 1 and 6 and implementing local pressurization at positions 4 and 6. X-ray inspection results for the optimized process showed that the shrinkage defects are eliminated at positions 1, 5 and 6, and there are no apparent defects at positions 2 and 4. The optimized squeeze casting process leads to a higher integrity of casting with less shrinkage defects.

Conclusions

1. A coupled thermo-mechanical simulation method for three-dimensional squeeze casting components has been developed, in which the effect of volume shrinkage, the effect of pressure on the latent heat release and the mutual dependence of interfacial heat transfer and casting deformation are taken into account.
2. A step-shaped trial casting was simulated, which demonstrated the ability of the method to simulate the pressure transmission and decline inside the casting as well as the distribution and evolution of the interfacial heat transfer coefficient.
3. The method was applied to simulate the solidification of an automotive sub-frame component, based on which the squeeze casting process of the component was optimized.

Acknowledgement

The research is funded by the National Natural Science Foundation of China (Project No. 51175291), the State Key Laboratory of Automotive Safety and Energy, Department of Automotive Engineering, Tsinghua University and Suzhou Automotive Research Institute, Tsinghua University under the contract of No. 2013XC-A-01.

References

1. M. R. Ghomashchi and A. Vikhrov, "Squeeze casting: An overview", *Journal of Materials Processing Technology*, 101 (2000), 1-9.
2. H. Hu, "Squeeze casting of magnesium alloys and their composites", *Journal of Materials Science*, 33 (1998), 1579-1589.
3. H. Hu and A. Yu, "Numerical simulation of squeeze cast magnesium alloy AZ91D", *Modeling and Simulation in Materials Science and Engineering*, 10 (2002), 1-11.
4. J. H. Lee, H. S. Kim, C. W. Won and B. Cantor, "Effect of the gap distance on the cooling behavior and the microstructure of indirect squeeze cast and gravity die cast 5083 wrought Al alloy", *Materials Science and Engineering A*, 338 (2002), 182-190.
5. W. Zhu, Z. Q. Han and B. C. Liu, "Thermomechanical modeling of solidification process of squeeze casting Ⅰ. Mathematic model and solution methodology", *Acta Metallurgica Sinica*, 3 (2009), 356-362.
6. C. S. Li and B. G. Thomas, "Thermomechanical Finite-Element Model of Shell Behavior in Continuous Casting of Steel", *Metallurgical and Materials Transactions B*, 6 (2004), 1151-1172.
7. S. Koric and B.G. Thomas, "Efficient thermo-mechanical model for solidification processes", *International Journal for Numerical Methods in Engineering*, 12 (2006), 1955-1989.
8. Y. Xu, J. W. Kang and T. Y. Huang, "Numerical simulation of the casting process with coupled thermal and mechanical effects", *Journal of Tsinghua University (Science & Technology)*, 5 (2008), 769-772.

Advances in the Science and Engineering of Casting Solidification
Edited by: Laurentiu Nastac, Baicheng Liu, Hasse Fredriksson, Jacques Lacaze, Chun-Pyo Hong, Adrian V. Catalina, Andreas Buhrig-Polaczek, Charles Monroe, Adrian S. Sabau, Roxana Elena Ligia Ruxanda, Alan Luo, Subhayu Sen, and Attila Diószegi
TMS (The Minerals, Metals & Materials Society), 2015

DOUBLE OXIDE FILM DEFECTS IN Al CASTINGS AND THE EFFECT OF DIFFERENT ELEMENT ADDITIONS

Q. Chen[1], A. J. Caden[1], W.D. Griffiths[1]

[1]School of Metallurgy and Materials, College of Engineering and Physical Science
University of Birmingham, Edgbaston, Birmingham, UK. B15 2TT

Keywords: Al castings, Oxide film defect, Element addition

Abstract

Double oxide film defects consist of doubled-over oxide films containing a gas-filled crevice, and are reported to cause both reductions in mechanical properties and increases in the scatter of properties in Al alloy castings. However, the gas entrapped in a double oxide film defect during its formation may be consumed by reaction with the surrounding melt. The defect might then be closed and its harmful effects might be reduced. In the experiments reported here, an air bubble was trapped inside an Al melt for up to 1 hour. The change in the volume of the bubble was determined and the oxide film created was investigated using SEM/EDX. The experiment was conducted with additions of Ti, Zr, Mo, Hf, Sc to commercial pure aluminum and 2L99 Al alloy, and it was found that the Mo addition affected the formation of the oxide layer and might therefore accelerate the consumption of the entrapped gas. Tensile testing of sand-cast 2L99 alloy with an addition of Mo suggested that with the Mo addition the Weibull modulus for Ultimate Tensile Strength was increased. Investigation of the fracture surfaces of the test bars suggested mechanisms of how this addition may affect double oxide film defects.

Introduction

An oxide film is readily formed on the surface of liquid aluminum. During casting, the molten metal is likely to experience surface turbulence leading to its oxide skin folding over upon itself entrapping a layer of gas, mostly air. This doubled over oxide film, when entrained into the molten metal, will form a double oxide film defect [1] (see Fig. 1). Double oxide film defects have been reported to be responsible for both variation and a reduction in mechanical properties. However, the entrapped gas in a double oxide film defect may be consumed by reaction with the surrounding molten metal, through rupture points or discontinuities in the oxide layer. Nyahumwa *et al.* [2] suggested that there may be a healing mechanism for this defect, as follows.

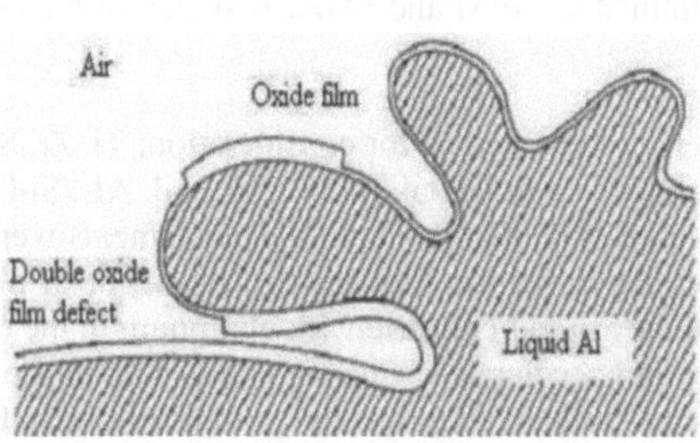

Fig. 1. Sketch showing the formation of a double oxide film defect.

The first surface oxide layer to form is probably in an amorphous state, which after a period of time transforms to γ-Al_2O_3 and finally α-Al_2O_3. This process might take from several minutes to several hours depending on the temperature. The γ-to-α transformation was accompanied by a 24 % volume increase, which might cause the oxide layer to be ruptured at some points. The molten metal would then come into contact with the internal atmosphere in the double oxide film defect. First oxygen and then nitrogen in the atmosphere would be consumed and then, after the majority of the gas was consumed, the two layers might bond together and the deleterious effect of the double oxide film defect might be reduced.

Based on these assumptions, El-Sayed et al. [3] used a Pore Gas Analyzer, (a device designed to analyze the gas extracted from a pore using a mass spectrometer), to investigate the change in the composition of an air bubble with time. Their results showed that in a casting of commercial purity aluminum, the majority of O_2 was consumed in the first 8 minutes before the nitrogen started to be consumed. While the consumption of nitrogen continued, the hydrogen content in the bubble increased.

The aim of the work reported here was to find an addition to an aluminum melt which could either compete with aluminum in forming an oxide or nitride, or somehow change the structure of the oxide layers in a bifilm defect, to allow the entrapped gas come into contact with the surrounding molten metal more easily. In both cases, the consumption rate of the entrapped gas might be accelerated.

Experimental procedure

750g of molten commercial purity aluminum was held in a cylindrical crucible at 730°C, in a 2.5 kW resistance-heated furnace. In order to trap an air bubble inside the aluminum melt, a steel rod with a blind hole at one end was inserted into a melt. The steel rod was 150 mm in length and 19 mm in diameter with the hole being 40 mm deep and 12.5 mm in diameter. A thermocouple was inserted 85 mm from the top of the furnace, (about the position of the bubble), to monitor the temperature of the melt during the experiment.

At the start of the experiment, the steel rod was rotated at 90 rpm/min and then plunged 85 mm beneath the surface of the melt. A thin steel rod was also inserted into the melt from the side of the crucible to touch the rotating steel rod. This was intended to break any oxide layer which might have been attached to the surfaces of the steel rod as it was plunged into the melt, which might form a leak path connecting the entrapped gas to the external environment. The rotation of the rod was halted after 5 s, and the steel rod held at the same position for 1 hour, before the melt was allowed to solidify. After the experiment, the metal inside the blind hole in the steel rod was cut out and its surface examined by SEM and EDX. A sketch of the experiment is shown in Fig. 2.

In the experiment, 5 elements were selected for consideration; Ti, Zr, Sc, Hf and Mo, which were added to two different alloys, commercial purity Al and Al-7Si-0.3Mg, (2L99) alloy. The composition of the alloys used is shown in Table I. The elements were selected chiefly because they had lower Gibbs free energy for the formation of oxide, or of nitride, compared to that of aluminum. In other words, they might compete with aluminum in the consumption of nitrogen in a double oxide film defect and accelerate the consumption rate of the entrapped gases. The two different alloys were selected because they were expected to form different oxide layers, Al_2O_3 for the commercial purity Al, and $MgAl_2O_4$, spinel, for the 2L99 alloy.

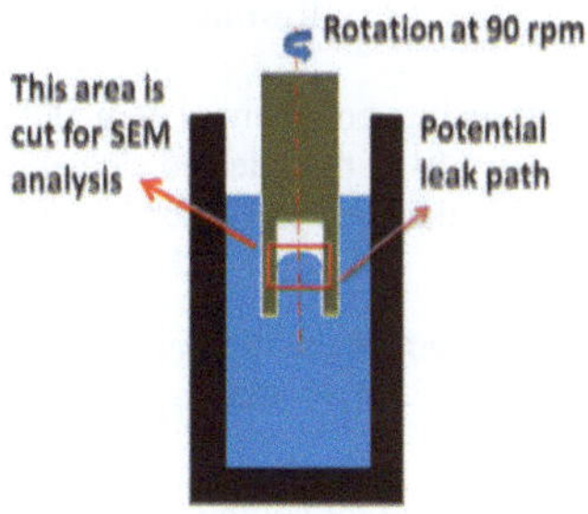

Fig. 2. A sketch of the experiment to hold a gas bubble in the Al melt for a period of time.

Table I. Chemical composition of the alloys.

Element		Si	Fe	Cu	Mn	Mg	Cr	Ni	Zn	Pb	Sn	Ti	Al
%	Pure Al	0.11	0.1	0.01	0.01	0.01	0.01	0.01	0.01	0.01	0.01	0.02	Bal.
	2L99	6.99	0.07	0.01	0.01	0.38	0.01	0.01	0.01	0.01	0.01	0.15	Bal.

Tensile tests were subsequently conducted, as the experimental results suggested that one of the element additions (Mo) might affect the behavior of bifilm defects in Al alloy castings. In this experiment, two sets of castings were prepared, using resin-bonded sand moulds (see Figure 3). (The moulds were prepared 1 week before casting to allow volatiles from the mould to be reduced.) Castings were made using 2L99 alloy and 2L99 + ~0.5 wt. % of Mo (both contain 20 test bars). The samples were heat treated to the T6 condition, (535°C solution treatment for 12 h, followed by quenching in 85°C hot water, followed by precipitation hardening for 4 h at 155°C.) The castings were then machined into tensile test bars with a gauge length of 37 mm and diameter in the gauge length of 6.75 mm, and subject to tensile testing with a strain rate of 1 mm min^{-1}.

After testing the Weibull modulus [4] was used to characterize the scatter of mechanical properties and the reproducibility of the castings.

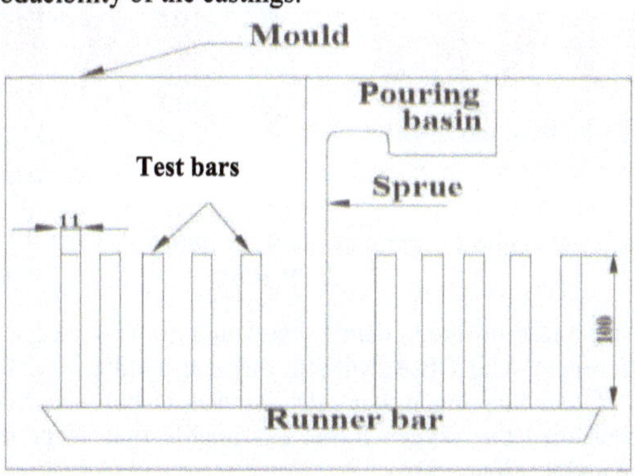

Fig. 3. A sketch of the test bar mould used in the experiment.

Results

The recovery of each element addition into commercial pure aluminum and 2L99 alloy are shown in Table II. For each case, 1 wt. % of the selected elements was initially added into the melt and the amount actually found in the melt varied for 15% to 70% recovery.

Table II. The recovery of each element addition in commercial purity aluminum and 2L99 alloy.

Element addition	Element Pickup for commercial purity aluminum (wt. %)	Element Pickup for 2L99 alloy (wt. %)
Mo	0.15%	0.37%
Zr	0.25%	0.41%
Ti	0.52%	0.51%
Hf	0.71%	0.35%
Sc	0.49%	---

Figure 4 (a) and (b) showed the amount of metal that had entered into the blind hole for the CP-Al and 2L99 alloy respectively. This was a measure of the rate of reaction of the alloy with air, following the element addition. For commercial purity aluminum, the Hf addition had the highest sample length, (constituting 75% of the hole depth). For 2L99 alloy, the Mo addition had the highest sample length, (constituting 45% of the hole depth).

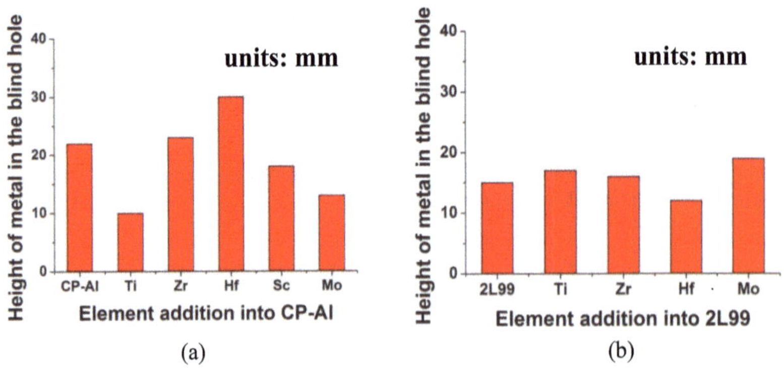

Fig. 4. (a) Sample depths for the element additions into commercial pure aluminum and (b) 2L99 alloy.

As shown in Fig. 5(a), a feather-like structure was found growing on the alumina layer from the bubble holding experiment with CP-Al without element additions. EDX results (Figure 5(b)) suggested that this feather-like shape was AlN, as was to be expected, as the liquid Al had reacted with and consumed the oxygen in the gas bubble first, to produce first alumina, (see Figure 5(c)), and then AlN (Figure 5(b)).

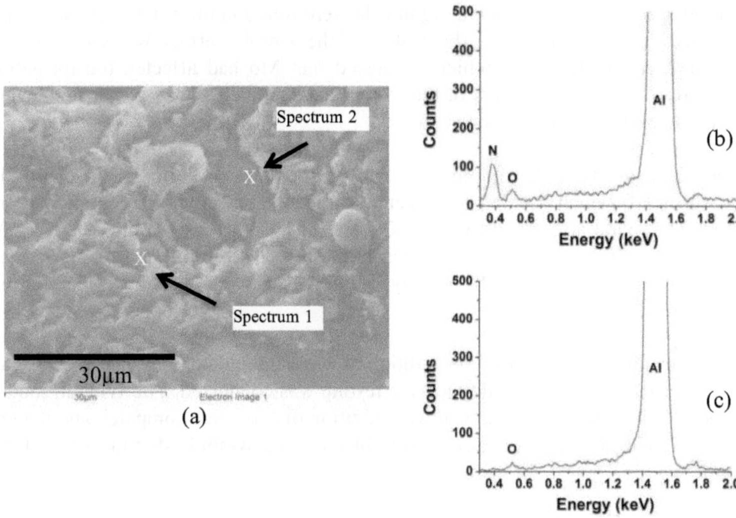

Fig. 5. (a) SEM image of the feather-like phase growing on an alumina layer and (b) the EDX result confirming the presence of nitrogen (c) the EDX result for spectrum 2.

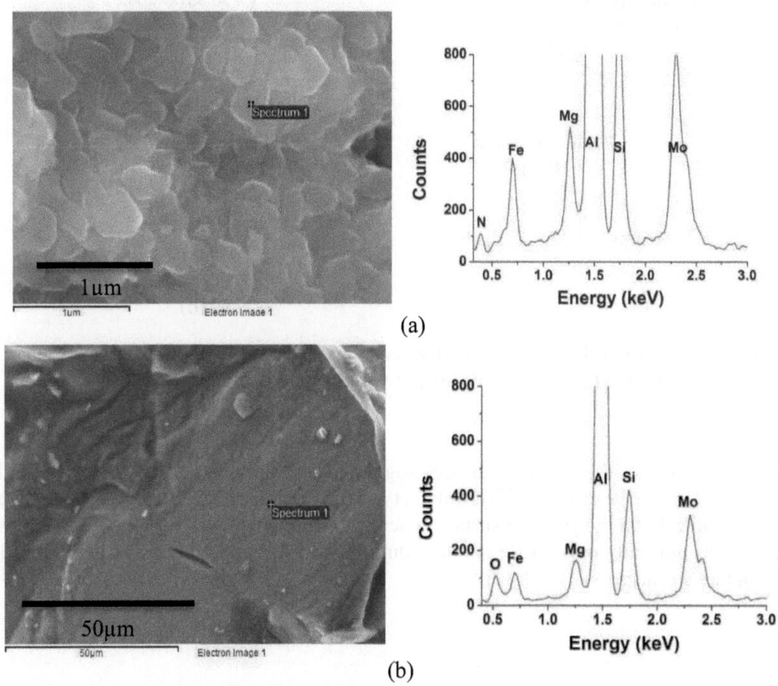

Fig. 6. (a) Mo containing nitride and (b) oxide layer found in bubble holding experiment with 2L99+Mo after holding for 1h.

Both Mo-containing oxide and Mo-containing nitride were found in the experiment with 2L99 + 0.5 wt. % Mo (see Fig. 6 (a) and (b)). About 80% of the sample surface was covered with an oxide and nitride layer of this kind, which suggested that Mo had affected the formation or structure of the oxide and nitride layers in the experiment, in such a way that it might help to accelerate the consumption of the entrapped gas.

No addition appeared to affect the morphology and composition of the oxide layer in CP-Al, even in the case of Hf, which seemed to consume more of the entrapped gas than the other elements added. The oxide layer found was Al_2O_3 and AlN in the case of nitride layer. When addition was added to 2L99 alloy, only Mo was found affect the formation of the oxide and nitride layer. For other additions into 2L99 alloy, $MgAl_2O_4$ and AlN was the oxide and nitride layer found respectively.

In order to further evaluate the effect of the addition of Mo on double oxide film defects, and the reproducibility of properties in 2L99 alloy, tensile testing was conducted. The Weibull moduli of the (heat-treated) 2L99 castings, without and with addition of Mo, were compared and the results are shown in Fig. 7. The addition of Mo seemed to increase the Weibull Modulus of the UTS by a factor of about two [5].

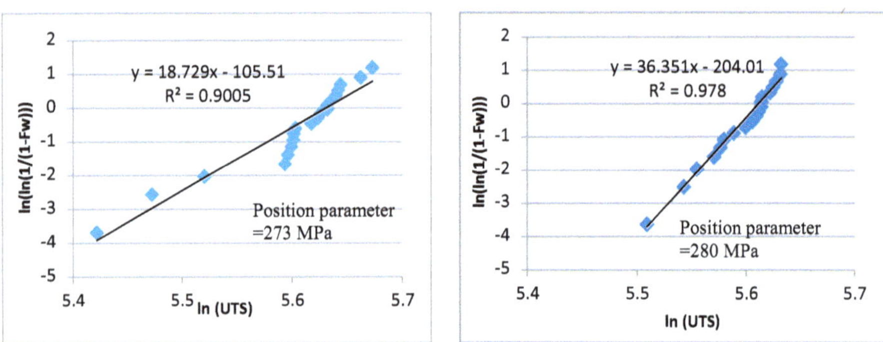

(a) (b)

Fig.7. (a) Weibull moduli of the UTS of the 2L99 castings without Mo addition, (b) Weibull moduli of UTS of the 2L99 casting with the Mo addition [5].

SEM/EDX examination of the fracture surfaces suggested that the presence of Mo affected the composition of the oxide layer in the bifilm defect (see Fig. 8), as 15% of the specimens were found to contain nitride on the fracture surface where a bifilm defect was observed (see Fig. 9). This is unusual, as normally only oxide is found during SEM examination of double oxide film defects on fracture surfaces.

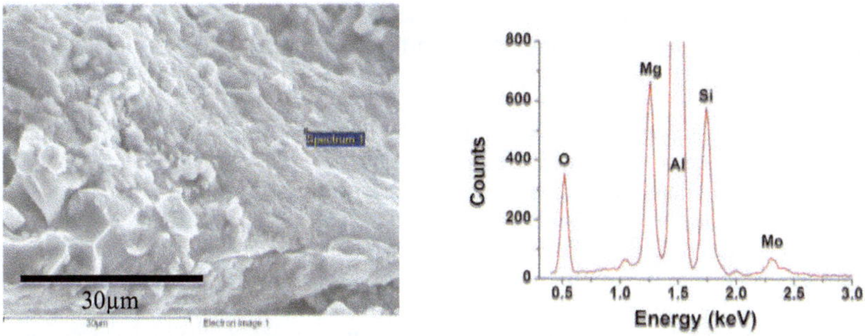

Fig.8. SEM/EDX results showing the presence of Mo in the oxide layer of a bifilm defect in the Mo containing tensile test bars.

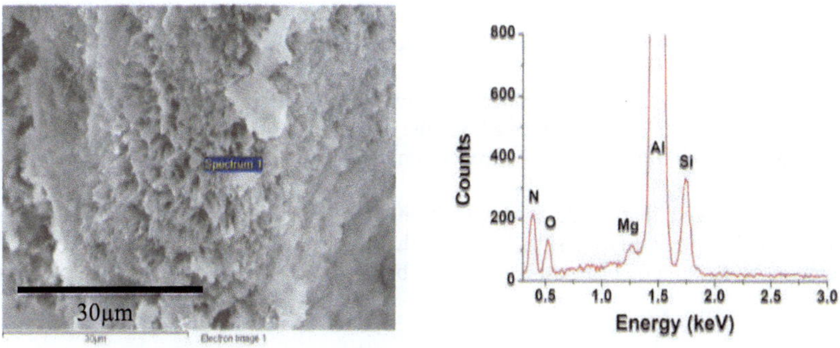

Fig.9. SEM/EDX results showing the occurrence of nitride in a bifilm defect in Mo containing tensile test bars.

Discussion

Figure 5 showed the presence of feather-like AlN growing on Al_2O_3, for the bubble-holding experiment with commercial purity aluminum. This could indicate the point where the majority of the oxygen in the entrapped atmosphere was depleted and nitrogen started to react. As the reaction between nitrogen and molten aluminum is difficult in the presence of a trace amount of oxygen [6], the presence of the nitride showed that any potential leak path which might connect the internal atmosphere of the bubble to the surroundings had been successfully disrupted in the experiment.

In the trapped bubble experiment, the Mo addition led to the majority of the sample surface being covered by a Mo-containing oxide or a Mo-containing nitride layer. This means that the Mo influenced the formation of both oxide and nitride, which could change the behavior of the oxide layer in a double oxide film defect. Fig. 8 showed that the Weibull Moduli of the 2L99 + Mo castings was approximately doubled compared with 2L99 castings without the Mo addition.

The improved Weibull moduli suggests that less variation of mechanical properties could be obtained, despite the bifilm defect occurring, due to surface turbulence during mould filling.

For the Mo containing castings, SEM examination of the fracture surface showed nitride found on the the fracture surface of the cast test bars. As discussed above, this indicates that the entrapped gas in the bifilm defect should have lost the majority of its oxygen by reaction with the liquid metal. The occurrence of nitride on the fracture surface of a test bar has not been reported in the literature, as bifilm defects are usually a few seconds old before solidification, in which case, it would be expect that the oxygen in the double oxide film defect would not have been completely consumed. The presence of nitride on the test bar fracture surfaces suggests that the reaction rate between the melt and the entrapped gas has been accelerated, due to the presence of Mo, perhaps reducing the deleterious effect of the defect.

Conclusions

1. The entrapped bubble experiment showed that none of the element additions, (Hf, Sc, Ti, Zr, or Mo) appeared to affect the oxide or nitride layer in commercial purity aluminum.

2. For the entrapped bubble experiment with 2L99 alloy, Mo-containing oxide and nitride occurred on the sample surface. This suggested that the presence of Mo in the alloy could accelerate the consumption of gas in a double oxide film defect.

3. Tensile testing showed that, with a Mo addition, the Weibull moduli of the UTS was approximately doubled. SEM/EDX results indicated that the Mo addition may have accelerated the consumption of the internal atmosphere in the bifilm defects, reducing their deleterious effects.

Acknowledgement

The authors would like to acknowledge the EPSRC Centre - LiME for providing financial support under Grant EP/H026177/1.

References

1. J. Campbell, *Castings*, 2nd Ed., (Butterworth-Heinemann, 2003), 17-69.
2. C. Nyahumwa, N.R. Green, and J. Campbell, “Effect of Mold-Filling Turbulence on Fatigue Properties of Cast Aluminium Alloys,” *AFS Transactions*, 58 (1998), 215-223.
3. M. A. El-Sayed, W.D. Griffiths, “Determination of the lifetime of a Double-Oxide Film in Al Castings,” *Metallurgy and Material Transactions B*, 45B (2014), 1398-1406.
4. Green, N.R. and Campbell,J., “Statistical distribution of fracture strengths of cast Al-7Si-Mg alloy,” *Material Science and Engineering: A*, 173(1-2) (1993), 261-266.
5. W. D. Griffiths, Q. Chen and A. J. Caden, “Aluminium Alloy and Method,” GB1411826.9, patent filed 02/07/14.
6. Q. Zheng, R. G. Reddy., “Mechanism of *in situ* formation of AlN in Al melt using nitrogen gas,” *Journal of Materials Science*, 39 (2004), 141-149.

Advances in the Science and Engineering of Casting Solidification
Edited by: Laurentiu Nastac, Baicheng Liu, Hasse Fredriksson, Jacques Lacaze, Chun-Pyo Hong, Adrian V. Catalina, Andreas Buhrig-Polaczek, Charles Monroe, Adrian S. Sabau, Roxana Elena Ligia Ruxanda, Alan Luo, Subhayu Sen, and Attila Diószegi
TMS (The Minerals, Metals & Materials Society), 2015

THE DESIGN OF NEW SUBMERGED ENTRY NOZZLES FOR BEAM-BLANK CONTINUOUS CASTING

Miaoyong Zhu, Mianguang Xu, Sen Luo

School of Materials and Metallurgy, Northeastern University, No. 11 Wenhua Road, Heping District, Shenyang, China

Keywords: Beam-blank continuous casting; Submerged entry nozzle; Transport phenomena; Numerical simulation; Water modelling

Abstract

Two types of radial flow submerged entry nozzle (SEN) were proposed to overcome the shortcomings of the straight SEN which has been widely used since the first beam-blank caster was installed in Algoma Steel over four decades ago. One radial flow SEN was designed with the purpose of "self-braking", with which fluid flow velocity in the mold can be reduced and the inclusion removal ratio can be increased. However, the "self-braking" gives a rise to a negative effect of confining too much superheat to a small region in the mold, which will cause serious quality problems and increase the risk of breakouts. Thereby another radial flow SEN was designed, which could avoid the heat concentrating by an asymmetrical flow pattern. In this paper, advantages and disadvantages of these three SENs for beam-blank continuous casting were discussed in detail by using numerical simulation, water modeling and industrial trials.

Introduction

Recently, as a near-net-shape continuous casting technique, the need of beam-blank steel has been experiencing a dramatic rapid increase and more than ten beam-blank continuous casting machines have been put into production in China. However, there is a common problem having to face, that is the quality of continuously cast beam-blank, especially the crack of the strand.

As it is known, most of defects affecting the quality of continuously cast are associated with the transport phenomena in the mold which is largely determined by the SEN structure. However, the complicated geometry of beam-blank mold creates great difficulties for the design of a suitable SEN and the simple straight SEN is still widely used. With this nozzle, molten steel is direct to pour into two ceramic funnels, which are located in the central flange regions of mold, as shown in **Figure 1**. Such pouring method can give rise to serious problems on the quality of final products, especially in the production of some new steel grades, so new types of SEN are necessary to be proposed to meet the demand for the product with higher quality.

Mathematical modeling of solidification during the continuous casting process has had a wide application as a powerful tool in the design of SEN and the analysis of the transport phenomena, and the use of mathematical model for investigating the suitable structure of SEN is very convenient. In beam-blank continuous casting, such models can be used to investigate the effects of different structures of SEN and operating conditions and predict their effects on the formation of the solidifying shell and on the levels of turbulence at the meniscus, which will have a major effect on the quality of final products. Unfortunately, most of mathematical models about beam-blank continuous casting were based on the thermal-mechanical analysis [1-3], water channel design [4,5] and secondary cooling strategy [6], a few literatures found are mainly on the straight SEN [7].

In the current study, the fluid flow, heat transfer and solidification in a beam-blank slab casting process were investigated. With the understanding of shortcomings of the straight SEN and the limitations of previous studies, two new designed three-port radial flow SEN geometries and proper relevant casting parameters were discussed.

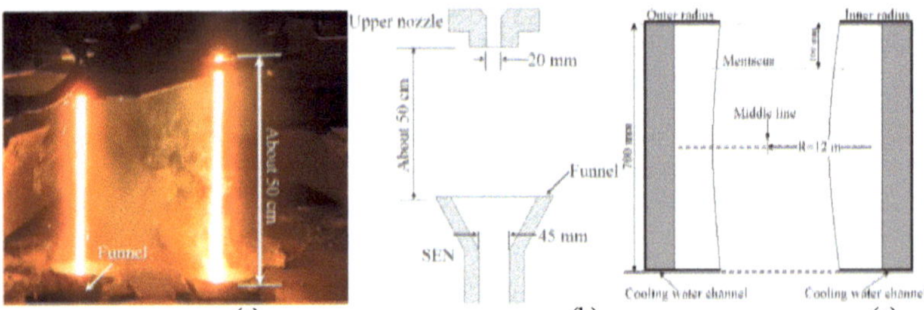

(a) (b) (c)

Figure 1. Molten steel open poured into ceramic funnels: (a) industrial trial, two nozzles per mold are used for uniform steel feeding; (b) schematic of upper nozzle and SEN; (c) vertical sectioning of the mold.

Mathematical models and casting conditions

The mathematical models and their boundary conditions could be found in the reference [8]. The research object of present work is a five-strand bloom/beam-blank continuous casting machine with the length about 37 m. The single-port straight SEN used in steelworks is shown in **Figure 2(a)** and the new designed three-port radial flow SENs are shown in **Figures 2(b)** and **2(c)**.The aspect ratio of thickness to height for the port of radial flow SEN was selected to be 0.83 in order to improve the effectiveness of the SEN having a better direction of molten steel flow to ensure its "self-braking" in the mold. The geometry for the port of radial flow SEN is shown in **Figures 2(b)** and **2(c)**, and the total area of ports is larger than that of the original straight SEN to enhance the inclusion removal ratio [8].

(a) (b) (c)

Figure 2. Schematic drawing of beam-blank mold and SEN vertical section: (a) original single-port straight SEN; (b) symmetry radial flow SEN; (c) asymmetry radial flow SEN.

The typical kinds of computation domains are presented in **Figure 3**. The geometry, casting conditions, materials properties and computational conditions are summarized in **Table 1**, and SEN submergence depth is defined as the distance of SEN bottom from meniscus.

Items	Values
Geometry	
Mold Length, m	0.8
Effective mold length, m	0.7
Casting parameter	
Casting speed, m/min	0.6, 1.0
Superheat, K	20
Straight SEN submergence depth, cm	6-10
Radial flow SEN submergence depth, cm	11.5
Thermo-physical properties of steel	
Liquids temperature, K	1791
Solidus temperature, K	1753
Specific heat of steel, J/kg/K	700
Latent heat of solidification of steel, J/kg	272000
Thermal conductivity of steel, W/m/K	32
Density of steel, kg/m^3	7020
Density of flux, kg/m^3	3000
Steel chemistry in mass %: C=0.17, Si=0.19, Mn=0.51, P=0.017, S=0.016.	

Table 1. Properties and conditions of the simulations

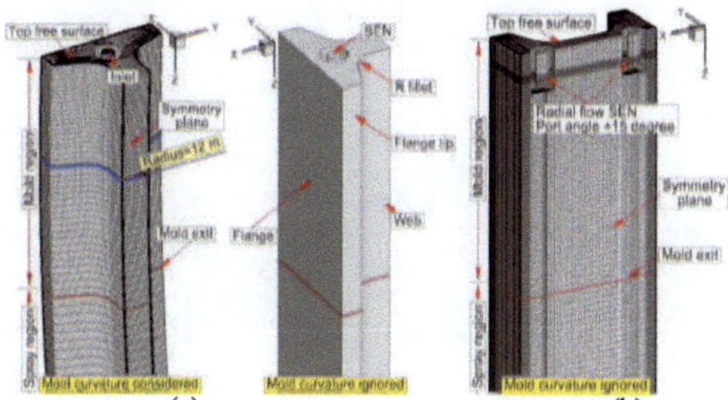

Figure 3. Geometrical model and mesh systems: (a) mold with the single-port straight SEN; (b) mold with the new designed three-port radial flow SEN.

Results and discussion

Meniscus behavior

Figure 4 shows the predicted level fluctuation, velocity magnitude and temperature distribution on the top free surface. With the straight SEN, the level fluctuation is less than 1 mm and the velocity magnitude is far from a reasonable interval. One of the reasons for the weak and calm top free surface is that such uniform steel feeding can reduce the impact of molten steel stream, and Lee et al [7] also have suggested that the flow near the meniscus is characterized by a low turbulence. With the radial flow SEN, the "self-braking" effect caused by two radial flow SENs provides a good flow stability at the web region, as shown in **Figure 4(a)**, and the level fluctuation, velocities and temperature at the top free surface of the mold are obviously larger than that of the straight SEN. The higher temperature and larger velocity magnitude are benefit for the function of mold flux. It implies the radial flow SEN has a better performance. However, the wave crest near the flange tip as shown in **Figure 4(a)** which generates the thinnest liquid flux layer may prevent the mold powder from penetrating into the gap between the mold and the solidifying shell, which is not favorable for the mold lubrication, and the maximum temperature of the top free surface appears at the web center as shown in **Figure** 4(c). At such high temperature locations, mold flux might be burned, which will increase the danger of more reoxidation and reducing the steel cleanliness. Considering the possible disadvantages, the submergence depth of the radial flow SEN with a positive angle should be deeper than 11.5 cm and the radial flow SEN with a negative angle could be used at a faster casting speed.

Figure 5 shows that the mold curvature causes the asymmetrical distributions of velocity and temperature at the top free surface. The top-surface near the outer radius is more active, which indicates that there is a stronger recirculation near the outer radius, because mold curvature and the gravitational force induce the molten steel stream trajectory closer to the outer radius.

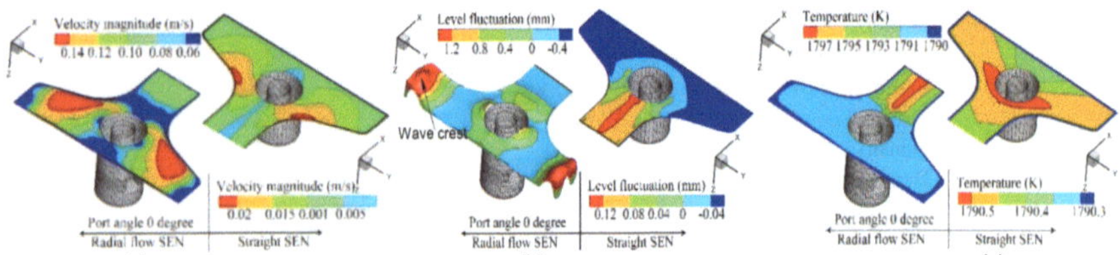

Figure 4. Top free surface characteristics: (a) level fluctuation; (b) velocity magnitude; (c) temperature.

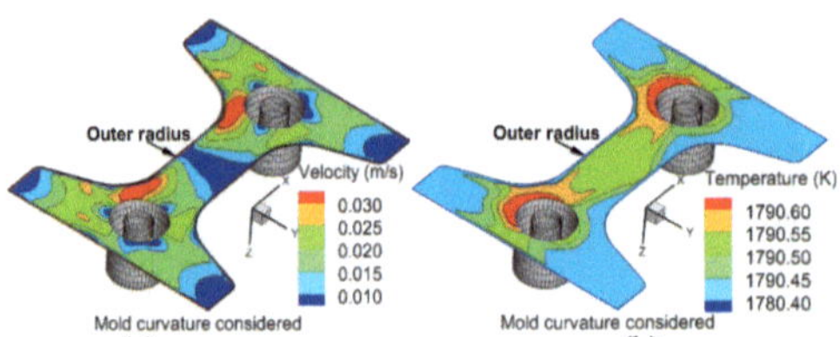

Figure 5. Velocity magnitude distribution and temperature distribution at the top free surface with considering mold curvature: (a) velocity magnitude distribution; (b) temperature distribution.

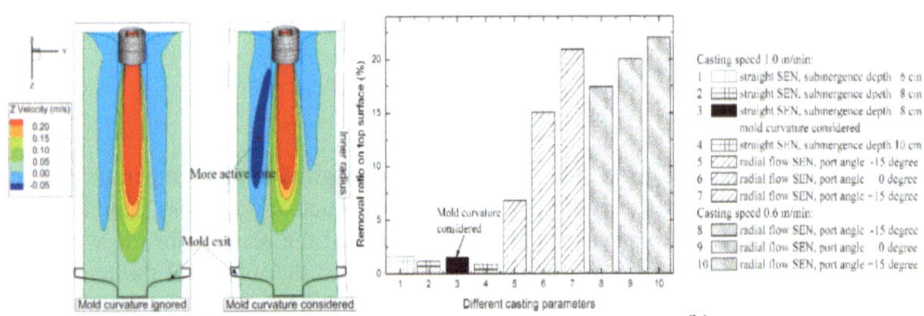

Figure 6. (a) Z-velocity distribution on the slice X=0.15 m without/with considering mold curvature; (b) Inclusion removal ratio with different casting parameters.

In **Figure 6(a)**, it can be seen that the molten steel jet is a slightly closer to the outer radius of the curved mold, which causes a more active zone near the outer radius and a possible dead zone near the inner radius. In the beam-blank mold, dead zone will lead to the aggregation of inclusions which are harmful to the final product. The inclusion removal ratio at the top free surface is shown in **Figure 6(b)**. With the straight SEN, thanks to the molten stream carries most inclusions into the deep mold, the removal ratio is very low and the fast flow from the nozzle also inhibits the flotation of inclusions. While the radial flow SEN appears to provide a more favorable conditions for the flotation of inclusions. With the port angles varying from -15 to +15 degree, the removal ratio of inclusions increases. With the same port angle, lower casting speed leads to a better inclusions removal efficiency.

Flow field in the mold

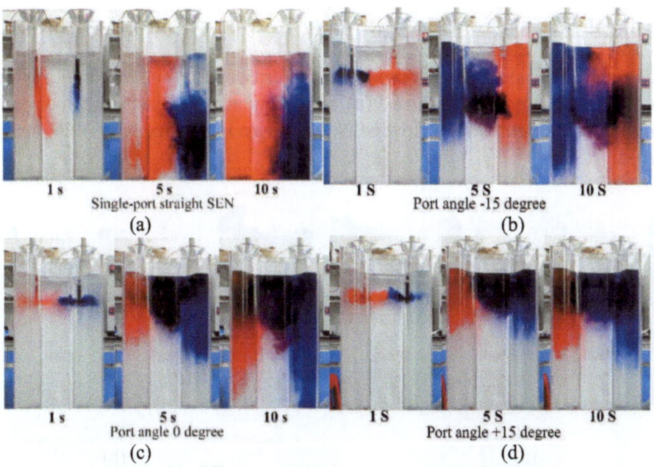

Figure 7. Flow patterns obtained with different nozzles at casting speed 1.0 m/min: (a) straight SEN: (b) radial flow SEN, port angle -15 degree; (c) radial flow SEN, port angle 0 degree; (d) radial flow SEN, port angle +15 degree.

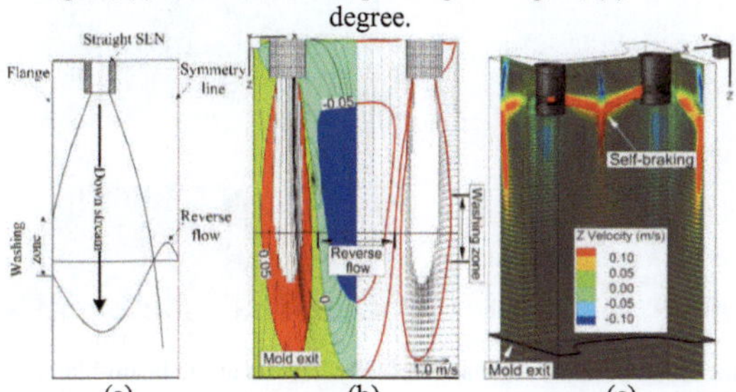

Figure 8. Flow pattern in the beam-blank mold: (a) characteristics of straight SEN, locating in an off-center position; (b) flow field obtained with straight SEN; (c) self-braking.

Figure 7 shows the flow patterns in a water model of the beam-blank continuous casting mold with different SENs at the same casting speed 1.0 m/min. It can be seen that using the radial flow SEN, with the port angles varying from -15 to +15 degree, the active region obviously moves upward to the meniscus. Zhang and his coworkers first proposed the "self-braking" for a single SEN with four ports [10], and in the present study, a clear "self-braking" effect as **Figure 8(c)**

shown has been achieved in the web region by two radial flow SENs. After the fluid leaves the SEN and enters the mold, the flow out from two SENs' ports will occur in the web center. After collision of the two streams, a stream will flow upward, and another downward. The upward stream will help to obtain an active meniscus, the upward and downward stream will augment the temperature of the meniscus and inside the mold.

In the beam-blank mold, the SEN is closer to the flange as **Figure 8(a)** indicates, so the "reverse flow" will not appear near the flange and there is a "washing zone". In **Figure 8(b),** on the wide face center-plane of the mold, two swirling centers at the same height can be found, which is a typical single-roll flow pattern and not favorable for the mold metallurgical behavior.

Temperature distribution and solidification

At the web center, the temperature is obviously higher because the "self-braking" effect confines superheat to a narrow region as **Figure 9(a)** shows, which increases the risk of breakouts and defects. Using the asymmetry radial flow SEN, as shown in **Figure 9(b)**, this phenomenon could be able to avoid.

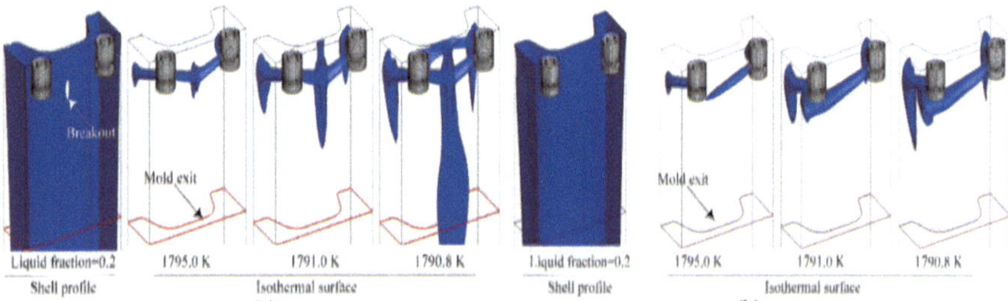

(a) (b)

Figure 9. Solidified shell profile and isothermal-surfaces: (a) symmetry; (b) asymmetry.

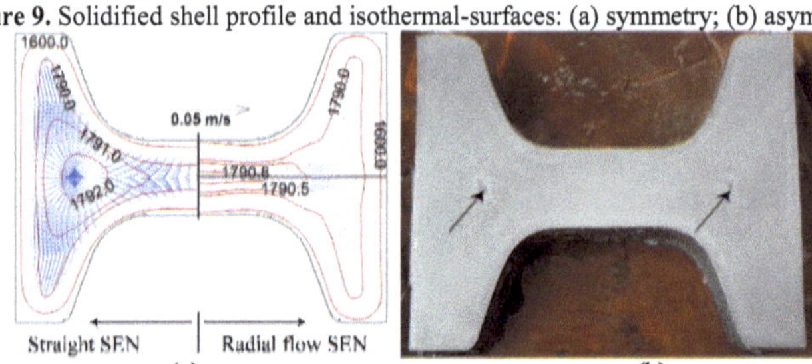

(a) (b)

Figure 10. (a) Contours of temperature field and velocity vectors at the beam-blank mold exit at casting speed 1.0 m/min, straight SEN with submergence depth 6 cm and radial flow SEN with port angle 0 degree; (b) Macroetch of a transverse section of beam-blank.

At the mold exit, the cross secion temperature distribution is presented in **Figure 10(a)**. With the straight SEN, there is a high temperature gradient, which could cause the center looseness shown in **Figure 10(b)**. With the radial flow SEN, the temperature at the mold exit is more uniform.

Figure 11 shows the solidified shell profile at the mold exit, and it can be clearly seen that there are many differences. With the straight SEN, the shell thickness is thinner at the fillet but thicker at the web, which can be validated by **Figure 12**. According to the experience in a few steelworks, the breakouts usually occurred at the fillet, which further confirms the present calculation results. The radial flow SEN could decrease the risk of breakouts caused by rupture of skin at the fillet. With the symmetry radial flow SEN, the shell thickness is thinner at the web center, as shown in **Figure 11(b)**, but it can be improved with the asymmetry radial flow SEN.

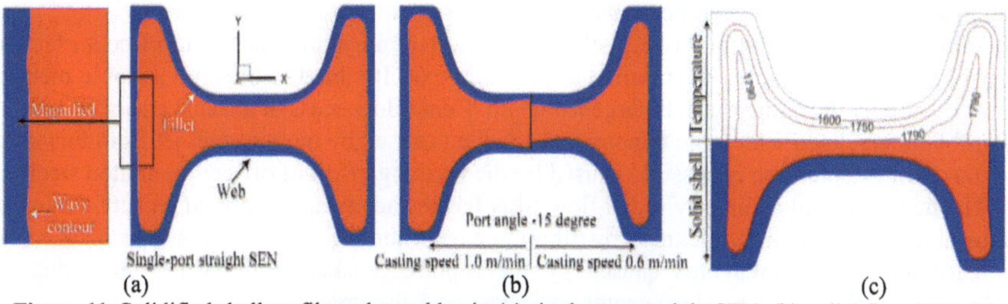

Figure 11. Solidified shell profile at the mold exit: (a) single-port straight SEN; (b) radial flow SEN; (c) asymmetry radial flow SEN.

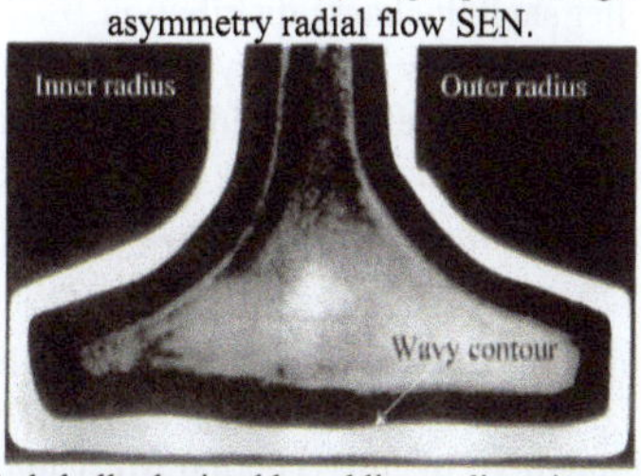

Figure 12. Photographs of solidified shell, obtained by adding radioactive tracer to the liquid pool [错误！未找到引用源。].

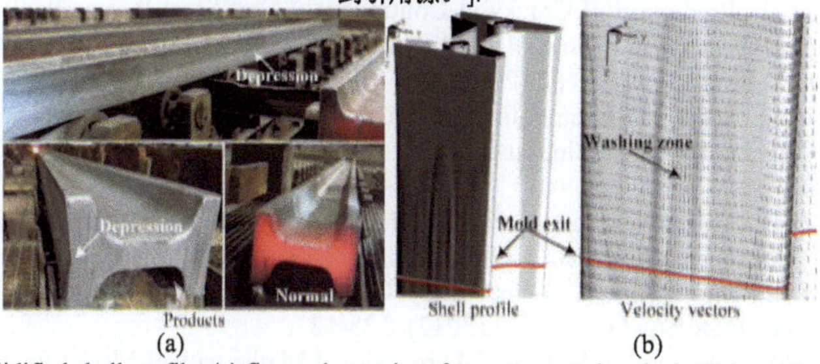

Figure 13. Solidified shell profile: (a) flange depression, from some steelworks in China; (b) calculated shell profile, an average value of solid fraction 0.5 was chosen in order to have a better view.

Using the single-port straight SEN, the depression sometimes occurs at the flange, as shown in **Figure 13(a)**, which should be caused by the structure of straight SEN. The characteristics of this nozzle locating in an off-center position as shown in **Figure 7(a)**, and the washing effect results in shell erosion in local thin spots and cause a "wavy contour" as shown in **Figures 11(a)** and **12(b)**. The flange depression indicates that the "wavy contour" shell could be unable to withstand the stress of solidified shell shrinkage process and could be pushed back by the mold inner surface or the retaining rolls in the secondary cooling zone. The simulated shell profile in the beam-blank mold is shown in **Figure 13**(b), and how the impinging molten steel jet pouring from the straight SEN affects the shell distribution of the flange can be found. Using the radial flow SEN, the shell thickness at the flange is very uniform and may avoid the flange depression.

Conclusions

The following conclusions can be drawn from this research:
(1) With the single-port straight SEN, the major shortcomings are inactive top free surface and non-uniform shell thickness distribution at the mold exit. It's hard to obtain a suitable meniscus status by simply changing the submergence depth. Center looseness of a transverse section of the beam-blank is related with large temperature gradient caused by molten steel jet from this nozzle and the flange depression is possibly caused by the washing effect of off-center molten steel jet
(2) The new designed symmetry radial flow SEN has the advantage to obtain an active meniscus status and a uniform shell thickness distribution except the web center. This type nozzle could decrease the risk of breakouts caused by the rupture of skin at the fillet and reduce the probability of flange depression, center looseness and the crack of strand. The new designed asymmetry flow SEN has a better shell thickness distribution.
(3) The "self-braking" effect caused by symmetry radial flow SEN provides a good stability of flow at the meniscus, but this braking effect could confine a large amount of superheat to a small region, which can be controlled with asymmetry radial flow SEN.

References

1. L.C. Hibbeler et al., Thermal-Mechanical Modeling of Beam Blank Casting, *Iron Steel Tech.*, 6 (2009), 60-73.
2. W. Chen, Numerical Simulation of the Thermo-Mechanical Process for Beam Blank Continuous Casting, *Acta Metall. Sin.*, 20 (2007), 241-250.
3. J.-E. Lee et al., Prediction of Cracks in Continuously Cast Steel Beam Blank through Fully Coupled Analysis of Fluid Flow, Heat Transfer, and Deformation Behavior of a Solidifying Shell, *Metall. Mater. Trans. A*, 31 (2000), 225-237.
4. H.L. Xu et al., Thermal Behaviour of Moulds with Different Water Channels and Their Influence on Quality in Continuous Casting of Beam Blanks, *Ironmaking Steelmaking*, 37 (2010), 380-386.
5. W. Luo et al., Improvement of Water Slot Design for Beam Blank Casting Mould, *Ironmaking Steelmaking*, 40 (2013), 582-560.
6. W. Luo et al., Improvement to Secondary Cooling Scheme for Beam Blank Continuous Casting, *Ironmaking Steelmaking*, 39 (2012), 125-133.
7. J.-E. Lee, J.-K. Yoon and H.N. Han, 3-Dimensional Mathematical Model for the Analysis of Continuous Beam Blank Casting Using Body Fitted Coordinate System, *ISIJ Int.*, 38 (1998), 132-141.
8. M.G. Xu and M.Y Zhu, Transport Phenomena in a Beam-Blank Continuous Casting Mold with Two Types of Submerged Entry Nozzle, *ISIJ Int.*, submitted.
9. F.M. Najiar, B.G. Thomas and D.E. Hershey, Numerical Study of Steady Turbulent Flow through Bifurcated Nozzles in Continuous Casting, *Metall. Mater. Trans. B*, 26 (1995), 749-765.
10. Y.F. Chen et al., Water Modeling of Self-Braking Submerged Entry Nozzle Used for Steel Continuous Casting Mold, *JOM*, 64 (2012), 1080-1086.

Advances in the Science and Engineering of Casting Solidification
Edited by: Laurentiu Nastac, Baicheng Liu, Hasse Fredriksson, Jacques Lacaze, Chun-Pyo Hong, Adrian V. Catalina, Andreas Buhrig-Polaczek, Charles Monroe, Adrian S. Sabau, Roxana Elena Ligia Ruxanda, Alan Luo, Subhayu Sen, and Attila Diószegi
TMS (The Minerals, Metals & Materials Society), 2015

Effects of Ce on the Thermal Stability of the Ω Phase in a Cast Aluminum Metal Matrix Composite

F. Melotti[1], A. Dustan[2], T. Hirst[3], W. D. Griffiths[1]

[1] School of Metallurgy and Materials, University of Birmingham, Edgbaston, Birmingham, UK, B152SE
[2] Aeromet International PLC, Cosgrove Close, Worcester, UK, WR38WA
[3] Controls and Data Services, Shaftmoor Lane, Hall Green, Birmingham, UK, B288SW

Keywords: cerium, aluminum alloys, metal matrix composite

Abstract

This work assesses the effect of Ce additions on the thermal stability of the Ω phase in an aluminum alloy matrix composite. Several additions of Ce, up to 1.5 wt. %, were made to an AA205 alloy, a novel MMC, and the thermal stability of the Ω phase tested at 230 °C, for up to 100 hours. Microstructural evaluation was carried out with optical microscopy and scanning electron microscopy (SEM). In addition, hardness tests were carried out during the 100 hours. The results showed that when the reinforcement particles were present in the alloy, the Ce did not have any beneficial effects on mechanical properties. In particular, the Ce poisoned the grain refinement of the alloy by nucleating around the TiB_2 particles. Energy dispersive X-ray spectroscopy (EDS) suggested the Ce-phase to be $CeTi_2Al_{20}$.

Introduction

Aluminum matrix composites (Al-MMC) are widely used in aerospace industries for their high strength and low density. Currently, the working temperatures of aluminum alloys are below 200 °C, so the goal of several researchers is to find an alloy which is able to operate at higher temperatures. Additions of transition metals such as scandium or titanium have been reported [1, 2] to improve the thermal stability of the alloy.

Rare earth metals, such as cerium and erbium, have also been recently studied with promising results [3-6]. Cerium has been reported to act as a grain refiner, it segregates into the dendrite arms giving strength and delaying recrystallization [7]. However, so far, the effects of cerium have never been studied in an Al-MMC.

AA205 is an Al-Cu alloy with additions of reinforcing particles such as TiB_2. The strength of a 2xxx series alloy with a specific Cu:Mg ratio derives mostly from the Ω phase. Studies have shown that the stability of this phase is reduced by exposure to temperatures above 200 °C. In particular, Hutchinson et al. [8], using HRTEM, defined the stability of the phase to be proportional to the number of ledges; the more ledges, the faster the flux of Cu diffusing into the phase, transforming it into an incoherent phase.

This aim of this work was to study the effects of Ce on an Al-MMC, both at room temperature and at an elevated temperature of 230 °C. Following this the microstructure was analyzed to assess the effect of Ce on grain refinement.

Experimental procedure

Four different alloys were cast by varying the composition of an AA205 alloy, (denoted alloy A). Alloys B, C and D were cast with additions of Ce of 0.2 wt. %, 0.5 wt. % and 1.5. %. Alloy E was cast without TiB_2 but with the addition of 0.5 wt. % of Ce. The compositions of the five alloys were analyzed by X-ray Fluorescence (XRF) and are reported in Table 1.

Table I: Chemical composition, given as ratios, of an Al-Cu alloy.

Alloy	Ce	TiB_2	Al
A	0	1	rem
B	0.2	1	rem
C	0.5	1	rem
D	1.5	1	rem
E	0.5	0	rem

The alloys were melted in an induction furnace and cast into sand molds to obtain test bars of 10 mm diameter and 100 mm length. A solution heat treatment was carried out, followed by quenching into cold water to achieve a super saturated solid solution, followed by precipitation hardening to a peak-aged condition.
Samples were also machined to 10 mm height and 10 mm diameter cylinders and held at 230 °C for up to several hundred hours in an air circulating furnace. Samples were stored in a refrigerator, mounted in bakelite and then polished. Optical microscopy (Zeiss Axioskop-2) and SEM (Jeol 6060, Phillips XL 30) were used to investigate microstructural changes occurring during the high temperature holding period. Kellers reagent and nitric acid were used to etch the samples.
Micro-hardness analysis was performed using a Zwick/Roell Indentec at 10 kg load; with more than five measurements performed on each sample. Means and standard deviations were calculated for each sample.
Differential scanning calorimetry (DSC) samples were machined to 5 mm diameter and 0.5 mm height, with around 28 mg mass. A Perkin-Elmer DSC 7 was used to measure heat evolution for these samples over a temperature range of 50 °C to 400 °C.

Results

Age hardening

Aluminum metal matrix composites are characterized by their fast response to age hardening, compared to standard aluminum alloys [9]. Figure 1 shows the variation in hardness during the precipitation process of the different alloys (A to E), aged at 170 °C. The response curves show that all the alloys increase in hardness by about 20 to 40 %. Alloy A at peak-aged condition had a hardness of ~162 VHN. With an addition of 0.2 wt. % and 0.5 wt. % of Ce, the peak hardness was ~155 VHN. Addition of 1.5 wt. % of Ce decreased the peak-age hardness to 135 VHN. Alloy E, which was the TiB_2-free alloy with Ce addition, showed a hardness of ~155 VHN. As mentioned previously, the reinforced alloys showed a faster response to the age hardening process followed by a plateau where the value for the hardness was constant. However, the non-reinforced alloy showed a slow response, with a peak age time of 24 hours, followed by a rapid drop in properties.

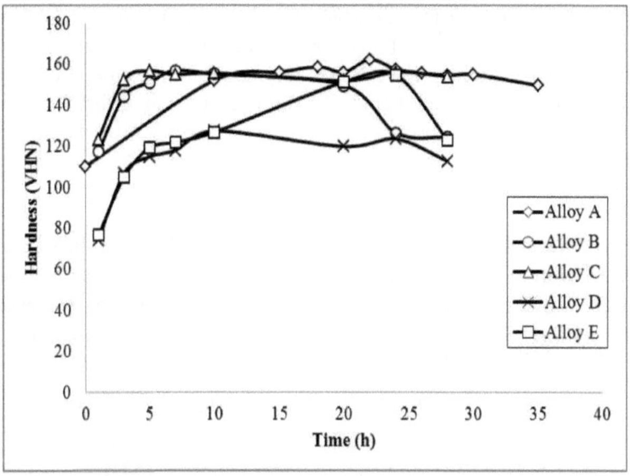

Figure 1: Hardness curves for alloy A, B, C, D, and E during the precipitation hardening process.

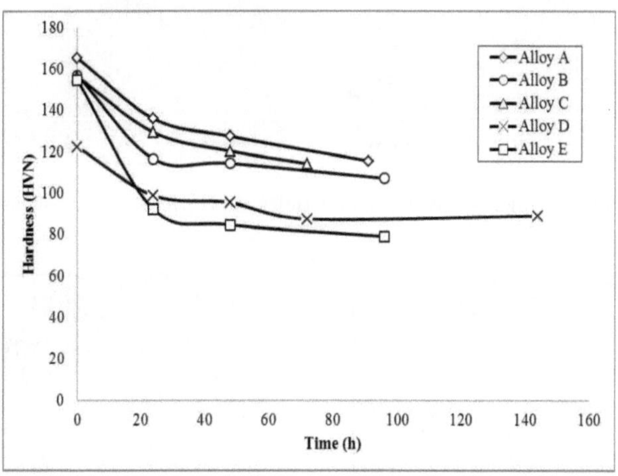

Figure 2: Hardness curves for alloy A, B, C, D and E during the holding at 230 °C

Elevated temperature exposure

Samples, in their peak-aged condition, were held at 230 °C, and their hardness was then measured for up to 100 hours holding time. Figure 2 shows the variation in hardness due to

holding at elevated temperature. The response curves showed that all alloys lost up to 30 % in hardness in the first 24 hours, suggesting that the main strengthening precipitate significantly decreased in all of the alloys studied.
The properties of alloys A, C and D decreased after 24 hours by almost 18 %. The hardness of alloy B decreased by 25%. Alloy E, however, showed a 40 % reduction in properties. This suggested that the reinforced alloys have better thermal stability compared to a non-reinforced alloy. However, this plot also showed that the Ce did not have any effects on the thermal stability of a reinforced alloy.

Microstructural analysis

Figure 3 shows the difference in the microstructures of alloys A and B. In alloy A, the grain sizes were around 30 µm, with a globular shape. On the contrary, alloy B showed a dendritic shape microstructure, typical of a non-reinforced alloy. The grain size greatly increased, from 30 µm to ~500 µm, suggesting that the grain refinement mechanism had been poisoned. In particular, in alloy B, a precipitate was observed which was not present in alloy A.

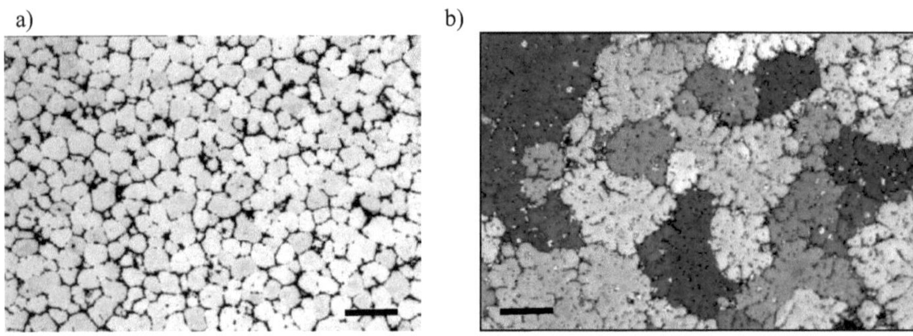

Figure 3: Optical microscopy of Alloy A (left) and B (right). Scale bars are 100 µm.

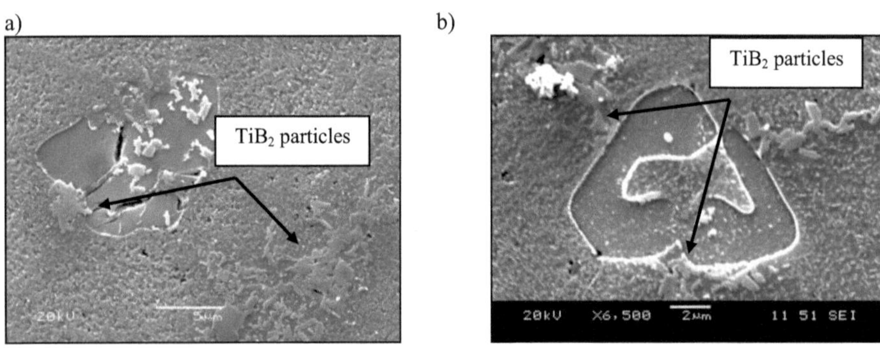

Figure 4: SEM micrographs of the CeTi2Al20 phase in alloy B.

Figure 4 shows SEM images of the Ce-bearing phase found in alloy A to D. This precipitate nucleated around the TiB_2 particles, apparently preventing their action as a grain refinement. The

size distribution of the phase was between 5 to 15 μm. EDS analysis was carried out over three samples, and ten samples each which suggested that the stoichiometry of the phase was $CeTi_2Al_{20}$. (The mean of the ten measurements has been reported in Table II.)
Figure 5 shows a DSC scan of alloys A and C. The plot shows that the dissolution temperature of the phase is around 590 °C and also that the Ce-bearing precipitate did not go into solution during the solution heat treatment process.

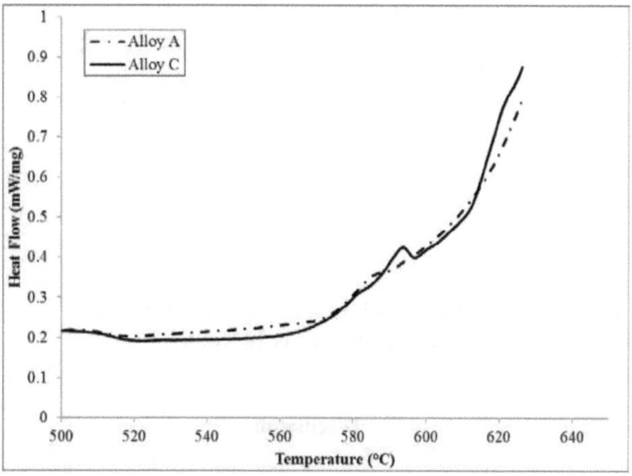

Figure 5: DSC plots for alloy A and C after solution heat treatment condition. The scanning rate used is 20 $Kmin^{-1}$

Table II: Chemical composition of the intermetallic phase found in alloys A, B, C and D. Results are the mean between 30 analyses.

Element	Weight (%)	Atomic (%)
Al	66	84
Ti	12	9
Ce	17	4
Cu	1	<0.5

<u>Differential scanning calorimetry</u>

Figure 6 shows the DSC traces of alloys A and B after solution heat treatment. As suggested in the literature, the peaks were related to the nucleation and dissolution of the GP zones and the Ω phase [10]. The addition of Ce in alloy B shifted the exothermic peak associated with the Ω nucleation to lower temperatures. In alloy A the shift was from 320 °C to 290 °C, which suggests that the precipitation of the Ω phase had been promoted.

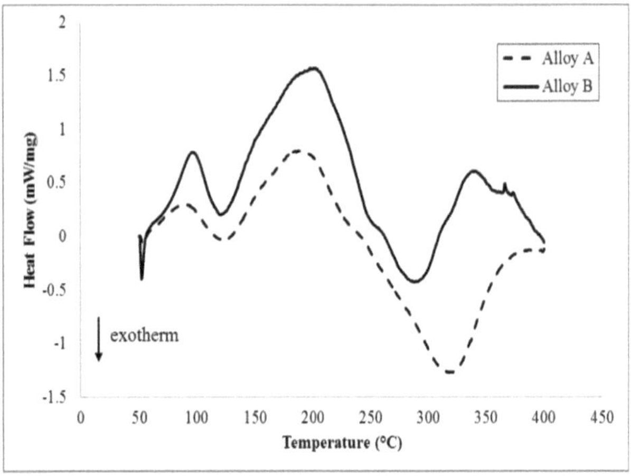

Figure 6: DSC plots for alloys A and B after solution heat treatment. The scanning rate used is 20 $Kmin^{-1}$.

Discussion

Stability of the Omega phase

In the Ce-bearing alloys, not all of the Ce formed intermetallic precipitates, as some of it was still in solution and helped to nucleate the Ω phase. This was suggested by the shift of the nucleation peak of the Ω phase to a lower temperature (see Figure 6), which indicated a promotion of the phase. The mechanism behind this process is still uncertain, but it can be deduced that the Ce addition promoted the nucleation of the GP zones and it acted as a heterogeneous nucleation site on the {1,0,0} and {1,1,1} planes to create finer and denser Ω and θ' phases.

The stability of the Ω phase has been the topic of many researches in the past years, and still it is unresolved. Several authors showed that the Ω phase was not stable at temperatures above 200 °C [11], while others state that it was stable for a long period of time at the same temperature [12].

It is suggested that the Ω phase loses stability because a flux of Cu atoms diffuses into the precipitate through ledges, and so the Ω phase increases in size becoming semi-coherent and losing strength. Previous work suggested that the effect of Ce was to restrict or prevent this Cu diffusion, so the Ω phase remains coherent for a longer time or at elevated temperatures. In this work it has been shown that the Ω phase completely disappears after a few hours at 230 °C and that the Ce, even if a small quantity is in solution, does not provide any high temperature stability.

The Ce- bearing phase

In this work, a new ternary precipitate has been found. After statistical analysis, the atomic composition was determined to be $CeTi_2Al_{20}$, with only a trace amount of Cu. Previously, Xiao

et al. [13] reported a Ce-Cu-bearing precipitate, $Al_xTi_6Ce_3Cu$. The Ce-bearing phase reported here was predicted and prepared by Niemann and Jeitschko [14] for the first time in 1995. However, since then, only this work was able to show the nucleation of this ternary phase in an alloy. The TiB_2 particles act as grain refiners because they nucleate the α-aluminum due to the formation of a $TiAl_3$ on the TiB_2 face. The poisoning mechanism is still unclear, however, the intermetallic prevents the globular growth of the alloy, creating bigger (~ 200%) grain size.
The peak-aged hardness of all alloys, besides alloy D, was similar, (with a difference of 3%), which suggested that the TiB_2 particles had only minor influence on the hardness. As suggested in the literature, MMC reinforced with TiB_2 showed a segregation of Cu atoms and Cu bearing phases to the ceramic particles [9, 15]. This segregation decreases the Cu atoms in solid solution, decreasing the volume fraction of strengthening precipitates. However, the reason for such a minor influence of the TiB_2 particles on hardness is still unclear.

Conclusions

Additions of Ce, up to 0.5 wt. %, to an Al metal matrix composite with TiB_2 reinforcing particles do not have any effects on the hardness. At higher amounts, such as 1.5 wt. %, the Ce addition does not have any beneficial effect on hardness. At elevated temperatures, above the temperature for stability of the Ω phase, the Ce does not provide any benefit to the alloy. In conclusion, the Ce poisons the microstructure of the alloy, causing larger and dendritic grains. The poisoning is due to a new phase, of composition $CeTi_2Al_{20}$.

Acknowledgments

The authors are grateful for financial support from the Technology Strategy Board. The authors also wish to acknowledge the technical assistance of by A. J. Caden.

References

1. Yin, Z., et al., "Effect of minor Sc and Zr on the microstructure and mechanical properties of Al–Mg based alloys." *Materials Science and Engineering: A*, 2000. 280(1): p. 151-155.
2. Fuller, C.B., D.N. Seidman, and D.C. Dunand, "Mechanical properties of Al (Sc, Zr) alloys at ambient and elevated temperatures." *Acta materialia*, 2003. 51(16): p. 4803-4814.
3. Chen, W., J. Kong, and W.J. Chen, "Effect of rare earth Ce on the microstructure, physical properties and thermal stability of a new lead-free solder." *Journal of Mining and Metallurgy Section B-Metallurgy*, 2011. 47(1): p. 11-21.
4. Wen, S.P., et al., "Synergetic effect of Er and Zr on the precipitation hardening of Al-Er-Zr alloy." *Scripta materialia*, 2011. 65(7): p. 592-595.
5. Xiao, D., et al., "Effect of rare earth Yb addition on mechanical properties of Al–5· 3Cu–0· 8Mg–0· 6Ag alloy." *Materials science and technology*, 2007. 23(10): p. 1156-1160.
6. Yu, X.X., et al., "Effects of Trace Addition of Cerium on Microstructure and Mechanical Properties of Novel Al-Cu-Li Alloy." *Rare Metal Materials and Engineering*, 2014. 43(2): p. 495-500.
7. Xiao, D.H., et al., "Effect of rare earth Ce addition on the microstructure and mechanical properties of an Al–Cu–Mg–Ag alloy." *Journal of Alloys and Compounds*, 2003. 352(1–2): p. 84-88.

8. Hutchinson, C., et al., "On the origin of the high coarsening resistance of Ω plates in Al–Cu–Mg–Ag Alloys." *Acta materialia*, 2001. 49(14): p. 2827-2841.
9. Siddhalingeshwar, I.G., et al., "Effect of mushy state rolling on age-hardening and tensile behavior of Al–4.5Cu alloy and in situ Al–4.5Cu–5TiB2 composite." *Materials Science and Engineering: A*, 2011. 528(3): p. 1787-1798.
10. Zhang, J.-b., et al., "Identification of thermal effects involved in DSC experiment on Al-Cu-Mg-Ag alloys with high Cu:Mg ratio." *International Journal of Minerals, Metallurgy, and Materials*, 2011. 18(6): p. 671-675.
11. Ringer, S.P., et al., "Precipitate stability in Al-Cu-Mg-Ag alloys aged at high temperatures." *Acta Metallurgica et Materialia*, 1994. 42(5): p. 1715-1725.
12. Garg, A. and J.M. Howe, "Nucleation and growth of Ω phase in Al-4.0 Cu-0.5 Mg-0.5 Ag alloy—An in situ hot-stage TEM study." *Acta Metallurgica et Materialia*, 1991. 39(8): p. 1925-1937.
13. Xiao, D.H., J.N. Wang, and D.Y. Ding, "Effect of minor cerium additions on microstructure and mechanical properties of cast Al-Cu-Mg-Ag alloy." *Materials science and technology*, 2004. 20(10): p. 1237-1240.
14. Niemann, S. and W. Jeitschko, "Ternary Aluminides AT2Al20 (A = Rare Earth Elements and Uranium; T = Ti, Nb, Ta, Mo, and W) with CeCr2Al20-Type Structure." *Journal of Solid State Chemistry*, 1995. 114(2): p. 337-341.
15. Pal, S., R. Mitra, and V.V. Bhanuprasad, "Aging behaviour of Al-Cu-Mg alloy-SiC composites." *Materials Science and Engineering a-Structural Materials Properties Microstructure and Processing*, 2008. 480(1-2): p. 496-505.

Advances in the Science and Engineering of

CASTING SOLIDIFICATION

An MPMD Symposium Honoring Doru Michael Stefanescu

Solidification Processing IV

Session Chair:

Attila Diószegi

Advances in the Science and Engineering of Casting Solidification
Edited by: Laurentiu Nastac, Baicheng Liu, Hasse Fredriksson, Jacques Lacaze, Chun-Pyo Hong, Adrian V. Catalina, Andreas Buhrig-Polaczek, Charles Monroe, Adrian S. Sabau, Roxana Elena Ligia Ruxanda, Alan Luo, Subhayu Sen, and Attila Diószegi
TMS (The Minerals, Metals & Materials Society), 2015

NON-METALLIC Ti OXIDES AND MnS/FeS_2 COMPLEX PRECIPITATION IN Ti-KILLED STEEL

Jieyun Chen, Dan Zhao, Huigai Li, Shaobo Zheng*

Shanghai Key Laboratory of Modern Metallurgy & Materials Processing, Shanghai University, Shanghai 200072, China
*Corresponding Author: sbzheng@staff.shu.edu.cn

Keywords: precipitation, sub-rapid solidification, Ti oxide, MnS/FeS_2

Abstract

Titanium deoxidized experiments can be carried in vacuum induction furnace by adding Ti-Fe alloy in molten steel to simulate strip casting. Sub-rapid solidification samples were obtained in the method of suing copper mold. The morphology, the chemical composition and the structures of nanometer precipitations were carried out to investigate by transmission electron microscope (TEM) with Energy Dispersive X ray Spectrum (EDX) and by collecting diffraction patterns with carbon extraction specimens. It has been found that titanium oxides were TiO monoclinic, Ti_4O_7 anorthic and TiO_2 orthogonal structure in one nanometer inclusion, as the composite oxide was precipitated MnS/FeS_2 cubic structure during sub-rapid solidification. Thermodynamic calculation analysis showed that it was possible to precipitate different kinds of nonstoichiometric TiOx. The solid solution between MnS/FeS_2 will precipitate on the surface of titanium oxides because of good coherency relationship.

Introduction

Microstructure what is influenced by the inclusions or precipitations is related to the mechanical performances closely in steel. It has been found that the process of deoxidation equilibrium establishes very rapidly, which provided a method to promote small size particle generation but restrict the particle growth [1].The product quality of rapid or sub-rapid solidification of molten metal depends mostly on the microstructure of the solidified materials, which is determined significantly by the solidification rate [2]. The simulated strip casting process with more than 10 K/s cooling rate is a non-equilibrium solidification process. Jacobson LA et al. produce supersaturated solid solution and nano-crystalline structure in this way [3].

Titanium has been used as a deoxidizer due to its strong affinity with oxygen. Considering compositional influences, additions of titanium often offer beneficial effect to metal, which effect has been assumed to be the formation of titanium oxides or nitride [4,5]. Industrially, Fe-Ti alloys are used as sources for Ti addition and the precipitation of inclusions in Ti deoxidized steel is a hot topic. In previous work, it has been found that TiO_2 and Fe_2SiO_4 grow together, and MnS precipitated on the outermost layer of the complex inclusions, which has been confirmed by crystallography analysis [6]. Microstructure and inclusion in Si-Mn-Ti deoxidized steel, and that MnS precipitated on the nucleus of oxide and surrounded has been investigated by Zhuo et al. [7].

The equilibrium between steel and inclusion can be estimated by thermodynamic calculation, which makes inclusions composition established [8]. The reaction in molten steel is quite different between common and sub-rapid solidification. But to some extent, it would be allowed to approximate sub-rapid solidification process to the thermodynamic equilibrium.

The aim of the present study is to investigate different kinds of titanic oxide inclusions with manganese sulfide. With TEM technology, various structures of titanic oxide inclusions and MnS/FeS_2 were got, and thermodynamic analysis for their formation sequence.

Experimental

The 5.18 kg industrial pure iron was melted at 1560℃ in a crucible with an induction arc heater under vacuum environment using an induction furnace. Ferro-titanium was added to the melt for the sake of deoxidization that declined the dissolved oxygen from 188.9 ppm to 17.9 ppm and component adjustment, which process made the temperature of molten to 1550 ℃. After five minutes of stabilization of the melt, a copper mold was used to get the 2mm thickness sample, then quenched in cold water quickly, which could solidify melt rapidly. With the help of Oxford PMI-MASTER PRO Spark Emission Spectrometer, the chemical composition of solidified as-cast steel in the present work is listed in Table 1. Moreover, the total N and O in steel were determined to 59ppm and 43ppm by LECO TC-436 N/O analyzer.

Table 1. Chemical Compositions of Sample by Ti Deoxidization

Ele.	Fe	C	Mn	Ti	Si	Al	Cr	Ni	Cu	S	P
wt%	99.74	0.0275	0.0897	0.0204	0.0050	0.0020	0.0569	0.0055	0.0030	0.018	0.017

The specimen for transmission electron microscopy (TEM) analysis was produced by cutting slices with the size of 10×10 mm^2, grinding and polishing to metallographic specimen at room temperature. Next the specimen was etched with the alcohol solution containing 3% nitric acid until the iron matrix was dissolved and surface grain structure could be observed. Carbon extraction replicas using the SBC-2 surface treating machine was prepared for FE-TEM observation. To avoid the possible detection of copper from the grid and the specimen holder, nitride net with diameter of 3 mm was adopted to load the carbon film containing inclusions and Be holder of TEM was used in this experiment. The dark-field scanning transmission electron microscopy (STEM) experiments and Energy Dispersive

X-ray Detector (EDX) mapping for investigating chemical compositions were conducted on a FE-TEM (TECNAI G2) instrument, operating at 200 kV with TEM point of 0.23nm and STEM resolution 0.19nm. Selected area diffraction (SAD) was used to distinguish different phases and to characterize the precipitates in the specimen and the orientations of the diffraction patterns were defined with respect to the electron beam.

Results and Discussion

The cooling rate was more than 10^2 K/s in the work which was calculated by measuring the second dendrite spacing in previous work [6]. In order to identify the morphology, compositions and constitute of precipitation were inspected for the presence of precipitates using TEM. Figure 1 shows a bright field image of typical complex precipitation and maps of different elements.

(a) (b) Ti (c) O

(d) Mn (e) Fe (f) S

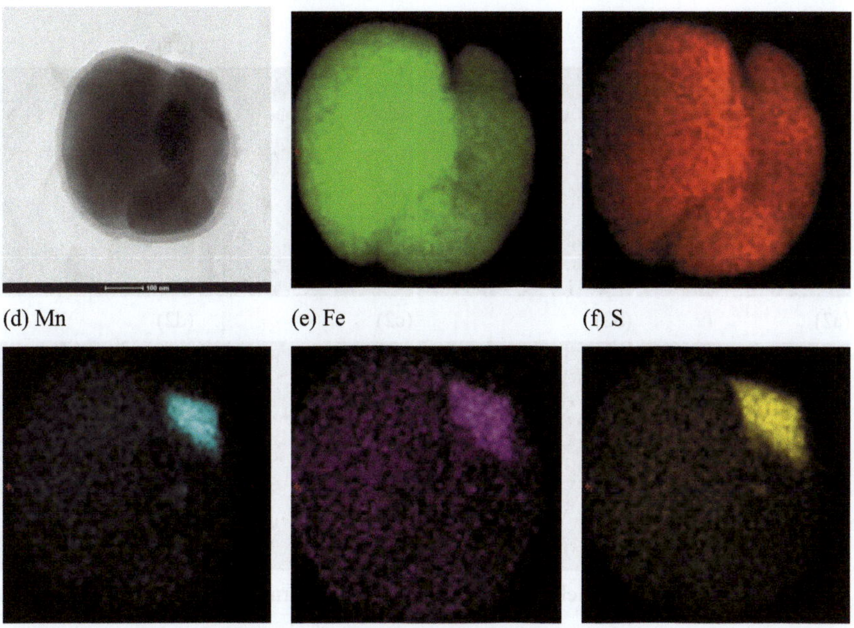

Figure 1. (a) Bright field image of precipitation;
(b-f) Element Ti, O, Mn, Fe, S of EDX spectrum in precipitation

Figure 1 shows the typical micrograph and EDX spectrum of a typical nonmetallic inclusion. It can be known, from the EDX spectrum, the size of this particle is about 400 nm and more than one part existed in this inclusion. Three parts of this particle are titanium oxide and one is sulfide. According to manganese or ferric sulfide is around the titanium oxide in the bright field image, it can be judged that manganese or ferric sulfide precipitated preferentially in association with Ti oxide phase. It is important to note that element carbon, has been removed for carbon film.

To get more information about this precipitation in detail, whose all diffraction patterns were got, which can confirm different structures through calibration. Dark field FE-TEM image and diffraction patterns of ellipsoids are shown in figure 2. Combined with the EDX mapping results and chemical analysis methods, the structure of the left part is Ti_4O_7 with a triclinic cell, whose beam direction is $[0\bar{1}0]$. Dark field image of the wrapping film and diffraction pattern are shown in Figure 2(a1) and 2(a2) respectively. In the same way, the other parts of this particle are orthorhombic TiO_2 in beam direction of $[01\bar{2}]$, monoclinic TiO in beam direction of $[\bar{1}\bar{1}1]$, cubic MnS or FeS_2 in beam direction of $[0\bar{1}0]$. Dark field images and diffraction patterns are shown in Figure 2(b1) to 2(d1) and 2(b2) to 2(d2) respectively.

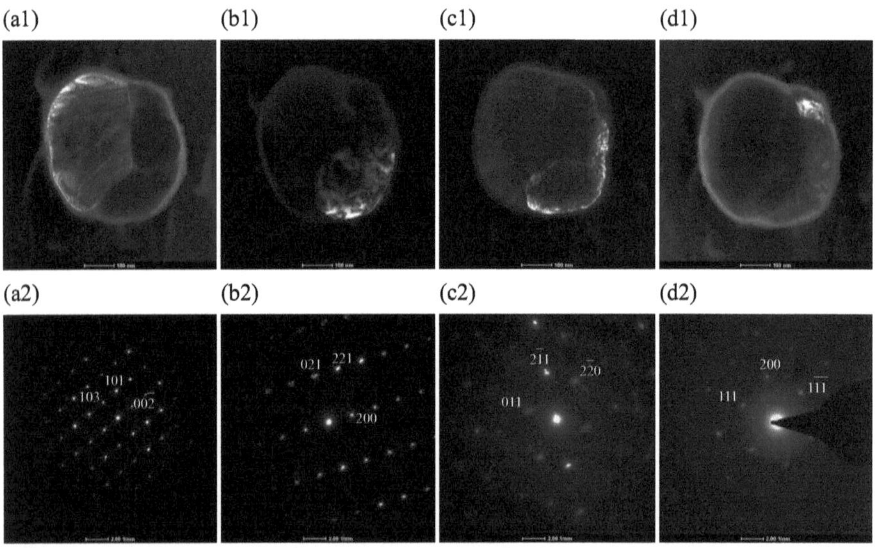

Figure 2. Dark field images and diffraction patterns of precipitation

(a1,2) image and patterns of Ti_4O_7 in beam direction of $[0\bar{1}0]$; (b1,2) image and pattern of TiO_2 in beam direction of $[01\bar{2}]$; (c1,2) image and pattern of TiO in beam direction of $[\bar{1}\bar{1}1]$; (d1,2) image and pattern of MnS or FeS_2 in beam direction of $[0\bar{1}0]$.

Based on the elements content shown in Table 1, the solidus and the liquidus of experimental low carbon steel in this article lead to the calculation that is 1807K for liquidus and 1781K for solidus by adopting the Equation (1) and (2)[9].

$$t_l = 1811-(65[\%C]+30[\%P]+25[\%S]+8[\%Si]+5[\%Mn]+1.5[\%Cr]+4.3[\%Ni]+2[\%Al]+90[\%N]+80[\%O]) \quad (1)$$

$$t_s = 1811-(175[\%C]+280[\%P]+575[\%S]+20[\%Si]+30[\%Mn]+7.5[\%Al]+6.5[\%Cr]+5[\%Mo]+1600[\%O]) \quad (2)$$

The precipitation temperature of various structure inclusions is important for judging precipitation sequence and thermodynamic calculation can provide a great reference. Previous study has summarized the Gibbs free energy of the reactions in the liquid steel. The relationship between the amount of precipitation of titanic oxide and manganese sulfide were shown in Equation (3), (4) and (5), respectively. [10][11] Where T is temperature (K).

$$[Ti]+[O]=TiO(s) \qquad \Delta G^0 = -367752+122.4T \text{ (J/mol)} \quad (3)$$
$$[Ti]+2[O]=TiO_2(s) \qquad \Delta G^0 = -678132+235.0T \text{ (J/mol)} \quad (4)$$
$$[Mn]+[S]=MnS(s) \qquad \Delta G^0 = -168822+98.87T \text{ (J/mol)} \quad (5)$$

The Gibbs free energy of the reactions in the liquid steel can be calculated with Equation (6).

$$\Delta G = \Delta G^0 + RT\ln K \quad (6)$$

Where the R is the gas constant, and the value is 8.314 J • mol^{-1} • K^{-1}. Combined with the data in reference [9][12], the activity coefficient and activity of the elements can be calculated with Equation (7) and (8). It should be mentioned that the activity coefficients was regarded remain unchanged in the certain range of temperature in this work.

$$lgf_i = \sum e_i^j [\%j] \quad (7)$$

$$a_i = f_i \cdot [\%i] \quad (8)$$

The precipitated temperature T can be obtained when count the pure substance with the activity of 1 in the reasonable extent. Assumes that ΔG=0, the precipitation temperature can be calculated as follow.

$$T_{TiO} = 1827\text{K} \qquad T_{TiO_2} = 1884 \qquad T_{MnS} = 1108\text{K}$$

According to the above result, precipitation temperature of TiO_2 is the highest and following the TiO, both of which prior to nucleate in liquid steel. Precipitation temperature of MnS is the lowest, which leads to the nucleation of MnS behind that of TiO_2 and TiO. Therefore, it can be known that the precipitation order is from TiO_2, TiO to MnS. Numbers of experiments showed that MnS could not precipitate in liquid low carbon steel, but only in the solid state. The reactions in actual liquid steel are in dynamic process, so different environments could impact the result of reactions.

The diffraction patterns of Mn-Fe-S were calibrated to cubic MnS or FeS_2 in beam direction of $[0\bar{1}0]$. The atoms of Fe and Mn are similar in size and property, and the lattice constants of cubic FeS_2 (a=b=c=5.405, α=β=γ=90°) are close to MnS (a=b=c=5.222, α=β=γ=90°). Combined with the distribution of three kinds of elements as Figure 1 (edf) shown, it could be considered that Fe atoms replace the Mn of cubic MnS in the cooling process after solidification reasonably.

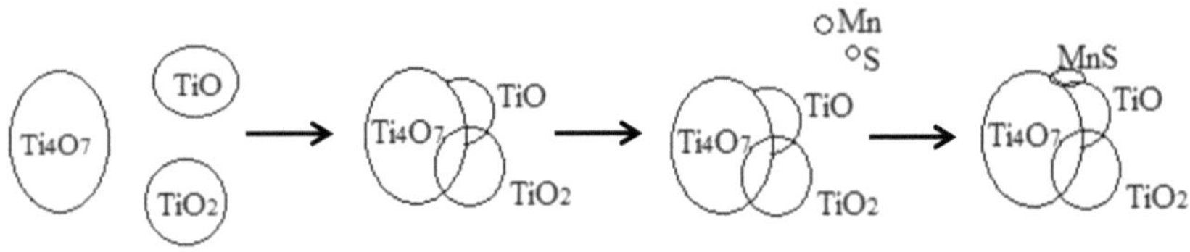

Figure 3. The speculated schematic of process about different kinds titanium oxides and MnS

In addition, crystal axis of Ti_4O_7 and MnS/FeS_2 are fitted perfectly, which can be estimated that MnS/FeS_2 attach to titanic oxide in the same crystal axis more easily, for lower interfacial energy and the good coherency relationship between them. It could be speculated that different kinds of small titanium oxides clumped together in molten steel based on thermodynamic analysis, as Figure 3 shown. But there did not show the evidences of Ti_4O_7, TiO_2 and TiO deformation process, which should be studied further.

Conclusions

(1) Based on the titanium deoxidation, various structure titanic oxides were existed in one inclusion with sub-rapid solidification, which are monoclinic TiO, anorthic Ti_4O_7 and orthogonal TiO_2, respectively.
(2) According to the thermodynamic calculation, it can be known that TiO_2 has priorities to nucleate, then following the TiO, MnS precipitated on the side of Ti_4O_7 due to low precipitation temperature.
(3) It could be estimated that MnS attach to the cluster after various titanic oxides clumping in the same crystal axis. But clear relationship and mechanism should be studied further.

Acknowledgements

The authors wish to express their grateful thanks to the National Natural Science Foundation of China (No.51104099) and (No. U1460103) for the financial support of this work.

References

1. H. Suito and H. Ohta, "Characteristics of Particle Size distribution in Early Stage of Deoxidation." *ISIJ International,* 46(1) (2006), 33-41.
2. F.J. Hong and H.H. Qiu, "Experimental Study on Rapid Solidification Process Using a Novel Ultrasound Technique." *Experimental Thermal and Fluid Science,* 30(1) (2005), 17-26.

3. L.A. Jacobson and J. McKittrick, "Rapid Solidification Processing." (Paper presented at the Materials Science and Engineering,11(8),1994), 355-408.
4. J.M. Gregg and H.K.D.H. Bhadeshia, "Solid-state Nucleation of Acicular Ferrite on Minerals Added to Molten Steel." *Acta Materialia,* 45(2) (1997), 739-748.
5. N. Kikuchi et al., "Effect of Ti De-oxidation on Solidification and Post-solidification Microstructure in Low Carbon High Manganese Steel." *ISIJ International*, 47(9) (2007), 1255-1264.
6. F.F. Sun et al., "Thermodynamic and Crystallographic Analysis on Complex Inclusions in Ti Deoxidized Low Carbon Steel." *TMS 2014 Supplemental Proceedings*, 511-516.
7. X.J. Zhuo et al., "Thermodynamic Calculation and MnS Solubility of Mn-Ti Oxide Formation in Si-Mn-Ti Deoxidized steel." *Journal of Iron and Steel Research, International*, 17(2)(2010), 10-16.
8. H. Suito and R Inoue, "Thermodynamic assessment of manganese distribution in hot metal and steel." *ISIJ international*, 35(3) (1995), 266-271.
9. J. X. Chen, *Handbook of Common Diagram and Data for Steelmaking* (Metallurgy Industry Press 1984).
10. S.S. Babu et al., "Development of Macro-and Microstructures of Carbon–manganese Low Alloy Steel Welds: Inclusion Formation." *Materials Science and Technology,* 11(2)(1995), 186-199.
11. O. Kubaschewski et al., *Spenscer: Materials Chemistry*, (6th, Pergamon Press, Oxford, 1993), 619.
12. X. G. Huang, *Theory of Metallurgy*, (Metallurgical Industry Press of China, 1998).

Advances in the Science and Engineering of Casting Solidification
Edited by: Laurentiu Nastac, Baicheng Liu, Hasse Fredriksson, Jacques Lacaze, Chun-Pyo Hong, Adrian V. Catalina, Andreas Buhrig-Polaczek, Charles Monroe, Adrian S. Sabau, Roxana Elena Ligia Ruxanda, Alan Luo, Subhayu Sen, and Attila Diószegi
TMS (The Minerals, Metals & Materials Society), 2015

Prediction of Surface Porosity Defects in High Pressure Die Casting

Mahdi Saeedipour[1], Simon Schneiderbauer[1], Stefan Pirker[1], Salar Bozorgi[2]

[1]Department of Particulate Flow Modelling, Johannes Kepler University;
Altenbergerstrasse 69; Linz, 4040, Austria
[2]LKR Leichtmetallkompetenzzentrum Ranshoden GmbH, Austrian Institute of Technology;
Postfach 26; Ranshofen, 5282, Austria

Keywords:
High pressure die casting (HPDC), Porosity, Jet breakup, Droplet impact, Droplet solidification.

Abstract

High pressure die casting (HPDC) is a novel manufacturing method with capability of mass production with higher accuracy. Porosity is one of the challenging defects in final product and may be affected by jet instability and atomization during injection phase. In case of atomization a large number of droplets with high velocity impinges the colder confining walls of the casting mold and might solidify consecutively. Different time scales of the impingement of the droplets and their solidification may result in heterogeneous structures near the surface of final product. A numerical framework using volume of fluid method (VOF) and an Eulerian-Lagrangian approach is established to simulate the liquid metal jet breakup and droplet formation during the injection phase. An analytical model for droplet impact on mold walls and solidification is studied and implemented in the numerical framework. The latter enables the prediction of porosity formation near the surface of final product.

Introduction

The high pressure die casting (HPDC) process is widely used for mass production of components based on aluminum, magnesium or zinc alloys. In this process molten metal is injected into the die at high speed (30-100 m/s) leading to a less physically understood filling behavior. Due to the turbulent nature of the flow, liquid jet may undergo severe instability, breakup and even atomization each of which could culminate to one or more typical casting defects. On the one hand, wavy disintegration and breakup of the liquid jet might result in cold shut defect in the final product; on the other hand a high degree of atomization strongly increases the porosity defect [1]. Although the influence of process parameters on the physics of HPDC is not fully understood so far [2], from an analytical viewpoint, the typical two phase flow with highly turbulent behavior is the main specification of its filling process. The numerical simulation of injection and filling of liquid metal in HPDC has been a challenging area of research for decades [1]. The global behavior of two phase flow in HPDC process is studied comprehensively in our previous works [1, 2, 3] in which the instability of the liquid jet, breakup and droplet formation around the surface of the jet have been investigated analytically, experimentally and numerically. But the focus of present study is on the porosity formation during HPDC process. Porosity is the formation of small voids in the final product of casting and is known as a challenging casting defect, which should be preferably avoided. Typically, there are two kinds of porosity defects in HPDC: (I) gas porosity and (II) shrinkage porosity. The former is a matter of fluid dynamics occurring as a result of air entrapment during the injection and filling phase while the latter occurs due to the variation of metal density in solid and liquid states during solidification process. The latter is not in the scope of present study.

Instabilities play an undeniable role in HPDC defects. It is revealed in [1] that due to the high inlet velocities in HPDC process, flow is in either so called "second-wind-induced" or "atomized" regime. There is no distinct boundary between these two regimes. Increasing the inlet velocity during injection leads to higher porosity because of the atomized nature of the flow and the droplet formation near the surface of the unstable liquid jet. The influence of other process parameters and global behavior of liquid jet on porosity formation has been studied in different works in the last decade.

In 2004, Lauki [4] studied the formation of microstructures during the injection phase in HPDC using optical microscopy and image analysis and suggested the thermal insulation during the injection phase to reduce porosity. Behoi [5] investigated the different flow regime in HPDC in a lab-scale set up using real time radiography. Lee et al. [6] did set of experiments to study Effect of process parameters on porosity distributions in HPDC product. It is demonstrated that by decreasing the inlet velocity and molten metal temperature, the total amount of porosity decreases. Rzychon and Kielbus [7] took the effect of wall thickness into account and showed that the increasing wall thickness results in reduced porosity. In 2012, Sadeghi and Mahmoudi [8] investigated the effect of die temperature on porosity formation. They showed that a decrease in the temperature range of die increases the probability of cold flow defects and air porosity.

From the numerical point of view, probably the earliest scientific effort to directly consider HPDC was conducted in early 1990s by Hu et al. [9] in which a computational model has been developed to simulate the flow, heat transfer and solidification during pressure die casting simultaneously. Recently by development of computational resources and due to the importance of turbulence modelling, CFD simulation of multiphase flow in HPDC process has been further developed. Homayonifar et al. [10] modified a VOF-based algorithm by using a concentration transport equation for calculating the air porosity distribution due to microbubbles entrapment and a coupled VOF-Lagrangian method for filling and splashing in HPDC. Later on, in 2010, Li et al. [11] have developed a numerical method to predict the air entrapment defects in high pressure die casting using VOF to track the micro bubbles entrapped in the liquid metal. Schneiderbauer et al. [2] studied the flow and filling characteristics during injection of liquid aluminum in HPDC using water analogy to capture the free jet regime changes and jet breakup for different nozzle types and inlet velocities. It is also revealed that the global flow behavior of the two phase flow in high pressure die casting can be captured by numerical simulation; however the simulation cannot resolve the drop formation with an acceptable computational effort. Following this work, Saeedipour et al. [1, 12] have developed an Eulerian-Lagrangian coupling framework to model the regime change and capture droplet formation around the liquid jet.

In present study the behavior of droplets around the liquid jet is studied with the focus on droplet-wall interaction as the main source of porosity formation near the surface. It is assumed that the droplets formed due to (i) instability and (ii) breakup might impinge the confining walls of the die. According to their local properties they are prone to be splashed or deposited at the wall depending on whether they are immediately solidified or not (?). In case of a large number of droplet impact on the wall with different time scales for solidification, heterogeneous microstructure can be formed over the die walls with air entrapped inside them. This is studied analytically as the probability of porous structure formation near the walls in HPDC process. In accordance with this analytical study for droplet impact and solidification, numerical simulations have been done to show the effect of droplet-wall interaction on the porosity formation in HPDC process.

Analytical model

A free liquid jet is always prone to encounter instabilities. The growth of disturbances along with the jet would cause instability, breakup and atomization. This leads to characterize the jet instability regimes according to flow properties in terms of dimensionless numbers: Reynolds (Re), Weber (We) and Ohnesorge (Oh) numbers.
As shown in Figure 1, based on instability theories and these dimensionless numbers, the free jet instability and breakup could form three main regimes: (I) Rayleigh breakup regime, (II) First and second wind-induced regimes and (III) Atomization regime.

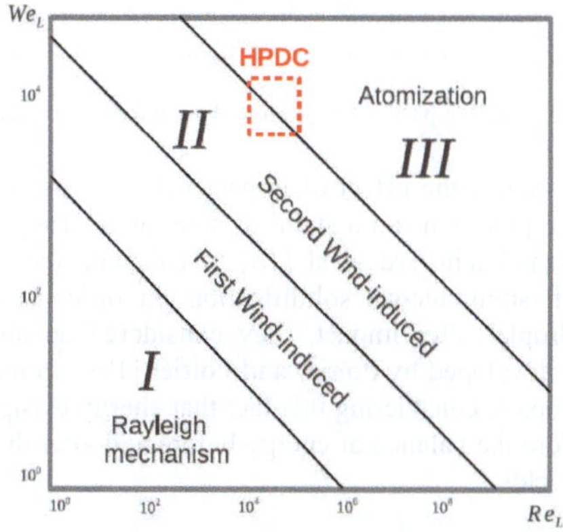

Figure 1. Liquid jet instability regimes and operation scope for HPDC process.

It can be deduced from the previous works [2, 13, 14] that the HPDC process seems to typically occur in a regime between second wind-induced and atomization in accordance with flow properties and process parameters. As stated in introduction, there is no distinct boundary between these two regimes and one may characterize the molten metal flow in HPDC process in a so-called "HPDC window" as depicted in Figure 1.
Considering the mechanisms of primary breakup of liquid jets, the higher inlet velocities and higher degree of turbulence lead to higher number of droplets formed near the jet surface and make the simulation and measurement a challenging issue. In addition, the larger number of produced droplets would increase the porosity defects in the final cast. These facts have motivated us to study an analytical model for porosity formation near the surface of the final product. For this purpose, an analytical study on droplet behavior after formation has been done as well as droplet-wall interaction toward the solidification process.
According to a precise study by Yarin [15], there are six different outcomes of a droplet impact on a dry surface: Deposition, Prompt splash, Corona splash, Receding breakup, Partial rebound and Rebound. There are also several works on the experimental side to characterize these different outcomes based on flow properties. Mundo et al. [16] have introduced splash parameters as a function of dimensionless numbers, Re and We as follows:

$$K = \sqrt{We\sqrt{Re}} \qquad with \;\; K_{cr} \cong 57.7 \tag{1}$$

This parameter could only predict whether the droplet is going to deposit or splash, but cannot distinguish between the other outcomes when the droplet is prone to splash. In other words, if the value of this splash parameter for a droplet approaching the nearest wall is below the critical value, the droplet will be deposited at the wall; otherwise it will be prone to other regimes such as splashing which are not in the frame of this work. Once deposited, the droplet usually spreads on the wall until its maximum diameter as shown in Figure 2 (left).

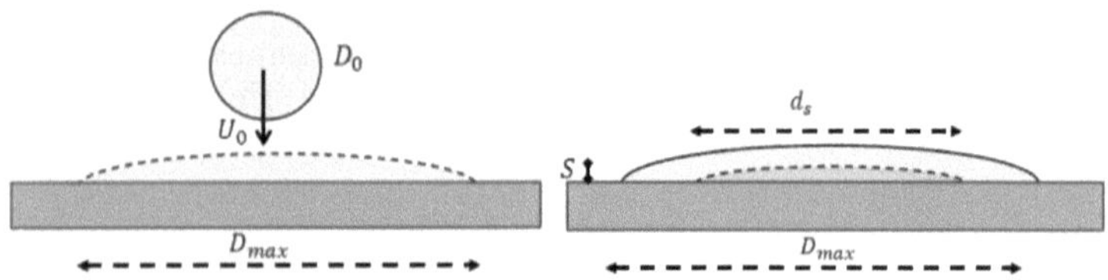

Figure 2. Schematics of droplet-wall impact and deposition (left) and droplet spread and solidified layer (right).

Aziz and Chandara [17] studied the effect of impact velocity and wall temperature during the impact of molten metal droplet (Tin) on a stainless steel plate. They also used an energy-based model developed by Passandideh-Fard et al [18] to calculate the maximum diameter of the droplet after impact with simultaneous solidification. In order to estimate the thickness of solidified layer in the droplet after impact, they considered an analytical solution for one-dimensional heat transfer developed by Poirier and Poirier [19]. In their model they equated the energy before and after impact, considering this fact that energy dissipates during droplet impact and solidification. Therefore the balance of energy before and after droplet impact (at the instant of maximum extension) yields:

$$KE_1 + SE_1 = KE_2 + SE_2 + W + \Delta KE \tag{2}$$

where (KE) is the kinetic energy, (SE) is the surface energy of the droplet, (W) is the work done in deforming the droplet against viscosity and (ΔKE) is the energy loss due to solidification which restricts the droplet spread. Subscripts 1 and 2 denote before and after impact respectively. These terms are described in [17] and only the solidification term is presented here again:

$$\Delta KE = \frac{1}{2}mu^2 = \left(\frac{1}{2}\rho U_0{}^2\right)(AS) = \left(\frac{1}{2}\rho U_0{}^2\right)\left(S\frac{\pi}{4}d_s{}^2\right) \tag{3}$$

where (S) is the thickness of solidified layer and (d_s) is its diameter of varying from zero to (D_{max}) during droplet spread as depicted in Figure 2 (right). A reasonable estimate of (d_s) is the half of D_{max} [17] and (S) could be calculated using the analytical model for one-dimensional heat conduction with the assumption of semi-infinite wall with no thermal contact resistance [19] and formulated as a function of Stefan number ($Ste = c_p(\frac{T_D - T_{wall}}{H_f})$, Peclet number ($Pe = Pr\,Re$), dimensionless time ($t^* = t\frac{U_0}{D_0}$) and material thermal inertia ($\gamma = c_p\rho k$) as follows:

$$S = D_0 \frac{2}{\sqrt{\pi}} Ste \sqrt{\frac{t^*}{Pe} \times \frac{\gamma_w}{\gamma_d}} \tag{4}$$

Substituting equation (4) into (3) and after doing some mathematical rearrangements [17], the maximum diameter of droplet after impact with simultaneous solidification is given in the equation (5), where subscripts (w) and (d) denote the properties corresponding to wall and droplet properties, respectively. (θ_A) is the advancing contact angle and is a material property of the wall.

$$\frac{D_{max}}{D_0} = \sqrt{\frac{We + 12}{3(1 - cos\theta_A) + 4\left(\frac{We}{\sqrt{Re}}\right) + We\,Ste\,\sqrt{\frac{3\gamma_w}{2\pi\,Pe\,\gamma_d}}}} \tag{5}$$

Comparing the relative magnitude of the terms in denomination, one can deduce that solicitation effects during droplet spread are negligible if [17]:

$$\Phi = \frac{Ste}{\sqrt{Pr}}\sqrt{\frac{\gamma_w}{\gamma_d}} < 1 \tag{6}$$

All the terms in equation above are specified according to either material properties or known process parameters. Therefore it is deduced that this (ϕ) could be a plausible criterion to predict whether the liquid metal droplet is rapidly solidified once deposited or it will form a liquid film over the wall.

Another analytical study has been done by Chiu and Ma [20] to investigate the time scales during droplet impact and solidification. They also used the same terminology presented in current work. They developed the model to obtain the solidification velocity for solidified layer, and then by dividing the solidified thickness by the solidification velocity, the dimensionless time of solidification reads:

$$t_s{}^* = \frac{2\sqrt{2}}{3(\frac{D_{max}}{D_0})^2}\sqrt{\frac{Ste}{Pe}t_r{}^*} \tag{7}$$

This relation shows the dependency of dimensionless solidification time on maximum spread factor, the thermal properties and temperature profile of the walls (Ste number) and flow properties at the instant of impact (Pe number) and the dimensionless time of the droplet spread (t_r). This also gives a reasonable estimate for time scales during solidification of droplets and shows whether the droplet will immediately be solidified once impinges or not. This could also interpret the effect of temperature difference between liquid metal and die walls on the porosity formation.

Numerical modelling

Computational fluid dynamics simulation of two phase flow in HPDC process is performed by a VOF-based Eulerian-Lagrangian approach which considers incompressible Navier-Stokes equations with two-phase mixture assumption. This Eulerian-Lagrangian coupling approach is implemented in the frame of OpenFOAM [21] and used here for simulation of liquid jet breakup and droplet formation during the filling phase. This method is extended with modeling the behavior of Lagrangian droplets after formation. The analytical model for droplet-wall interaction described in current work is applied in the numerical framework presented in [12], according to local properties of the droplets while approaching the near walls. As mentioned in the introduction, the focus of this work is on droplet-wall interactions, therefore the splash parameter (K) is implemented in the numerical model to decide whether the droplet is going to be deposited at the wall or not. Then according to the analytical model for droplet spreading and simultaneous solidification, droplets are considered to be solidified with the same mass and momentum on the impact point. This would result in the solidified droplet pile up and formation of porous structures with increasing temperature inwardly from the walls.

Results and discussion

For numerical simulation of high pressure die casting, a liquid Aluminum round jet with the diameter of 1 mm is studies with three different inlet velocities (30, 50 and 70 m/s) located in different instability regimes in HPDC window. The specifications of each case are listed in Table 1 and the Re-We instability diagram is depicted in Figure 3 (left). In order to use the analytical model for droplet-wall interaction presented in this work, a 3D geometry of a simple round jet is created with an obstacle (with wall boundary condition) positioned in the middle of the die cavity.

Table 1. Aluminum liquid jet with different inlet velocities (d= 1mm).

U(m/s)	*Re*	*We*	*Oh*	*Regime*
30	64360	2470	0.0007	2nd wind-induced
50	107000	6900	0.0007	in between
70	150000	13000	0.0007	Atomization

This obstacle increases the complexity of the problem by means of changing the flow direction and imposing droplet-wall interaction. It is assumed that this cavity is made of stainless steel at room temperature (300 K); therefore the thermal properties of the walls could be taken into account for our analytical model. The ϕ factor in equation (6) is about 9 at the beginning of the process and might get to 4 due to the temperature increase in the walls. It means that the solidification during the droplet spread is not negligible and as the droplet impinges the wall it starts solidifying. The solidification time scale as a function of spreading time and wall temperature is depicted in Figure 3 (right). It reveals that the first front of Aluminum droplets with temperature just above the melting point (934 K) will get immediately solidified after impact to the cold walls because the dimensionless (t_s) is one order of magnitude smaller than (t_r) at the beginning. When the following droplet fronts arrive, the solidified layers are accumulated consecutively and play the role of walls for the next coming droplets with the different thermal properties. This might end up with the temperature increase inward the cavity and consequently increase the solidification time scale in comparison to spreading time scale (the time scales get to the same order of magnitude as temperature increases). This temperature increase is taken into account for every 100 K in Figure 3 (right). It means that the next droplet fronts may not be solidified as they impinge the newly formed representative walls (the thickness of accumulated layers) and the porosity formation could be decreased. This also agrees with the results in [8] which concluded that any decrease in die temperature results in increasing the porosity.

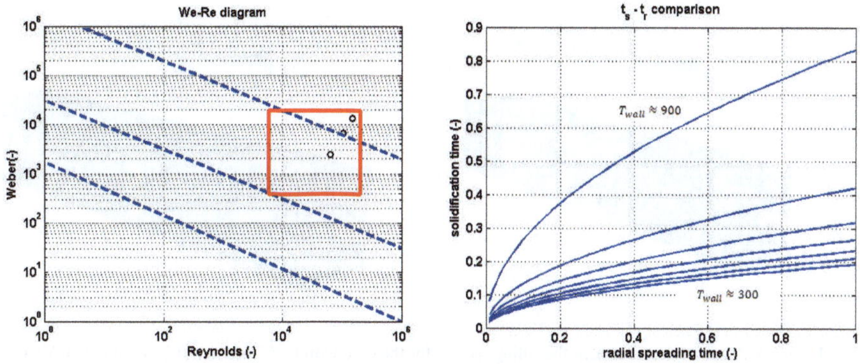

Figure 3. We-Re instability diagram for different velocities of the Aluminum jet (left), comparison of solidification and radial spread time scales for impinging droplets.

A computational grid is generated with the minimum grid spacing of 250 micron as shown in Figure 4. As mentioned in previous section, three dimensional Navier-Stokes equations are solved using the Eulerian-Lagrangian approach. The pressure-velocity coupling is carried out with PIMPLE algorithm and for turbulence modelling the Reynold stress model (RSM) is applied. The simulation results for each case are shown in Figures 4 (right) and 5. Figure 4 (right) shows the general liquid jet behavior for the case with (U=50 m/s). The instability regime transition and droplet formation could be seen around the jet at (t=0.5 ms). To show the capability of this analytical model in prediction of porosity formation, a comparison is done for this case in Figure 5. The filling pattern obtained by this droplet-wall model is compared to the one obtained by simple VOF simulation. It can be observed that, the droplet-wall interaction and prompt solidification in the middle of the cavity, where the obstacle is located, end up in porosity formation which cannot be seen in the standard VOF simulation with the same grid spacing.

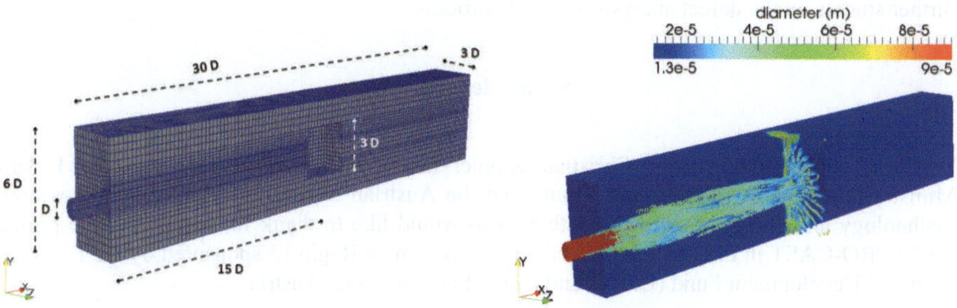

Figure 4. The geometry and grid network of die cavity with an obstacle (left), Numerical simulation of the liquid jet for the case with U=50 m/s at t=0.5 ms.

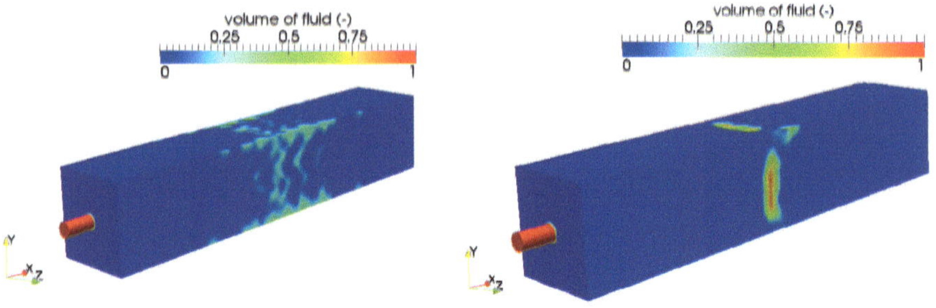

Figure 5. A comparison between the filling pattern for the case with U=50 m/s at t=0.5 ms: standard VOF simulation (left) simulation including the droplet-wall interaction model (right).

Conclusion

In this paper, the high pressure die casting flow characteristics are studied analytically and numerically. The CFD simulations of the general instability, breakup and droplet formation are performed using the numerical solver presented in our previous works. An analytical model based on droplet-wall interaction and 1D solidification analysis is developed and implemented in the numerical solver to predict the porosity formation. It shows that for Aluminum HPDC process, the generated droplets due to the jet breakup will get immediately solidified at the instant of impact to the walls. This also agrees with the observations from experimental works in the literature and reveals that the droplet impact on the die walls and solidification are probably the main mechanism for porosity formation near the surface of final casting. The simulation results also show the feasibility of porosity prediction by this model. To conclude, this study introduces a new method for the numerical modelling of HPDC and could be a strong basis for further studies on the defect analysis in HPDC process.

Acknowledgements

This work was funded by the Christian-Doppler Research Association, the Austrian Federal Ministry of Economy, Family and Youth, and the Austrian National Foundation for Research, Technology and Development. The authors also would like to thank for supporting the project MEEEPRO-CAST in the framework of the EU-programme Regio 13 sponsored by the European Regional Development Fund (EFRE) and by the State of Upper Austria.

References

[1] M. Saeedipour et al., "A Numerical and Experimental Study of Flow Behavior in High Pressure Die Casting", Proceedings of TMS annual meeting: Magnesium Technology, San Diego, USA, 2014.
[2] S. Schneiderbauer et al., "Studies on Flow Characteristics at High-Pressure Die-Casting", *IOP Conference series, Material Science and engineering,* 27 (2011).

[3] C. Chimani et al., "Studies on Flow Characteristics at High-Pressure Die-Casting", Proceedings of TMS annual meeting: Light Metals, Orlando, USA, 2012.
[4] H. I. Lauki, "High Pressure Die Casting of Aluminum and Magnesium Alloys - Grain Structure and Segregation Characteristics", PhD Thesis, Norwegian University of Science and Technology, 2004.
[5] A. Bhoi, "Evaluation of Metal Flow in High Pressure Die Casting via Radiography", Master Thesis, The Ohio state university, 2005.
[6] S. G. Lee et al., "Effect of process parameters on porosity distributions in high-pressure die-cast M50 Mg-alloy", *Materials Science and Engineering*, A 427 (2006).
[7] T. Rzychon and A. Kielbus, "The influence of wall thickness on the microstructure of HPDC AE44 alloy", *Achieves of Materials Science and Engineering,* 28 (2007).
[8] M. Sadeghi and J. Mahmoudi, "Experimental and Theoretical Studies on the Effect of Die Temperature on the Quality of the Products in High-Pressure Die-Casting Process", *Advances in Materials Science and Engineering* (2012).
[9] E. R. G. Eckert et al., "A complete solution for the flow, heat transfer, solidification and porosity distribution in pressure die casting", (Report UMSI 92/56, University of Minnesota Supercomputer Institute Research report, 1992).
[10] P. Homayounifar et al., "Numerical modelling of splashing and air entrapment in high-Pressure die casting", *International Journal of Advanced Manufacturing Technology*, 39 (2008).
[11] S. Li et al., "Numerical Simulation of Flow-Induced Air Entrapment Defects in the High Pressure Die Casting Process", Proceedings of TMS annual meeting: Magnesium Technology, Seattle, USA, 2010.
[12] M. Saeedipour et al., "Numerical simulation of turbulent liquid jet breakup using a sub-grid criterion with industrial application", Proceedings of ILASS-Europe annual conference on liquid atomization and spray systems, Bremen, Germany, 2014.
[13] K. Kuwana et al., "Assessment of Computer Simulation Software and Process Data for High Pressure Die Casting of Magnesium", (Report ORNL 0596, University of Kentucky, 2007).
[14] C. Chimani et al., "Investigations on Microstructure Effect of Changing Fluid Flow Characteristic in High Pressure Die Casting", *Materials Science Forum*, 782 (2014).
[15] A. L. Yarin, "Drop Impact Dynamics: Splashing, Spreading, Receding, Bouncing...", *Annual Review Fluid Mechanics*, 38 (2006).
[16] C. Mundo et al., "Droplet-wall collisions: Experimental studies of the deformation and breakup process", *International Journal of Multiphase Flow,* 21 (1995).
[17] S. D. Aziz and S. Chandra, "Impact, recoil and splashing of molten metal droplets", *International Journal of Multiphase Flow,* 43 (2000).
[18] M. Pasandideh-Fard et al., "Deposition of Tin droplets on a steel plate: simulations and experiments", *International Journal of Heat and Mass Transfer,* 41 (1998).
[19] D.R. Poirier and E.J. Poirier, "Heat Transfer Fundamentals for Metal Casting", Proceedings of TMS annual meeting , Warrendale, USA, 1994.
[20] H. R. Chiu and H. K. Ma, "Solidification Models of an Impinging Metal Droplet", *Particulate Science and Technology,* 27 (2009).
[21] OpenFOAM, The open source CFD toolbox, http://www.OpenFOAM.org .

Advances in the Science and Engineering of Casting Solidification
Edited by: Laurentiu Nastac, Baicheng Liu, Hasse Fredriksson, Jacques Lacaze, Chun-Pyo Hong, Adrian V. Catalina, Andreas Buhrig-Polaczek, Charles Monroe, Adrian S. Sabau, Roxana Elena Ligia Ruxanda, Alan Luo, Subhayu Sen, and Attila Diószegi
TMS (The Minerals, Metals & Materials Society), 2015

ENGINEERED COOLING PROCESS FOR HIGH STRENGTH DUCTILE IRON CASTINGS

Simon N. Lekakh, Anthony Mikhailov, and Joseph Kramer

Missouri University of Science and Technology; Rolla, MO. 65409, USA

Keywords: Ductile Iron, Strength, Cooling, Structure, Mechanical Properties

Abstract

Professor Stefanescu contributed fundamentally to the science of solidification and microstructural evolutions in ductile irons. In this article, the possibility of development of high strength ductile iron by applying an engineered cooling process after casting early shake out from the sand mold was explored. The structures in industrial ductile iron were experimentally simulated using a computer controlled heating/cooling device. CFD modeling was used for process simulation and an experimental bench scale system was developed. The process concept was experimentally verified by producing cast plates with 25 mm wall thickness. The tensile strength was increased from 550 MPa to 1000 MPa in as-cast condition without the need for alloying and heat treatment. The possible practical applications were discussed.

Introduction

The mechanical and thermo-physical properties of cast iron castings produced in sand mold are mainly controlled through the development of desired primary solidification structure by: (i) variation of carbon equivalent for changing the primary austenite/graphite eutectic ratio, (ii) inoculation for promoting graphite nucleation and decreasing chill tendency, and (iii) magnesium treatment for modifying graphite shape (flake in GI, vermicular in CGI, and spherical in SGI). Professor Doru Stefanescu made fundamental contributions to the science of cast iron solidification [1] and his studies bring a light on the mechanisms of the many liquid/solid transformations.

Cooling rate adjustment and alloying with additions of Cu, Mo, Ni are other methods which are used in practices for metal matrix structure control of the cast iron that is formed during the solid/solid eutectoid reaction. However, these methods have serious limitations, because it is difficult to achieve significant change in the heat extraction rate from the casting into the sand mold during the eutectoid reaction, the high cost of additions, and a limited ability to increase strength in the as-cast condition. Currently, if high strength is needed, an additional austempering heat treatment is used to produce ADI castings. In this heat treatment, rapid undercooling of austenite combined with isothermal holding at 350-420°C forms an ausferrite, or bainite structure with increased strength and toughness.

Several different ideas involving integrating rapid cooling into the metal casting process in order to increase strength without requiring an additional heat treatment have been discussed during the last few decades in the metal casting community. Recently, Stefanescu and co-authors [2] studied a process for direct development of high strength ductile iron with an ausferrite structure in the as-cast condition by a combination of alloying by 3-5% Ni, early shake out, and air cooling to the isothermal bainitic transformation temperature [2]. In this article, the further possibility of development of high strength ductile iron by applying an engineered cooling to the casting after early shake out from the sand mold was explored. The objective was development of a process for the production of high strength ductile iron in the as-cast condition, eliminating both alloying and additional heat treatment.

Process simulation

Employing a specially designed cooling schedule (engineered cooling) during solid state transformations allows control of the structure without needing to alter the alloy chemistry[3]. The various high strength products of the solid state reaction could be formed in lean ductile iron during the decomposition of undercooled austenite. The combination of high carbon concentration in austenite and the suppression of carbon diffusion by high cooling rate stabilizes the undercooled austenite. Under these conditions, carbon has a major role as an alloying element. The key feature of the engineered cooling process[3] is a seamless integration of the desired cooling profile into the casting process, combining early shake out (at a temperature above eutectoid transformation) and controlled cooling after that to maximize strengthening.

Experimental simulations of achievable structures

In order to experimentally simulate the different engineered cooling scenarios, a special device called the "Thermal Simulator" was developed[3]. Small test specimens (50x6x4 mm), machined from industrial ductile iron castings, were subjected to a heating/cooling cycle. The specimen heating was performed by a computer controlled high ampere DC current power supply. Temperature measurement was done by a thermocouple welded onto the hot zone and a high precision infrared pyrometer with a 1 mm spot size. The compressed air used in the cooling loop was controlled by a proportional electromagnetic valve. These two controlling loops (heating and cooling) in combination with the small thermal inertia of the test specimen allowed reproduction of any cycle with up to 80 C/sec heating and cooling rates. The "Thermal Simulator" measured the electrical resistivity (ρ) of the specimen (a structure sensitive physical property) and the supplied electrical power (W) at constant heating or cooling rate (a parameter sensitive to the heat of phase transformation, similar to scanning calorimetry test). A combination of ρ and W measurements was used to determine the phase transformation temperatures and kinetics (Figure 1).

These ductile iron specimens were subjected to heating and cooling cycles designed to simulate engineered cooling. Two types of heating cycles were studied: (a) heating to austenization temperature (920^{0}C), with 5-30 minutes holding for carbon saturation in the austenite, and continuous cooling with 0.3-20 C/sec cooling rate to room temperature and (b) isothermal treatment, including the same austenization heating schedule followed by 2-20 C/sec cooling, 60 minutes isothermal hold at 380^{0}C, and fast cooling to room temperature.

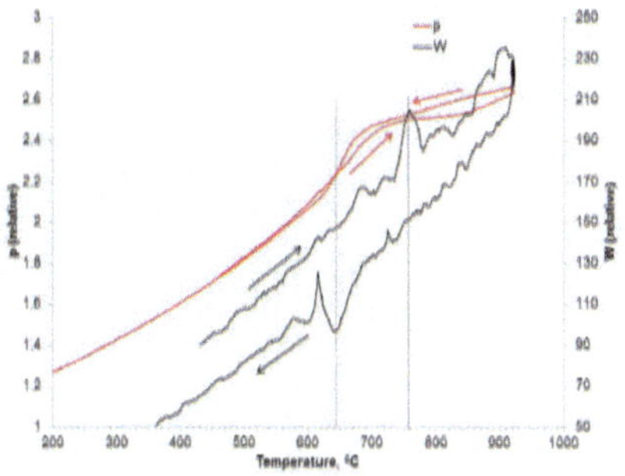

Figure 1. Example of electrical resistivity and power curves during heating and 2 C/s continuous cooling cycle (shown by arrows)[3].

Cooling rate has a significant effect on the macro-hardness of the ductile iron (HB, black lines in Figure 2a) by changing the volumes of phases and the phases' internal structure and microhardness (HV, black lines in Figure 2b). Cooling rates up to 2 C/s increase the volume and microhardness of pearlite. Cooling rates from 2 C/s to 10 C/s exhibit a sharp increase of hardness, mainly because of the formation of quenched martensite with 550-600 HV microhardness. These continuous cooling experiments showed the limitations since too high of a cooling rate resulted in an undesirable martensitic transformation.

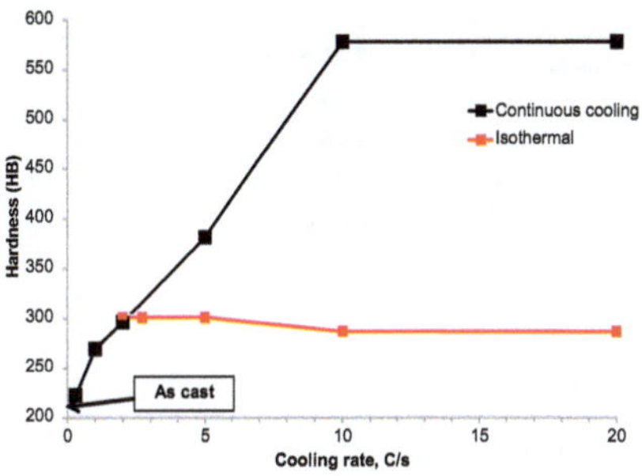

(a)

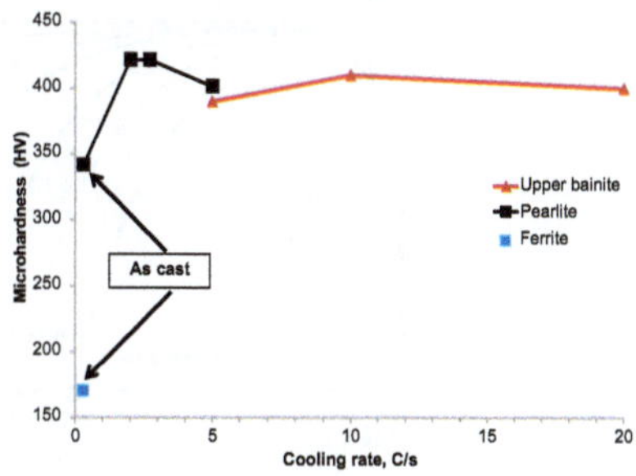

(b)

Figure 2. (a) Effect of cooling rate on hardness and (b) the microhardness of individual phases [3].

In order to develop microstructures that would provide a combination of high strength and toughness, isothermal heat treatments were investigated using the "Thermal Simulator"[3]. The specimens, taken from cast ductile iron plates, were heated to 920^{0}C, cooled to 380^{0}C at different cooling rates, and isothermally held for 60 minutes. A change in electrical resistivity indicated that ausferrite formation was complete at 25 minutes during 380^{0}C isothermal holding (Figure 3a). The different mixtures of fine pearlite and ausferrite can be developed by changing the cooling rate (Figure 3b). At 10 C/s cooling rate an ausferrite structure with small local pearlite spots around graphite nodules developed (Figure 3c).

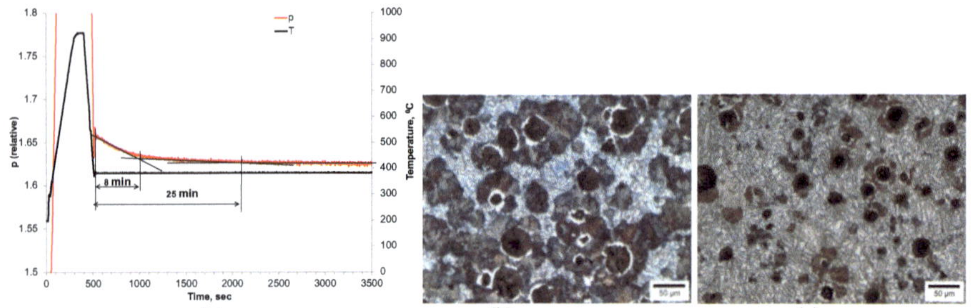

(a) (b) (c)

Figure 3. (a) Electrical resistivity during isothermal holding at 380^{0}C (5 C/s cooling from 920^{0}C) and microstructures of ductile iron cooled: (b) 5 C/s and (c) 10 C/s[3].

Figure 4 summarizes the achievable micro-structures after continuous cooling to room temperature and cooling to isothermal hold temperature (380^{0}C) at different cooling rates. A minimum cooling rate of 2 C/s is required to achieve the fine pearlite structure in the ductile iron investigated. At higher cooling rates, a mixture of fine pearlite and ausferrite can be formed by isothermal holding above the Ms temperature. Based on these experimental studies, a range of process parameters for engineered cooling were suggested[3].

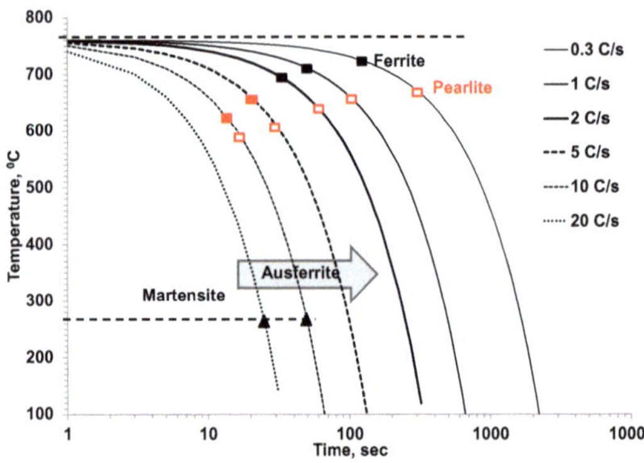

Figure 4. Achievable structures by applying continuous cooling to room temperature and to isothermal holding temperature (380^{0}C) at different cooling rates[3].

Simulation and design engineered cooling

Computational fluid dynamic (CFD) simulations and experimental tests were used to design the engineered cooling parameters[3]. CFD simulations (FLUENT software) were used to predict the effect of different cooling methods on the temperature profiles in the center and on the surface of a 25x150x200 mm plate casting. Three critical parameters were considered:

- cooling rate (>2 C/s, to develop desired structure of metal matrix (Figure 4))
- temperature gradient in the casting wall (<100 - 150^{0}C, to avoid thermal cracks and casting distortion), and

- final surface temperature above Ms (>260 - 300^{0}C, to avoid martencite formation).

In order of increasing surface heat flux, the simulated cooling methods included: (a) still air, (b) forced air, (c) wide angle water/compressed air atomizer nozzle, and (d) wide angle water spray nozzle (Figure 5).

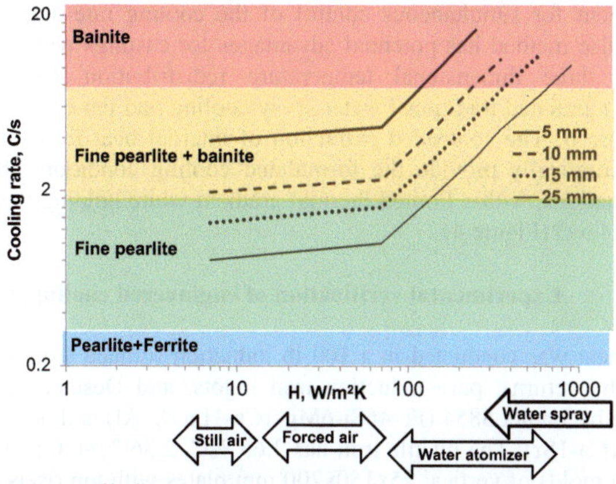

Figure 5. Ductile iron structure-cooling intensity diagram for plates with different wall thicknesses.

The first two, "soft" cooling methods: still air and forced air without water, do not provide the required cooling rate to achieve the ausferrite structure in ductile iron castings with 25 mm wall thickness. On the contrary, the mild water/compressed air atomizer and the more intensive water spray cooling methods both provide a high enough cooling rate, but significantly increases the temperature gradient in the casting wall and quickly decreases the surface temperature below the Ms temperature while solid transformations were not finished in the center (Figure 6).

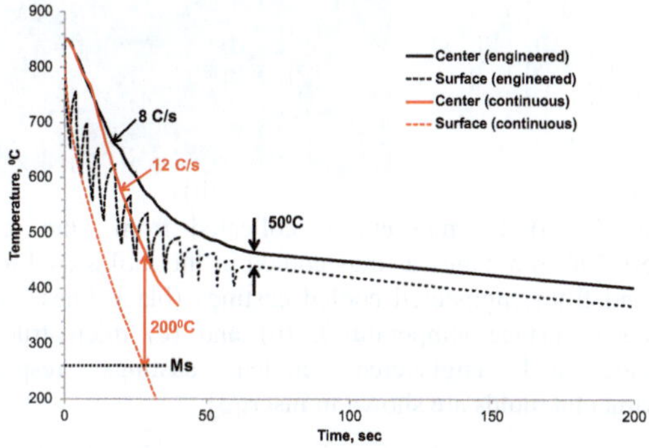

Figure 6. CFD simulated cooling 25x150x200 mm ductile iron plate applying continuous and engineered (impulse) water spray cooling.

To satisfy the critical cooling parameters and optimize the casting cooling, a computer assisted engineered cooling method was applied. In this method, cooling intensity was controlled using feedback from a surface temperature input. The different controlling scenarios were virtually simulated for a 25 mm wall thickness plate. These scenarios included proportional control of water flow in water-air atomizer for changing cooling efficiency and impulse on/off control of water flow in water spray nozzle. The simulations showed that the both proportional and impulse methods are efficient for simultaneous control of the cooling rate and the thermal gradient; however, the impulse method has potential advantages for castings with complicated geometry allowing internal three dimensional temperature redistribution during each cycle. The comparison of the simulated traditional water spray cooling and the engineered impulse cooling are shown in Figure 6. The controlled extraction of internal heat from the casting after early shake out can theoretically provide the formulated cooling conditions, based on kinetics of structure transformations, with a limited thermal gradient while holding the surface temperature above the required level (Figure 4).

Experimental verification of engineered cooling[3]

An experimental heat was conducted in a 100 lb induction furnace with a charge consisting of ductile iron foundry returns, pure induction iron ingots, and Desulco carbon. The melt was treated in the ladle by Lamet 5854 (Fe-46Si-6Mg-1Ca-1La-0.7Al) and inoculated by Superseeed (Fe-70Si-0.4Al-0.1Ca-1Sr). The ductile iron has 3.65 %C, 2.36 %Si, 0.55 %Mn and 0.55 %Cu. Four no-bake sand molds of vertical 25x150x200 mm plates with top risers were poured (Figure 7). Two reference plates had K-type thermocouples (protected by a quartz tube) in the casting
The reference plates were cooled in the molds (base process). The two other molds had an investment ceramic coated 12 mm rod in the riser sleeve for transferring castings to the cooling device. These two castings were shaken out early and subjected to engineered cooling.

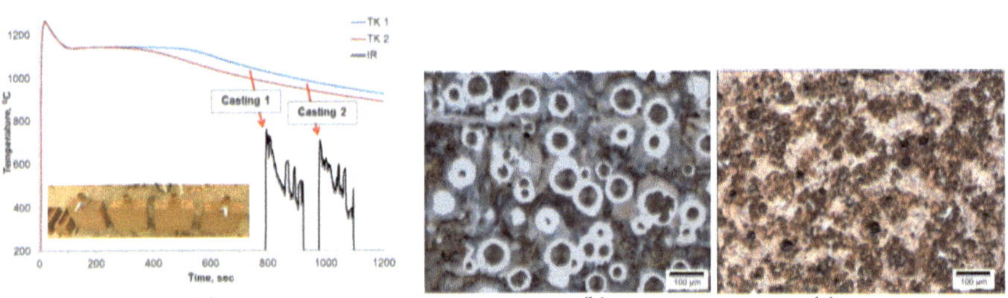

(a) (b) (c)

Figure 7. (a) Thermal curves collected from the reference 25x150x200 mm plate castings cooled in the molds (red and blue line) and from engineered cooled castings (black lines - infrared pyrometer surface temperature), (b) and (c) microstructure of reference and engineered cooled castings respectively. Experimental molds are shown in insert[3].

Base casting had a structure of lamellar pearlite with 10-15% ferrite. After the application of engineered cooling, the structure became a mixture of ausferrite and fine pearlite. Engineered cooling nearly doubled the tensile strength of ductile iron from 550-600 MPa at 8 % elongation to 1000-1050 MPa at 4% elongation in the as-cast condition. Described engineered cooling method has a wide process parameter window and authors are planning future experimental studies to optimize ductile iron properties.

Conclusions

A process for the production of high strength ductile iron in the as-cast condition by applying engineered cooling was discussed. The process includes an early shake out and the specially designed engineered cooling to develop the desired structure. The "Thermal Simulator" experimental technique and CFD simulations were used to formulate the process parameters.. The process was experimentally verified by pouring 25 mm thick plate castings and subjecting them to engineered cooling after early shake out. The tensile strength of ductile iron was increased from 550-600 MPa for mold cooled castings to 1000 – 1050 MPa for castings subjected to engineered cooling. The process is under further development in accord with Phase 2 of the project funded by American Foundry Society.

References

1. D. Stefanescu, *Science and Engineering of Casting Solidification. Second Addition* (Springer Science+business Media, LLC, 2009).
2. U. de La Torre, D.M. Stefanescu, D. Hartmann, and R. Suarez, "As-cast Austenitic Ductile Iron," *Keith Millis Symposium on Ductile Iron, AFS* (2013), 223-243.
3. S. Lekakh, "High Strength Ductile Iron Produced by the Engineered Cooling: Process Concept," Int'l J. of Metalcasting, vol. 9, issue 2 (2015).

Acknowledgement

Authors gratefully acknowledge:
- the American Foundry Society for funding of this project and support from AFS Steering Committee: Mike Riabov (Elkem), Matt Meyer (Bremen Casting), Eric Nelson (Dotson Iron Castings), and Don Craig (Selee)
- Missouri University of Science and Technology personal, including Professor Von Richards and students: Seth Rummel, Mingzhi Xu, and Jingjing Quin for support experiments.

Advances in the Science and Engineering of Casting Solidification
Edited by: Laurentiu Nastac, Baicheng Liu, Hasse Fredriksson, Jacques Lacaze, Chun-Pyo Hong, Adrian V. Catalina, Andreas Buhrig-Polaczek, Charles Monroe, Adrian S. Sabau, Roxana Elena Ligia Ruxanda, Alan Luo, Subhayu Sen, and Attila Diószegi
TMS (The Minerals, Metals & Materials Society), 2015

NUMERICAL SIMULATION OF SOLIDIFICATION STRUCTURE FOR YQ450NQR1 STEEL BLOOM IN CONTINUOUS CASTING PROCESS

Kun Dou[1], Lei Wang[1], Jiasheng Qing[1], Xiaofeng Zhang[1], Bao Wang[1], Bo Liu[1], Qing Liu[1] ✉

[1]State Key Laboratory of Advanced Metallurgy, University of Science and Technology Beijing; 30 Xueyuan Rd.; Haidian District, Beijing, 100083, P. R. China

Keywords: Continuous casting bloom, Solidification structure, Cellular Automaton-Finite Element, Numerical simulation, YQ450NQR1 steel.

Abstract

The solidification structure of continuous casting bloom has a significant impact on quality and performance of final products. In this paper, numerical simulation for a 360mm×450mm YQ450NQR1 steel continuous casting bloom is carried out to study the impact of superheat and secondary cooling water flowrate on solidification structure of the bloom using coupled Cellular Automaton-Finite Element (CA-FE) method. The results show that increase of superheat and secondary cooling water flowrate leads to dendrites growth acceleration and coarsening from bloom surface to the center and decrease in equiaxed grain ratio. When the superheat increases from 10℃to 40℃, the equiaxed grain ratio drops from 45.6% to 22.3% and the average grain diameter increases from 1.63mm to 1.92mm. When secondary cooling water flowrate increases from 0.35L/kg to 0.65L/kg, the equiaxed grain ratio drops from 37.7% to 23.1% and the average grain diameter increases from 1.68mm to 1.98mm.

Introduction

As an important semi-product, the quality of steel continuous casting bloom has a significant impact on final products and the proportion of columnar to equiaxed grains during solidification process could reflect the distribution uniformity of solute elements[1,2,3]. Therefore, it is necessary to study the solidification structure of steel. In this paper, numerical simulation for a 360mm×450mm YQ450NQR1 steel continuous casting bloom is carried out to study the impact of superheat and secondary cooling water flowrate on solidification structure, which will provide a theoretical foundation for the fine production of YQ450NQR1 steel bloom.

Mathematical Model

A heat transfer model is firstly established according to actual production process of YQ450NQR1 steel bloom, on this basis, nucleation and grain growth models are coupled to study the bloom microstructure evolution.

Macro Heat Transfer

Equation (1) is the two dimensional heat transfer governing equation for bloom continuous casting[4,5].

$$\rho c\frac{\partial T}{\partial t} = \frac{\partial}{\partial x}\left(k_{eff}\frac{\partial T}{\partial x}\right) + \frac{\partial}{\partial y}\left(k_{eff}\frac{\partial T}{\partial y}\right) + S \tag{1}$$

Initial Condition. At the beginning of casting process, molten steel temperature equals to pouring temperature, which is illustrated as $T(x, y, 0) = T_c$, where T_c is the pouring temperature.

Boundary Condition. In the mold,

$$-k_{eff}\frac{\partial T}{\partial n} = \rho_w c_w W \frac{\Delta T}{A_m} \tag{2}$$

In secondary cooling zone,

$$-k_{eff}\frac{\partial T}{\partial n} = h(T - T_w) + \sigma\varepsilon[(T + 298)^4 - (T_{ext} + 298)^4] \tag{3}$$

In air cooling zone,

$$-k_{eff}\frac{\partial T}{\partial n} = \sigma\varepsilon[(T + 298)^4 - (T_{ext} + 298)^4] \tag{4}$$

In bloom center,

$$-k_{eff}\frac{\partial T}{\partial n} = 0 \tag{5}$$

In all the equations above, T_c is the pouring temperature, k_{eff} is effective thermal conductivity, ρ is density of molten steel, ρ_w is water density, c_w is specific heat capacity, W is mold cooling water flow rate, ΔT is the temperature difference between water that flows in and out of the mold, A_m is the surface area of the mold, T_w is temperature of cooling water, h is the heat transfer coefficient between mold and bloom, σ is Stefan-Boltzmann constant, ε is radiation coefficient, T_{ext} is external temperature.

Heterogeneous Nucleation

A continuous nucleation distribution function $dn/d(\Delta T)$ can be used to describe the change in grain density, dn, which is induced by an increase in undercooling $d(\Delta T)$ [6,7,8]. The distribution function $dn/d(\Delta T)$ is described by the following equation (6):

$$\frac{dn}{d(\Delta T)} = \frac{n_{max}}{\sqrt{2\pi}\Delta T_\sigma}\exp\left[-\frac{1}{2}\left(\frac{\Delta T - \Delta T_{max}}{\Delta T_\sigma}\right)\right] \tag{6}$$

Where ΔT is the calculated local undercooling, ΔT_{max} is the mean undercooling, ΔT_σ is the standard deviation, n_{max} is the maximum nucleation density which can be reached when all the nucleation sites are activated while cooling.

Dendrite Tip Growth Kinetics

The total undercooling of the dendrite tip, ΔT, is considered as the sum of four contributions[9]:

$$\Delta T = \Delta T_c + \Delta T_t + \Delta T_r + \Delta T_k \tag{7}$$

Where ΔT_c, ΔT_t, ΔT_r and ΔT_k are the undercooling contributions associated with solute diffusion, thermal diffusion, solid-liquid interface curvature and attachment kinetics, respectively. In order to accelerate the computation course, the model is fitted and the following equation is gained:

$$v(\Delta T) = a_2 \Delta T^2 + a_3 \Delta T^3 \tag{8}$$

Where a_2 and a_3 are the coefficients of the multinomial of dendrite tip growth velocity, ΔT is the total undercooling of the dendrite tip.

Determination of Calculation Parameters

Simulation for solidification structure of YQ450NQR1 steel bloom is based on precise determination of related continuous casting process parameters and nucleation parameters.

Calculation of Thermal-Physical Properties

Main chemical compositions of YQ450NQR1 steel are given in Table I. Temperature dependence of enthalpy, density, thermal conductivity and solid fraction are calculated using thermodynamic database from JMatPro software, as shown in Figure 1. The turning points from thermal-physical properties curves correspond to phase transformation temperatures, e.g., the range of 1460~1510℃ and 700~800℃ is in correspondence with 1 →γ and γ→α, respectively.

Table I. Main Chemical Compositions of YQ450NQR1 Steel

C	Si	Mn	P	S
0.12	0.34	1.28	0.025	0.010

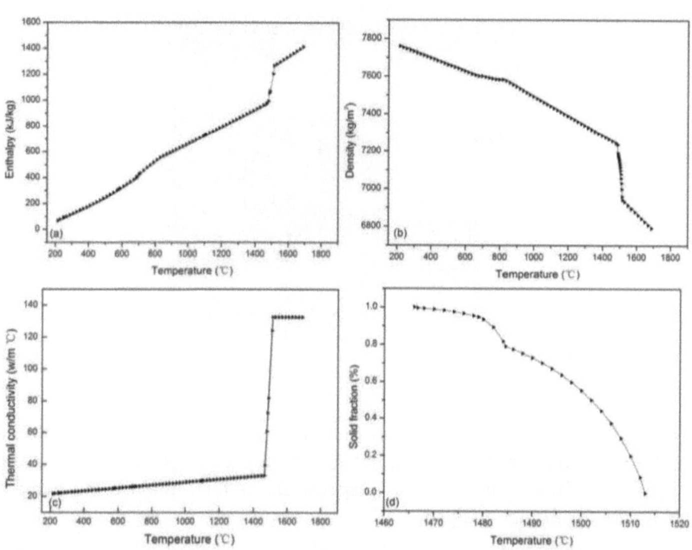

Figure 1. Calculated thermo-physical properties with temperature variation
(a) Enthalpy, (b) Density, (c) Thermal conductivity, (d) Solid fraction

Selection of Nucleation Parameters

Due to complexity of steel continuous casting process and limitation in experiment condition, research about selection of nucleation parameters based on given steel compositions is rare. In this paper, nucleation parameters from casting process of Fe-C binary alloy is firstly adopted and then revised according to comparison between simulation and experiment results until well agreement are obtained. The revised nucleation parameters for YQ450NQR1 steel are listed in Table II.

Table II. Nucleation Parameters for YQ450NQR1 Steel

Mold Surface			Bulk of Liquid		
$n_{max,s}$ (m^{-2})	$\Delta T_{n,s}$ (K)	$\Delta T_{\sigma,s}$ (K)	$n_{max,b}$ (m^{-3})	$\Delta T_{n,b}$ (K)	$\Delta T_{\sigma,b}$ (K)
(1.2~2.2) $\times 10^8$	1.0	0.1	(1.6~3.2) $\times 10^9$	8~15	1.5

Model Validation

In order to verify the accuracy of the model established above, the solidification structure of transverse section for YQ450NQR1 steel bloom is etched by H_2O-50% HCl regent. The related continuous casting parameters are shown in Table III and comparison between experiment results and corresponding simulated results are illustrated in Figure 2.

Table III. Operation Parameters for YQ450NQR1 Steel Bloom

Item	Cross sectional dimension	Pouring temperature	Superheat	Casting speed	Secondary cooling water flowrate
Value	360mm×450mm	1550℃	37℃	0.5m/min	0.35 L/kg

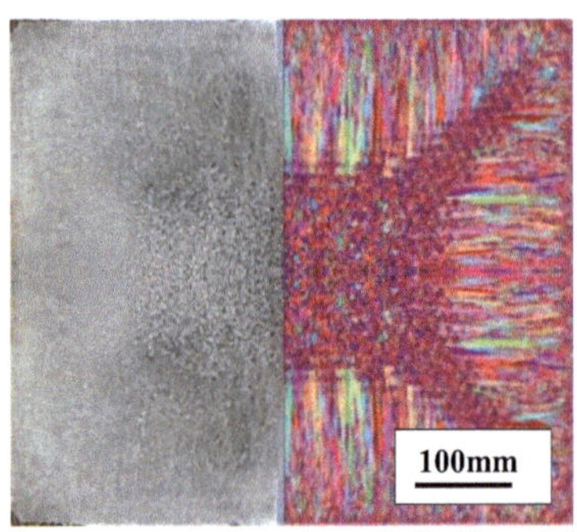

Figure 2. Comparison between experiment result (left) and simulation result (right) (1/2 cross section)

As is compared in Figure 2, the dendrites orientations and columnar to equiaxed transition trend are quite similar and further measurements indicate the equiaxed grain ratio for experiment and simulation results is 26.3% and 27.1%, respectively, which proves the reliability of the mathematical model for further analysis.

Results and Discussions

To quantitatively describe the influence of operation parameters on solidification structure of YQ450NQR1 steel bloom, the equiaxed grain ratio and average grain diameter (D_m) are adopted as quantify indexes. The average grain diameter (D_m) is defined as the following equation based on the assumption that all dendrites are spherical.

$$D_m = 2\sqrt{S/(N \cdot \pi)} \tag{9}$$

Where S is the area of calculation, N is the nuclei number in the calculation area.

Effect of Superheat on Solidification Structure Characteristics

The effect of different superheats on solidification structure morphology of YQ450NQR1 steel bloom is simulated as shown in Figure 3. It can be found that increase of superheat leads to dendrites growth acceleration and coarsening from bloom surface to the center and decrease in equiaxed grain ratio. From Figure 4, it is evident that when the superheat increases from 10℃ to 40℃, the equiaxed grain ratio drops from 45.6% to 22.3% and the average grain diameter increases from 1.63mm to 1.92mm. This is because the increase of superheat leads to nucleation sites number reduction and decrease of undercooling at solid/liquid interface, which promotes dendrites growth and inhibits the development of equiaxed grain. Besides, the excessive growth of dendrites due to superheat addition prevents the effective diffusion of solute elements and causes severe micro-segregation. Therefore, low superheat is recommended to obtain fine grains and high equiaxed grain ratio.

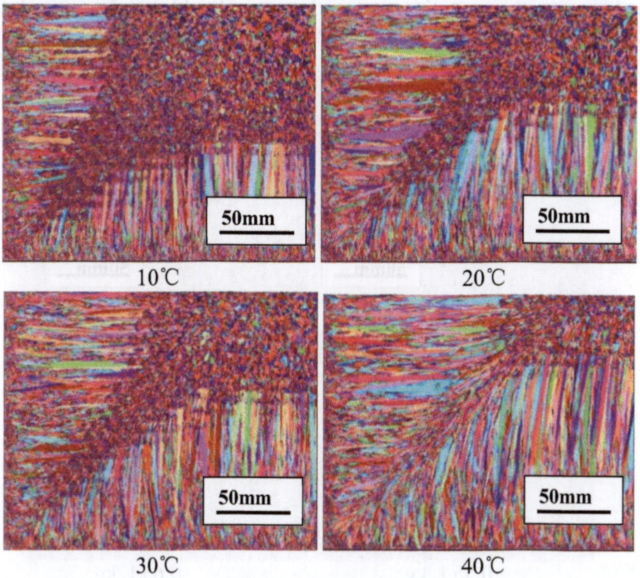

Figure 3. Simulation results for solidification structure of YQ450NQR1 steel bloom under different superheats (1/4 cross section)

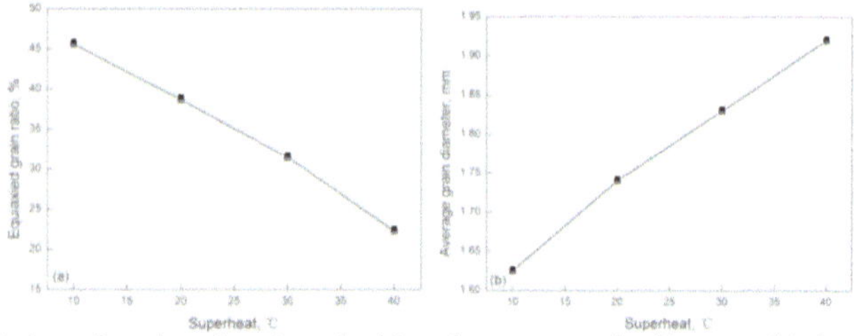

Figure 4. Variation of equiaxed grain ratio (a) and average grain diameter (b) for YQ450NQR1 steel bloom under different superheats

Effect of Secondary Cooling Water Flowrate on Solidification Structure Characteristics

The influence of secondary cooling water flowrate on solidification structure morphology of YQ450NQR1 steel bloom is simulated as shown in Figure 5, where the superheat is 20℃. It could be observed that increase of secondary cooling water flowrate results in rising of undercooling at solid/liquid interface, which is beneficial to dendrites growth. Moreover, the secondary cooling water flowrate variation has obvious impact on central equiaxed grains formation compared with those in the chill zone of bloom surface, which usually forms in the mold. Statistics from Figure 6 quantitatively describe the effect of secondary cooling water flowrate on equiaxed grain ratio and average grain diameter. As secondary cooling water flowrate increases from 0.35L/kg to 0.65 L/kg, the equiaxed grain ratio decreases from 37.7% to 23.1% and the average grain diameter increases from 1.68mm to 1.98mm. Based on the above analysis, suitable secondary cooling water flowrate should be determined thus obtaining reasonable equiaxed grain ratio and grain size in actual steel continuous casting process.

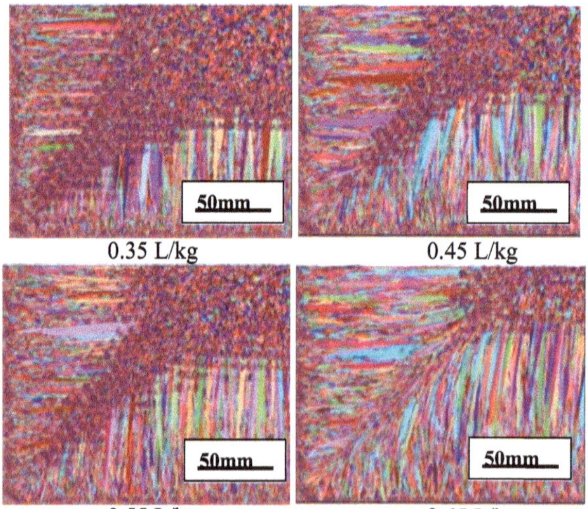

0.35 L/kg 0.45 L/kg

0.55 L/kg 0.65 L/kg

Figure 5. Simulation results for solidification structure of YQ450NQR1 steel bloom under different secondary cooling water flowrate (1/4 cross section)

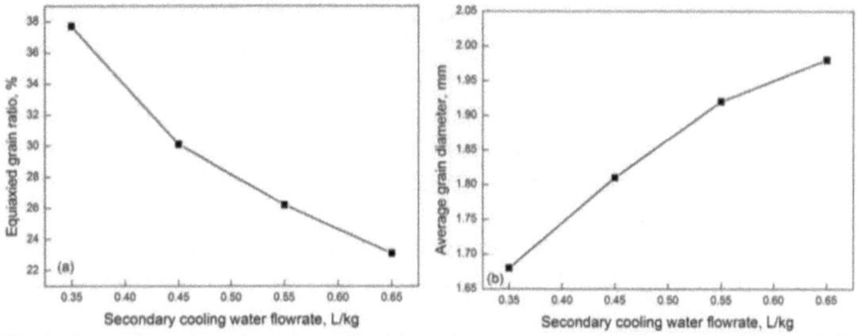

Figure 6. Variation of equiaxed grain ratio (a) and average grain diameter (b) for YQ450NQR1 steel bloom under different secondary cooling water flowrate

Conclusions

In this paper, the solidification structure of YQ450NQR1 steel bloom is simulated with Cellular Automaton-Finite Element (CA-FE) method and accuracy of the model is verified with sample from actual production. Relevant research results are as follows.

(1) With the increase of superheat, dendrites growth accelerate from bloom surface to center and dendrites zone enlarges evidently while the equiaxed zone reduces accordingly. When the superheat increases from 10℃ to 40℃, the equiaxed grain ratio drops from 45.6% to 22.3% and the average grain diameter increases from 1.63mm to 1.92mm.

(2) The increase of secondary cooling water flowrate results in rising of undercooling at solid/liquid interface thus promoting dendrites growth and prohibiting the formation and growth of central equiaxed grains. When the secondary cooling water flowrate increases from 0.35L/kg to 0.65 L/kg, the equiaxed grain ratio decreases from 37.7% to 23.1% and the average grain diameter increases from 1.68mm to 1.98mm.

(3) In the actual steel continuous casting process, low superheat and suitable secondary cooling water flowrate should be selected to inhibit dendrites growth and to promote the formation of equiaxed grains thus improving the quality of continuous casting blooms.

Acknowledgements

The authors would like to acknowledge the financial support for this study provided by Independent Research and Development Foundation of State Key Laboratory of Advanced Metallurgy, University of Science and Technology Beijing (No. 41602023).

References

1. F. Yin, L. Hua et al., "Microstructural Modeling and Simulation for GCr15 Steel During Elevated Temperature Deformation," *Materials & Design*, 2014, no. 55: 560-573.

2. M. Wu, J. Domitner, and A. Ludwig, "Using A Two-Phase Columnar Solidification Model to Study the Principle of Mechanical Soft Reduction in Slab Casting," *Metallurgical and Materials Transactions A*, 43(3) (2012), 945-964.

3. T. Carozzani, C.A. Gandin and H. Digonnet, "Optimized Parallel Computing for Cellular Automaton-Finite Element Modeling of Solidification Grain Structures," *Modeling and Simulation in Materials Science and Engineering*, 22(1) (2014), 015012.

4. Z.B. Hou, F. Jiang, and G.G. Cheng, "Solidification Structure and Compactness Degree of Central Equiaxed Grain Zone in Continuous Casting Billet Using Cellular Automaton-Finite Element Method," *ISIJ International*, 52(7) (2012), 1301-1309.

5. C.L. Jing, X.H. Wang, and M. Jiang, "Study on Solidification Structure of Wheel Steel Round Billet Using FE-CA Coupling Model," *Steel Research International*, 82(10) (2011), 1173-1179.

6. Y.L. Zhang et al., "Effects of Alloying Elements on the Microstructure of 22CrMoH Gear Steels Billets," *Advanced Materials Research*, 239 (2011), 3-14.

7. S. Luo, M.Y. Zhu, and S. Louhenkilpi, "Numerical Simulation of Solidification Structure of High Carbon Steel in Continuous Casting Using Cellular Automaton Method," *ISIJ International*, 52(5) (2012), 823-830.

8. S. Vernède, M. Rappaz, "A Simple and Efficient Model for Mesoscale Solidification Simulation of Globular Grain Structures," *Acta materialia*, 55(5) (2007), 1703-1710.

9. Z.B. Hou et al., "Compactness Degree of Longitudinal Section of Outer Columnar Grain Zone in Continuous Casting Billet Using Cellular Automaton-Finite Element Method," *ISIJ International*, 53(4) (2013), 655-664.

Advances in the Science and Engineering of

CASTING SOLIDIFICATION

An MPMD Symposium Honoring Doru Michael Stefanescu

Microstructure Evolution I

Session Chairs:

Hideyuki Yasuda

Laurentiu Nastac

Advances in the Science and Engineering of Casting Solidification
Edited by: Laurentiu Nastac, Baicheng Liu, Hasse Fredriksson, Jacques Lacaze, Chun-Pyo Hong, Adrian V. Catalina, Andreas Buhrig-Polaczek, Charles Monroe, Adrian S. Sabau, Roxana Elena Ligia Ruxanda, Alan Luo, Subhayu Sen, and Attila Diószegi
TMS (The Minerals, Metals & Materials Society), 2015

MODELING OF MICROSTRUCTURE EVOLUTION DURING ALLOY SOLIDIFICATION

Mingfang Zhu[1], Shiyan Pan[1,2], Dongke Sun[1,3]

[1]Jiangsu Key Laboratory for Advanced Metallic Materials, School of Materials Science and Engineering, Southeast University; Jiangning, Nanjing, Jiangsu, 211189, China
[2]Engineering Training Center, Nanjing University of Science and Technology; Nanjing, Jiangsu, 210094, China
[3]School of Materials Science and Engineering, Shanghai Jiao Tong University; Shanghai, 200240, China

Keywords: Solidification, Microstructure, Modeling, Cellular automaton, Lattice Boltzmann method.

Abstract

In recent years, considerable advances have been achieved in the numerical modeling of microstructure evolution during solidification. This paper presents the models based on the cellular automaton (CA) technique and lattice Boltzmann method (LBM), which can reproduce a wide variety of solidification microstructure features observed experimentally with an acceptable computational efficiency. The capabilities of the models are addressed by presenting representative examples encompassing a broad variety of issues, such as the evolution of dendritic structure and microsegregation in two and three dimensions, dendritic growth in the presence of convection, divorced eutectic solidification of spheroidal graphite irons, and gas porosity formation. The simulations offer insights into the underlying physics of microstructure formation during alloy solidification.

Introduction

Solidification is a complex process controlled by the interplay of heat, solute, capillary, thermodynamics, and kinetics. Computational modeling, which enables extensive use of mathematics for solving complicated problems, has emerged as an important tool to study the underlying physics of microstructural formation during solidification [1]. The cellular automaton (CA) approach can reproduce various microstructure features with an acceptable computational efficiency, indicating the considerable potential for practical applications. It has, therefore, drawn great interest in academia and achieved remarkable advances in the modeling of microstructures. The CA approach was first applied to simulate grain structure at the meso-scale [2]. It has been extended to the micro-scale to simulate the microstructures, involving two and three dimensional dendritic growth in binary and ternary alloys [3-11], non-dendritic or globular structure evolution in semi-solid process [12], dendritic growth in a fluid flow [13-14], the formation of microstructures in regular, irregular and divorced eutectic solidification [15-17], and dendritic growth with microporosity formation [18].

Recently, a kinetic-based lattice Boltzmann method (LBM) has developed rapidly as a new powerful technique of computational fluid dynamics (CFD) [19]. Compared to the conventional CFD, LBM has the attractive merits of simplicity of coding, high computational performance associated with time-efficiency, and good numerical stability for the calculation of fluid flow in complex geometries. Since LBM describes fluid motion at the level of pseudo-particles represented by the distribution functions, it can be naturally incorporated with the related numerical techniques to simulate crystal growth in a fluid flow [20-23].

This paper briefly describes some microstructure simulation models proposed in our latest work, which are developed within the frameworks of CA and LBM. Some simulation examples are presented to illustrate the capabilities of the models.

Model Description and Numerical Method

Cellular Automaton Approach.

The computation domain is divided into uniform square cells for two dimensions (2D) or cubic cells for three dimensions (3D). Each cell is characterized by the variables of concentration, temperature, solid fraction, and crystallographic orientation. The state of a cell can be liquid (f_s=0), solid (f_s=1) or interface (0<f_s<1). The kinetics of the moving solid/liquid (S/L) interface can be calculated from the Gibbs-Thomson equation by [13]

$$V/(\mu_k \cdot f_1(\varphi,\theta_0)) = T_l^{eq} - T^* + m\cdot[C_l^* - C_0] - \Gamma \cdot f_2(\varphi,\theta_0)\cdot \mathrm{K} \tag{1}$$

where V is the interface growth velocity and μ_k is the kinetic coefficient. $f_1(\varphi,\theta_0)$ is a function accounting for the kinetic anisotropy and is calculated by $f_1(\varphi,\theta_0)=1+\delta_k\cos[4(\varphi-\theta_0)]$, where δ_k is the degree of kinetic anisotropy, φ is the growth angle, calculated by $\varphi = \arccos(\partial_x f_s/[(\partial_x f_s)^2+(\partial_y f_s)^2]^{1/2})$, and θ_0 is the preferential growth orientation. T_l^{eq} is the liquidus temperature at the initial composition C_0, and m is the liquidus slope. Γ is the Gibbs-Thomson coefficient. $f_2(\varphi,\theta_0)$ is a function accounting for the anisotropy of the surface energy and is calculated by $f_2(\varphi,\theta_0)=1-15\varepsilon\cos[4(\varphi-\theta_0)]$, where ε is the degree of anisotropy of the surface energy. K is the local interface curvature that can be calculated using the counting-cell method [3] or the finite difference method according to the gradient of local solid fraction at the S/L interface [6]. T^* and C_l^* are the interface temperature and interface liquid composition, determined by numerically solving the following governing equations of transient heat conduction and solute diffusion using the finite diffidence method (FDM).

$$\rho C_p \partial_t T = \nabla\cdot(\lambda\nabla T) + \rho\Delta H\partial_t f_s, \quad \partial_t C_i = \nabla\cdot(D_i\nabla C_i) + C_l(1-k)\partial_t f_s \tag{2}$$

where ρ is the density, C_p is the specific heat, f_s is the solid fraction, λ is the thermal conductivity, ΔH is the latent heat, D is the solute diffusivity. The subscript i indicates the liquid or solid, and k is the partition coefficient.

For the calculation of growth kinetics using Eq. 1, a kinetic coefficient, μ_k, must be properly determined for quantitative simulations. Zhu and Stefanescu (ZS) [8] proposed a solution to the growth kinetics based on a local composition equilibrium approach. According to this approach, the increase in solid fraction, Δf_s, of an interface cell at one time step interval, Δt, is evaluated from the difference between the equilibrium composition and the actual liquid composition by

$$\Delta f_s = (C_l^{eq} - C_l^*)/(C_l^{eq}(1-k)) \text{ with } C_l^{eq} = C_0 + (T^* - T_l^{eq})/m_l + \Gamma(\theta)K/m_l \tag{3}$$

where C_l^{eq} is the interface equilibrium liquid composition. The normal growth velocity of the S/L interface is obtained by $V_n = \Delta f_s \cdot \Delta x / \Delta t$. The ZS model allows reasonable calculation of solutal dendritic growth without the need of introducing a kinetic coefficient.

The 2D ZS model was extended to 3D [10]. The most difficult issue for this extension is the description of different crystallographic orientations in 3D. The extended 3D model calculates the 3D weighted mean curvature (*wmc*) that is incorporated with the anisotropy of surface energy based on the surface divergence of the Cahn-Hoffman ξ-vector [24] by $wmc = -\nabla_s \cdot \xi(\hat{n})$ with $\xi(\hat{n}) = \tilde{\nabla}(|\nabla f_s| \cdot a(\hat{n}))$, where $a(\hat{n})$ is an anisotropy function of surface energy.

Lattice Boltzmann Method.

The Lattice Boltzmann method (LBM) describes fluid flow by the particle distribution function. According to the Bhatnagar-Gross-Krook (BGK) approximation with a single relaxation time scheme [25], the evolution equation of the particle distribution function is expressed as:

$$f_i(\mathbf{x} + \mathbf{e}_i \Delta t, t + \Delta t) - f_i(\mathbf{x}, t) = -\left[f_i(\mathbf{x}, t) - f_i^{eq}(\mathbf{x}, t) \right] / \tau + F_i(\mathbf{x}, t) \tag{4}$$

where $f_i(\mathbf{x},t)$ is the particle distribution function representing the probability of finding a pseudo-particle at location $\mathbf{x}$ and time t, $\mathbf{e}_i$ is the discrete velocity of the pseudo-particle, Δt is the time step, τ is the relaxation time, $f_i^{eq}(\mathbf{x},t)$ is the equilibrium distribution function, and $F_i(\mathbf{x},t)$ is the force term caused by the external fields or internal interaction such as the buoyancy effect.

A D2Q9 topology [25], where the 2D space is discretized into a regular square lattice including nine velocities, the equilibrium particle distribution function, $f_i^{eq}(\mathbf{x},t)$, and the force term, $F_i(\mathbf{x},t)$, in Eq. (4) can be expressed as

$$f_i^{eq}(\mathbf{x}, t) = w_i \left(\sum_{i=0}^{8} f_i(\mathbf{x}, t) \right) \left[1 + 3(\mathbf{e}_i \cdot \mathbf{u})/c^2 + 4.5(\mathbf{e}_i \cdot \mathbf{u})^2/c^4 - 1.5\mathbf{u}^2/c^2 \right] \tag{5}$$

$$F_i(\mathbf{x}, t) = (1 - 0.5/\tau) w_i [3(\mathbf{e}_i - \mathbf{u})/c^2 + 9(\mathbf{e}_i \cdot \mathbf{u}/c^4)\mathbf{e}_i] \cdot \mathbf{F} \Delta t \tag{6}$$

where w_i are the weight coefficients with $w_0 = 4/9$, $w_{1\text{-}4} = 1/9$ and $w_{5\text{-}8} = 1/36$, $c = \Delta x / \Delta t$ is the lattice speed, $\mathbf{u}$ is the macroscopic flow velocity, and $\mathbf{F}$ is the buoyancy force.

The LBM can also be used to calculate heat and solute transport controlled by convection and diffusion. The relevant LB equations are similar to Eqs. 4 and 5 with $f_i(\mathbf{x},t)$ substituted by $g_i(\mathbf{x},t)$ and $h_i(\mathbf{x},t)$, $f_i^{eq}(\mathbf{x},t)$ substituted by $g_i^{eq}(\mathbf{x},t)$ and $h_i^{eq}(\mathbf{x},t)$, τ substituted by τ_D and τ_α, and $F_i(\mathbf{x},t)$ substituted by $G_i(\mathbf{x},t)$ and $H_i(\mathbf{x},t)$, where $G_i(\mathbf{x},t) = w_i \Delta f_s C_l (1-k)$ and $H_i(\mathbf{x},t) = w_i \Delta f_s \Delta H / C_p$ are the source terms due to rejected solute and released latent heat during solidification, respectively. The quantities of momentum, $\rho\mathbf{u}$, fluid density, ρ, concentration, C, and temperature, T, can be computed from the relevant PDFs by $\rho\mathbf{u} = \sum_{i=0}^{8} \mathbf{e}_i f_i + \mathbf{F}\Delta t/2$, $\rho = \sum_{i=0}^{8} f_i$, $C = \sum_{i=0}^{8} g_i$, and $T = \sum_{i=0}^{8} h_i$. The kinematic viscosity, ν, and the solutal and thermal diffusivities, D and α, are related to the relevant relaxation times, τ, τ_D, and τ_α, by $\nu = c^2\Delta t(2\tau - 1)/6$, $D = c^2\Delta t(2\tau_D - 1)/6$, and $\alpha = c^2\Delta t(2\tau_\alpha - 1)/6$, respectively.

The coupled CA-LBM models are developed to simulate dendritic growth with convection[21,23] and gas bubble formation [26]. In the coupled models, the CA approach is for the simulation of dendritic growth, and the LBM is for numerically solving the thermal, solutal and flow fields.

Simulation Examples

3D Dendritic Growth.

The modeling of single dendrite was performed for a Ni–5 wt.% Nb alloy using the 3D ZS model [10]. The physical properties of the alloy studied are: m_l=−7.5 K/(*wt%*), k=0.81, Γ=3.65×10^{-7}*mK*, $T_{l,0}^{eq} = 1417.5K$, D_l=3.0×10^{-9}m^2/s, D_s=1.0×10^{-12}m^2/s. The calculations were carried out on a cubic domain of a 250×250×250 mesh with Δx=0.8 μm. Figure 1 presents the simulated dendrite morphologies for a Ni–5 wt.%Nb alloy at undercoolings of 2 K and 3.5 K. The degree of anisotropy of the surface energy, ε, was chosen as 0.02 and 0.04 for Figure 1 (a) and (b), respectively. It can be seen that at a lower undercooling and a smaller value of ε, the dendrite exhibits a branchless needle shape. With a higher undercooling and a larger value of ε, dendrite with well developed side branches is produced. Primary dendritic trunks grow along the preferential crystallographic directions. Secondary dendritic arms with random spacing are formed approximately perpendicular to the primary trunks. Near the dendritic tip, there is an incubation length along the tip before branching secondary arms. These simulated dendritic features are observed in the experimental dendrite microstructures.

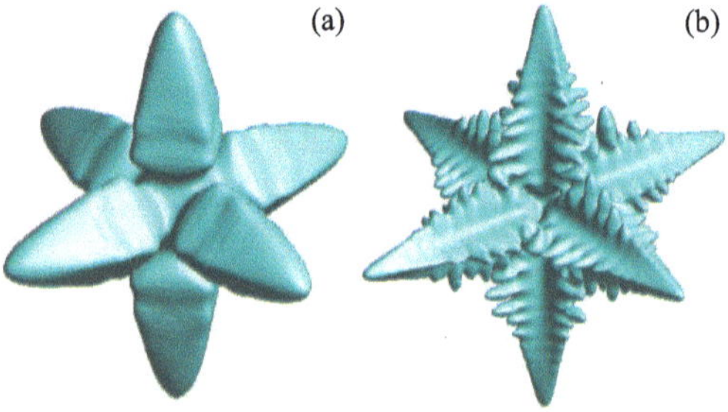

Figure 1. Simulated 3D single dendrite morphologies of a Ni-5wt.% Nb alloy at melt undercoolings and the degree of anisotropy of the surface energy: (a) ΔT=2K, ε=0.02; (b) ΔT=3.5K, ε=0.04.

Dendritic Growth under Natural Convection.

The CA-LBM model [23] was applied to simulate dendritic growth with natural convection. The physical parameters of succinonitrile-acetone (SCN-Ace) alloys are used for this simulation: k=0.1; m=−2.16*K/mol.%*; Γ=6.52×10^{-8}*mK*; ΔH=4.62×10^4J/m^3; C_p=1940 $J/(m^3{\cdot}K)$; T_m=331.23K; ν=2.6×10^{-6}m^2/s; D=1.27×10^{-9}m^2/s; and α=1.12×10^{-7}m^2/s. Figure 2 presents the evolution of the composition, temperature, and flow fields, and the morphology of multiple dendritic growth with natural convection for a SCN-0.5*mol.%*Ace alloy. It can be seen that the dendrites grow along various preferential crystallographic orientations. The natural flow is complex with several developed rotating vortices. The flow flux in the vicinity of the dendrites is, however, almost directly upwards. The growth of the dendritic arms in the inner region is mostly suppressed by the nearby dendrite and not affected much by convection, while the growth of the dendrites in the outer region is evidently influenced by natural convection. The dendrite arms in the lower region are more developed compared to the ones in the upper region. The upward natural flow transports the released heat and solute from the lower region to the upper region, providing a relative larger driving force for the growth of the dendrite arms in the upstream region.

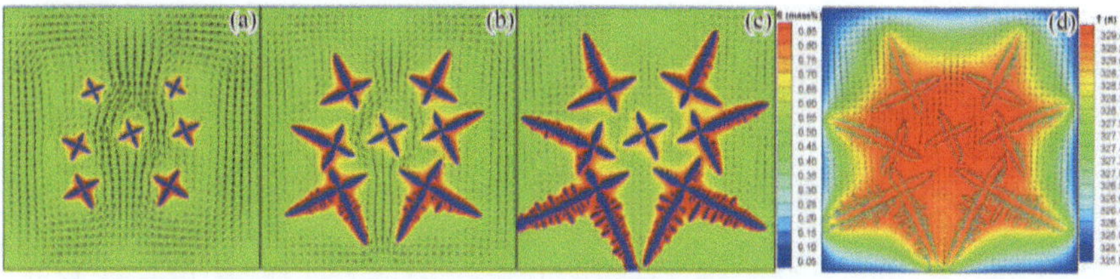

Figure 2. Evolution of the multiple dendrites of a SCN-0.5*mol.*%Ace alloy growing in a melt with ΔT=1 *K*, the cooling rate of 20 K/s, and Rayleigh numbers of $Ra_T=Ra_C=5\times10^3$: (a) f_s =3%, (b) f_s =10%, and (c) & (d) f_s =17% (f_s: total solid fraction): (a)-(c) showing the concentration field, and (d) showing the temperature field (domain: 500×500 with Δx=1 μm).

Divorced Eutectic Solidification of Spheroidal Graphite Irons

The ZS model was extended to the multi-phase system to simulate the divorced eutectic solidification [17]. Figure 3 presents the evolution of the microstructure and solute field for a hypoeutectic spheroidal graphite (SG) iron with C_0=4.1wt.%C. As shown, the solidification of a hypoeutectic SG iron starts with the nucleation and growth of primary austenite dendrites. When the temperature reaches the eutectic temperature, graphite nuclei precipitate from the liquid and grow. Once austenite dendrites get in contact with graphite nodules, they quickly engulf the nodules. Then, the austenite shells that surround graphite grow from the liquid in an isotropic manner. The enveloped graphite nodule grows within the austenite by carbon diffusion from the liquid through the austenite shell. After fully solidified, the graphite nodules continue growing by consuming the supersaturated carbon in the austenite phase. The final microstructure exhibits several graphite nodules encapsulated in each austenite grain.

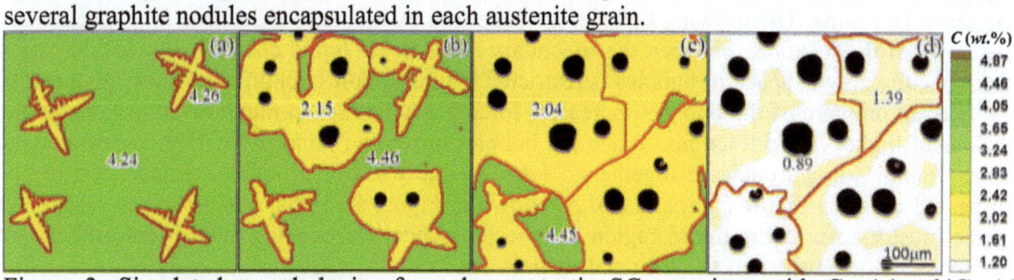

Figure 3. Simulated morphologies for a hypoeutectic SG cast iron with C_0=4.1wt.%C: (a) f_s=7%,T=1163°*C*; (b) f_s=40%, T=1147°*C*; (c) f_s=84%; T=1146°*C*; (d) f_s=100%; T=740°*C*. (Numbers on the figures show the local carbon concentration, and f_s is the total solid fraction).

Dendrite Growth and Microporosity Formation.

Figure 4 presents the evolution of porosities and dendrites of an Al-7wt.%Si alloy simulated by the CA-FDM model [18]. It is seen that the concentrations of Si and H in liquid increases with dendritic growth and Si is obviously enriched at the S/L interface, while H concentration is less enriched at the S/L interface. When hydrogen concentration in liquid is increased to be higher than the supersaturation for porosity nucleation, one pore precipitates from the liquid. To better match the experimental microstructure, the location of pore nucleus was artificially assigned. It is noticed that in the early stage, the pore grow spherically due to the effect of the gas/liquid interfacial tension. After the pore contacts and impinges on dendrites, however, its shape becomes non-spherical. Since the growing pores become sinks for the supersaturated hydrogen in liquid, the local hydrogen concentration close to the pore is lower than that far away from it.

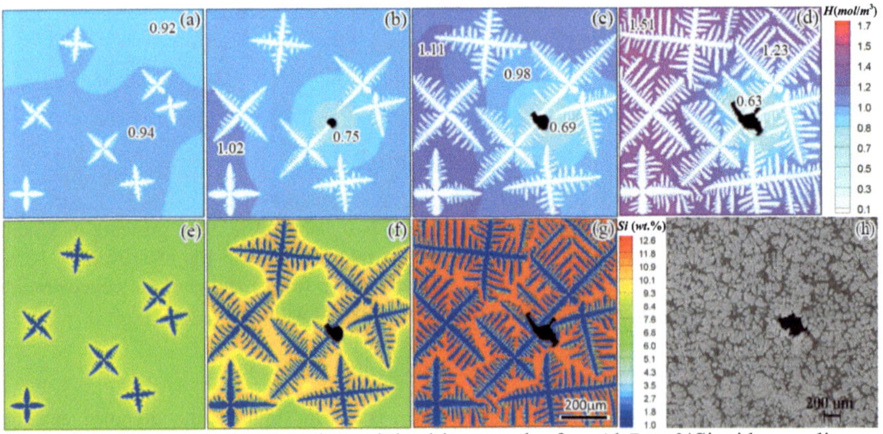

Figure 4. Evolution of microporosity and dendrite growth of an Al-7 wt.%Si with a cooling rate of 10°*C/s* and H_0=0.9 *mol./m*^{3}: (a, e) *T*=606°*C*; (b) 601°*C*; (c, f) T=597°*C*, (d, g) T=580°*C*, (a, b, c, d) H concentration field, (e, f, g) Si concentration field, (h) experiment [27].(domain: 300×300 with Δ*x*=3 *μm*, and numbers on the figures show the local hydrogen concentration).

Figure 5 presents the evolution of columnar dendrites and gas bubbles of an Al-4*wt.*%Cu alloy in directional solidification simulated by the CA-LBM model [26]. As shown, after initial competition, three dendrites overgrow the others. When the solid fraction reach around 14%, the supersaturated gas element is separated to form tiny bubbles mostly at the S/L interface between the secondary arms. The growing bubbles are squeezed by the narrow dendrite arm spacing and escape from the locations between the secondary arms to the liquid channels. As solidification proceeds, the growth space of bubbles is restricted by the complex dendrite network. As a result, the internal pressure of the bubbles increases, forcing the bubbles to move or jump towards the regions of higher liquid fractions. When the bubbles move and jump in the liquid channels, the nearby bubbles might coalesce together to form a large worm-like bubble, as indicated by the circle and arrow in Figure 5 (b) and (c). As shown in Figure 5 (d) that some bubbles escape out of the mushy zone to the liquid region and become spherical bubbles, while some others are entrapped in the mushy zone by the dendrite network.

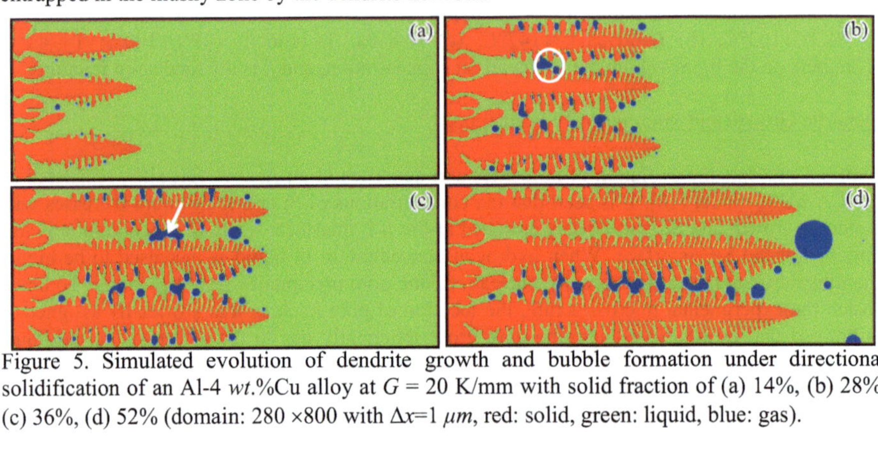

Figure 5. Simulated evolution of dendrite growth and bubble formation under directional solidification of an Al-4 *wt.*%Cu alloy at *G* = 20 K/mm with solid fraction of (a) 14%, (b) 28%, (c) 36%, (d) 52% (domain: 280 ×800 with Δ*x*=1 *μm*, red: solid, green: liquid, blue: gas).

Conclusions

The cellular automaton (CA) and lattice Boltzmann method (LBM) based models for the simulations of solidification microstructures are presented. The CA models can reasonably describe the kinetics of phase transformation, and the thermal and solutal transport during dendritic growth, eutectic solidification, and dendritic growth with pore formation. The coupled models have been proposed by incorporating CA with LBM to simulate dendritic growth with melt convection, as well as the formation and movement of gas bubbles during alloy solidification. Efforts will be made continuously to further improve the present models, develop new models to simulate microstructure formation of multi-phases in multi-scales, and extend the applications of microstructure simulation to practical material processes.

Acknowledgments

We wish to thank Professor Doru M. Stefanescu at the Ohio State University, and Professor Dierk Raabe at Max-Planck-Institut für Eisenforschung, Germany, for helpful discussions. The work was supported by the National Natural Science Foundation of China (Grant No. 51371051).

References

1. M. Asta et al., "Solidification Microstructures and Solid-State Parallels: Recent Developments, Future Directions," *Acta Mater*, 57 (2009), 941-971.

2. M. Rappaz and Ch.-A. Gandin, "Probabilistic Modelling of Microstructure Formation in Solidification Processes," *Acta Metall Mater*, 41(1993), 345-360.

3. L. Nastac, "Numerical Modeling of Solidification Morphologies and Segregation Patterns in Cast Dendritic Alloys," *Acta Mater*, 47(1999), 4253-4262.

4. M. F. Zhu and C. P. Hong, "A Modified Cellular Automaton Model for the Simulation of Dendritic Growth in Solidification of Alloys," *ISIJ Int*, 41(5) (2001), 436-445.

5. Z. J. Liang, Q. Y. Xu, and B. C. Liu, "3D Modeling and Simulation of Dendritic Growth during Solidification," *J. Mater Sci Technol*, 19 (2003), Suppl. 1-4.

6. L. Belteran-Sanchez and D. M. Stefanescu, "A Quantitative Dendrite Growth Model and Analysis of Stability Concepts," *Metall Mater Trans A*, 35A (2004): 2471-2485.

7. H. B. Dong and P.D. Lee, "Simulation of the Columnar-to-equiaxed Transition in Directionally Solidified Al–Cu Alloys," *Acta Mater*, 53 (2005), 659-668.

8. M.F. Zhu and D.M. Stefanescu, "Virtual Front Tracking Model for the Quantitative Modeling of Dendritic Growth in Solidification of Alloys," *Acta Mater*, 55 (2007), 1741-1755.

9. L. Wei et al., "A Cellular Automaton Model for the Solidification of a Pure Substance," *Appl Phys A* 103 (2011), 123-133.

10. S.Y. Pan and M.F. Zhu, "A Three-Dimensional Sharp Interface Model for the Quantitative Simulation of Solutal Dendritic Growth," *Acta Mater*, 58(2010), 340-352.

11. X. Zhang et al., "A Three-Dimensional Cellular Automaton Model for Dendritic Growth in Multi-Component alloys," *Acta Mater*, 60 (2012), 2249-2257.

12. M.F. Zhu et al., "Computational Modeling of Microstructure Evolution in Solidification of Aluminum Alloys," *Metall Mater Trans B*, 38B (8) (2007), 517-524.

13. M.F. Zhu, S. Y. Lee, and C.P. Hong, "Modified Cellular Automaton Model for the Prediction of Dendritic Growth with Melt Convection," *Phys Rev E*, 69 (2004), 061610.

14. L. Yuan and P. D. Lee, "Dendritic Solidification under Natural and Forced Convection in Binary Alloys: 2D Versus 3D Simulation," *Modelling Simul Mater Sci Eng*, 18 (2010), 055008

15. M. F. Zhu and C.P. Hong, "Modeling of Microstructure Evolution in Regular Eutectic Growth," *Phys Rev B*, 66 (2002), 155428.

16. M.F. Zhu and C.P. Hong, "Modeling of Irregular Eutectic Microstructures in Solidification of Al-Si Alloys," *Metall Mater Trans A*, 35A (2004), 1555-1563.

17. M. F. Zhu et al., "Numerical Simulation of Microstructure Evolution during Alloy Solidification by using Cellular Automaton Method," *ISIJ Int*, 50(12) (2010), 1851-1858.

18. M. Zhu et al., "Cellular Automaton Modeling of Microporosity Formation during Solidification of Aluminum Alloys," *ISIJ Int*, 54 (2) (2014), 384-391.

19. D. Raabe, "Overview of the Lattice Boltzmann Method for Nano- and Microscale Fluid Dynamics in Materials Science and Engineering," *Modelling Simul Mater Sci Eng*, 12 (2004), R13-R46.

20. W. Miller, S. Succi, and D. Mansutti, "Lattice Boltzmann Model for Anisotropic Liquid-Solid Phase Transition," Phys. Rev. Lett., 86 (2001), 3578-3581.

21. D. K. Sun et al., "Lattice Boltzmann Modeling of Dendritic Growth in a Forced Melt Convection," *Acta Mater*, 57 (2009), 1755-1767.

22. H. Yin, S.D. Felicelli, and L. Wang, "Simulation of a Dendritic Microstructure with the Lattice Boltzmann and Cellular Automaton Methods," *Acta Mater*, 59 (2011), 3124-3136.

23. M. Zhu et al., "Modelling of Dendritic Growth during Alloy Solidification under Natural Convection," *Modelling Simul Mater Sci Eng*, 22 (2014), 034006.

24. J. E. Taylor, "II—Mean Curvature and Weighted Mean Curvature," *Acta Metall Mater*, 40(7) (1992), 1475-1485.

25. Y. H. Qian, D. d'Humibres, and P. Lallemand, "Lattice BGK Models for Navier-Stokes Equation," *Europhys Lett*, 17(6) (1992), 479-484.

26. W. Wu et al., "Modelling of dendritic growth and bubble formation," *IOP Conf Ser Mater Sci Eng*. 33(2012), 012103.

27. L. Yao et al., "Study of Microporosity Formation under Different Pouring Conditions in A356 Aluminum Alloy Castings" (Paper presented at 140th TMS Annual Meeting, San Diego, CA, 27 February, 2011), 783.

Advances in the Science and Engineering of Casting Solidification
Edited by: Laurentiu Nastac, Baicheng Liu, Hasse Fredriksson, Jacques Lacaze, Chun-Pyo Hong, Adrian V. Catalina, Andreas Buhrig-Polaczek, Charles Monroe, Adrian S. Sabau, Roxana Elena Ligia Ruxanda, Alan Luo, Subhayu Sen, and Attila Diószegi
TMS (The Minerals, Metals & Materials Society), 2015

A LATTICE BOLTZMANN MODEL FOR DENDRITIC GROWTH UNDER NATURAL CONVECTION

Mohammad Hashemi[1], Mohsen Eshraghi[2], Sergio Felicelli[1]

[1] Department of Mechanical Engineering, The University of Akron, Akron, OH 44325
[2] Department of Mechanical Engineering, California State University Los Angeles, Los Angeles, CA 90032

Keywords: Solidification, Modeling, Lattice Boltzmann Method, Cellular Automaton, Natural Convection

Abstract

A two-dimensional (2D) lattice Boltzmann (LB) and Cellular Automaton (CA) model is developed to study the buoyancy-induced flow pattern during equiaxed and columnar dendritic solidification in a binary alloy. The complex flow pattern during solidification has a significant influence on redistribution of solute and heat. In the present work, lattice Boltzmann method (LBM) is used to solve for solute diffusion, heat transfer and fluid flow, while CA is employed to capture the solidification interface. The results show that natural convection has a significant effect on morphology of both columnar and equiaxed dendrites.

Introduction

Segregation defects formed during solidification of metallic alloys are of utmost importance for casting industries, as they degrade the mechanical properties of the castings. Of these defects are the channel-like macrosegregation defects, also known as freckles, often observed during directional solidification of metallic alloys. The channels form in the mushy zone between the dendritic arms, declining the mechanical properties and causing subsequent rejection of the casting products. Complex thermosolutal convection induces the formation of these defects. Studying the flow pattern during solidification can provide a more comprehensive picture of this phenomenon. Flemings and co-workers [1-3] considered the inerdendritic fluid flow through a fixed dendritic solid network and achieved "local solute redistribution function (LSRE)", as a mathematical model to explain the behavior of flow in solidification process. Mehrabian et al. [4], suggested Darcy's law to calculate interdendritic flow velocity which induces microsegregation. They assumed the interdendritc spacing between dendrites to be a porous media. Fujii et. al [5] extended the work of Flemings and Mehrabian [6] to ternary alloys to study macrosegregation in multicomponent low alloy steels. They solved a coupled set of equations given by Darcy's law and LSRE. However, they took temperature from measurement. In the mid-1980s, Benonn and Incropera [7], and Beckermann and Viskanta [8] derived continuum or volume-averaged models for macrosegregation in alloy solidification. They solved solute concentration, temperature and fluid flow for solid, mushy and liquid regions of the solidification system. Ramirez and Beckermann [9] suggested a criterion to predict the formation of freckles in Pb-Sn and Ni-based superalloys based on a maximum value for Rayleigh number. The numerical results obtained in these studies reveal the existence of complex thermo-solutal convection patterns in the melt. In particular, Bennon and Incropera [10] presented the first direct numerical simulations of freckles. Felicelli et al. [11] explained the emergence and survival of channels by a two dimensional mathematical model for Pb-10wt%Sn alloy. Also, Felicelli et al. [12] developed a three dimensional finite element model using a thermodynamic function to express the solidification path. Recently, Yuan and Lee [13] developed a three dimensional microsclae model for freckling in Pb-Sn alloys.

Lattice Boltzmann Method (LBM) is a numerical technique that offers many advantages over conventional computational fluid dynamics methods including: ability to handle arbitrarily complex geometry, local structure, and good scalability for parallel computations. LBM has been used in various fields of science and engineering, including phase change and solidification. Sun et al. [14] introduced a two-dimensional (2D) LB–CA model for dendritic solidification. Eshraghi and Felicelli [15] suggested a new implicit variation of LBM to solve the heat conduction problem with phase change. Eshraghi et al. [16] also developed a three dimensional LB-CA model for dendrite growth. However, the convection effects were not included in the above-mentioned studies. Jelinek et al. [17] presented a scalable parallel LB model for dendritic growth.

In the present work, a two dimensional LB-CA model is developed to study the buoyancy-induced flow pattern during equiaxed and columnar dendritic solidification in a binary alloy.

Model Description

Governing Equations

In the present model, we assume that molten metal is an incompressible fluid and the dendrite growth is calculated by the difference between the local actual and local equilibrium solute compositions at the local temperature. The governing equations for mass, momentum, energy and solute are given by:

$$\nabla.(f_l\vec{u}) = 0 \tag{1}$$

$$\frac{\partial}{\partial t}(f_l\vec{u}) + \nabla.(f_l\vec{u}\vec{u}) - \nabla.\left[\frac{\mu}{\rho}\nabla(f_l\vec{u})\right] = -\frac{f_l}{\rho}\nabla P + S \tag{2}$$

$$\frac{\partial T}{\partial t} + \nabla.(\vec{u}T) = \frac{\lambda}{\rho c_p}\nabla^2 T - \frac{L}{c_p}\frac{\partial f_l}{\partial t} \tag{3}$$

$$\frac{\partial C_e}{\partial t} + \nabla.(\vec{u}C_l) = \nabla.(D_e\nabla C_l) \tag{4}$$

where $\vec{u}$ is velocity vector in liquid, t is time, P is pressure, μ is viscosity, c_p is the specific heat, λ is thermal conductivity, ρ is the liquid density, f_l is the fraction of liquid, L is the latent heat, C_s and C_l are the average solute concentrations of the solid and liquid, T is the temperature, and S is the source term for buoyancy flow. C_e and D_e are the equivalent concentration and solute diffusion coefficient, respectively. Natural buoyancy force can be characterized by this formula:

$$F = -\rho_0 g \beta_T (T - T_{ref}) - \rho_0 g \beta_C (C_l - C_{ref}) \tag{5}$$

where g is the gravitational vector, ρ_0 is the fluid density at temperature T_{ref} and composition C_{ref}, β_C and β_T are the expansion coefficients for composition and temperature. The direction of gravitational vector is perpendicular and toward south wall in Figure 1 and 2.

Lattice Boltzmann formulation

According to Bhatnagar–Gross–Krook (BGK) approximation [13], the discrete form of the LB equation satisfying the momentum conservation, including a force term can be written as:

$$f_i(x + c_i \Delta t, t + \Delta t) - f_i(x,t) = -\frac{f_i(x,t) - f_i^{eq}(x,t)}{\tau_V} + \Delta t F_i(x,t) \tag{6}$$

where $f_i(x,t)$ is the particle distribution function (PDF) representing the probability of finding a particle at location, x, at time, t, c_i is the discrete moving velocity of the pseudo-particle, Δt is the time step, τ_V is the relaxation time for fluid flow model, $f_i^{eq}(x,t)$ is the equilibrium PDF (EPDF), and $F_i(x,t)$ is the force term caused by natural buoyancy force.

$$F_i = -3 w_i \rho c_i F / c^2 \tag{7}$$

Where w_i are the weight coefficients given by $w_0 = 4/9$, $w_{1-4} = 1/9$ and $w_{5-9} = 1/36$. $c = \Delta x / \Delta t$ is lattice speed. The energy equation involves heat diffusion, convection flow and latent heat release. The latent heat is added to the LB equation as a source term:

$$h_i(x + e_i \Delta t, t + \Delta t) - h_i(x,t) = -\frac{h_i(x,t) - h_i^{eq}(x,t)}{\tau_T} + \Delta t H_i(x,t) \tag{8}$$

where $h_i(x,t)$ and $H_i(x,t)$ are the particle distribution function for temperature and latent heat source term, respectively. $h_i^{eq}(x,t)$ represents EPDF for heat transfer and τ_T is the relaxation time for the heat transfer model. The latent heat source term can be calculated as:

$$H_i = w_i \Delta \emptyset_s L / C_p \tag{9}$$

$\Delta \emptyset_s$ is the volumetric solid fraction change. The LB equation for solute diffusion can be written in a similar way:

$$g_i(x + e_i \Delta t, t + \Delta t) - g_i(x,t) = -[g_i(x,t) - g_i^{eq}(x,t)/\tau_C] \tag{10}$$

where $g_i(x,t)$ and $g_i^{eq}(x,t)$ are the PDF and EPDF for solute transport. τ_D is taken as the relaxation time for the diffusion LB equation. The macroscopic density, velocity, solute composition and temperature can be calculated as:

$$\rho = \sum_i f_i \tag{11}$$

$$u(x,t) = \frac{1}{\rho}\sum_i c_i f_i(x,t) \tag{12}$$

$$C_l = \sum_i g_i \tag{13}$$

$$T = \sum_i h_i \tag{14}$$

The macroscopic velocity $u(x,t)$ calculated by Eq. (12) will be used in the equilibrium function. The equilibrium functions for fluid flow, temperature and solute transport models can be presented as:

$$f_i^{eq}(x,t) = w_i\rho\left[1 + \frac{3(e_i.u)}{c^2} + 4.5\frac{(e_i.u)^2}{c^2} - 1.5\frac{u^2}{c^2}\right] \tag{15}$$

$$h_i^{eq}(x,t) = w_i T[1 + \frac{3(e_i.u)}{c^2} + 4.5\frac{(e_i.u)^2}{c^2} - 1.5\frac{u^2}{c^2}] \tag{16}$$

$$g_i^{eq}(x,t) = w_i C\left[1 + \frac{3(e_i.u)}{c^2} + 4.5\frac{(e_i.u)^2}{c^2} - 1.5\frac{u^2}{c^2}\right] \tag{17}$$

Evolution of the solid-liquid interface

In the present model, it is assumed that the driving force for solidification is the difference between the local equilibrium and local actual concentrations. The local equilibrium concentration can be calculated by:

$$C_l^*(x,t) = C_0 + \frac{T_l^*(x,t) - T_l^{eq} + \Gamma K[1 - 15\varepsilon cos4(\theta - \theta_0)]}{m_l} \tag{18}$$

where T_l^{eq} stands for the equilibrium liquidus temperature at the initial concentration C_0, T_l^* is the interface temperature, ε is the degree of anisotropy of surface energy, Γ is the Gibbs-Thomson coefficient, K is defined as the curvature of solid/liquid interface, θ is the growth angle between normal vector to the interface and the x-axis, and θ_0 represent the preferential growth direction with respect to x-axis. m_l is the slope of the liquidus line in the phase diagram. The local actual concentration and temperature are obtained using LBM. The change in the fraction of solid in each interface cell after one time step can be calculated by:

$$\Delta\emptyset_s = (C_l^* - C_l)/[C_l^*(1-k)] \tag{19}$$

where k is the solute partition coefficient. As the solid fraction increases, solute is rejected to the interface and distributed equally to the nearest interface cells:

$$C_i(x) = w_i \Delta\emptyset_s C_l(x)(1-k) \tag{20}$$

The interface curvature can be calculated as:

$$K = \left[2\partial_x\emptyset_s\partial_y\emptyset_s\partial_{xy}^2\emptyset_s - (\partial_x\emptyset_s)^2\partial_y^2\emptyset_s - (\partial_y\emptyset_s)^2\partial_x^2\emptyset_s\right].\left[(\partial_x\emptyset_s)^2 + (\partial_y\emptyset_s)^2\right]^{-3/2} \qquad (21)$$

A Cellular Automaton (CA) scheme is implemented to capture new interface cells.

Results and Discussion

Equiaxed dendrite growth

The evolution of an equiaxed dendrite is demonstrated and the effect of natural convection on the dendrite growth is investigated. The calculation domain consists of a 600×600 grid which is equivalent to a 200μm×200μm domain. Figure 1 presents the simulated morphology of a single equiaxed dendrite for a binary alloy with $C_0 = 0.4\%$ in an undercooled melt with ΔT=0.8 K with and without natural convection. The growth time is 1.25 s for all cases presented in this figure. Parts (a) and (b) show the solutal field, and parts (c) and (d) show the thermal field. The velocity vector plots indicate the strength and direction of natural flow.The Rayleigh numbers are taken as $Ra_C = Ra_T = 5 \times 10^3$. The material properties are given in [18]. The following boundary conditions are implemented to compute the unknown PDFs at the boundary cells of a 2D Square domain. All four sides of domain are assumed to be solid walls. Therefore, bounce-back rule is applied for unknown PDFs at all four boundaries. A similar boundary condition is also applied at the solid/liquid and a zero-flux boundary condition is implemented on all boundaries of the calculation domain for solute transport and heat transfer.

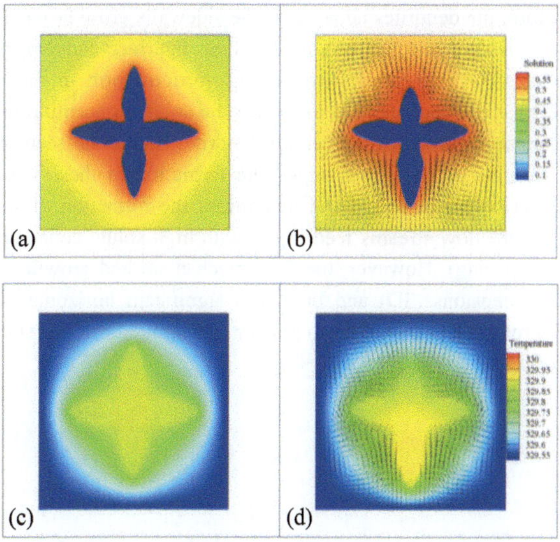

Figure 1. Simulated morphologies of an equiaxed dendrite freely growing in an undercooled melt (ΔT = 0.8 K) without convection: (a) and (c), and with natural convection: (b) and (d).

Solute is rejected to the interface during solidification and simultaneously this phenomenon occurs with release of heat. Solute and temperature gradients lead to natural convection flows which induce some vortex flows as it can be observed in Figure 1. The downward arm grows faster in comparison with the others, since the buoyancy-induced flow washes the solute from the interface of that arm and leads to a higher difference between local actual and local equilibrium concentrations. Consequently, more liquid transforms to solid around the downward arm, according to Eq. (19). It should be noted that the effect of convection is not considered in the heat transfer model.

Columnar dendrite Growth

At the beginning of simulation, five dendrite seeds were placed at the bottom of the domain. The simulation domain consists of 300×300 lattice points that represents a 60μm×60μm domain. The Rayleigh number was taken as $Ra_C = 10^5$. Figure 2 shows the simulated morphologies of the columnar dendrites at three different time steps. The material considered for the simulation was Fe-0.36%C steel alloy, but a lower viscosity value of 0.042×10^{-3} Pa.s was used to amplify the fluid flow effects. The boundary conditions for flow field and solute transport were similar to the equiaxed case. The temperature gradient (G) and casting speed (R) were assumed to be 3,700 m.$s^{-1}$and 0.054 m.s^{-1}, respectively.
As can be seen in Figure 2(a), the streams arising between dendrites due to buoyancy force combine to form an upward flow field above the central dendrites. The direction of these convection flows increases the solute concentration at the center of the domain. As the solidification proceeds, solute accumulates between dendritic arms. The dendrites adjacent to the walls experience less solute accumulation, because there is not solute rejection from the wall's side. Therefore, the dendrites adjacent to the sidewalls grow faster. Due to this higher growth rate, these dendrites reject more solute to their interface and suppress the growth of neighboring dendrites. Over the time, two strong vortices form in the mid-width of the domain (Figure 2(b)). The vortices help to reduce the solute concentration in the middle of the domain, accelerating growth of the central dendrite. As can be seen in Figure 2(c), the streams become stronger at later time steps. The formation of these vortices significantly influences the redistribution of solute and determines the formation of segregation defects such as freckles, where flow streams feed liquid with high solute concentration in channels (interdendritic arm spacing). However, the flow mechanism and growth kinetics are totally different in three-dimensions (3D) and involve a significant horizontal flow between the dendritic arms. Therefore, it is necessary to perform the simulations in 3D in order to capture the physics correctly.

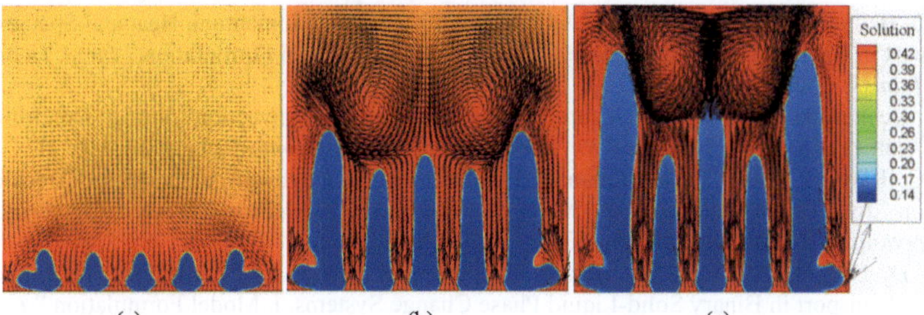

Figure 2. Simulated morphologies of columnar dendrites freely growing under the effect of natural convection at different time steps: (a) 6.67 ms, (b) 27.34 ms, and (c) 37.32 ms.

Conclusions

In summary, a two dimensional LB-CA model is developed for simulation of dendrite growth in the presence of natural convection. Calculations were performed for equiaxed and columnar dendritic solidification. For the equiaxed dendrite, it was observed that the downward arm grow faster in comparison with the others under the effect of free convection. In the case of columnar dendrites, channels with high solute composition form during solidification. A buoyancy-induced vortex stream may form during solidification. These vortices alter redistribution of solute and energy, and significantly affect the kinetics of dendrite growth. The fluid flow induced by natural convection determines the mechanism of solute redistribution and eventual formation of segregation defects such as freckles. Nevertheless, 2D simulations are not capable of capturing the correct physical phenomena, since the fluid regimes and growth kinetics are completely different in 3D. Considering the advantages offered by LBM, extension of the current model to 3D can provide a great tool for simulation of fluid flow and microsegregation during dendritic solidification of metallic alloys, which is the next step we are pursuing in this research project.

References

1. M.C. Flemings, and G.E. Nereo, "Macrosegregation: Part I," *Trans. AIME*, 239 (1967), 1449-1461.

2. M.C. Flemings, R. Mehrabian, and G.E. Nereo, "Macrosegregation: Part II," *Trans. AIME*, 242 (1968), 41-49.

3. M.C. Flemings, and G.E. Nereo, "Macrosegregation: Part III," *Trans. AIME*, 242 (1968), 50-55.

4. R. Mehrabian, M. Keane, and M.C. Flemings, "Interdendritic Fluid Flow and Macrosegregation; Influence of Gravity," *Metall. Trans.*, 1 (1970), 1209-1220.

5. T. Fujii, D.R. Poirier, and M.C. Flemings, "Macrosegregation in a Multicomponent Low Alloy Steel," *Metall. Trans. B*, 10(1979), 331-339.

6. R. Mehrabian, and M.C. Flemings, "Macrosegregation in Ternary Alloys," *Metall. Trans.*, 1 (1970), 455-464.

7. W.D. Bennon, and F.P. Incropera, “A Continuum Model for Momentum, Heat and Species Transport in Binary Solid-Liquid Phase Change Systems. I. Model Formulation,” *Int. J. Heat Mass Transfer*, 30 (1987), 2161-2170.
8. C. Beckermann, and R. Viskanta, “Double-Diffusive Convection during Dendritic Solidification of a Binary Mixture,” *PhysicoChemical Hydrodynamics*, 10 (1988), 195-213.
9. J. C. Ramirez, and C. Beckermann, “Evaluation of a Rayleigh-number-based freckle criterion for Pb-Sn alloys and Ni-base superalloys,” *Metallurgical and Materials Transactions A*, *34*(7) (2003), 1525-1536.
10. W.D. Bennon and F.P. Incropera, “A Continuum Model for Momentum, Heat and Species Transport in Binary Solid-Liquid Phase Change Systems. I. Model Formulation,” *Int. J. Heat Mass Transfer*, 30 (1987), 2161-2170.
11. S. D. Felicelli, J. C. Heinrich, and D. R. Poirier, ”Simulation of freckles during vertical solidification of binary alloys,” *Metallurgical Transactions B*, *22*(6) (1991), 847-859.
12. S.D. Felicelli, D. R. Poirier, and J. C. Heinrich, “Modeling freckle formation in three dimensions during solidification of multicomponent alloys,” *Metallurgical and Materials Transactions B*, *29* (4) (1998), 847-855.
13. L. Yuan, and P. D. Lee, “A new mechanism for freckle initiation based on microstructural level simulation,” *Acta Materialia*, *60*(12) (2012), 4917-4926.
14. D. Sun, M. Zhu, S. Pan, and D. Raabe, “Lattice Boltzmann modeling of dendrite growth in a forced melt convection,” *Acta Materialia*, *57*(6) (2009), 1755-1767.
15. M. Eshraghi, and S. D. Felicelli,”An implicit lattice Boltzmann model for heat conduction with phase change,” *International Journal of Heat and Mass Transfer*, *55*(9) (2012), 2420-2428.
16. M. Eshraghi, S. D. Felicelli, B. Jelinek”Three dimensional simulation of solutal dendrite growth using lattice Boltzmann and cellular automaton methods,” *Journal of Crystal Growth*, *354*(1) (2012), 129-134.
17. B. Jelinek, M. Eshraghi, S. Felicelli, and J. F. Peters, “Large-scale parallel lattice Boltzmann–cellular automaton model of two-dimensional dendritic growth,” *Computer Physics Communications*, *185*(3) (2014), 939-947.
18. D. K. Sun, M. F. Zhu, S. Y. Pan, C. R. Yang, and D. Raabe, ”Lattice Boltzmann modeling of dendritic growth in forced and natural convection,” *Computers & Mathematics with Applications*, *61*(12) (2011), 3585-3592.

Advances in the Science and Engineering of Casting Solidification
Edited by: Laurentiu Nastac, Baicheng Liu, Hasse Fredriksson, Jacques Lacaze, Chun-Pyo Hong, Adrian V. Catalina, Andreas Buhrig-Polaczek, Charles Monroe, Adrian S. Sabau, Roxana Elena Ligia Ruxanda, Alan Luo, Subhayu Sen, and Attila Diószegi
TMS (The Minerals, Metals & Materials Society), 2015

MODELING OF DENDRITIC STRUCTURE AND MICROSEGREGATION IN SOLIDIFICATION OF Al-RICH QUATERNARY ALLOYS

Ting Dai[1], Mingfang Zhu[1], Shuanglin Chen[2], Weisheng Cao[2]

[1] Jiangsu Key Laboratory of Advanced Metallic Materials, School of Materials Science and Engineering, Southeast University, Nanjing, Jiangsu, 211189, China
[2]CompuTherm LLC; 437 S. Yellowstone Dr., Suite 217; Madison; WI 53719, USA

Keywords: solidification, modeling, cellular automaton, PanEngine, dendritic growth, microsegregation, Al-rich quaternary alloy

Abstract

A two-dimensional cellular automaton (CA) model is coupled with a CALPHAD tool for the simulation of dendritic growth and microsegregation in solidification of quaternary alloys. The dynamics of dendritic growth is calculated according to the difference between the local equilibrium liquidus temperature and the actual temperature, incorporating with the Gibbs–Thomson effect and preferential dendritic growth orientations. Based on the local liquid compositions determined by solving the solutal transport equation in the domain, the local equilibrium liquidus temperature and the solid concentrations at the solid/liquid (SL) interface are calculated by the CALPHAD tool. The model was validated through the comparisons of the simulated results with the Scheil predictions for the solid composition profiles as a function of solid fraction in an Al-6wt%Cu-0.6wt%Mg-1wt%Si alloy. It is demonstrated that the model is capable of not only reproducing realistic dendrite morphologies, but also reasonably predicting microsegregation patterns in solidification of Al-rich quaternary alloys.

Introduction

Microstructure and microsegregation of solidified metallic materials are closely associated with the properties of final products. Since most commercial metallic materials are multi-component alloy systems, research of microstructure and microsegregation evolution in solidification of multi-component alloys has received increasing interest [1]. With the development of powerful computers and advanced numerical techniques, numerical modeling has emerged as an important and indispensable tool of solidification research. Various numerical approaches are recently proposed to handle the complex patterns in microstructure and microsegregation evolution which involves complicated phenomena, such as thermal/solutal transport, dendrite coarsening, the curvature and kinetic effects on the moving solid/liquid (SL) interface [2-4]. The models based on the cellular automaton (CA) technique can reproduce a wide range of microstructure features observed experimentally with an acceptable computational efficiency, indicating the excellent potential for engineering applications. The CA models have thus recently achieved a considerable interest and advances in modeling of microstructure evolution during solidification [5-7].

In this paper, a two-dimensional (2D) cellular automaton (CA)-PanEngine (a CALPHAD tool) coupled model, which was developed for microstructure simulation of ternary alloys [8], is further extended to apply to quaternary alloy systems. The model is applied to simulate dendrite growth morphology and microsegregation in an Al-6wt%Cu-0.6wt%Mg-1wt%Si alloy. The simulated results are validated by the comparison with the predictions of the Scheil model.

Model Description and Numerical Algorithm

It is well known that for most commercial multi-component alloys the solute partition coefficient and the liquidus slope vary significantly with temperature and compositions. Therefore, it is necessary to couple microstructure simulation models with thermodynamic calculations in the modeling of microstructure and microsegregation of multi-component alloys. The present work was performed based on a CA approach coupled with PanEngine which is the kernel part of a thermodynamic and phase equilibrium calculation software package PANDAT [9]. To improve computational efficiency, a data tabulation coupling strategy between CA and PanEngine was adopted. Before starting the CA microstructure simulation, a data file is generated by PanEngine, in which the equilibrium liquidus temperatures and the equilibrium solid compositions are tabulated with respect to the relevant liquid compositions for a uniform grid spacing of 1at%. During the CA simulation, the needed data of equilibrium liquidus temperature and interface solid compositions are interpolated from the data stored in the tabulation grids.

To describe the evolution of microstructure and microsegregation in a quaternary alloy system, three solute fields must be calculated. The governing equation for solute diffusion is given by

$$\frac{\partial C_i(m)}{\partial t} = \nabla \cdot \left[D_i(m)\nabla C_i(m)\right] \qquad (i = s, l \text{ and } m = 1,2,3) \tag{1}$$

where $C_i(m)$ is the concentration and $D_i(m)$ is the diffusion coefficient of solute element (m) in phase (i). All diffusion coefficients $D_i(m)$ are considered to be independent of composition but temperature dependent. Three solutes are considered to diffuse independently and cross diffusion is neglected. Eq. (1) is solved using an explicit finite difference scheme. The zero-flux boundary condition is applied for the cells located at the four walls of the calculation domain.

The growth of the dendrite is driven by the local undercooling. The local undercooling at time tn, $\Delta T(t_n)$, is considered to be the difference between the local equilibrium liquidus temperature $T^*(t_n)$ and local actual temperature $T(t_n)$, incorporating the effect of curvature and it is given by

$$\Delta T(t_n) = T^*(t_n) - T(t_n) - \Gamma(\theta)K(t_n) \tag{2}$$

where $\Gamma(\theta)$ is the Gibbs-Thomson coefficient, and $K(t_n)$ is the local interface curvature. The local equilibrium liquidus temperature $T^*(t_n)$ is obtained with the aid of PanEngine based on the local liquid compositions of three solutes. In the present work, the temperature in the domain is assumed to be uniform and cool down at a certain cooling rate. Thus, $T(t_n)$ in Eq.(2) is determined by an imposed cooling rate. The local interface curvature $K(t_n)$ is calculated using the following equation.

$$K(t_n) = \left[\left(\frac{\partial f_s}{\partial x}\right)^2 + \left(\frac{\partial f_s}{\partial y}\right)^2\right]^{-3/2} \left[2\frac{\partial f_s}{\partial x}\frac{\partial f_s}{\partial y}\frac{\partial^2 f_s}{\partial x \partial y} - \left(\frac{\partial f_s}{\partial x}\right)^2 \frac{\partial^2 f_s}{\partial y^2} - \left(\frac{\partial f_s}{\partial y}\right)^2 \frac{\partial^2 f_s}{\partial x^2}\right] \tag{3}$$

The growth velocity and local undercooling is related by the classical sharp interface model [10]

$$V_g = \mu_k(\theta) \cdot \Delta T(t_n) \tag{4}$$

where $\mu_k(\theta)$ is the interface kinetic coefficient.

To simulate dendrite growth with specific crystallographic orientations, it is necessary to consider anisotropy in either the surface energy or interfacial attachment kinetics (or both) in the models of dendrite growth [11]. The present model accounts for the anisotropy in both surface energy and interfacial kinetics. It is known that fcc-lattice crystals of Al-rich alloys exhibit the fourfold anisotropies of the surface energy and kinetics at the SL interface. The Gibbs-Thomson coefficient $\Gamma(\theta)$ and the interface kinetics coefficient $\mu_k(\theta)$ are thus given by

$$\Gamma(\theta) = \overline{\Gamma}\{1 - \delta_t \cos[4(\theta - \theta_0)]\} \tag{5}$$

$$\mu_k(\theta) = \overline{\mu}_k \{1 + \delta_k \cos[4(\theta - \theta_0)]\} \tag{6}$$

where $\bar{\mu}_k$, δ_k, $\bar{\Gamma}$, and δ_t are the average interface kinetic coefficient, the degree of the kinetic anisotropy, the average Gibbs-Thomson coefficient, and the degree of the surface energy anisotropy, respectively. θ is the angle between the normal of the SL interface and the horizontal direction, and θ_0 is the preferred dendrite growth orientation. The angle θ can be calculated according to the gradient of solid fraction at the SL interface using the following equation.

$$\theta = \arctan\left(\frac{\partial f_s}{\partial y} \bigg/ \frac{\partial f_s}{\partial x} \right) \tag{7}$$

The growth velocities of interface cells were calculated with Eqs. (2)~(7). The change rate of solid fraction of an interface cell can thus be evaluated from the growth velocity V_g as follows

$$\frac{\partial f_s}{\partial t} = G \frac{V_g}{\Delta x} \tag{8}$$

where Δx is the cell spacing. G is a geometrical factor related to the state of neighbor cells, which is defined by

$$G = \min\left[1, \frac{1}{2}\left(\sum_{m=1}^{4} s_m^I + \frac{1}{\sqrt{2}} \sum_{m=1}^{4} s_m^{II} \right) \right]$$

where S_I and S_{II} indicate the state of the nearest neighbor cells and the second-nearest neighbor cells, respectively. According to the state of a neighbor cell, S_I and S_{II} are determined by

$$s^I, s^{II} = \begin{cases} 0 & (f_s < 1) \\ 1 & (f_s = 1) \end{cases} \tag{9}$$

It is assumed that the local thermodynamic equilibrium is maintained at the SL interface and thus the solidified cells always adopt the equilibrium solid compositions. As dendrite grows, the growing cells reject solutes at the SL interface. The rejected amount of solute element *(m)* is evaluated by

$$\Delta C(m) = \Delta f_s \left(C_l^*(m) - C_s^*(m) \right) \tag{10}$$

where Δf_s is the solid fraction increment of the interface cell at one time step, which is evaluated by Eq.(8),$C_l^*(m)$ and$C_s^*(m)$ are the interface liquid and solid compositions of solute element *(m)*, respectively. The rejected $\Delta C(m)$ is added to the remaining liquid in the same cell and its surrounding neighbor cells. The interface solid composition$C_s^*(m)$in Eq. (10) and interface equilibrium liquidus temperature $T^*(t_n)$ in Eq. (2) are obtained by interpolation according to the local interface liquid compositions of three solutes $C_l^*(1)$, $C_l^*(2)$and$C_l^*(3)$ which are determined by solving Eq.(1). The physical parameters used in the present work are given in Table 1.

Table 1 Physical parameters used in the present work

Symbol	Definition and unites	Value	Refs.
$D_l(Cu)$	Liquid diffusion coefficient of Cu (m^2/s)	$1.05 \times 10^{-7}exp(-2856/T)$	[3]
$D_s(Cu)$	Solid diffusion coefficient of Cu (m^2/s)	$4.8\times10^{-5}exp(-16069/T)$	[3]
$D_l(Mg)$	Liquid diffusion coefficient of Mg (m^2/s)	$9.9\times 10^{-5}exp(-8610/T)$	[3]
$D_s(Mg)$	Solid diffusion coefficient of Mg (m^2/s)	$6.23\times10^{-6}exp(-13831/T)$	[3,12]
$D_l(Si)$	Liquid diffusion coefficient of Si (m^2/s)	$0.11\times10^{-4}exp(-5696/T)$	[12]
$D_s(Si)$	Solid diffusion coefficient of Si (m^2/s)	$2.02\times10^{-4}exp(-16069/T)$	[3,12]
δ_t	Degree of the surface energy anisotropy	0.3	[8]
δ_k	Degree of the kinetic anisotropy	0.3	[8]
$\overline{\Gamma}$	Average Gibbs-Thomson coefficient (mK)	1.7×10^{-7}	[7,8]

Results and Discussion

The CA-PanEngine model is applied to simulate the microstructure and microsegregation in solidification of an Al-6wt%Cu-0.6wt%Mg-1wt%Si quaternary alloy. Figure 1 shows the simulated dendrite morphology and composition maps of copper, magnesium and silicon with a cooling rate of 20*K/s*. The simulation was performed in a domain consisting of 200×200 meshes with a mesh size of 3μm. It can be noted from Figure 1 that the dendrite morphology exhibits well developed side branches. Besides, the center region of the dendrites shows lower compositions than the outside shell. Since the simulation stopped at the end of primary dendrite solidification, the remaining interdendritic liquid will transform to eutectics in the subsequent solidification.

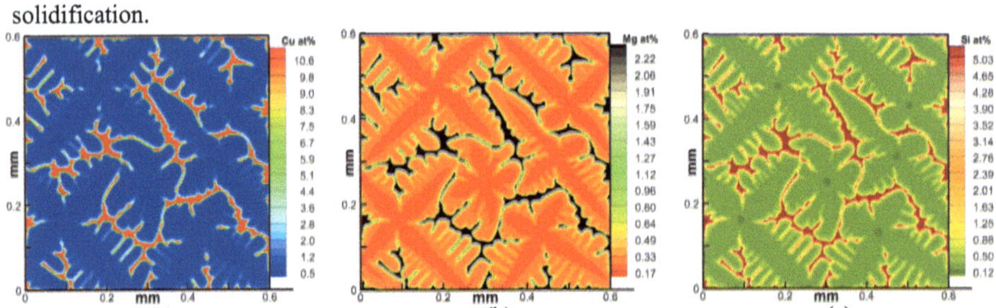

(a) (b) (c)

Fig.1 Simulated equiaxed dendrite morphology and solute maps of (a) Cu, (b) Mg, and (c) Si for an Al-6wt%Cu-0.6wt%Mg-1wt%Si alloy solidified at a cooling rate of 20 K/s.

During dendritic growth, the solute concentrations in the liquid phase increase because of the solute partition at the SL interface, leading to a continuous increase of solute concentrations in the newly formed solid. The variation of solid composition with the increase of solid fraction obtained from the model calculation was measured and compared with the predictions of the Scheil model. The CA simulation involves non-zero solid diffusivity and the limited liquid diffusivity using the realistic solute diffusivities, whereas the Scheil model is derived based on the assumptions of no diffusion in the solid and complete mixing in the liquid. To analyze the effect of diffusion in liquid and solid on microsegregation respectively, the comparison with the Scheil model was made in two steps. First, the simulations were carried out with the condition of zero solid diffusivity (*Ds(m)=0*) and realistic liquid diffusivity. The simulated solid composition profiles of three solutes with the cooling rates of 0.25, 20 and 100*K/s* are compared with the Scheil predictions in Figure 2. It can be noted that the solid compositions of all solutes predicted by the Scheil model are relative lower at the early solidification stage but higher at the late stage. With decreasing of solidification rate, the difference between the simulated data and the Scheil profiles decreases. In case of very low cooling rate of 0.25 K/s, the CA calculated data are nearly superposed on the Scheil profiles.

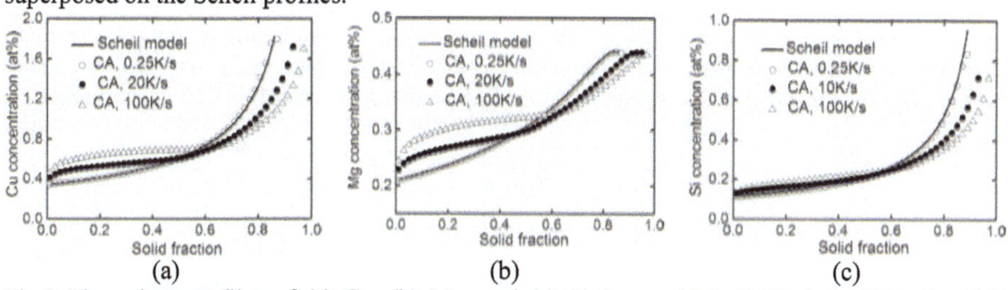

Fig.2 The solute profiles of (a) Cu, (b) Mg, and (c) Si for an Al-6wt%Cu-0.6wt%Mg-1wt%Si alloy solidified with various cooling rates predicted by the Scheil model and the CA-PanEngine model with zero solid diffusivity and realistic liquid diffusivities.

Figure 3 indicates the simulated dendritic morphologies and Cu composition maps with various cooling rates when solid fraction is 0.2. It can be noted from Fig. 3(a) that low cooling rate provides enough time for solute diffusion in liquid, resulting in an almost uniform liquid composition field which is very close to the Scheil condition. However, with an increase in cooling rate, solute gets accumulated in front of the SL interface, which not only enhances the interface instability to cause side branches, but also results in solute trapping in the early solidified region and thus relatively lower composition will be produced in the late solidified region. The higher the cooling rate, the heavier the solute buildup in front of the SL interface as shown in Fig. 3 (b) and (c). Similar results were observed for solutes Mg and Si. Consequently, it is understandable that the difference between the simulated data and the Scheil profiles will increase with solidification rate as indicated in Fig. 2.

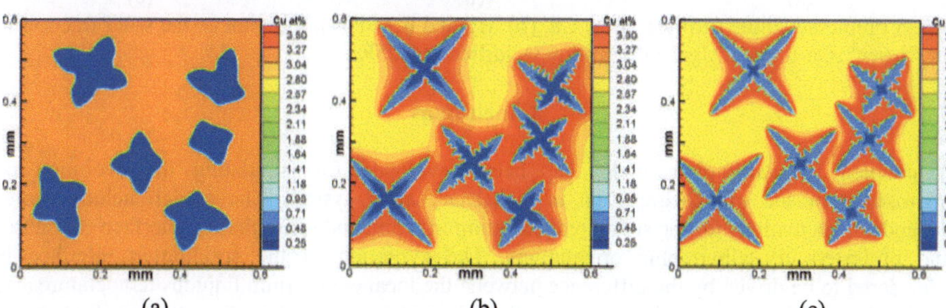

Fig. 3 Simulated dendritic morphologies and Cu composition maps for an Al-6wt%Cu-0.6wt%Mg-1wt%Si alloy (fs=0.2) with various cooling rates of (a) 0.25K/s, (b) 20K/s, and (c)

100K/s.

As the second stage, the simulations were performed with zero and non-zero solid diffusivities to investigate the effect of solid diffusion on the solute microsegregation. Realistic liquid diffusivities were adopted for all simulations. The results with the cooling rates of 0.25 and 50K/s are presented in Figures 4 and 5. It can be seen from Fig. 4 that for the case of 50K/s cooling rate, two sets of data calculated from zero and non-zero solid diffusivity are almost superposed for all solutes. Since in this case the total solidification time is only about 4 seconds, it is not long enough to generate a detectable back diffusion. On the other hand, when the cooling rate slows down to 0.25K/s, the composition profiles of all solutes with non-zero solid diffusivities are higher at the early stage and lower later than those obtained with zero solid diffusivity, indicating the evident effect of back diffusion. This is understandable that the slow solidification process provides longer time for solid diffusion. Accordingly, the present model permits not only reproducing realistic-looking dendritic morphologies, but also predicting reasonably the microsegregation patterns depending on the involved solidification conditions.

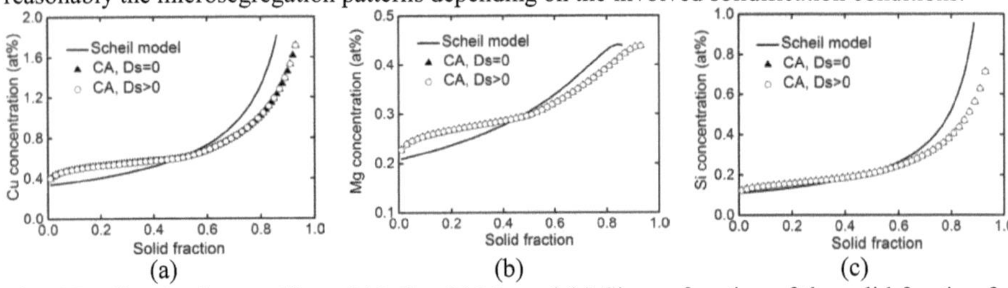

Fig. 4 Predicted solute profiles of (a) Cu, (b) Mg and (c) Si as a function of the solid fraction for an Al-6wt%Cu-0.6wt%Mg-1wt%Si alloy solidified at 20K/s.

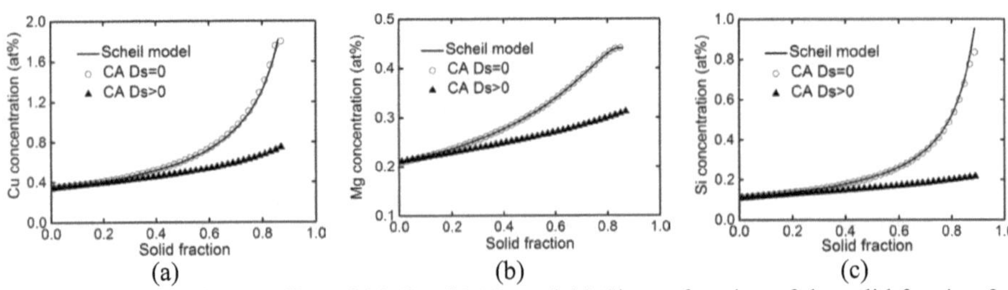

Fig. 5 Predicted solute profiles of (a) Cu, (b) Mg and (c) Si as a function of the solid fraction for an Al-6wt%Cu-0.6wt%Mg-1wt%Si alloy solidified at 0.25K/s.

Conclusions

A coupled CA-PanEngine model has been developed for simulating the evolution of microstructure and microsegregation in quaternary alloy system. The model includes time-dependent calculations for the solute redistribution in liquid and solid phases, interface curvature, and preferred crystallographic orientations. The evolution of the solid/liquid interface is considered to be driven by the difference between the local equilibrium liquidus temperature and the local actual temperature, incorporating the effect of curvature. Based on the interface liquid compositions of three solutes, which are determined by numerically solving the mass transport equation of three solutes in the whole domain, the local equilibrium liquidus temperature and the interface equilibrium solid compositions of three solutes are obtained with the aid of PanEngine. The simulated composition profiles of three solutes in solidified dendrites were compared with the predictions of the Scheil model for an Al-6wt%Cu-0.6wt%Mg-1wt%Si alloy. The results show that the simulated solid compositions with the zero solid diffusivity and the realistic liquid diffusivities are higher than the Scheil predictions in the early stage and lower in the late solidification stage. With the decrease of cooling rate, the simulated solid composition profiles were found to increasingly approach to the Scheil profiles. Moreover, at lower solidification rates, the calculated solid composition profiles with finite diffusion in solid are initially higher and later lower than those obtained with no diffusion in solid, illustrating the effect of back diffusion. However, the effect of back diffusion is not evident when solidification rate increases. The simulation results demonstrate that the present model can predict reasonably the microsegregation patterns in solidification of quaternary alloys depending on the involved cooling conditions. The model is also able to predict dendritic growth morphology with various orientations and well developed side branches.

Acknowledgments

This work was supported by the National Natural Science Foundation of China under Grant No. 50671025.

References

1. U. Hecht et al., "Multiphase solidification in multicomponent alloys," *Mater Sci Eng*, 46(R)(2004),1-49.

2. F.-Y. Xie et al., "Microstructure and microsegregation in Al-rich Al-Cu-Mg alloys," *Acta Mater.*, 47(1999), 489-500.

3. X. Yan et al., "Computational and experimental investigation of microsegregation in an Al-rich Al–Cu–Mg–Si quaternary alloy," Acta Mater., 50(2002), 2199-2207.

4. P.D. Lee et al., "Multiscale modelling of solidification microstructures, including microsegregation and microporosity in an Al-Si-Cu alloy,"Mater Sci Eng,365(A)(2004), 57-65.

5. M. F. Zhu, C. P. Hong, "A modified cellular automaton model for the simulation of dendritic growth in solidification of alloys," ISIJ Int, 41(2001), 436-445.

6. W. Wang, P. D. Lee and M. Mclean, "A model of solidification microstructures in nickel-based superalloys: predicting primary dendrite spacing selection," Acta Mater, 51(2003), 2971-2987.

7. L. Belteran-Sanchez and D. M. Stefanescu, "A quantitative dendrite growth model and analysis of stability concepts," Metall Mater Trans A, 35(2004), 2471-2485.

8. M.-F. Zhu et al., "Modified Cellular Automaton Model for Modeling of Microstructure and Microsegregation in Solidification of Ternary Alloys," Trans. Nonferrous Met. Soc. China, 16 (S2) (2006), s180-s185.

9. S.-L. Chen et al., "The PANDAT software package and its applications," CALPHAD, 26(2)(2002), 175-188.

10. A. A. Wheeler, W. J. Boettinger, and G. B. McFadden, "Phase-field model for isothermal phase transitions in binary alloys," Phys. Rev. A, 45(1992), 7424-7439.

11. L. Granasy et al., "Growth of 'dizzy dendrites' in a random field of foreign particles," Nature Materials, 2(2003), 92-96.

12. Brandes EA. In: Brook GB, ed., Smithells Metals Reference Book. 7 ed. (Berlin, Springer Verlag: Heidelburg; 1992).

Advances in the Science and Engineering of

CASTING SOLIDIFICATION

An MPMD Symposium Honoring Doru Michael Stefanescu

Microstructure Evolution II

Session Chair:

Adrian S. Sabau

Advances in the Science and Engineering of Casting Solidification
Edited by: Laurentiu Nastac, Baicheng Liu, Hasse Fredriksson, Jacques Lacaze, Chun-Pyo Hong, Adrian V. Catalina, Andreas Buhrig-Polaczek, Charles Monroe, Adrian S. Sabau, Roxana Elena Ligia Ruxanda, Alan Luo, Subhayu Sen, and Attila Diószegi
TMS (The Minerals, Metals & Materials Society), 2015

IN SITU STUDY ON THE EVOLUTION OF DENDRITE MORPHOLOGY AFFECTED BY ELECTRIC FIELD IN Sn-Bi ALLOY

Fenfen Yang[1], Fei Cao[1], Rengeng Li[1], Huijun Kang[1], Yanan Fu[2], Tiqiao Xiao[2], Tongmin Wang[1]

[1] Key Laboratory of Materials Modification by Laser, Ion and Electron Beams (Ministry of Education), School of Material Science and Engineering, Dalian University of Technology, Dalian 116024, China
[2] Shanghai Institute of Applied Physics, Chinese Academy of Sciences, Shanghai 201204, China

Keywords: Microstructure, Electric field, Dendrite growth, Synchrotron radiation

Abstract

With high energy, high brilliance and eminent collimation, synchrotron X-ray imaging enables in situ observation of dendrite growth during the solidification of Sn-Bi alloy under an applied electric field such as direct current (DC) and electric current pulse (ECP). In present study, the application of DC leads to the suppression of dendrite growth and the degeneration of the secondary and higher order dendrites. Due to the induced Joule heat, the sharp dendrite tips are modified to be round or flat. By applying ECP to the sample during the whole solidification process, a significant grain refinement is observed.

Introduction

It is well known that the solidification structure plays a vital role in the mechanical properties of metals and alloys [1]. With an isotropic finer-grained structure, the cast alloys could provide an excellent comprehensive mechanical properties [2]. Therefore, a majority of research groups dedicated to the study on grain refinement and controlling of microstructures in order to improve the materials performance [3-6].The application of electric field (such as DC, ECP) to metallic liquid can achieve significant grain refinement, which is proved by previous studies on the effect of external fields [7-10]. However, it is worth noticing that microstructure evolution of solidification is usually investigated through the post-mortem analysis, which results in the lack of some detailed dynamic information in the solidification process. In recent years, a lot of studies indicate that the synchrotron X-ray imaging techniques enables real-time observation of solidification in metallic systems [11-13]. Consequently, it is of major interest to study the time evolution of dendrites under the electric field. In this paper, we investigated the evolution of dendrite morphology of Sn-12wt.%Bi alloy under Direct current (DC) and electric current pulse (ECP) using synchrotron X-ray radiography technique. Based on our experiments, the suppression mechanism of DC and the refinement mechanism of ECP were discussed.

Experimental procedure

The Sn-12wt.%Bi alloy was prepared by melting high-purity Sn (99.99%) and pure Bi (99.99%) in a ceramic crucible. Sn-Bi alloy was selected since the absorption difference between Sn and Bi is sufficient and a good contrast can be obtained. The 100 μm thick sample was cut into a rectangular slice of 10 × 20 mm^2, and then sandwiched between two ceramic plates. A 100 μm Teflon sheet was also placed between the ceramic plates, and used to fix the sample. Two pieces of aluminum foil acted as electrodes, were brought in direct contact with the upper and lower parts of sample, respectively. A special furnace system consists of two heated zones was designed to melt the sample during the experiment. The temperature of each furnace could be controlled independently. Fig. 1 shows the schematic of synchrotron radiation imaging experimental setting. DC and ECP were applied by connecting the sample to an electric current generator.

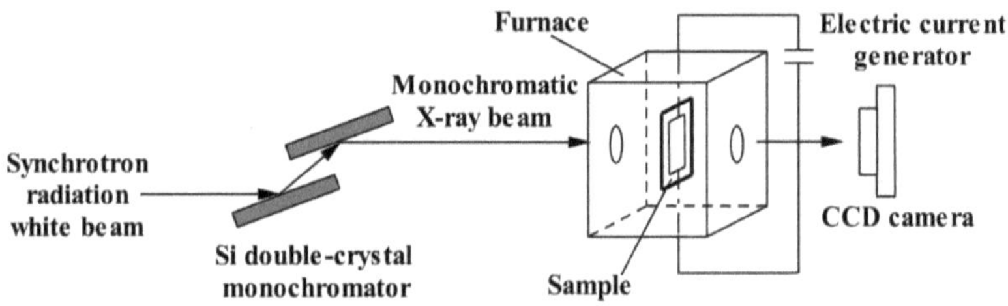

Fig. 1 Schematic of synchrotron radiation imaging experimental setting.

In situ synchrotron X-ray radiography experiments were carried out on BL13W1 beamline of Shanghai Synchrotron Radiation Facility (SSRF) in China. The main surface of the sample (10 × 20 mm^2) was set perpendicular to the incident monochromatic X-ray beam with energy of 20 keV. The images during the solidification of sample were recorded with a fast readout, low noise charge coupled device (CCD) camera. The field of view was 3.5 × 3.5 mm^2 with a resolution of 1.85 μm/pixel. The sample-to-detector distance was 15 cm. The sample was heated to a temperature at which the superheat degree was 5 ℃ and then cooled at a rate of 1.5 ℃/min. DC and ECP were applied to the melt during the whole solidification process. Fig. 2 shows the output waveform of ECP. Δt_1 and Δt_2 refer to the impulse width and the impulse cycle time, respectively. In present work, impulse frequency was set to 1000 HZ (impulse cycle time is 0.001 s), impulse duty raio ($\Delta t_1/\Delta t_2$) is 60%, voltage value (U) is 7 V, impulse current density is 270 A/cm^2 and reversing time (τ) is 10 s.

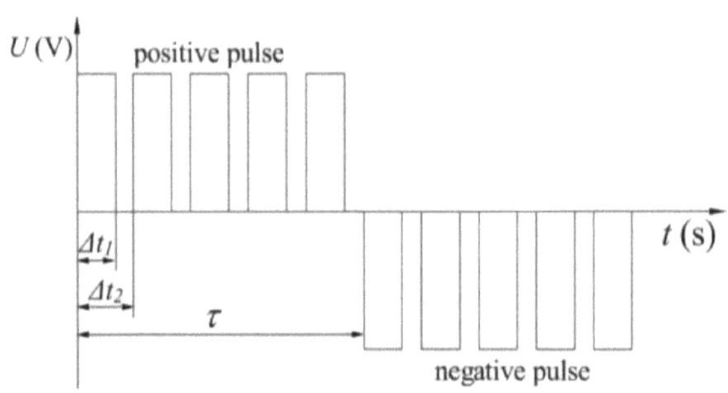

Fig. 2 Schematic diagram of output waveform of ECP

Results and discussion

Fig. 3 shows the dendrite growth morphology in Sn-12wt.%Bi alloy without and with DC. The solid phase mainly constituted of Sn appears in grey while the Bi enriched liquid is dark. During the Sn-dendrites growth, excess Bi is rejected to the melt by diffusion. Comparing the Fig. 3a and Fig. 3b, a remarkable difference can be directly observed. Without DC treatment, a dendritic columnar morphology was presented because of an unavoidable temperature gradient in the vertical direction (Fig. 3a). In addition, the thin dendrites with sharp tips as denoted by white arrows were clearly visible. When a DC with a density of 3 A/cm^2 was applied, the equiaxed dendrites can be seen in Fig. 3b. The dendrite branching and arm growth were significantly suppressed due to the effect of DC, the primary dendrite arms grew remarkably "fatter", the second dendrite arms were severely restrained and the higher order dendrite arms disappeared. Moreover, the sharp tips of the primary and the secondary dendrites were suppressed so that they tended to be round or flat as denoted by white arrows (Fig. 3b). According to our previous work [14], this is mainly ascribed to the Joule heat effect and current crowding effect. Since the electrical resistivity of the melt is greater than that of the solid phase, the electric current prefers to penetrate the sharp dendrite tip first when it is passing through the solid-liquid interface, leading to the rising temperature in the tip caused by the much more Joule heat generated there. As a consequence, the growth of dendrite tip was suppressed and the sharp tips tended to be round or even flat to reduce the effect of dense electric streamlines.

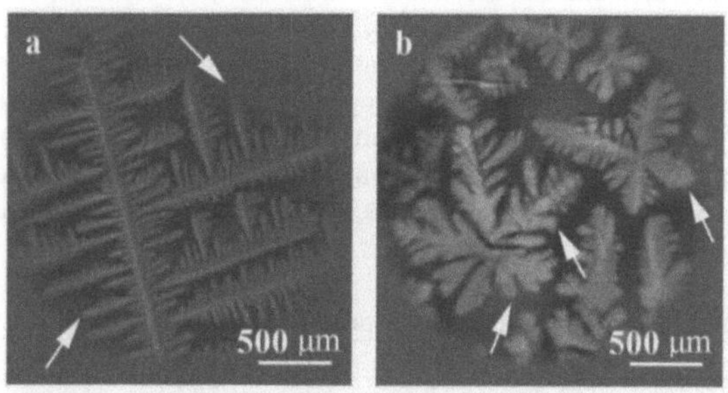

Fig. 3. Dendrite growth morphology in Sn-12wt.%Bi alloy sample: (a) without DC and (b) with DC, the current density is 3 A/cm^2.

Fig. 4 shows the dendrite growth morphology without and with ECP applied during the whole solidification process. The coarsening columnar dendrites with a wide mean primary dendrite arm spacing (PDAS) are found in Fig. 4a while the thin dendrites with a considerable narrow PDAS are shown in Fig. 4b. It is obvious that the dendrites are effectively refined by applying ECP during the entire solidification process.

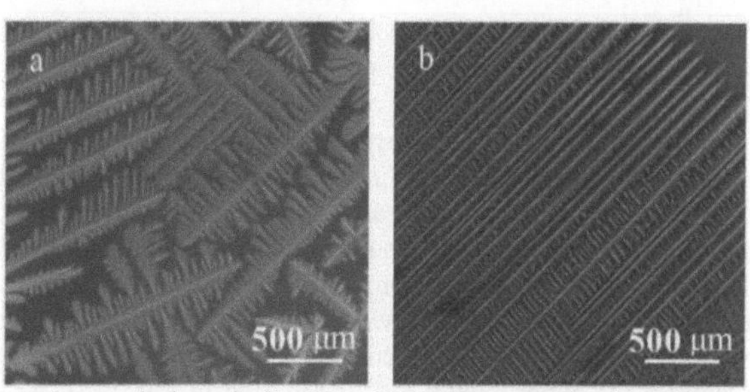

Fig. 4 Dendrite growth morphology in Sn-12wt.%Bi alloy sample: (a) without ECP and (b) with ECP (U=7 V, I=2.7 A and τ=10 s).

The mean PDAS and growth rate of primary dendrites over time without and with ECP are listed in Table I and Table II, respectively. Comparing Table I and Table II, it is found that without ECP the solidification starts at 1569 s and with ECP the solidification starts at 2165 s, which indicates that the incubation time increases when ECP is applied. Without ECP treatment, the temperature of the sample is determined by q_{out} which is caused by the thermal transfer between the sample and furnace:

$$q_{out} = hA\Delta T \tag{1}$$

where h is the heat transfer coefficient, A is heat exchange area and ΔT is the temperature difference between sample and furnace. When ECP is applied, there is a heat input q_{in} in addition to the q_{out}:

$$q_{in} = UI \tag{2}$$

where U is the voltage value and I is the current value. The heat input q_{in} is stable as the voltage value and current value are constant, while q_{out} is linearly related to the ΔT. Without ECP treatment, there is no heat input, the temperature of sample decreases as the solidification proceeds. As soon as the sample temperature drops to the liquidus, the initiation of solidification occurs. However, the temperature of sample under ECP is determined by both q_{in} and q_{out}. The temperature of sample cannot go down immediately due to the heat input, as a consequence, it takes longer to cool the sample to the liquidus, in other words, the incubation time increases when ECP is applied.

Table I. Mean PDAS and growth rate of primary dendrites over time without ECP.

Time (s)	1569	1579	1589	1599	1609	1619	1629
PDAS (μm)	442.9	527.3	585	587.5	589.1	589.5	590*
Growth rate (μm/s)	—	29.8	40	44.9*	33.1	10.6	0.5

* denotes the maximum value, t=0 represents the time when cooling is applied and t=1569 s represents the time when solidification starts.

Table II. Mean PDAS and growth rate of primary dendrites over time with ECP.

Time (s)	2165	2170	2175	2180	2195	2210	2225
PDAS (μm)	116.4	116.8	117.2	117.5	117.9	118.2	118.4*
Growth rate (μm/s)	—	287.8*	203.4	58.1	—	—	—

* denotes the maximum value, t=0 represents the time when cooling is applied and t=2165 s represents the time when solidification starts.

As shown in Table I, the growth rate of primary dendrites increased at first and decreased later with time, and the peak in growth rate (44.9 μm/s) occurred at 1599 s. That is because in the initial stage of solidification, the dendrites are almost isolated from the surrounding dendrites in the direction of growth, the freely growth without the influence of other dendrites results in the increase of growth rate. As the solidification proceeds, the solute-enriched layers overlap, which gradually evens out the Bi concentration between the two dendrite tips. Finally, the dendrite growth stopped when the concentration gradient between dendrites decays to zero. These real-time observations support the experimental results reported in Ref. [15]. As shown in Table II, the dendrites grew at a higher growth rate (nearly 290 μm/s) in the very beginning of solidification (2170 s) then decreased rapidly when ECP was applied. Longer incubation time as detailed above leads to the undercooling accumulation, so the cooling intensity is stronger than that without ECP. In this case, the dendrites break out to grow simultaneously at a high growth rate (almost six times of the growth rate without ECP treatment) at the very beginning of solidification. The primary dendrite arm spacing $\overline{\lambda}$ (PDAS) and the dendrite growth rate v_d are simply related by [16]:

$$\overline{\lambda} \propto v_d^{-a} G_l^{-b} \tag{3}$$

where G_l refers to the temperature gradient in the alloy melt, a and b are constants. Because a is a positive constant, $\overline{\lambda}$ decreases with the increasing dendrite growth v_d. Considering that v_d with ECP is much higher than that without ECP, the PDAS should be much smaller compared to that without ECP, which is proved by the data in Table I and Table II. It is found that the mean PDAS varies from 442.9 μm to 590 μm without ECP, while the maximum mean PDAS is even below 120 μm when ECP is applied. It follows that the mean PDAS decreases obviously due to the high growth rate under ECP, and a remarkable refinement is achieved.

Conclusions

1. Due to the induced Joule heat, the application of DC to the sample led to the significantly suppression of dendrite growth, the degeneration of the secondary and higher order dendrites. Moreover, the sharp dendrite tips were modified to be round of flat for exactly the same reason.

2. When ECP is applied to the sample during the whole solidification, the incubation time notably increases, the growth velocity improved significantly and a refined microstructure with narrow PDAS was observed because of the undercooling accumulation.

Acknowledgements

The authors gratefully acknowledge the supports of National Natural Science Foundation of China (Nos. 51274054, U1332115, 51271042, 51375070), the Key grant Project of Chinese Ministry of Education (No. 313011), the China Postdoctoral Science Foundation (2014M551075). The authors wish to thank all the staff members of the BL13W1 beamline of SSRF and 4W1A beamline of Beijing Synchrotron Radiation Facility (BSRF) for corresponding experiments and discussion.

References

1. G. Reinhart et al., "Investigation of columnar-equiaxed transition and equiaxed growth of aluminium based alloys by X-ray radiography," Materials Science and Engineering A 413-414 (2005), 384-388.

2. E. Liotti et al., "A synchrotron X-ray radiography study of dendrite fragmentation induced by a pulsed electromagnetic field in an Al-15Cu alloy," Acta Materialia 70 (2014), 228-239.

3. D. Räbiger et al., "The relevance of melt convection to grain refinement in Al-Si alloys solidified under the impact of electric currents," Acta Materialia 79 (2014), 327-338.

4. J. Zhu et al., "Real time observation of equiaxed growth of Sn-Pb alloy under an applied direct current by synchrotron microradiography," Materials Letters 89 (2012), 137-139.

5. C.X. Jiang et al., "Grain refining effect of magnetic field on Mg2Ni0.8Mn0.2 hydrogen storage alloys during rapid quenching," Electrochimica Acta 112 (2013), 535-540.

6. A.L. Greer et al., "Modelling of inoculation of metallic melts: application to grain refinement of aluminium by Al-Ti-B," Acta Materialia 48 (2000), 2823-2835.

7. F. Li, L.L. Regel, and W.R. Wilcox, "The influence of electric current pulses on the microstructure of the MnBi/Bi eutectic," Journal of Crystal Growth 223 (2001), 251-264.

8. X.L. Liao et al., "Refining mechanism of the electric current pulse on the solidification structure of pure aluminum," Acta Materialia 55 (2007), 3103-3109.

9. T.M. Wang et al., "In situ synchrotron X-ray imaging on morphological evolution of dendrites in Sn-Bi hypoeutectic alloy under electric currents," Applied Physics A (2014), DOI 10.1007/s00339-014-8537-6.

10. J. Li et al., "Research on solidification structure refinement of pure aluminium by electric current pulse with parallel electrodes," Materials Science and Engineering A 490 (2008), 452-456.

11. R.H. Mathiesen et al., "In situ X-ray video microscopy as a tool in solidification science," Journal of Metals, 64 (2012), 76-82.

12. H. Nguyen-Thi et al., "On the interest of synchrotron X-ray imaging for the study of solidification in metallic alloys," Comptes Rendus Physique 13 (2012), 237-245.

13. T. Schenk et al., "Application of synchrotron X-ray imaging to the study of directional solidification of aluminium-based alloys," Journal of Crystal Growth 275 (2005) 201-208.

14. T.M. Wang et al., "Evolution of dendrite morphology of a binary alloy under an applied electric current: An in situ observation," Physical Review E 81 (2010), 0421601.

15. A. Bogno et al., "In situ analysis of dendritic equiaxed microstructure formation in Al-Cu alloys by synchrotron X-ray radiography," Transactions of The Indian Institute of Metals 62 (4-5) (2009), 427-431.

16. R. Trivedi, "Theory of dendritic growth during the directional solidification of binary alloys," Journal of Crystal Growth 49 (1980), 219-232.

Advances in the Science and Engineering of Casting Solidification
Edited by: Laurentiu Nastac, Baicheng Liu, Hasse Fredriksson, Jacques Lacaze, Chun-Pyo Hong, Adrian V. Catalina, Andreas Buhrig-Polaczek, Charles Monroe, Adrian S. Sabau, Roxana Elena Ligia Ruxanda, Alan Luo, Subhayu Sen, and Attila Diószegi
TMS (The Minerals, Metals & Materials Society), 2015

HETEROGENEOUS STRIP ORIGINATED FROM INCLUSIONS: CHARACTERIZATION AND PHYSICAL MECHANISM

Xiaoping Ma, Dianzhong Li

Shenyang National Laboratory for Materials Science, Institute of Metal Research, Chinese Academy of Sciences, 110016 Shenyang, China

Keywords: heterogeneous strip, mushy zone, inclusion movement, interface tension, phase transition.

Abstract

The heterogeneity, such as well-known macrosegregation, is a major problem in the casting of steel ingots. Most heterogeneity generally originated from the solute partition and solute distribution in the solidification process with interdendritic convection. In this article, a new heterogeneous phenomenon originated from inclusions is discovered by detailed experimental characterization in a steel ingot. A strip characterized as ferrite chains and inclusions chains is revealed by macro-etching. The formation mechanism for the heterogeneous strip is proposed. In the mushy zone of the ingot, a large amount of separate MnS inclusions move laterally and upwards. Some MnS inclusions will remain in the moving trace. Such residual MnS inclusions appear as a large amount of separate MnS inclusion chains. In the subsequent solid phase transition process, promoted by the MnS chains, ferrite prefers to be transited from the austenite near the MnS inclusions and shows as a large amount of separate ferrite chains. A large amount of ferrite chains align in a strip-like zone, which results in the heterogeneous strip phenomenon in macro-etching process. The physical model about the driving force for the MnS movement is further theoretically analyzed. In the mushy zone, the interface tension resultant applied on the MnS inclusions can act as the drive force for the lateral movement of MnS inclusions. And the buoyance applied on the MnS inclusions acts as the drive force for the upwards movement of inclusions.

1 Introduction

The heterogeneity of steel ingots significantly damages the mechanical properties of forgings produced from ingots [1-3]. For example, some disasters happened in crucial nuclear plant equipment could be induced by the non-uniform properties of the ingots. Generally, such heterogeneity of steel ingots is known as macrosegregation. Because the macrosegregation can hardly be eliminated by forging and heat treatment process, the macrosegregation [2, 4-8] is considered as one of the most severe problems in steel ingots. Many researches have been done about formation mechanism of macrosegregation. Generally, the macrosegregation is considered as the result of microsegregation and interdendritic flow in solidification process. And the interdendritic flow is induced by density difference and gravity [9-13]. Therefore, the interdendritic flow was highly evaluated for the macrosegregation, and much numerical simulations about macrosegregation were performed based on thermosolutal flow calculation [2, 12, 14-18]. Besides, some experimental physical simulations with model alloy system were also performed [19, 20]. In these experimental works, the thermosolutal convection and the formation of macrosegregation were directly observed [20].

Interestingly, before establishing the mechanism of interdendritic thermosolutal flow, in the earlier research period, the researchers simply accepted the understanding that the solidification begins at the outside of the ingot. And the advancing solidification frontier layer will reject the impurities into the interior liquid melt [21]. This early understanding about impurity rejection by solidification frontier layer is rebutted by the mush zone model proposed in 1960's. According to the mush zone model, only negligible solute buildup can occur in front of the mush zone, and solute rejection from the mush zone is impossible [8].

However, several questions should be clarified. Firstly, the concept of impurities is different from the partitioned solutes. The inclusions are also possible impurity produced in the solidification process in the mushy zone. The rejection of solute from the mush zone is impossible. But the rejection of inclusions is fundamentally different from the rejection of solute. Indeed, the rejection of inclusions in front of the solid-liquid interface has been reported by the in-situ experimental observation [22]. Secondly, there may be no contradiction between the impurity rejection and negligible solute buildup at the solidification frontier. The solute buildup at the solidification frontier is impossible according to the current mushy zone model. But it should be realized that the inclusion rejection by the mushy zone may not demand for the solute buildup at the solidification frontier. Thus, reevaluation about the impurity rejection by the mushy zone should be performed.

If the inclusions can be rejected by the mushy zone, the movement of solute and the movement of inclusions may be independent and different. In this situation, besides the macrosegregation, the heterogeneity of steel ingots actually includes the distribution of inclusions. Investigations seldom focus on the distribution of inclusions and relating mechanism in actual steel ingot.

In order to clarify above questions and possibility, the reliable basic information about heterogeneity must be acquired firstly by the detailed experimental characterization in steel ingot. And further, theory model about the phenomenon provided by the experimental characterization should be analyzed and established.

2 Experimental Procedures

In this article, the heterogeneity is carefully characterized in a 500kg sand mold steel ingot, which nominal chemical compositions (weight percent) is C 0.30, Si 0.09, Mn 0.35, S 0.03, P 0.03, Cr 0.36, Ni 0.22, Cu 0.03, and Fe balanced. The steel was melted at 1600℃ by induction furnace, and was bottom filled at 1550℃ in the atmosphere after Al-killed process. The solidified ingot was cut in half along the longitudinal axle. The sectioned slice was further grinded and polished. In order to reveal the heterogeneity, the traditional macro-etching (etched by 20vol%HNO_3-5vol%H_2SO_4-H_2O solution, 20vol%HNO_3-H_2O solution and 5vol%HNO_3-H_2O solution, rinsed by hot water and C_2H_5OH, and subsequently dried by hot air) was carried out. And in order to reveal further information about the channel segregation, a new macro-etching technology was adopted (etched by 20vol%HNO_3-5vol%H_2SO_4-H_2O solution, 20vol%HNO_3-H_2O solution and 5vol%HNO_3-H_2O solution, wiped with C_2H_5OH and cotton, rinsed by hot water and C_2H_5OH, and finally dried by hot air). The macrographs for the whole ingot and amplified macrograph for specific channel segregation position were observed. In order to reveal the further microstructural information about the heterogeneity, the specific section of the channel segregation was etched by the 4vol%HNO_3-C_2H_5OH solution, and was further observed by optical microscope and scan electron microscope.

3 Experimental Results

3.1 The Macrograph of the Heterogeneity

The macrograph revealed by traditional macro-etching technology is shown in Figure 1(a), which displays an obvious heterogeneous strip. The heterogeneous strip shows as a continuous strip-like dark zone with about 50mm away from the ingot surface. The improved macro-etching technology can further reveal the appearance and additional information about the heterogeneous strip, as shown in Figure 1(b). Some distinct features are displayed compared with the macrograph revealed by traditional macro-etching technology. Revealed by the new macro-etching technology, the heterogeneous strip is actually a strip-like zone with a large amount of special dark lines intermingled in the common matrix. It should be noted that the special dark lines are actually separate and very slim. The special dark lines can be observed more clearly in the amplified part of macrograph, as shown in the square 1 and 2 of Figure 1(c). The slim dark lines start from the outer position and stretches to the interior and upward position in a local zone.

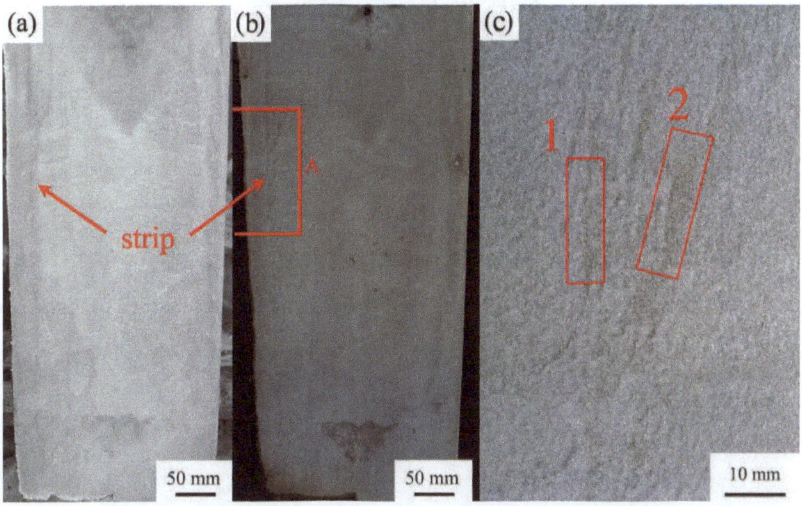

Figure 1 Macro-etching pictures of the steel slice sectioned from the ingot. (a) with the common macro-etching technology; (b) with the improved macro-etching technology; (c) amplified picture in the red square area A of (b).

3.2 The Microstructures about the Special Dark Lines

In order to reveal the innate character of the slim dark lines, we etched the sample with the 4vol%HNO_3-C_2H_5OH solution. Figure 2 show the optical microscopic pictures of dark line 2 located in Figure 1(c). In the Figure 2, the white phase is the ferrite, and the black phase is the pearlite. We can see that the normal microstructures of the ingot matrix are consisted by the ferrite on original austenite grain boundary, the pearlite and the acicular intragranular ferrite. The the dark line 2 is constituted by the continuous distributed ferrite arranged and displayed as chains.

In order to further reveal the innate character of the slim dark line, we observed the A and B zones in Figure 2 with SEM. As shown in Figure 3, we can see that the common matrix surrounding the slim dark lines is the microstructures constituted by the acicular ferrite, the pearlite and the grain boundary ferrite. In the dark line zones, the acicular ferrite reduces dramatically and granular ferrite appears. The ferrite chains coincide with the positions of the slim dark lines.

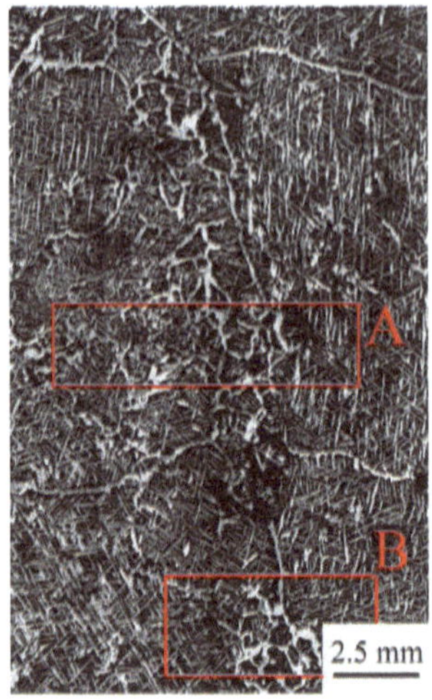

Figure 2 Microstructures observed with optical microscope in the red squares 2 of Figure 1(c).

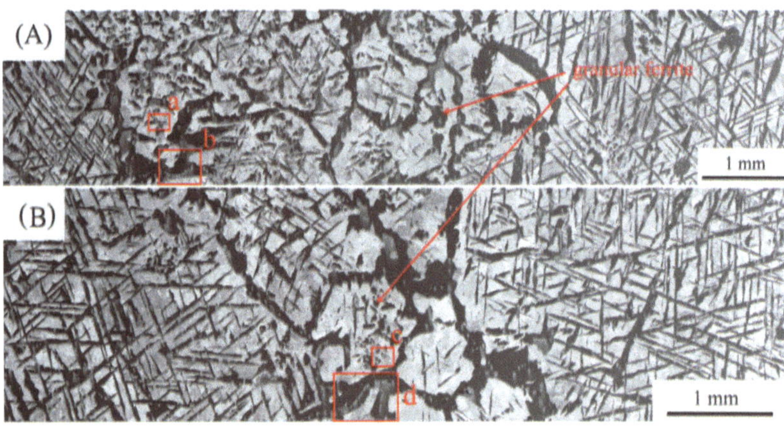

Figure 3 Microstructures observed with scan electrical microscope in the red squares A and B of Figure 2. (A) microstructures in the red square A of Figure 2; (B) microstructures in the red square B of Figure 2.

3.3 The MnS Inclusions in the Special Dark Lines

We further observed the ferrite in Figure 3 by SEM, as shown in Figure 4. The important information lies in that MnS inclusions always appear in the inner of the ferrite chains, the granular ferrite. As shown in Figure 4 (a) and (c), the separate MnS inclusions appear in the inner of granular ferrite. And as shown in Figure 4 (b) and (d), the MnS chains appear in the interior of the ferrite chains.

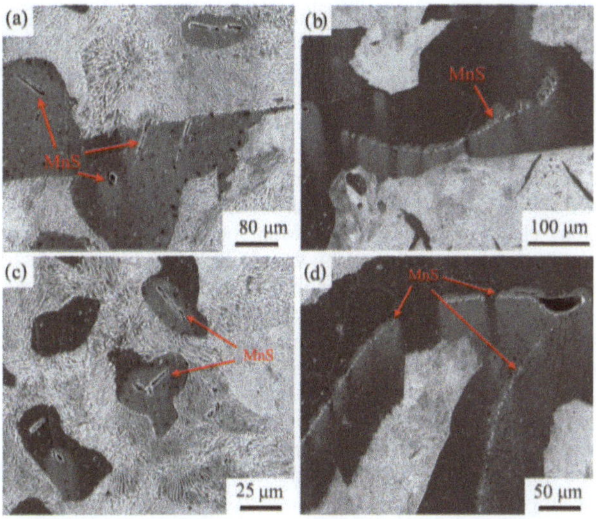

Figure 4 Microstructures observed with scan electrical microscope in the red squares a-d of Figure 3. (a), (b), (c) and (d) are the microstructures in the red squares a-d of Figure 3 respectively.

4 Discussions

4.1 The Innate Character of Heterogeneous Strip

Based on the macrograph observation and the microstructures analysis, important information should be realized about the innate character of heterogeneous strip. First, the heterogeneous strip is not a uniform and whole zone. Instead, the heterogeneous strip is a zone with common matrix and separate slim dark lines. This fact implies that the separate dark lines are the innate character of heterogeneous strip. Second, the slim dark lines are separate and independent with each other, which implies that the heterogeneous strip is a macro-phenomenon simultaneously constituted by a large amount of separate micro-events. According to the experiment results, the micro-events related with the slim dark lines are the formation of ferrite chains. Thus, the heterogeneous strip is the simultaneous formation process of a large amount of separate slim ferrite chains.

This heterogeneous strip should be distinguished from a well-known phenomenon "the channel segregation". The channel segregation is a continuous channel zone characterized with higher solute content. While, the heterogeneous strip in this paper is consisted by a large amount of separate ferrite chains located along a specific strip-like zone. Moreover, the ferrite chains intermingled with the normal microstructures will contain lower solute contents. Besides, the position of the heterogeneous strip is rather close to the surface of the ingot, which is different from conventional channel segregation. Thus, the heterogeneous strip is a new phenomenon different from channel segregation. Accordingly, the mechanism of enriched interdendritic melt flow for channel segregation obviously fails to explain the formation of heterogeneous strip.

4.2 The Formation Mechanism of Ferrite Chains

It is quite interesting that why the continuous or discrete ferrite chains can form. Based on our observation, the MnS inclusions appear in the inner of the ferrite, and MnS chains appear along with the ferrite chains. It has been well recognized in the oxide metallurgy research field that the MnS inclusion can promote the formation of ferrite in the solid phase transition process [23-25]. Thus, an interesting conclusion is proposed that the innate character of the ferrite chain is the MnS inclusions chain. Accordingly, the origin of the heterogeneous strip can now be further traced back to a large amount of separate MnS inclusions chains.

4. 3 The Physical Mechanism for MnS Inclusions Chains in the Mushy Zone

Based on all above experimental observation and analysis, we propose that the innate character of the heterogeneous strip is the large amount of separate ferrite chains. And the separate MnS inclusions chains are the origin of the ferrite chains. The further question is how the MnS inclusions chain is produced. The MnS inclusions chain may be induced by the interdendritic thermosolutal convection or by the solidification frontier rejection. According to the common understanding about interdendritic thermosolutal convection pattern, the shape of the MnS inclusions chains is obviously opponent to the thermosolutal convection pattern. The MnS inclusions chain starts from the outer position and stretches to the interior and upward position in a local zone. Moreover, if the MnS inclusions chain is produced by interdendritic thermosolutal convection, the solute enrichment should co-exist with the inclusions. This inference is disapproved by the experimental characterization. Thus, we would like to suggest that the MnS inclusions move under the influence of upward buoyance F_b and lateral solidification layer rejection force F_{ix} in the mushy zone. The physical model for the MnS inclusions movement is shown in Figure 5.

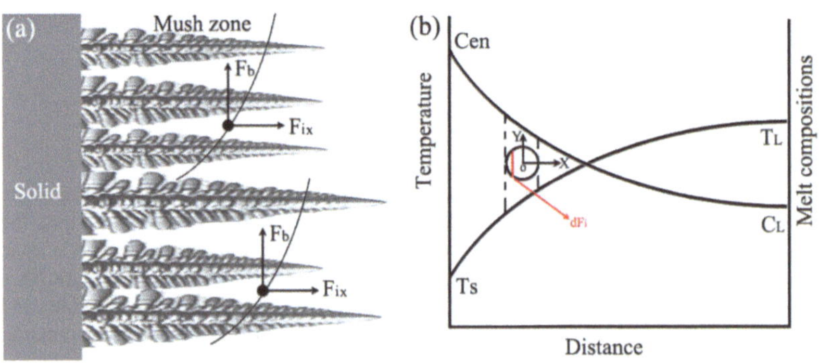

Figure 5 The sketch for the MnS inclusions movement in the mushy zone (a) and the physical model about lateral driving force for the inclusions in the mush zone (b). The sphere means the inclusions. F_b, the buoyant force; F_i, the interface tension; F_{ix}, the horizontal component of the the interface tension; Ts and Cen, the temperature and solute content at the left side of the mushy zone; T_L and C_L, the temperature and solute content at the right side of the mushy zone.

Although the formation of separate MnS chains, ferrite chains, dark lines and heterogeneous strip is clarified, the question that why a large amount of separate MnS chains simultaneous appear along a specific strip-like zone still remains. To answer this question, we have to consider the value of F_{ix}. If the F_{ix} is large enough to excite rapid movement of MnS inclusions, and if the movement of inclusions is faster than the growth of crystals, more MnS inclusions will move to the frontier of mush zone. In the frontier of mush zone, the F_{ix} is negligible and the MnS inclusions are free for upwards floating without dendrite obstruction. Therefore, we suggest that the specific strip-like zone is a just specific zone providing large lateral solidification layer rejection force F_{ix}. But, the exact and quantitive critical value about F_{ix} remains unsolved, because whether the MnS can move to the frontier of mush zone or not also depends on the diameter of MnS and the growth velocity of mushy zone. Especially, this movement of MnS will also be affected by the dynamic dendritic morphology evolution, which makes the solution more complicated.

5 Conclusions

By experimental characterization and theory analysis, the new phenomenon of heterogeneous strip is clarified. As the solidification layer push forwards in the steel ingot, specific strip-like zone can provide the enough interface tension gradient. And the interface tension resultant motivates the lateral movement of MnS inclusions. Assisted by the gravity, buoyance and thermosolutal convection, the MnS moves upwards simultaneously and was finally entrapped by the solidification process. The entrapped MnS further promotes the intragranular ferrite formation in the process of solid phase transition. The entrapped MnS inclusions and induced intragranular ferrite are finally revealed by macro-etching, which is the truth for heterogeneous strip.

Acknowledgements

This research is financially supported by the Youth Innovation Promotion Association of Chinese Academy of Sciences.

References

1. C. Maidorn, D. Blind, “Solidification and segregation in heavy forging ingots”, *Nucl Eng Des*, 84 (1985), 285-296.
2. G. Lesoult, “Macro segregation in steel strands and ingots: characterisation, formation and consequences”, *Mater Sci Eng A*, 413-414 (2005) 19-29.
3. J.Q. Wang, P.X. Fu, H.W. Liu, D.Z. Li, Y.Y. Li, “Shrinkage porosity criteria and optimized design of a 100-ton 30Cr2Ni4MoV forging ingot”, *Mater Des*, 35 (2012) 446-456.
4. K. Tashiro, T. Todoroki, S. Kimura, “Consideration on the formation mechanisms of macroscopic segregation in large steel ingot or casting", *Tetsu-to-Hagane*, 57 (1971) 1654-1675.
5. K. Suzuki, T. Miyamoto, “Study on formation of A-segregation in steel ingot”, *Trans ISIJ*, 18 (1978) 80-89.
6. S. Asai, H. Nakato, I. Muchi, “Theoretical analysis and model study on the structure of A-type segregation”, *Trans ISIJ*, 22 (1982) 190-197.
7. N. Mori, K. Ogi, “Study on formation of channel-type segregation”, *Metall Trans A*, 22 (1991) 1663-1672.
8. M.C. Flemings, “Our understanding of macrosegregation: past and present” *ISIJ Int*, 40 (2000) 833-841.
9. C. Beckermann, “Modelling of macrosegregation: applications and future needs”, *Int Mater Rev*, 47 (2002) 243-261.
10. M. Simpson, M. Yerebakan, M.C. Flemings, “Influence of dendrite network defects on channel segregate growth”, *Metall Trans A*, 16 (1985) 1687-1689.
11. G.H. Goldsztein, J.C.Santamarina, “Solute transport during cyclic flow in saturated porous media”, *Appl Phys Lett*, 85 (2004) 2432-2434.
12. M. Zaloznik, H. Combeau, “Thermosolutal flow in steel ingots and the formation of mesosegregates”, *Int J Thermal Sci*, 49 (2010) 1500-1509.
13. M.G.Worster, “Convection in mushy layers”, *Annu Rev Fluid Mech*, 29 (1997) 91-122.
14. J.J. Moore, N.A. Shah, “Mechanisms of formation of A- and V-segregation in cast steel”, *Int Metals Rev*, 28 (1983) 338-356.
15. T.P. Battle, “Mathmatical modeling of solute segregation in solidifying materials”, *Int Mater Rev*, 37 (1992) 249-270.
16. L. Yuan, PD. Lee, “A new mechanism for freckle initiation based on microstructural level simulation”, *Acta Mater*, 60 (2012) 4917-4926.

17. J.C. Heinrich, U.K. Sajja, S.D. Felicelli, D.G.Westra, "Projection method for flows with large local density gradients: application to dendritic solidification", *Int J Numer Meth Fluids*, 57 (2008) 1211-1226.
18. M. Wu, A. Ludwig, "Modeling equiaxed solidification with melt convection and grain sedimentation-I: model description", *Acta Mater*, 57 (2009) 5621-5631.
19. J.R. Sarazin, A. Hellawell, "Channel formation in Pb-Sn, Pb-Sb, and Pb-Sn-Sb alloy ingots and comparison with the system NH_4Cl-H_2O", *Metall. Trans. A*, 19 (1988) 1861-1871.
20. N. Shevchenko, S.Boden, G.Gerbeth, S.Eckert, "Chimney formation in solidifying Ga-25wt pct in alloys under the influence of thermosolutal melt convection", Metall. Trans. A, 44 (2013) 3797-3808.
21. B. Stoughton, The Metallurgy of Iron and Steel (New York, NY: McGraw-Hill, 1911).
22. S. Kimura, Y. Nabeshima, K. Nakajima, S. Mizoguchi, "Behavior of nonmetallic inclusions in front of the solid-liquid interface in low-carbon steels", *Metall. Trans. B*, 31 (2000) 1013-1021.
23. G. Miyamoto, T. Shinyoshi, J. Yamaguchi, T. Furuhara, T. Maki, R. Uemori, "Crystallography of intragranular ferrite formed on (MnS + V(C, N)) complex precipitate in austenite", *Scripta Mater*, 48 (2003) 371-377.
24. T. Yokomizo, M. Enomoto, O. Umezawa, G. Spanos, R.O. Rosenberg, "Three-dimensional distribution, morphology, and nucleation site of intragranular ferrite formed in association with inclusions", *Mater Sci Eng A*, 334 (2003) 261-267.
25. T. Pan, Z.G. Yang, C. Zhang, B.Z. Bai, H.S. Fang, "Kinetics and mechanisms of intragranular ferrite nucleation on non-metallic inclusions in low carbon steels", *Mater Sci Eng A*, 438-440 (2006) 1128-1132.

Advances in the Science and Engineering of Casting Solidification
Edited by: Laurentiu Nastac, Baicheng Liu, Hasse Fredriksson, Jacques Lacaze, Chun-Pyo Hong, Adrian V. Catalina, Andreas Buhrig-Polaczek, Charles Monroe, Adrian S. Sabau, Roxana Elena Ligia Ruxanda, Alan Luo, Subhayu Sen, and Attila Diószegi
TMS (The Minerals, Metals & Materials Society), 2015

AN INVESTIGATION OF DENDRITIC SEGREGATION IN DIRECTIONALLY SOLIDIFIED CMSX-4 SUPERALLOY

Gheorghe Matache[1], Doru Michael Stefanescu[2], Cristian Puscasu[1], Elvira Alexandrescu[1]

[1]National Research and Development Institute for Gas Turbines COMOTI;
220D Iuliu Maniu Ave.; 061126 Bucharest 6, Romania
[2]The Ohio State University and The University of Alabama, USA
182 S High Street, Dublin OH 43017

Keywords: Single-Crystal Superalloy, Directional Solidification, Segregation

Abstract

Dendritic segregation of directionally solidified columnar and single crystal CMSX-4® superalloy in as-cast and heat-treated condition was investigated. The partition coefficients of alloying elements derived from EDS measurements for various secondary dendrite arm spacings (SDAS) exhibited strong segregation in the dendrite cores (Re, W) or in the interdendritic regions (Al, Ti, Ta). Titanium segregation is more sensitive to the SDAS than that of the other elements. In the final stages of the interdendritic solidification chromium has a tendency to accumulate in a supersaturated γ phase layer around the coarse gamma prime islands.
The solidification partition coefficients calculated with Pandat™ and the PanNi thermodynamic database showed a good agreement with the experimental data. Pandat™ calculations using the local chemical composition of segregated areas were also used for the optimization of the solutionizing heat treatment window.

Introduction

The chemistry of different generations of single-crystal (SX) Ni-base superalloys includes a high content of refractory elements (W, Ta, Re, Ru) and a reduced content of Cr, Co, Mo compared with the previous generations of equiaxed cast superalloys. The main γ'-forming elements (Ti, Al) are still used, but the Ti content is reduced to low level, while Al is increased [1]. After the introduction of directional solidification (DS), some of the elements traditionally used for grain boundary strengthening (B, C) were partially or totally removed from the alloy chemistry [2]. SX alloys design is focused on optimizing the strengthening γ' fraction, high temperatures mechanical properties, oxidation and corrosion resistance [1, 3].
A major concern in the development of SX superalloys is the avoidance of defects produced by solute segregation during solidification. Co, Cr, Re, W segregate in the dendrite core, while Al, Ta, Ti segregate in the interdendritic regions [4, 5]. Strong partition tendency of high density refractory elements (W, Re, Ta) between the dendrites and the remaining liquid affect thermosolutal convection leading to density inversion in the mushy zone [6, 7]. During DS, the buoyancy forces related to the density inversion lead to the formation of unacceptable grain defects termed freckles [8, 9]. Freckles occur as long vertical chains of random equiaxed grains roughly aligned parallel to the gravity direction. The challenge in alloys chemistry optimization is to establish an optimum content of refractory elements (Ta, Re, W) in order to reduce defects tendency [10].
The high segregation of some alloying elements in as-cast superalloys cannot be completely removed by post-processing after solidification, but can be diminished by high temperature homogenization heat treatments [4, 11]. Thus, a first objective of this research was to obtain a quantitative description of the segregation of the various elements in SX and columnar as-cast

alloys, and to evaluate the effect of homogenization heat treatment on the compositional map. A second objective was to determine experimentally the partition coefficients for the elements of the multicomponent alloys, and to compare them with those obtained through thermodynamic calculations. This is particularly important as the interaction between the various elements of the alloy affects the partition coefficient as demonstrated for Rhenium [12].

Experimental Methodology

Single-crystal and columnar structure cylindrical rods of second generation Ni-base single-crystal superalloy CMSX-4® (registered trademark of Cannon-Muskegon Corporation) were used in this work. The typical chemical composition (wt.%) of this alloy is as follows: 5.6Al, 1Ti, 6.5Ta, 6W, 3Re, 0.6Mo, 6.5Cr, 9Co, 0.1Hf, balance Ni. Some of the alloying elements (Re, W) significantly increase the liquidus (T_L) and solidus (T_S) temperatures of the alloy, and strongly segregate in the γ dendrites during solidification [1]. Other γ forming alloying elements (Co, Cr, Mo) tend to lower T_L and have a moderate tendency to segregate in the γ-phase during solidification [13]. A high fraction of γ' precipitates is generated due to the Al, Ti, and Ta content.
The experimental cylindrical rods (16 mm dia., 200 mm length) were produced in an industrial Bridgman type vacuum induction melting and casting furnace at the Foundry-Institute, RWTH Aachen University. The casting and solidification conditions were as follows: melt and heating resistance temperature 1550°C, temperature gradient 3K/mm and withdrawal rate 3mm/min. Each shell mold included six rods and was designed to obtain either SX or columnar structure. Samples were cut from the rods transverse and parallel to the solidification direction. Metallographic samples were prepared by grinding, polishing and then etching (Marble etchant: 10 g $CuSO_4$, 50 ml HCl, 50 ml H_2O) to dissolve the γ' phase. Microstructure qualitative investigations, as well as quantitative local chemical analysis, were performed using a FEI Inspect F50 SEM equipped with an energy dispersive X-ray spectrometer EDAX APEX 2i with SDD Apollo X detector. To improve the accuracy of the quantitative EDS results, the known bulk chemical composition was used as standard, targeting values close to actual values for elements that are over 1%, and not expecting high accuracy for small amounts, as for Mo and Hf. Solute segregation was investigated on both transverse and longitudinal sections using backscattered electron imaging. Thermodynamic calculations were performed using the commercial Pandat™ software (registered trademark of CompuTherm LLC, USA) with PanNi multi-component thermodynamic database to investigate the solidification path and phase formation during solidification.

Results and Discussion

Experimental Measurements

Freckle formation can be predicted by simple criteria based on temperature gradient *G*, local solidification velocity *V*, or local solidification time t_f [9]. More appropriate criteria are based on the Rayleigh number by taking into account primary (PDAS) and secondary (SDAS) dendrite arm spacing. The length-scale of PDAS and SDAS is a function of *V* and *G*. Higher *G* produce finer spacing, reduced segregation and less freckles [9, 14].
During DS, γ-dendrites with orientations close to the <001> crystallographic direction develop from the liquid (Figure 1a). The as-cast microstructure of both SX and columnar grain structure in cross section (Figure 1b) are similar and consist of an array of cut dendrite trunks with a fine γ-γ' structure, interdendritic regions with a coarser γ-γ' structure, coarse γ' islands, and pools of γ-γ' eutectic (Figure 2). The coarse γ' islands are coarse cellular γ' particles precipitated within the γ channels developed in the interdendritic regions during solidification [15]. The partition of

the alloying elements at the solid/liquid (S/L) interface and the reduced solid diffusivity of the refractory elements generates highly segregated structures [1, 16, 17].

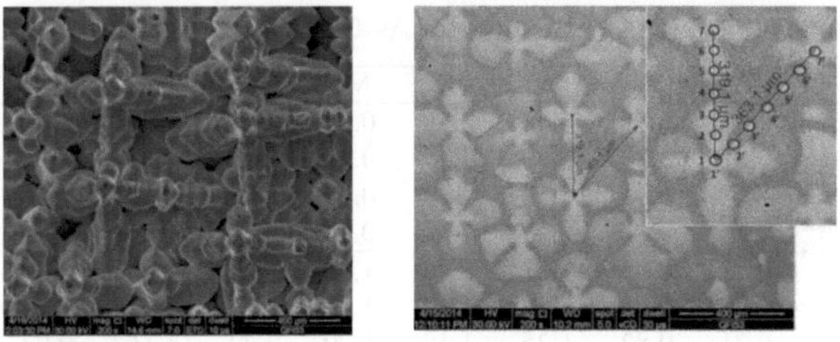

a) Spatial 3-D top view; interrupted solidification.

b) SEM backscattered electron image showing high segregation.

Figure 1. SEM images of SX dendrites in the as-cast microstructure in transverse section.

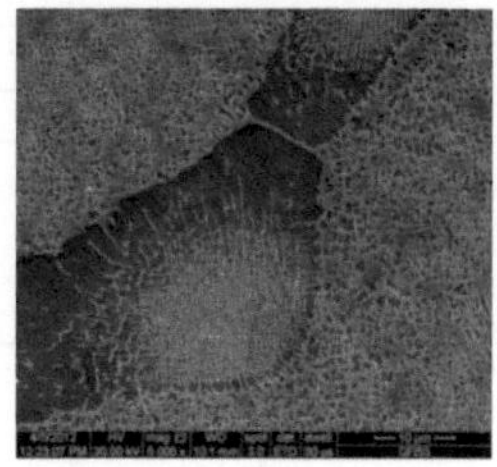

γ/γ'eutectic pool morphology.

Figure 2. CMSX-4 single crystal interdendritic microstructure.

EDS quantitative analysis was done between neighboring dendrites along the side and on the diagonal of the array which they form (see window in Figure 1b). As expected, the refractory elements (W, Re) are present in higher quantities in the dendrite cores and secondary arms than in the interdendritic regions. The γ' forming elements (Al, Ta, Ti) exhibit an opposite behavior, showing higher concentrations in the interdendritic regions as shown in Figure 3 for the EDS results on the diagonal of dendrites array.

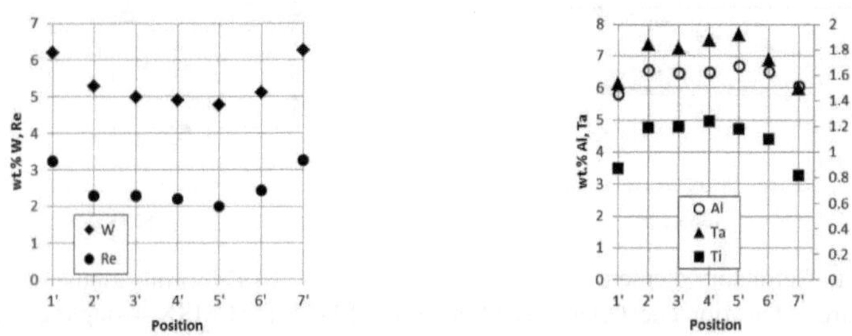

a) Heavy elements. b) γ' forming elements.

Figure 3. Concentration of the alloying elements on the diagonal of dendrites array.

The segregation level of the elements in as-cast structures can be expressed by the partition coefficient between the dendrite core and the interdendritic region: $k_I = C_D/C_I$, where C_D and C_I are the concentration of the element in the dendrite core and in the interdendritic region, respectively. Similarly, a partition coefficient between the dendrite core and the eutectic pools (C_E) can be defined as $k_E = C_D/C_E$. Because the interdendritic regions in the as-cast structure consist of both γ-γ' areas and eutectic pools, another partition coefficient ($k_{IE} = C_D/C_{IE}$) was calculated where C_{IE} is the average between the measured elements concentration in the γ-γ' interdendritic regions and in the eutectic pools. A partition coefficient $k > 1$ suggest segregation to the dendrite core, while $k < 1$ to the interdendritic regions. The experimental EDS chemical compositions and the calculated k are summarized in Table I for single-crystals. For both SX and columnar structures, W and Re strongly segregate to the dendrite cores (k_{IE} >>1). Al, Ta, and Ti accumulate in the interdendritic regions and in the eutectic. Some elements that segregate to the dendrites (Co, Cr, Mo) exhibit k_{IE} close to 1, but are depleted in the eutectic pools.

Table I. EDS Results and Calculated Partition Coefficients for As-Cast CMSX-4 Single Crystal

	Al	Ti	Ta	W	Re	Mo	Cr	Co	Ni
Bulk composition, wt.%	5.64	0.95	6.25	6.46	2.89	0.59	6.24	9.27	61.62
Dendrite core, wt.%	5.33	0.82	5.80	6.99	3.35	0.60	6.33	9.68	60.98
Interdendritic, wt. %	5.92	1.13	6.80	5.72	2.29	0.57	6.48	8.93	62.07
Eutectic, wt.%	6.65	1.18	7.32	5.45	1.94	0.54	4.94	7.83	64.08
k_I	0.90	0.73	0.85	1.22	1.46	1.06	0.98	1.08	0.98
k_E	0.80	0.70	0.79	1.28	1.73	1.12	1.28	1.24	0.95
k_{IE}	0.85	0.71	0.82	1.25	1.59	1.09	1.11	1.16	0.97

Since most of the high temperature properties of SX superalloys are related to SDAS, additional measurements were done to evaluate the influence of SDAS on the segregation behavior. EDS quantitative analysis was performed in longitudinal sections of SX for SDAS in the range 50 – 100 μm. Partition coefficients calculated as $k_I = C_D/C_I$ show that they are sensitive to SDAS values. The partition coefficients of W and Re increase with SDAS, while those of Al, Ti, Ta decrease (Figure 4). Titanium segregation is more sensitive to the SDAS than that of the other elements (higher slope of the trend line than for Al and Ta). In general, smaller SDAS generates less segregation and should have a positive influence on avoidance of structure defects.

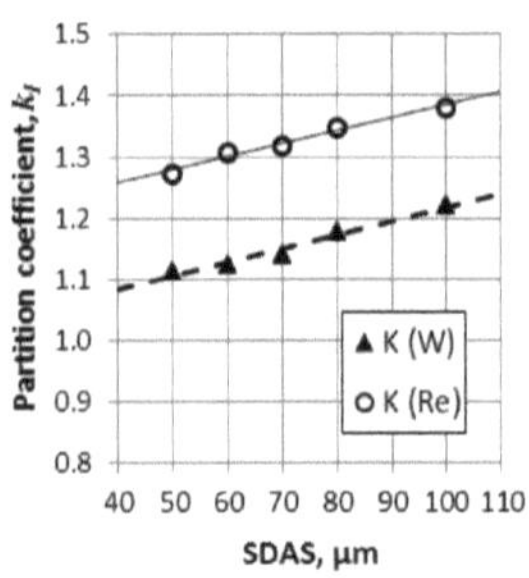

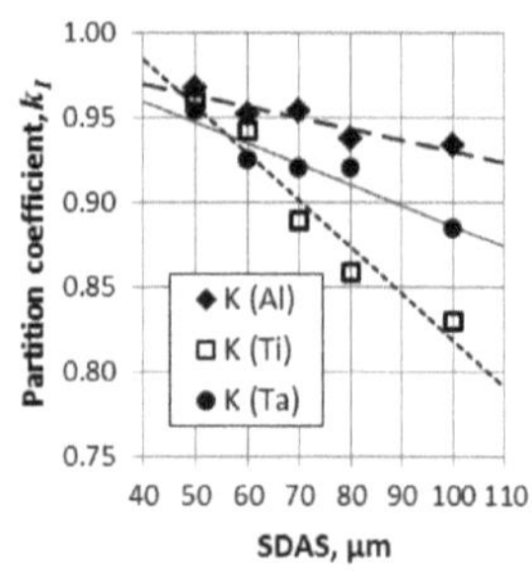

a) Refractory elements. b) γ' forming elements.

Figure 4. Partition coefficients as a function of SDAS for CMSX-4 single-crystal.

Observations on cross sections revealed that Cr strongly segregates around the coarse γ' islands for both SX and columnar structures. Such a behavior for Cr was also observed in other DS superalloys [11, 18]. Additional quantitative EDS measurements were done to investigate the alloying elements repartition in the Cr-rich layer that surrounds the coarse γ' islands for SX. The EDS results (Figure 5) show that solutes other than Cr are distributed differently at the interdendritic region/coarse γ' boundary. The concentration of γ' forming elements Al and Ta is smaller in the Cr rich layer by comparison with the rest of the interdendritic region (Figure 5b). For the elements that segregate to the dendrites, the layer is enriched in Cr, Co, and to some extent in Re, while W is slightly depleted (Figure 5c). Table II summarizes the chemical composition of different regions across the analyzed line where coarse γ-γ' is the average of the measurements in the γ-γ' interdendritic area (points 1, 2, 3, 8, 9, 10), Cr rich layer is the average of points 4 and 7, and the coarse γ' is the average of points 5 and 6. The Cr rich layer is enriched in Re, Mo, and

Co and depleted in γ' forming elements (Al and Ta), while the Ti content appears to remain at levels comparable to the interdendritic region.

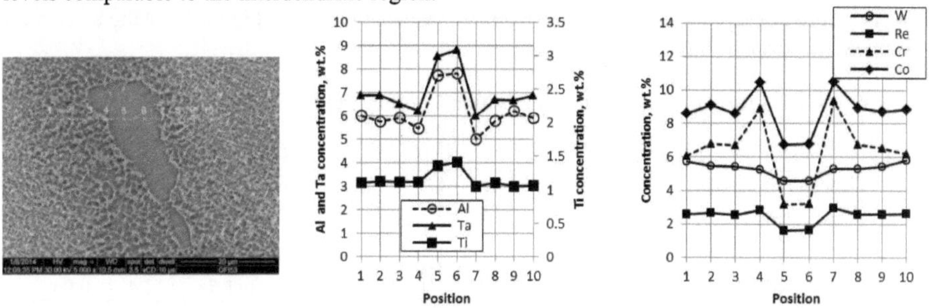

a) b) c)

Figure 5. Variation of alloying elements concentrations along the analyzed line; analysis points 4 and 7 correspond to the Cr rich layer as identified by Cr surface distribution;

Table II. Average Chemical Composition of Cr Rich Layer and of the Neighboring Regions

	Al	Ti	Ta	W	Re	Mo	Cr	Co	Ni
Coarse γ - γ'	5.92	1.09	6.74	5.52	2.60	0.67	6.50	8.82	62.05
Cr rich layer	5.23	1.08	6.11	5.27	2.91	0.80	9.13	10.51	58.91
Coarse γ'	7.77	1.39	8.67	4.59	1.65	0.71	3.20	6.76	65.19

Because the coarse γ' island is depleted in W, Re, Co and Cr, the enrichment of the surrounding layer in these elements can occur only if, in the final stages, the interdendritic regions γ' solidifies first and rejects these elements into the last fraction of liquid, which then solidifies between the interdendritic coarse γ-γ' and the coarse γ' islands. This explain also why the concentrations of Cr, Co and Re are higher in these layers than in the interdendritic γ-γ' and the dendrite cores. The thermodynamic calculations for the chemical compositions in Table II predict that the first γ' fraction is the result of an eutectic reaction occurring at 1350°C, slightly over the T_S of the Cr rich layer predicted at 1348°C. Even supersaturated in some solutes Cr rich layer is predicted to solidify as γ phase. Since the Cr solubility in the γ' phase is reduced, the solute difference is rejected to the coarse γ'-γ dendrites interface that contain also a lower amount of Cr. In a lesser extent the same behavior is exhibited by Co and Mo identified in higher quantities in the Cr-rich layer than in the neighboring regions.

To reduce the high compositional inhomogeneity, the as-cast alloy must be solution heat treated at temperatures above the γ' solvus. Subsequent precipitation heat treatment is required to optimize the cuboidal morphology of γ' precipitates. As-cast test samples were subject to a standard solution heat treatment under vacuum (heating to 1290°C with 15°C/min., 2 hrs. holding, heating to 1310°C and 6 hrs. holding, followed by cooling in argon stream). The heat-treated samples were investigated similar to the as-cast structures to quantify the effect of the solution heat treatment on the segregation of the alloying elements as summarized in Table III for SX. Similar data were obtained for the columnar structure.

The results show a good efficiency of the high temperature solution heat treatment on the chemical homogenization of the alloy. As compared with the as-cast structures, the *k* of the elements that strongly segregate in the dendrite cores tend to decrease, while that of the elements that partition to the interdendritic regions and eutectic increase.

One of the goals was to identify the Cr-rich layer response to heat treatment. Qualitative investigation by CrK_α radiation surface or line distribution established that most of the Cr segregated layers where homogenized by the heat treatment, as exemplified in Figure 7 for the as-cast and solution heat treated single crystal.

Table III. EDS Results and Calculated Partition Coefficients for Solution Heat Treated Single Crystal

	Al	Ti	Ta	W	Re	Mo	Cr	Co	Ni
Dendrite core, wt.%	5.78	0.93	6.16	6.59	3.30	0.69	6.69	9.45	60.32
Interdendritic, wt. %	6.17	1.02	6.44	5.75	2.57	0.64	6.16	8.64	62.51
Eutectic, wt.%	7.12	1.09	7.43	5.53	1.98	0.61	4.33	7.63	64.17
k_I	0.94	0.92	0.96	1.15	1.28	1.07	1.09	1.09	0.96
k_E	0.81	0.85	0.83	1.19	1.67	1.13	1.54	1.24	0.94

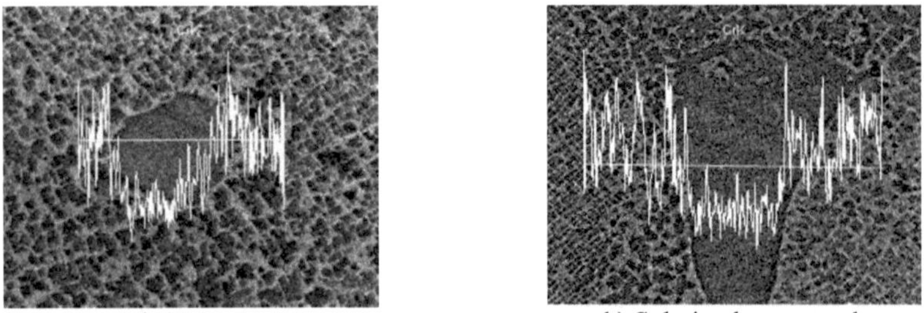

a) As-cast. b) Solution heat treated.

Figure 7. Line distribution of CrK_α radiation across the coarse γ' islands in single crystals.

Thermodynamic Calculations

Thermodynamic calculation of the solidification path of CMSX-4 using the commercial software Pandat predicts that under both equilibrium and Scheil solidification assumptions the solidification starts with the formation of the γ phase from the liquid. However, the similarities between the two solidification models stop here. The equilibrium calculation predicts that solidification ends when the γ phase solid fraction (f_S) reaches 1, and do not offer any information about the γ' phase. It is assumed that γ' precipitates later from the solid phase after reaching the γ' solvus temperature. The Scheil model predicts that primary γ solidifies until $f_S = 0.88$. Then, because of the depletion of the γ-forming refractory elements in the remaining liquid, and because of the enrichment in γ'-forming elements, the eutectic reaction ($L \rightarrow \gamma + \gamma'$) is favored. The predicted T_L is the same using both methods. The Scheil model can estimate the T_L but overestimates solute segregation in the liquid, and thus predicts a very low T_S [12, 19]. The calculated solidification sequence of CMSX-4 predicts a fraction of 72% γ' that precipitates from the γ phase below the γ' solvus temperature. This prediction is in very good agreement with the results reported by other authors for the same alloy [1, 17].

Thermodynamic calculations were also used to predict the variation of the concentration of alloying elements in different phases during solidification. The partition coefficients were calculated as $k_i = (C_S/C_L)_i$ and $k_i = (C_S/C_o)_i$ using the Scheil model for the predicted first fraction of solid as summarized in Table IV. C_S and C_L are the predicted compositions of the alloying elements in the solid and liquid at the S/L interface, respectively. C_o is the bulk composition of the alloy. The partition coefficients calculated using the predicted C_S compositions at the S/L interface are very close for both methods of calculation. The comparison between experiment-based and Pandat composition–based partition coefficients shows a good agreement for most of the alloying elements. Pandat predictions for the elements that have $k \cong 1$ (Ni, Cr, Co) are in very good agreement with the experimental results, with the exception of Mo that is underestimated.

Table IV. Partition Coefficients Calculated Based on the Predicted C_S and C_L at the S/L Interface

	Al	Ti	Ta	W	Re	Mo	Cr	Co	Ni
$k = C_S/C_L$	0.823	0.832	0.658	1.218	1.543	0.627	0.919	1.049	1.014
$k = C_S/C_0$	0.824	0.832	0.658	1.218	1.541	0.628	0.919	1.049	1.014

Pandat calculations were used to determine the solution heat treatment window (SHTW) (Table V). ΔT_{SHTW} is the solution heat treatment window calculated as the gap between T_S and the γ' solvus. To be efficient, the solution heat treatment should be done above the highest γ' solvus and the lowest T_S corresponding to the ΔT_{SHTW} of interdendritic regions of 70°C, between 1270 – 1340°C. This will ensure that all segregated areas are within the solubility range of the γ' during the heat treatment. A further optimization of the SHTW has to be done by taking into account the eutectic temperature in order to avoid local remelting during the heat treatment. For the EDS composition of eutectic pools, the predicted eutectic reaction ends at 1320°C. Thus, the SHTW for the investigated alloy shrinks to only 50°C between 1270 – 1320°C.

Table V. Predicted Solidus and γ' Solvus Temperatures

Phase	T_S, °C	γ' solvus, °C	ΔT_{SHTW}, °C
Dendrite core	1359	1250	109
Interdendritic	1340	1270	70
Cr-rich layer	1348	1220	128

Conclusions

1. Calculated partition coefficients based on EDS analysis have shown a strong segregation tendency of heavy elements (Re, W) to the dendrite core, while in the interdendritic regions mostly the γ' forming elements (Al, Ti, Ta) segregate. Other elements (Ni, Mo, Co, Cr) exhibit a moderate tendency to partition either to the dendrites or to the interdendritic areas.
2. The partition coefficients of the main alloying elements as a function of SDAS have shown that Ti segregation increases with the distance between the dendrites secondary arms with a higher slope than Al or Ta.
3. In the final stages of solidification, a highly segregated supersaturated γ phase rich in Cr, Co and Mo forms as a layer around the coarse γ' islands.
4. After a standard solution heat treatment, the chemical inhomogeneity between dendrites and interdendritic regions diminishes. The heat treatment had also a positive influence on the chemical homogenization of the as-cast Cr rich layers.
5. The calculated partition coefficients of solutes based on thermodynamic simulations using Pandat software show a good agreement with the as-cast experimental results. Pandat thermodynamic calculations predicted also an optimized solution heat treatment window of 50°C.

Acknowledgements

This work was supported by the Romanian National Authority for Scientific Research, POS CCE, code SMIS-CSNR 12607, financing agreement no. 174/16.06.2010. The authors are grateful to Professor A. Bührig Polaczek from the RWTH Aachen University - Foundry-Institute, Aachen, Germany for the casting of the experimental rods and generous technical assistance.

References

[1] R.C. Reed, *The Superalloys: Fundamentals and Applications*, Cambridge, UK: Cambridge UP, 2006
[2] M.G. Ardakani et al.,"Competitive Grain Growth and Texture Evolution during Directional Solidification of Superalloys", *Superalloys 2000*, (TMS, 2000), 219-228
[3] P. Caron, "High γ' solvus New Generation Nickel-based Superalloys for Single Crystal Turbine Blade Applications,", *Superalloys 2000*, (TMS, 2000), 737–746
[4] M.S.A. Karunaratne et al., "Modelling of the microsegregation in CMSX-4 Superalloy and its homogenization during Heat Treatment", *Superalloys 2000*, (TMS, 2000), 263–272
[5] M. Ganesan, D. Dye, and P.D. Lee, "A Technique for Characterizing Microsegregation in Multi-component Alloys and Its Application to Single-Crystal Superalloy Casting", *Metall. Mater. Trans. A*, 2005 36A:2191–2204
[6] T.M. Pollock and W.H. Murphy, *Metall. Mater. Trans. A*, 1996, 27:1081–1094
[7] R.A. Hobbs, S. Tin, and C.M.F. Rae, "A Castability Model Based on Elemental Solid-Liquid Partitioning in Advanced Nickel-Base Single-Crystal Superalloys", *Metall. Mater. Trans. A*, 2005 36A:2761-2773
[8] P. Auburtin *et al.*, "Freckle Formation and Freckle Criterion in Superalloy Castings", *Metall. Mater. Trans. B*, 2000 31B:801–811
[9] A.F. Giamei and B.H. Kear, "On the Nature of Freckles in Nickel Base Superalloys", *Metall. Trans.*, 1970, 1:2185-2192
[10] S. Tin, T.M. Pollock, and W.T. King, "Carbon Additions and Grain Defect Formation in High Refractory Nickel-base Single Crystal Superalloys", *Superalloys 2000*, (TMS, 2000), 201–210
[11] H.T. Pang et al., "Solution Heat Treatment Optimization of Fourth-Generation Single-Crystal Nickel-Base Superalloys", *Metall. Mater. Trans. A*, 2012 43A:3264-3282
[12] A. Heckl, R. Rettig, and R.F. Singer, "Solidification Characteristics and Segregation Behavior of Nickel-Base Superalloys in Dependence on Different Rhenium and Ruthenium Contents", *Metall. Mater. Trans. A*, 2010 41A:202-211
[13] R.A. Hobbs et al., "Solidification Characteristics of Advanced Nickel-base Single Crystal Superalloys", *Superalloys 2004*, (TMS, 2004), 819-825
[14] A. Wagner, B.A. Shollock, and M. McLean, "Grain structure development in directional solidi-fication of nickel-base superalloys", *Mater. Sci. Eng. A*, 374 (2004) 270-279
[15] H.S. Whitesell, L. Li, and R.A. Overfelt, "Influence of Solidification Variables on the Dendrite Arm Spacings of Ni-Based Superalloys", *Metall. Mater. Trans. B*, 2000 31B: 546–551
[16] H.T. Pang, et al., "Microstructure and Solidification Sequence of the Interdendritic Region in a Third Generation Single-Crystal Nickel-Base Superalloy", *Metall. Mater. Trans. A*, 2009 40A:1660–1669
[17] S.M. Seo et al., "A comparative Study of the γ/γ' Eutectic Evolution During the Solidification of Ni-Base Superalloys", *Metall. Mater. Trans. A*, 2011 42A:3150–3159
[18] S.M. Seo et al., "Solute Redistribution during Planar and Dendritic Growth of Directionally Solidified Ni-Base Superalloy CMSX-10", *Superalloys 2008*, (TMS, 2008), 277–286
[19] G.E. Fuchs and B.A. Boutwell, , "Calculating Solidification and Transformation in As-Cast CMSX-10", JOM, 2002 54 (1):45-48

Advances in the Science and Engineering of Casting Solidification
Edited by: Laurentiu Nastac, Baicheng Liu, Hasse Fredriksson, Jacques Lacaze, Chun-Pyo Hong, Adrian V. Catalina, Andreas Buhrig-Polaczek, Charles Monroe, Adrian S. Sabau, Roxana Elena Ligia Ruxanda, Alan Luo, Subhayu Sen, and Attila Diószegi
TMS (The Minerals, Metals & Materials Society), 2015

MODELING OF CASTING DEFECTS IN AN INTEGRATED COMPUTATIONAL MATERIALS ENGINEERING APPROACH

Adrian S. Sabau[1]
[1]Oak Ridge National Laboratory, Oak Ridge, Tennessee, 37831

Keywords: microporosity, simulation, modeling, nucleation, yield stress, hot tearing

Abstract

To accelerate the introduction of new cast alloys, the modeling and simulation of multiphysical phenomena needs to be considered in the design and optimization of mechanical properties of cast components. The required models related to casting defects, such as microporosity and hot tears, are reviewed. Three aluminum alloys are considered A356, 356 and 319. The data on calculated solidification shrinkage is presented and its effects on microporosity levels discussed. Examples are given for predicting microporosity defects and microstructure distribution for a plate casting. Models to predict fatigue life and yield stress are briefly highlighted here for the sake of completion and to illustrate how the length scales of the microstructure features as well as porosity defects are taken into account for modeling the mechanical properties. The data on casting defects, including microstructure features, is crucial for evaluating the final performance-related properties of the component.

INTRODUCTION

In castings, cavity defects can have regular, well-rounded shapes, or irregular, interdendritic shapes. For example, "hydrogen" microporosity consists of well-rounded and isolated voids while "shrinkage" microporosity and "hot-tear" defects are of irregular shape corresponding to the shape of the interdendritic region [1]. Over the last decades, industry has expanded the use of computer-aided engineering in reducing the manufacturing design and production cycle and cost, especially by implementing defect models into casting software. Integrated Computational Materials Engineering (ICME) has recently been established as a research protocol that combines theoretical analysis, design, and materials processing at the component level for materials design, including alloy development. In this study, the integration of casting defect modeling into an ICME framework is investigated.

For alloy design, it is desired to predict the resulting mechanical properties after heat treatment by taking into account casting defects and microstructure features (dendrite arm spacing, phases, amounts of phases, morphology, and lengthscales). In this context, for an ICME-based approach to the development of new cast alloys, the computational tools need to include the following models and data: (a) nucleation and growth models for defect during metal casting (e.g., microporosity, macroporosity, hot tearing), (b) process simulation models to obtain the size distribution within the casting of the microstructure length scales, phases, and defects, (c) microstructure/defects-to-mechanical property models to evaluate the resulting mechanical properties in as-cast condition, (d) precipitate nucleation and growth during heat treatment, and (e) microstructure/defects-to-mechanical property models to evaluate the resulting mechanical properties after heat treatment. In this study, the state-of-the art for microporosity defect prediction, including the integration of data on porosity defects and microstructure features in an ICME-based approach, is reviewed.

In the last decade, truly coupled multi-physics software were developed specifically for metal casting by coupling most of the phenomena relevant to metal casting: fluid dynamics, stress evolution, diffusion, microstructure evolution, and defect formation/evolution. With the exception of several success stories in truly applying ICME-like methodologies [2-5], most ICME studies were about combined experimental-modeling of individual phenomena, such as grain growth, grain recrystallization, and prediction of stress-strain response with only a few presentations showing some actual "integrated" computational materials science and engineering results.

For alloy development studies, thermodynamics-based simulations, which can provide data on phases, phase stability, and amount of constituents, require short computational times and are widely used. However, as these thermodynamics-based simulations are geometry-less, a generic cooling rate expected during casting or a generic time-temperature schedule that includes not only the casting process but also the heat treatment step is as required at input. On the other hand, casting defects are not a material property but rather a result of the combined effect of fluid dynamics, diffusion, and microstructure evolution [6]. The availability of constitutive models for porosity defects and the much longer computational times required for process simulations poses a challenge to the ICME models.

REVIEW OF MODELS FOR MICROPOROSITY AND HOT TEARING

The amount of gas porosity, of regular shaped, isolated voids, can be very well predicted using models that include fluid dynamics, solidification, and gas segregation in molten alloys [7]. However, the prediction of irregular shaped porosity has been challenging. Recently, models have been proposed to formulate computational methodologies that can predict the occurrence of irregular shaped void defects, such as hot-tears and shrinkage porosity (Table I). In Table I, P_g is the gas pressure, P is the liquid metal pressure, P_c is the cavitation pressure of Al, and $P_S \,\square\, 2\square/r$ is the pressure due to surface tension, σ, and f_g is the gas mass fraction. For the sake of completeness, it has to be mentioned that Sabau [8] proposed to estimate the degree of pore irregularity by the extent of pore growth after the cavitation pressure threshold has been reached.

Table III. Models for the onset of growth of irregular void defects [8].

Defect	*Fluid dynamics*	*Gas evolution*	*Stress*	*Porosity shape*	*Criteria and Reference*
microporosity	y	y	-	regular	$P_g > P + P_S$ [7]
shrinkage microporosity	y	y	-	*irregular*	$P_g > P + P_S$ and $P < P_c$ [1]
Hot tearing	y	-	y	irregular	$P < P_c$ [9]
Hot tearing	y	y	y	irregular	$P_g > P + P_S$ [10]
Microporosity and/or hot tearing	y	-	y	regular	$\partial f_g/\partial t > 0$ and/or $\sigma > \sigma_{\min}(d)$ [11]

One of the most important factors in microporosity growth is the pore curvature. Until recently the pore curvature was taken to be half of the secondary dendrite arm spacing. More accurate estimates for the cavity curvature, which is considered to be restricted by the solid dendrites, were developed by Pequet [12]. Recently, volatile elements were shown to affect the porosity formation in alloys since the nucleation and growth of pores in solidifying alloys was influenced by the partial vapor pressure of volatile solute elements [13]. However, these microporosity models predict only the average pore fraction at a given location in the casting, while the pore size distribution is needed for the prediction of fatigue resistance. Modeling the stochastic distribution of the pore size and density is a very active area of research [14-16]. Several submodels have been developed for the stochastic nucleation and diffusion-based growth of microporosity [17] by coupling Cellular Automata (CA) and finite element methods (FEM) in a similar manner as those developed solely for microstructure evolution [18, 19]. However, due to the complex phenomena involved, such as heterogeneous nucleation of microporosity and pore migration, no comprehensive model is yet available for microporosity shape prediction similar to the CAFE models for microstructure evolution [18, 19]. Direct numerical simulation CA-FE models for the prediction of microporosity defect distribution in entire shape castings are prohibitively computational intensive. Lee et al. [17] used physics-based correlations for predicting microporosity lengthscales in complex 319 Al alloy cast components. It has to be mentioned that these physics-based correlations, such as the one for the maximum pore length [17], were developed based on CA simulation results for a representative control volume for different process variables, including solidification time, hydrogen content, Cu content, and liquid pressure as opposed to empirically developed correlations that were used for decades with limited results when the casting shape was changed. However, only the proposed equation is available for the maximum pore length correlation in 319 Al alloy, as the actual parameters used in the correlation have not been published [17].

MICROSTRUCTURE COMPUTATIONAL RESULTS

Three alloys were considered in this study, as shown in Table II, in order to illustrate how microporosity depends on alloy composition in order to develop ICME-based practices for alloy design that take into account a realistic microstructure with intrinsic defects that lower their mechanical performance. Ni [%wt] is 0.029, 0.002, and 0.013; Pb[%wt] is 0.021, <0.001, and 0.007 for 319, A356, and 356, respectively, while Na, Sr, P, B, Ca, Sb were present at ppm levels. The thermodynamic simulations were conducted only for the as-cast condition, as the thermodynamic models of the microstructure evolution during heat treatment have not yet been implemented in ProCAST.

Table II. Composition of aluminum alloys considered [% wt].

Elem./ Alloy	Si	Cu	Fe	Mn	Zn	Ti	Cr	Sn
319	8.29	3.17	0.683	0.393	0.337	0.313	0.166	0.035
A356	7.32	0.002	0.1	0.044	0.402	0.005	0.156	<0.001
356	7.21	0.138	0.385	0.254	0.372	0.169	0.183	0.021

The typical microstructure for the alloys consisted of α-Al, Al_5FeSi□□β- AlFeSi), Al_2Cu (AlCu-θ), eutectic Si phases and porosity, where the terminology used in the ProCAST and/or CompuTherm is indicated in parenthesis [20, 21]. The microstructure model in ProCAST was used to conduct the thermodynamic simulations for the three alloys considered based on a back diffusion model for a constant cooling rate of 1 °C/s. The type of phases and their predicted amounts at the end of solidification for the three alloys considered are listed in Table III. The phase stability of precipitates is a very important factor in attaining adequate mechanical properties and the heat treatment effects on the as-cast microstructure need to be considered in further studies.

Table III. Phases and their calculated volumetric concentration [%] using the Al material database (Computherm) and microstructure module in ProCAST for the as-cast condition.

Phase/ Alloy	α-Al	*D-A4	Al_2Cu	$Al_{15}FeMn_3Si_2$	Al_5FeSi	$Al_5Cu_2Mg_8Si_6$	$Al_8FeMg_3Si_6$	Al_3Ti	Al_3Ni_1	Mg_2Si
319	84.90	6.53	2.9069	2.386	1.246	0.5445	-	0.376	0.039	-
A356	92.59	5.73	-	0.166	0.215	-	0.142	0.297	-	0.168
356	90.98	5.52	-	1.347	0.647	-	0.119	0.375	-	0.019

*Diamond-A4

EVALUATION OF SOLIDIFICATION SHRINKAGE

Using accurate material properties is paramount to the accuracy of process simulations. Through partnerships with thermodynamic database and software developers, the current-state-of the art metal casting software enables the evaluation of the following material properties: density in the liquid, mushy zone, and solid phase, thermal expansion, thermal conductivity, specific heat, viscosity of the molten metal, Young's modulus, and yield stress (YS). As an example of this emerging thermodynamic-based capability, results are presented in Figure 1 for the density. In order to illustrate the density variation in the latest stages of solidification, the density and temperature were shown in dimensionless form in Figure 1b. These data shows that there is significant variation in the shrinkage in the last stages of solidification among the three alloys considered. This last stage solidification shrinkage is very difficult to feed, as the liquid fraction is low, permeability becomes very low, resulting in large pressure drops in the mushy zone. Thus severe microporosity and hot tear defects are expected to occur, as discussed in the next section.

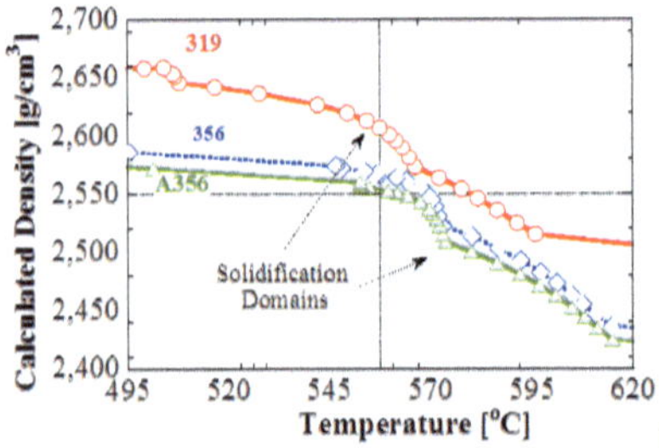

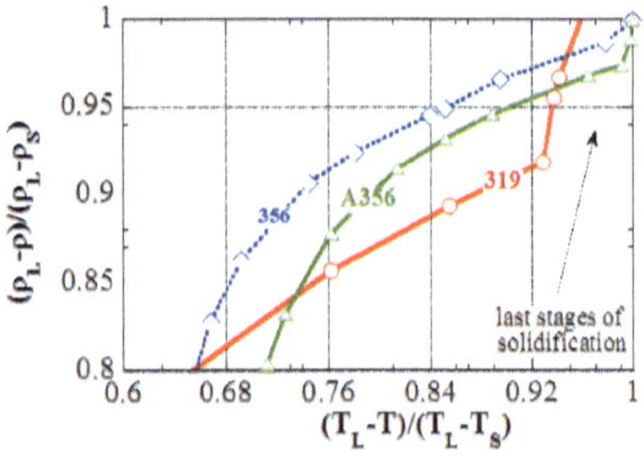

Figure 1. Calculated density using the Al material database Computherm and microstructure module in ProCAST for the alloys considered: (a) data (b) dimensionless data.

CASTING SIMULATION RESULTS

Numerical simulation results are presented in this section for a plate casting with cast iron chills (Sabau and Viswanathan, 2002). The mesh used in the computations is shown in Figure 2. The plate dimensions are 14 x 22.9 x 3.2 cm. The top and bottom chill dimensions were nominally 20 x 15 x 2.5 cm. The end-chill dimensions were nominally 6 x 15 x 6 cm. All the plates were contained in sand molds with nominal dimensions of 59 x 29 x 15.5 cm, respectively. ProCAST software was used in this study for metal casting simulations [22]. For Al alloys considered in this study, the model of the nucleation of the primary dendritic grains is that presented by Rappaz et al. [23, 24] based on a distribution of solid nuclei with undercooling. The nucleation of the eutectic grains is based on the model introduced by Oldfield [25]. In Figure 3, the following microstructure variables are shown: (a) secondary dendrite arm spacing (SDAS), (b) eutectic grain radius (EGR), which is the radius of the eutectic grains that nucleate in between the dendrites of the primary phase, and (c) eutectic lamellar spacing (ELS), which is the average distance between the eutectic lamellae or rods. Excluding the end plate regions (approx. last 4 cm on either side), these microstructural length scales for SDAS, EGR and ELS are uniform through the plate length. Near the plate top surface and bottom surface, the microstructure length scales are different from the core region.

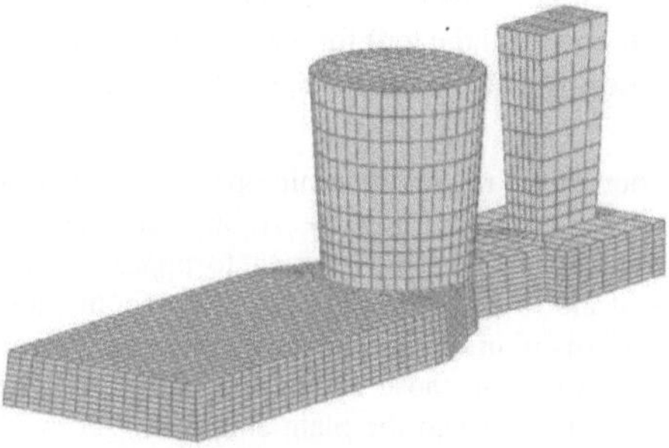

Figure 2. Finite element mesh used for metal casting simulations of a plate, including riser and sprue.

Numerical simulation results for microporosity, which were obtained with the advanced microporosity model available in ProCAST [12, 13], are shown for the three alloys in Figure 4a, b, and c (initial hydrogen content of 0.1cc/100g). The region with microporosity higher than 0.2% is referred in the remainder of this section as the high-microporosity region. For the A356 case (initial hydrogen content of 0.112cc/100g), the high-microporosity region is shown to appear slightly toward the plate end, extending from distances of 2.3 to 19.4 cm (from the plate end), while the experimental data presented in Sabau and Viswanathan [1] indicate that average porosity values larger than 0.2% covered a region from 3.9 to 13 cm (as measured from the plate end). Thus, the location of highest porosity region for A356 simulation is in good agreement with experimental data; however its predicted length is larger than that estimated from porosity measurements.

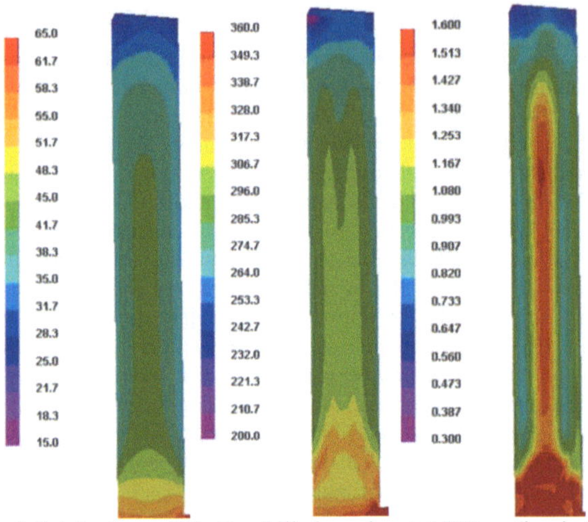

Figure 3. Computed distribution within the chilled-sand cast A356 casting in the vertical cross-section of the plate (plate end is on the top) for the: (a) SDAS [μm], (b) eutectic grain radius [μm], and (c) eutectic lamellar spacing [μm].

These results for the position of the region with microporosity higher than 0.2% are summarized for the alloys considered in Figure 5b. For the A356 case with initial hydrogen content of 0.1cc/100g, the high-microporosity region is shown to appear slightly toward the plate end, extending from distances of 2.6 to 18.1 cm. For the 356 case, the maximum porosity is higher than that for A356 and its distribution is quite different, e.g., the core along the centerline exhibits a lower porosity region than those above and below it. The predicted high-porosity region for 356 is located, with respect to the plate end, from 7.3 to 16.6 cm. For the 319, the length of the high-porosity region and the maximum microporosity value were found to be the largest among the three alloys considered. This is somehow expected, as shrinkage demand in the last stages of solidification is highest for 319, as shown in Figure 1b. The calculated high-porosity region for 319 alloy was found to be located at distances of 1.4 to 19.6 cm (with respect to the plate end).

The minimum pressure in the interdendritic liquid is shown in Figure 4 d, e, and f. Based on models presented in Pequet et al. (2002); Couturier and Rappaz (2006) the negative pressures in the liquid are allowed, as long as $P_g \square P \square P_S \square 0$. On the other hand, the pressure levels below the cavitation pressure of Al were associated with severe shrinkage regions, regions in which the microporosity is irregular in shape [1, 8]. Thus, the negative pressure values, as the cavitation pressure of Al is very small at the solidus temperature, can be used as an indicator of a change in the porosity morphology, from rounded porosity defects to irregular-shaped porosity defects. This severe shrinkage region is thus expected to extend from 2 to 17.6 cm, 5.4 to 18 cm, and 1.5 to 18.5 cm for A356, 356 and 319 alloys, respectively, with respect to the plate end.

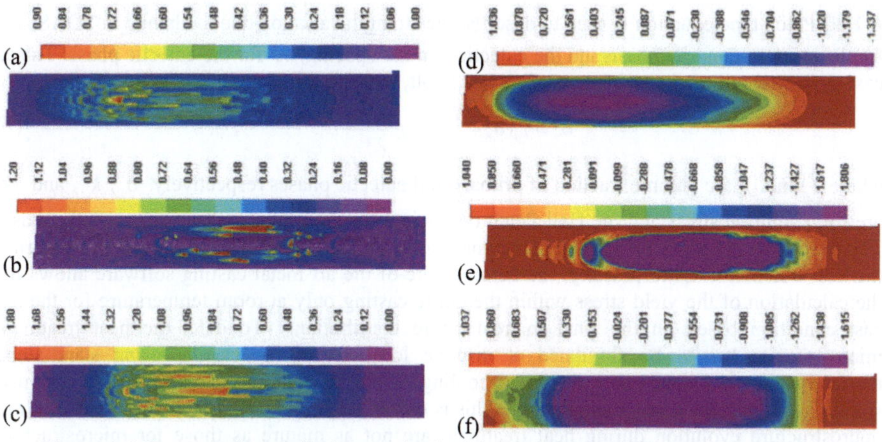

Figure 4. Microporosity prediction results for (a) A356, (b) 356, (c) 319. Corresponding minimum pressure in the liquid (d) A356, (e) 356, (f) 319 (plate end is located at the left-hand side).

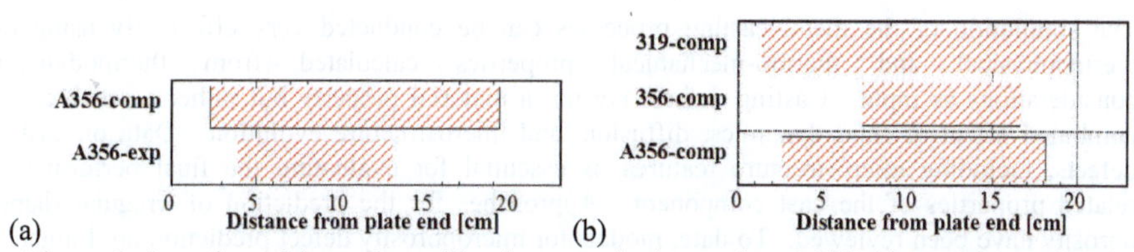

Figure 5. Location of the region with microporosity larger than 0.2%: (a) experimental and computed results for A356 alloy at an initial hydrogen content of 0.112cc/100g, and (b) for the alloys considered and an initial hydrogen content of 0.1cc/100g.

MICROSTRUCTURE/DEFECTS-TO-MECHANICAL PROPERTY MODELS

One of the most challenging steps in ICME-like approaches is the prediction of mechanical properties based on the calculated or measured distribution of both microstructure and void defects. Fatigue life and yield stress are briefly highlighted here for the sake of completeness and to illustrate how the length scales of the microstructure features as well as porosity defects are taken into account for modeling the mechanical properties. The following microstructure features can significantly affect the fatigue crack initiation and its growth behavior: SDAS, size and distribution of silicon particles, fracture resistance of silicon, strength of interface between silicon and aluminum matrix, intermetallics, and porosity [26]. A microstructure based-fatigue model is desired in order to be able to account for the effect of these stress concentrators [27].

A Hall-Petch type equation is used to describe the strength variation due to change in SDAS, λ_2, for the primary dendrite phase and the eutectic laminar spacing, λ, for the eutectic phase. The as cast yield strength can be written, as (Guo and Scott, 2002):

$$\sigma_c = f_p(\sigma_o + \frac{k_p}{\lambda_2^{1/2}}) + f_e \frac{k_e}{\lambda^{1/2}} + \sigma_P \tag{1}$$

where f_p and f_e are volume fraction of primary and eutectic phases respectively. σ_o, k_p, and k_e are solid solution stress and the Hall-Petch coefficient for the primary phase and eutectic phase. The third term accounts for the strengthening due to precipitates that nucleate and grow during the heat treatment. Unfortunately, the current-state-of the art metal casting software allows for the calculation of the yield stress within the entire casting only at room temperature for the as-cast condition based on the final microstructure distribution. To date, the heat treatment microstructure has to be simulated at discrete locations in the casting using stand-alone thermodynamic software based on actual cooling/heating/quenching curves that were obtained from a metal process casting simulation. This is due in large part to the fact that models for the microstructure evolution during heat treatment are not as mature as those for microstructure evolution during casting solidification. Microporosity, which obviously decreases mechanical properties, is not considered in the above equation.

DISCUSSIONS AND CONCLUSIONS

The simulation of the metal casting processes can be conducted very effectively using the thermophysical and thermo-mechanical properties calculated from thermodynamic considerations as input. Casting defects are not a material property but rather a result of the combined effect of fluid dynamics, diffusion, and microstructure evolution. Data on casting defects, including microstructure features, is essential for evaluating the final performance-related properties of the cast component. Approaches for the prediction of irregular-shaped porosity have been reviewed. To date, models for microporosity defect prediction are limited to the volumetric fraction of porosity. Models for the prediction of the stochastic variables that describe the distribution of the pore size and density are being developed and not yet available to the community at large.

The spatial distribution of the as-cast microstructure length-scales can be accurately predicted. To date, the distribution of the amount and morphology of strengthening precipitates following a *heat treatment* cannot be simulated in the entire casting. Microstructure-to-property models that would take into account the distribution within the component of the as-heat-treated microstructure and casting defects are being developed to predict both the yield stress at operating temperatures and in-service overall performance of the component.

ACKNOWLEDGEMENTS

This work was performed under a Cooperative Research and Development Agreement (CRADA) with the Nemak Inc., and Chrysler Co. for the project "High Performance Cast Aluminum Alloys for Next Generation Passenger Vehicle Engines." The author would also like to thank Amit Shyam for reviewing the paper and Andres Rodriguez of Nemak Inc. Research sponsored by the U. S. Department of Energy, Office of Energy Efficiency and Renewable Energy, Vehicle Technologies Office, as part of the Propulsion Materials Program under contract

DE-AC05-00OR22725 with UT-Battelle, LLC. Part of this research was conducted through the Oak Ridge National Laboratory's High Temperature Materials Laboratory User Program, which is sponsored by the U. S. Department of Energy, Office of Energy Efficiency and Renewable Energy, Vehicle Technologies Program.

REFERENCES

1. Sabau, A.S. and Viswanathan, S, "Microporosity Prediction in Aluminum Alloy Castings," Metallurgical and Materials Transactions B, 2002, Vol. 33B, pp. 243-255.
2. NRC, 2008, "Integrated Computational Materials Engineering (ICME): A Transformational Discipline for Improved Competitiveness and National Security", Committee on Integrated Computational Materials Engineering, National Research Council.
3. Miller, L.K.; "Simulation-Based Engineering for Industrial Competitive Advantage," Computing in Science & Engineering, vol.12, no.3, pp.14-21, 2010.
4. Allison J, Backman D, Christodoulou L, 2006, Integrated computational materials engineering: A new paradigm for the global materials profession, JOM, Vol. 58, pp. 25-27.
5. Allison, J., Mei Li, C. Wolverton, and Xu Ming Su, 2006, Virtual Aluminum Castings: An Industrial Application of ICME. JOM, Vol. 58, pp. 28-35.
6. A.S. Sabau, W.D. Porter, S. Roy, and A. Shyam, Process Simulation Role in the Development of New Alloys Based On An Integrated Computational Materials Engineering Approach, paper IMECE2014-37982, Proceedings of the ASME 2014 Int. Mech. Eng. Congress & Exposition IMECE2014, Nov. 14-20, 2014, Montreal, Quebec, Canada.
7. K. Kubo and R. D. Pehlke: *Metall. Trans. B*, 1985, vol. 16B, pp. 359-66.
8. Sabau, A.S. Predicting Interdendritic Cavity Defects during Casting Solidification, JOM, 2004, Vol. 56, pp. 54-56.
9. Rappaz M. Drezet, JM; Gremaud, M, "A new hot-tearing criterion," *Metall. Mater. Trans. A.*, Vol. 30, (1999), pp. 449-455.
10. J. F. Grandfield, C. J. Davidson, and J. A. Taylor, "Application of a New Hot Tearing Analysis in Horizontal Direct-Chill Cast Magnesium Alloy AZ91," Light Metals, (131st TMS Annual Meeting and Exhibition), Seattle (USA), 11-15 February 2001, pp. 207-213.
11. Suyitno, W.H. Kool, and L. Katgerman, *Mater. Sci. Forum*, Vols. 396-402 (2002), pp. 179-184.
12. C. Pequet, M. Gremaud, and M. Rappaz, "Modeling of Microporosity, Macroporosity, and Pipe-shrinkage Formation During the Solidification of Alloys Using a Mushy-zone Refinement Method: Applications to Aluminum Alloys," *Metall. Mater. Trans. A.*, Vol. 33, (2002) pp. 2095-2106.
13. G. Couturier and M. Rappaz, Modelling and Simulation in Materials Science and Engineering, Vol. 14, pp. 253, 2006.
14. P.D. Lee, A. Chirazi, D. See, Modeling microporosity in aluminum±silicon alloys: a review, Journal of Light Metals 1 (2001), pp. 15-30.

15. R.C. Atwood, P.D. Lee, Simulation of the three-dimensional morphology of solidification porosity in an aluminium–silicon alloy, Acta Materialia, Vol. 51, pp. 544–5466, 2003.
16. Wang, J.; Li, M.; Allison, J.; and Lee, P.D, Multiscale modeling of the influence of Fe content in a Al-Si-Cu alloy on the size distribution of intermetallic phases and micropores, Journal of Applied Physics, Vol. 107, Article No. 061804, 2010.
17. P. D. Lee, A. Chirazi, R. C. Atwood, and W. Wang, Multiscale modelling of solidification microstructures, including microsegregation and microporosity, in an Al–Si–Cu alloy, Mater. Sci. Eng., A 365, p. 57, 2004.
18. Ch.-A. Gandin and M. Rappaz, *Acta Metall. Mater.,* 42 (1994), pp. 2233–2246.
19. Ch.-A. Gandin, J.-L. Desbiolles, M. Rappaz and Ph. Thevoz, *Metall. Mater. Trans.*, 30A (1999), pp. 3153–3165.
20. J. Guo and M. T. Samonds T. "Property prediction with coupled macro-micromodeling and computational thermodynamics," Proc. of MCSP6, Taiwan: Kaohsiung, 2004, pp. 157-164.
21. J. Guo and M. T. Samonds, Alloy Thermal Physical Property Prediction Coupled Computational Thermodynamics with Back Diffusion Consideration, Journal of Phase Equilibria and Diffusion, 2007, Vol. 28, pp. 58-63.
22. J. Guo and M. T. Samonds, Modeling of alloy casting solidification, Journal of Metals, Vol. 63, 2011, pp. 19-28.
23. M. Rappaz and Ph. Thevoz, *Acta Metall.*, 1987, vol. 35, pp. 1487–97.
24. M. Rappaz and Ph. Thevoz, *Acta Metall.*, 1987, vol. 35, pp. 2929–33.
25. W. Oldfield, ASM Transactions, 1996, Vol. 59, pp. 945-61.
26. Ye H., An overview of the development of Al-Si-Alloy based material for engine applications, Journal of Materials Engineering and Performance, 2003, Volume 12, pp. 288-297.
27. D.L McDowell, K Gall, M.F Horstemeyer, J Fan, Microstructure-based fatigue modeling of cast A356-T6 alloy, Engineering Fracture Mechanics, Vol. 70, 2003, pp. 49–80.

Advances in the Science and Engineering of Casting Solidification
Edited by: Laurentiu Nastac, Baicheng Liu, Hasse Fredriksson, Jacques Lacaze, Chun-Pyo Hong, Adrian V. Catalina, Andreas Buhrig-Polaczek, Charles Monroe, Adrian S. Sabau, Roxana Elena Ligia Ruxanda, Alan Luo, Subhayu Sen, and Attila Diószegi
TMS (The Minerals, Metals & Materials Society), 2015

X-RAY OBSERVATIONS SHOWING THE EFFECT OF FLUID FLOW ON DENDRITIC SOLIDIFICATION IN Ga-In ALLOYS

Natalia Shevchenko, Olga Roshchupkina, Sven Eckert

Helmholtz-Zentrum Dresden-Rossendorf, P.O. Box 510119, 01314 Dresden, Germany

Keywords: dendritic solidification, melt convection, X-ray radioscopy, flow patterns, forced convection

Abstract

The directional solidification of Ga–25wt%In alloys within a Hele-Shaw cell has been studied by X-ray radioscopy. The investigations were focused on the influence of melt convection on the dendritic growth. Natural convection occurs during a bottom up solidification because a lighter solute is rejected during crystallization. Forced convection has been produced by a specific electromagnetic pump. The direction of forced melt flow is almost horizontal at the solidification front. Melt flow induces various effects on grain morphology caused primarily by convective transport of solute, such as facilitation of the growth of primary trunks or lateral branches, dendrite remelting, fragmentation or freckle formation depending on the dendrite orientation, the flow direction and intensity. Forced flow eliminates solutal plumes and damps local fluctuations of solute. A preferential growth of the secondary arms occurs at the upstream side of the dendrites, whereas high solute concentration at the downstream side inhibits the formation of secondary branches.

Introduction

Thermosolutal convection has been identified as the main reason for the development of solute-rich channels in the mushy zone during the solidification of metallic alloys [1-4]. The occurrence of such segregation channels in fully-solidified castings has to be considered as a serious casting defect.

A sufficient understanding of the strongly coupled interaction between the melt flow and the solidification requires an authentic knowledge of the velocity field especially in the vicinity of the solidification front. During an upwards solidification solutal convection may arise from an unstable density stratification at the solid-liquid interface if the solute component rejected during the freezing of the primary crystals is lighter than the initial composition of the melt. In turn, the melt convection redistributes the solute concentration in the boundary layer and the mushy zone leading to dramatic changes in growth rate and direction [5-8]. Further effects concern the selection of the secondary branches of growing dendrites [9]. In general, the flow impact might be different for different regions of the mushy zone. The growth rate is increased by a removal of the solute, whereas an influx of solute-rich melt causes the formation of segregation freckles by a remelting of already solidified crystals.

Many experimental investigations have been carried out using aqueous analogues as model liquids, where the convective flow pattern can be observed visually by diverse optical diagnostic techniques [10-12]. The dimensions of segregation channels occurring in such systems were shown to be comparable to those found by metallographic examinations in fully solidified metallic castings [2]. Although these similarities give point to extrapolate the findings from the transparent materials to metal alloys to a certain extent, corresponding measurements in real metallic materials would attract wide interest. Real-time observations of solidification processes in opaque metal alloys can be realized by means of the X-ray radioscopic method [13]. Only a few studies were made so far disclosing information with respect to the flow structure near the growth front. Koster et al. [14] studied natural convection in liquid gallium and gallium alloys. Qualitative images of the flow field were derived from precise measurements of local density differences from temperature gradients. However, the spatial resolution achieved in these experiments was not sufficient to perceive details of the flow structure being in the order of the typical dendrite spacing. More recent studies applied the optical flow approach to estimate the liquid flow field during the solidification of Ga-In alloys [7, 15-16]. In this paper, we focus on the impact of melt convection on the dendrite growth during the bottom-up solidification of Ga-In alloys.

Experimental setup

The directional solidification of alloys within a Hele-Shaw cell was visualized by X-ray radioscopy. The solidification experiments were carried out at HZDR using an experimental setup which is described in detail elsewhere [7, 17]. The nominal composition of the Ga–25wt%In alloy was prepared from 99.99% Ga and 99.99% In. The alloy was melted and filled into a Hele-Shaw quartz cell with a gap of 150 μm. The quartz cell was produced in two geometric versions: a square form of a surface area of about 30 x 30 mm^2 (Figure 1(a)) and a hexagonal shape with an area of 30 x 35 mm^2 (Figure 1(b)).

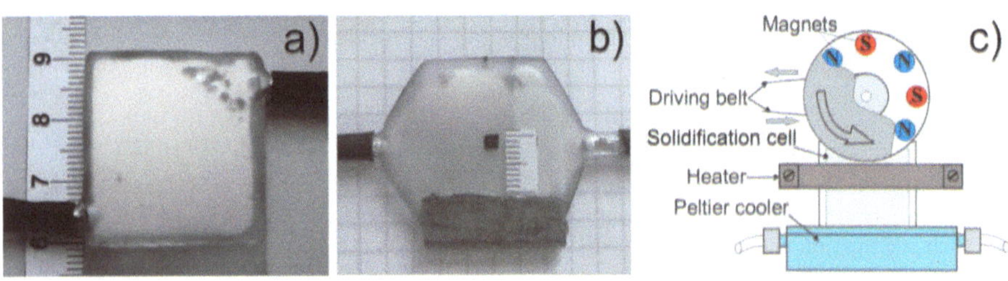

Figure1. (a) Hele-Shaw square quartz cell; (b) Hele-Shaw hexagonal shape cell; (c) schematic view of the solidification cell equipped with electric heater, Peltier elements and magnetic wheel.

Two Peltier elements acted as a cooler which was attached to the bottom part of the cell. A second pair of Peltier elements or an electric heater was mounted on the upper part of the solidification cell to heat up the alloy. In the course of the solidification the temperature gradient across the cell and the cooling rate were controlled by a simultaneous and synchronized operation of the heater and cooler. Four miniature K-type thermocouples were contacted to the surface of the cell to monitor the temperature and the temperature gradient. The experiments were carried out at constant cooling rate of 0.01 K/s and vertical temperature gradient of 1.3 - 1.5 K/mm. The electromagnetically driven flow was produced by a rotating wheel comprising two parallel iron disks (see Figure 1(c)). A set of permanent magnets (NdFeB) with alternating polarization was located at their inner sides. The top part of the solidification cell was positioned in the gap between the disks. The rotation speed of the magnetic wheel was 10 revolutions per minute (rpm). A microfocus X-ray source (Phoenix XS225D-OEM) was used to perform the radioscopy. X-ray observations were performed over a rectangular window of 9 × 12 mm^2. Images were captured at a scan rate of 50 half frames per second and a spatial resolution of ten microns. For reducing the noise level single images were integrated over a period of 1 s.

Data analysis and flow control

The local solute concentration was derived from the relative brightness in acquired images. Further information with respect to the experimental hardware and procedure can be found in [15, 17]. The analysis of the solidification front velocity, tip velocity and primary arm spacing follows the algorithms described in [15]. The Optical flow approach was applied to reconstruct the flow patterns in the melt. The algorithm description and specific results concerning the flow fields in the melt recovered by the Optical Flow approach can be found elsewhere [15].

We consider three different modes of melt convection during the solidification experiments. The conditions with respect to the temperature gradient and the cooling rate are similar in all experiments. The melt flow patterns in the liquid above the mushy layer are shown schematically in Figure 2.

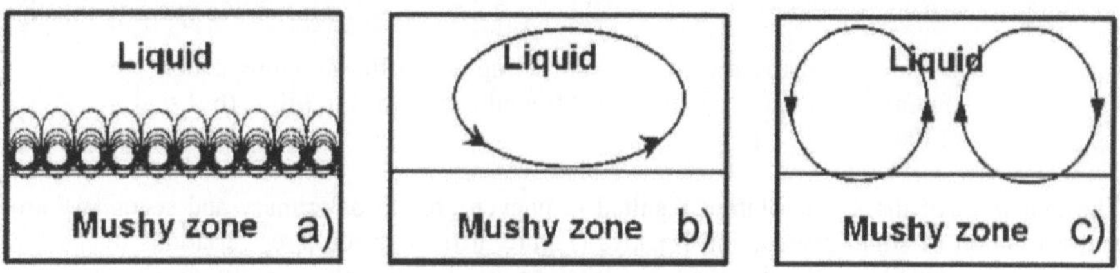

Figure 2. Streamlines for the three modes of convection: (a) thermosolutal convection (fine-scale convective cells); (b) single vortex (forced convection); (c) double vortex patterns.

The first case (Figure 2(a)) corresponds to a solidification process under the thermosolutal convection. A single vortex melt flow as shown in Figure 2(b) was produced by the electromagnetic pump. In this case the natural convection is superimposed by an electromagnetically driven flow. The solidification process in the hexagonal solidification cell is characterized by two dominating convection rolls in the liquid phase (Figure 2(c)), which are driven by the convex shape of the solidification front. Such shape of the solidification fronts can be related to nonuniform cooling of the cell [17].

Results and discussion

Natural convection

Thermosolutal convection occurs during bottom up solidification because lighter solute was rejected during crystallization at the solid-liquid interface. Typical snapshots of dendrite network obtained during solidification are shown in Figure 3. Dark grey Indium dendrites at the bottom of the solidification cell and plume-like patterns (light grey color) above the mushy layer were observed. Our previous study [18] showed that the plumes become more intense after an initial solidification stage and the size, position and number of the plumes change over process time. The velocity inside the ascending plumes was in the range of 50 – 100 μm/s. The rising plumes are accompanied by a downward flow of In-rich melt over the mushy zone. This leads to an inhomogeneous horizontal concentration distribution and formation of convective cells along the solidification front (Figure 3(a)). A diameter of convective roll approximated to few primary dendritic spacing.

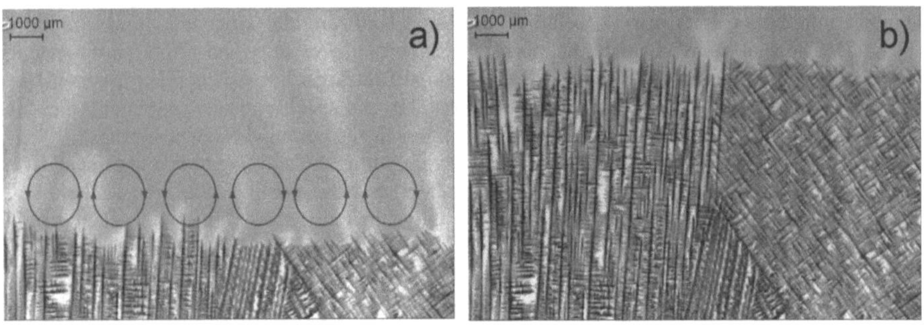

Figure 3. Snapshots of the solidifying dendritic structure under natural convection captured at different times: (a) t = 340 s, (b) t = 1300 s.

The dynamics of the solute plumes resulted in uneven growth of primary and secondary arms and formation of small segregation freckles (Figure 3(b)). Furthermore, variations of the local concentration profiles caused strong fluctuations in the growth velocity of the individual dendrites [15]. This led to the formation of a rough solid-liquid interface. The microstructure at the end of the experiment (Figure 3(b)) reveals the formation of three grains with different direction of dendrite growth. Assessment of the spacing between the primary trunks has a mean value of about 320 μm.

Forced convection

In the next experiment the natural convection was superimposed by an electromagnetically driven flow produced by a specific electromagnetic pump [7, 16]. Figure 4a captures the initial stage after the magnetic pump was activated. The magnetic wheel was rotated at a speed of 10 rpm in a clockwise direction. This leads to an almost horizontal flow of ~100 μm/s velocity in the contraclockwise direction along the solidification front as shown in Figure 4(a).

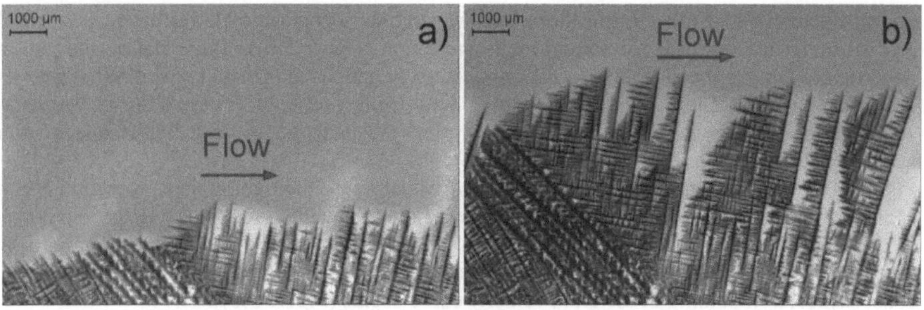

Figure 4. Snapshots of the solidifying dendritic structure under forced convection captured at different time steps: (a) t=380 s – just after the onset of the electromagnetic stirring, (b) t=1500 s.

The forced flow strongly changes the concentration field in the liquid phase: strong vertical plumes of natural convection are replaced by weak smeared plumes which are moving along the solidification front from left to the right (see Figure 4a). An uneven growth of primary dendrites due to the lateral inflow causes the formation of Ga-rich zones near the solidification front which develop into distinct segregation freckles (Figure 4(b) right). Preferential growth of the secondary arms is observed at the upstream side of the dendrites; high solute concentration at the downstream side prevents the formation of secondary arms. Competitive growth between secondary branches and primary dendrites leads to an increase in the primary arm spacing. An increase in the spacing between the primary trunks from about 300 µm in the case without forced convection up to about 700 µm was observed. Moreover, the inclination angles of the growing dendrites are slightly changing under the influenced of the incident flow. The consequences on the microstructure become apparent in Figure 4(b).

Figure 5 shows the vertical growth velocity of two grains being representative for natural and forced melt convection experiments.

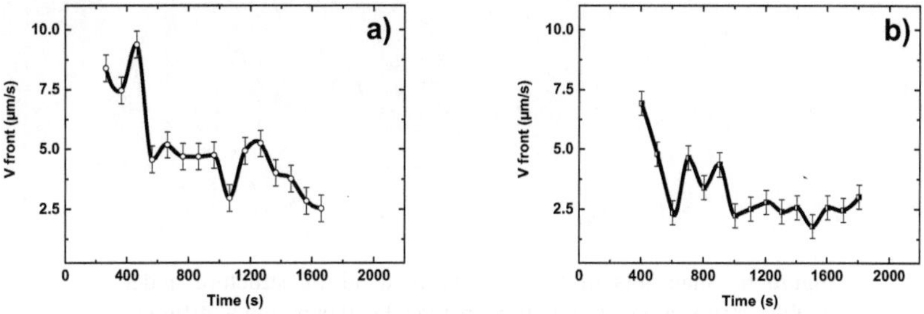

Figure 5. Velocity of the solidification front for two selected grains: (a) the left grain (as highlighted in Figure 3) under influence of the natural convection; (b) the right grain under influence of the forced convection (as highlighted in Figure 4).

Strong fluctuations of the solidification front velocity are a characteristic feature of the growth process under the influence of natural melt convection (Figure 5(a)). These fluctuations of the dendrite growth may be attributed to the pronounced variations in the solute distribution near the solidification front. The decreasing growth velocity from 9 μm/s to 2.5 μm/s during the entire solidification process can be explained in terms of the increase in the gallium concentration above the solidification front.

Figure 5(b) shows the growth velocity for an individual grain (the right grain as shown in Figure 4) in the forced convection mode. One can observe the fluctuating growth during the early stage of the solidification experiment. About 600 s after the outset of the forced convection, the flow reaches a steady state and damps the local fluctuations of solute concentration. As a consequence, a smoothing of the grain velocity curve occurs (Figure 5(b)).

Double-vortex convection

The third situation is the dendrite growth under the double-vortex convection. Double-vortex flow patterns arise from the specific hexagonal geometry of the solidification cell. The conical shape of the side walls leads to nonuniform cooling of the cell and, therefore, to an inhomogeneous solidification process [17]. The accelerated solidification in the central region results in a strong ascending Ga-rich flow, which generates a downward flow due to a suction effect. The resulting double vortex patterns may span over the entire cell volume.

The optical flow method identify a converging flow ahead of the mushy layer coming from the side walls and leading to accumulation of the rising plumes in the central area of the cell. As shown in Figure 6(a), there are two large convection rolls imposing a converging tangential flow along the solidification front. A lateral component of the flow velocity was determined to be in the range between 30 and 70 μm/s [17].

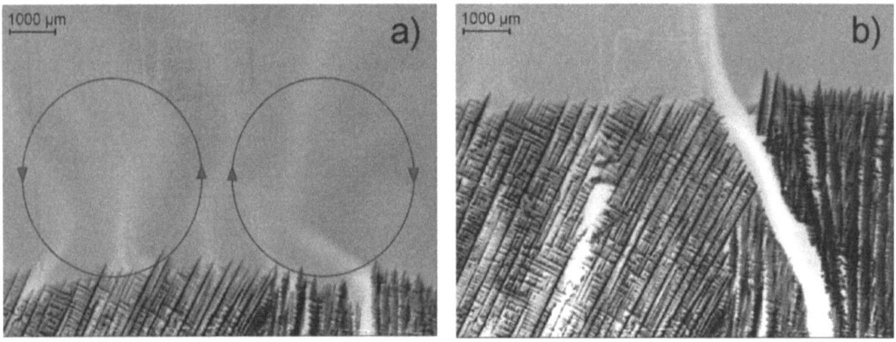

Figure 6. Snapshots of the solidifying dendritic structure under double-vortex convection in the hexagonal cell captured at different time steps: (a) t = 370 s, (b) t = 1650 s.

Figure 6(b) shows a representative microstructure with a strong segregation channel. In all fifteen solidification experiments performed under such conditions, at least one stable chimney in the central part of the solidification cell can be detected. The occurrence of such central chimneys is promoted by the converging flow ahead of the solidification front. The development of the chimneys is strongly coupled with a continuous accumulation of solute in the central part of the cell. For further details concerning the formation of segregation channels in the mushy zone see refs. [16, 17].

During the initial period of solidification the channels show a rather unstable behavior, in particular they can migrate or disappear and re-emerge at another location [17]. The kinetics of chimney development depends on fluctuations of the global melt flow. The solute-rich fluid rises through the channel and erodes surrounding dendrites (Figure 6(b)). In this way, the initiation and development of the channels affect the growth velocity and morphology of neighboring dendrites. The mean values of a solidification front velocity are 5 μm/s and 3 μm/s for the case of the natural and double-vortex convections, respectively. The decreasing growth rate of the grains is caused by continuous accumulation of solute above the solid-liquid interface in the central part of the cell.

Summary

An in situ X-ray study is presented focusing on the dendrite growth during the bottom-up solidification of Ga-In alloys. The visualization of the solidifying alloys provides information on both the structure of the dendrite fraction and the flow field in the vicinity of the solid-liquid interface. Three experiments have been chosen within this paper for examining the effect of melt convection on the growth behavior of the Indium dendrites. Melt flow effects on grain morphology caused primarily via convective transport of solute, such as acceleration of the growth of primary trunks or lateral branches, dendrite remelting or freckle formation depending on the dendrite orientation, the flow intensity and direction. Our visualization experiments reveal the way by which specific flow patterns generated by different types of melt convection provoke a solute transport in the mushy zone and the formation of spacious segregation freckles.

Acknowledgments

The research is supported by the German Helmholtz Association in form of the Helmholtz-Alliance "LIMTECH".

References

1. A. Hellawell, J.R. Sarazin, R.S. Steube, "Channel convection in partly solidified systems," *Philosophical Transactions: Physical Sciences and Engineering*, 345 (1993), 507-544.
2. A.K. Sample, A. Hellawell, "The mechanisms of formation and prevention of channel segregation during alloy solidification," *Metallurgical Transactions*, 15 (A) (1984), 2163-2173.
3. S.N. Tewari, R. Shah, "Macrosegregation during steady-state arrayed growth of dendrites in directionally solidified Pb-Sn alloys," *Metallurgical Transactions,* 23 (A) (1992), 3383-3392.
4. M.I. Bergman et al., "Convection and channel formation in solidifying Pb-Sn alloys," *Metallurgical and Materials Transactions*, 28 (A) (1997), 859–866.
5. K. Murakami et al., "Influence of melt flow on the growth directions of columnar grains and columnar dendrites," *Acta Metallurgica*, 31 (1983), 1425-1432.

6. M. Medina et al., "Channel segregation during solidification and the effects of an alternating traveling magnetic field," *Metallurgical and Materials Transactions*, 35 (B) (2004), 743-754.
7. S. Boden, S. Eckert, G. Gerbeth, "Visualization of freckle formation induced by forced melt convection in solidifying GaIn alloys," *Materials Letters*, 64 (2010), 1340-1343.
8. B. Willer et al., "The columnar-to-equiaxed transition in Pb–Sn alloys affected by electromagnetically driven convection," *Materials Science and Engineering*, 402 (A) (2005), 55-65.
9. S. Steinbach, L. Ratke, "The influence of fluid flow on the microstructure of directionally solidified AlSi-base alloys," *Metallurgical and Materials Transactions*, 38 (A) (2007), 1388-1395.
10. S.S.L. Peppin, E. Huppert, M.G. Worster, "Steady-state solidification of aqueous ammonium chloride," *Journal of Fluid Mechanics*, 599 (2008), 465-476.
11. B.T. Murray, A.A. Wheeler, M.E. Glicksman, "Simulations of experimentally observed dendritic growth behavior using a phase-field model," *Journal of Crystal Growth*, 154 (1995), 386-400.
12. M. Zhang, T. Maxworthy, "The interactive dynamics of flow and directional solidification in a Hele-Shaw cell Part 1. Experimental investigation of parallel shear flow," *Journal of Fluid Mechanics*, 470 (2002), 247-268.
13. R.H. Mathiesen, L. Arnberg, "X-ray radiography observations of columnar dendritic growth and constitutional undercooling in an Al–30wt%Cu alloy," *Acta Metallurgica*, 53 (2005), 947-956.
14. J.N. Koster, T. Seidel, R. Derebail, "A radioscopic technique to study convective fluid dynamics in opaque liquid metals," *Journal of Fluid Mechanics*, 343 (1997), 29-41.
15. S. Boden et al., "X-ray radioscopic visualization of the solutal convection during solidification of a Ga-30 wt pct in alloy," *Metallurgical and Materials Transactions*, 39 (A) (2008), 613-623.
16. N. Shevchenko et al., "In situ X-ray monitoring of convection effects on segregation freckle formation," *IOP Conference Series: Materials Science and Engineering*, 33 (2012), 012035
17. N. Shevchenko et al., "Chimney Formation in Solidifying Ga-25wt%In Alloys Under the Influence of Thermosolutal Melt Convection," *Metallurgical and Materials Transactions*, 44 (A) (2013), 3797-3808.
18. N. Shevchenko et al., "The effect of natural and forced melt convection on dendritic solidification in Ga-In alloys", *Journal of Crystal Growth* (2015), in press

Advances in the Science and Engineering of

CASTING SOLIDIFICATION

An MPMD Symposium Honoring Doru Michael Stefanescu

Cast Iron I

Session Chair:
Roxana Elena Ligia Ruxanda

Advances in the Science and Engineering of Casting Solidification
Edited by: Laurentiu Nastac, Baicheng Liu, Hasse Fredriksson, Jacques Lacaze, Chun-Pyo Hong, Adrian V. Catalina, Andreas Buhrig-Polaczek, Charles Monroe, Adrian S. Sabau, Roxana Elena Ligia Ruxanda, Alan Luo, Subhayu Sen, and Attila Diószegi
TMS (The Minerals, Metals & Materials Society), 2015

DEFECT FORMATION MECHANISMS IN LAMELLAR CAST IRON RELATED TO THE CASTING GEOMETRY

Attila Diószegi[1], Peter Svidró[1], Lennart Elmquist[2], Izudin Dugic[3]

[1]Jönköping University, School of Engineering, Department of Material and Manufacturing – Foundry Technology, P.O. Box 1026, SE-551 11 Jönköping, Sweden
[2]SinterCast AB, Technical Centre, SE-641 30 Katrineholm, Sweden
[3]Linnaeus University, Faculty of Technology, Department of Mechanical Engineering, SE-351 95 Växjö, Sweden

Keywords: Lamellar Cast Iron, Columnar Zone, Equiaxed Zone, Permeability, Material Transport

Abstract

Although lamellar cast iron has been used in advanced applications for about twenty years, our knowledge about the mechanisms affecting microstructure and defect formation is relatively limited. The present paper summarizes some solidification related phenomena from a series of recently published peer reviewed papers and scientific theses and suggests a mechanism of defect formation which is dependent on the shape of the solidifying casting geometry. When shrinkage porosity or metal expansion penetration occurs evidence of material transport in the intergranular zone of primary equiaxed austenite grains in the casting and in the intergranular regions between the sand grains in the mold material is seen. Material transport occurs across the casting-mold interface where the existence of or the permeability of the primary columnar zone determines if material transport can take place.

Introduction

Lamellar cast iron is the most often used cast material for engineering purposes. Despite the many advantages such high thermal conductivity, high vibration damping or good machinability being reasons for its widespread application there are some common manufacturing defects which detract from these excellent properties. These defects are shrinkage porosity (SP) [1] known to cause leakage in complex shaped castings intended to operate under pressure, and metal expansion penetration (MEP) [2], which causes extrusion of metal between sand grains of the mold, creating metal-sand phases protruding from the casting surfaces. These defects are reported especially from production of automotive castings. Massive efforts have been dedicated towards understanding the mechanisms of these defects formation [3-5]. Progress in understanding the solidification mechanisms in lamellar cast iron and which help us to understand the defect formation mechanisms was possible when a series of new investigation methods was introduced during the last two decades. These include color etching [6], direct austempering after solidification [7], electron back scattering diffraction [8], thermal analysis [9], and density and volume change measurements [10]. Castings with both primitive geometries and complex shapes have been used to investigate the solidification mechanisms. The aim of the present paper is to summarize the results in the literature which pertain to the solidification and defect formation mechanisms of SP and MEP in lamellar cast iron with respect to the casting geometry. The material included in the study is hypoeutectic lamellar cast iron which includes in addition to the standard alloying elements

Si, Mn, P and S, small quantities of perlite stabilizers including Cu, Mo and Cr. The melts were produced under industrial conditions in either induction or cupola furnaces. Molding materials included in the investigation are resin-bonded quartz sand mixtures for the special samples or a combination of green sand and resin-bonded sand mixtures for complex shaped automotive castings. Detailed information of the experimental processes is given in the literature references.

Literature

Introduction of DAAS and the interpretation of the primary austenite grains observed as columnar or equiaxed zones.

Solidification of lamellar cast iron starts with the nucleation and growth of primary austenite grains. The most obvious place to nucleate austenite grains are at the metal/mold interface. Foundrymen have for a long time ago observed this phase and named it “casting skin”. Unfortunately the primary crystal structure could not be observed at ambient temperature after solidification until a special heat treatment named Direct Austempering After Solidification was introduced [7], (DAAS). This method is developed to preserve the original primary grains, FCC structure by quenching the cast sample through the solid-state transformation and thereby avoiding the formation of the BCC structure of pearlite or ferrite. Figure 1 shows the primary austenite grain structure of a sample known from the literature as Quick-cup, treated by DAAS. Figure 2 was created by drawing the grain borders observed on the original sample with the help of variable inclination of the incident light. The grains observed to be in contact with the casting surface are represented by a darker color and are assumed to form the columnar zone (casting skin). The internal zone of the casting within the coherency line (red) is denoted as the equiaxed zone. A clear orientation between the outer, columnar zone and the casting surface is observable, in contrast to the equiaxed zone. A study of the morphology of the columnar and equiaxed zones was presented in the literature [11]. Further investigations [8] have confirmed the correctness of interpreting the DAAS treated samples using the variable angle of incident light by using electron backscattering diffraction (EBSD), which shows that each observed unit has a unique crystallographic orientation.

Figure 1. DAAS treated quick-cup sample on hypoeutectic LGI.

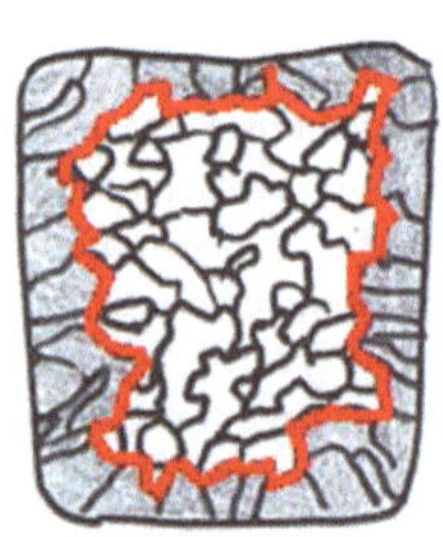

Figure 2. Columnar to equiaxed transition line in the quick-cup sample.

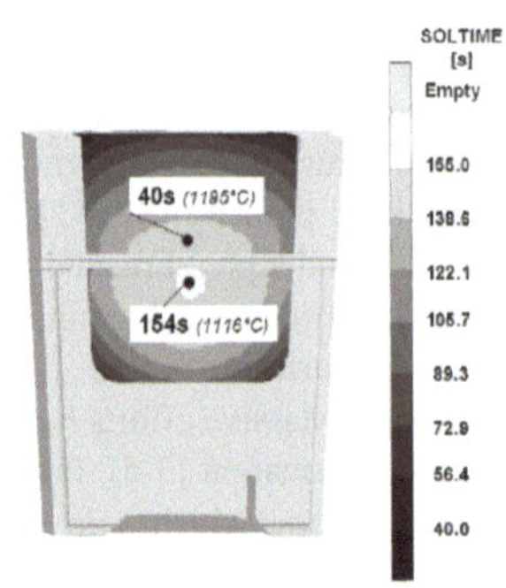

Figure 3. Simulated local solidification time of the quick-cup sample. The dots show the hot spot position at different times.

Study of the thermal field and its influence on the columnar zone

Investigation of complex shaped castings is useful for understanding the solidification and defect formation mechanisms under industrial conditions but it is a tedious work to investigate the local macro- and microstructure properties. The DAAS treatment also has limitations with respect to the size and the complexity of the cast sample which can be treated. By increasing the size and geometrical complexity of the casting the DAAS quenching effect decreases and observation of the primary grains cannot be made. For this reason a combination of different primitive geometrical volumes has been developed which successfully provoke the formation of MEP [12] and SP [13]. The MEP sample is shown in Figure 4 and the SP sample is shown in Figure 7. Simulated local solidification times of samples with different geometrical complexity are shown in Figure 3, Figure 6 and Figure 9, [14].

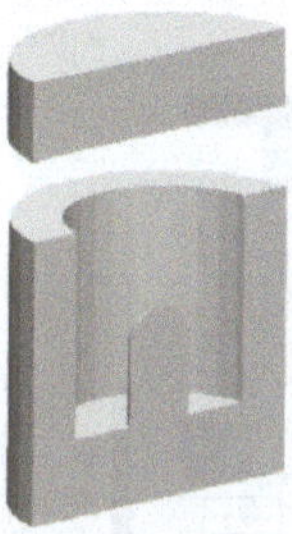

Figure 4. Cross-section of sand mold used for MEP sample production.

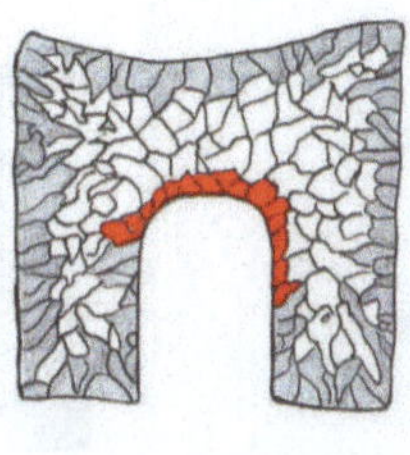

Figure 5. Columnar and equiaxed crystals from the DAAS investigated MEP sample. The red zone indicated the smallest primary grains.

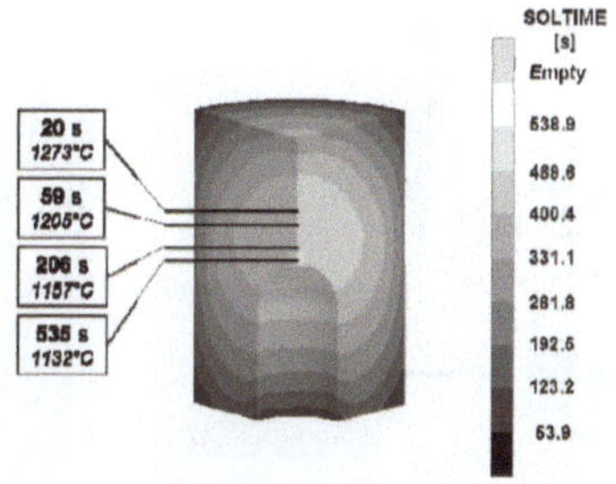

Figure 6. Simulated local solidification time of the MEP sample. The dots show the hot spot position at different times.

Simulation results showing the solidification time of the Quick-cup sample which is formed by a single primitive geometry (a cube) are given in Figure 3. The calculated solidification times for the MEP sample which is a combination of two cylindrical units where the smaller cylinder forms an internal core inside a large cylinder are shown in Figure 6. The SP sample is built-up of three plates of different thicknesses and three cylindrical columns in combination (shown schematically in Figure 7). The simulated solidification time for the cross section of the sample is shown in Figure 9. The hottest point and therefore the last spot of the primitive geometry to solidify is situated in the middle of the sample. A direct consequence of the thermal field obtained is the continuous and constant thickness columnar zone over all surfaces of the Quick-cup sample. The cooling conditions during solidification are nearly isotropic. The MEP and SP samples solidify last at the metal/mold interface. Consequently the thickness and size of the columnar zone in the last solidifying region is less developed, as can be seen in Figure 5 for the MEP sample and Figure 8 in the SP sample. The columnar zone marked red in Figure 5 is influenced by heat saturation in the internal core of the sample. The austenite grain structure of the SP sample taken from the lower cross-section of the middle cylindrical component of the sample is shown in Figure 8, where the thermal field is confined between a cylindrical column and a parallel plate. On the left side of the micrograph the columnar zone is thicker compared to the right side. The mold material in contact with the surface on the right side creates a heat buffer and disturbs the columnar zone development. Recent research results show the dependence between the intra-dendritic space and the local solidification time [15 – 16] which being a consequent phenomenon to the dynamic dendrite coarsening. Further investigations [17], showed that the dynamic coarsening is valid even for the MEP and SP samples, and a gradient of the increasing intradendritic space is observed from zones with short local solidification time to zones with long local solidification time. The longest local solidification time for the MEP and SP samples was found at the metal/mold interface, as shown in Figure 6 and Figure 9. These zones are consequently expected to have the dendritic network with the largest permeability.

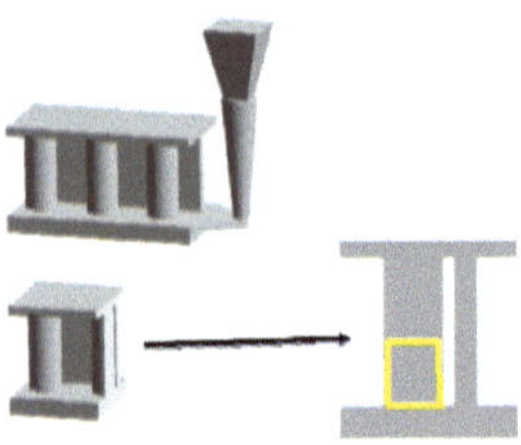

Figure 7. SP sample indicating where the macrostructure was inspected.

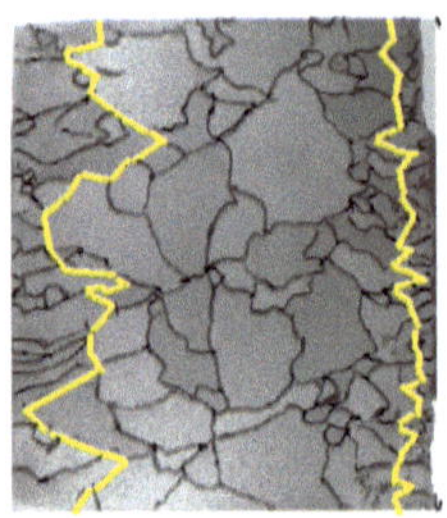

Figure 8. Columnar to equiaxed transition lines in the SP sample.

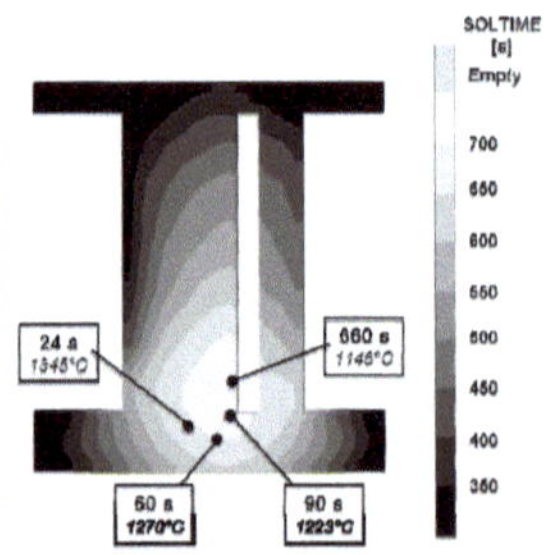

Figure 9.Simulated local solidification time of the SP sample. The dots show the hot spot position at different times.

Material transport path and SP formation

Shrinkage porosity in lamellar cast iron causing leakage in the casting during operation has been identified to have a 3D distribution enclosing several eutectic cells [1]. See Figure 10. After introducing the DAAS visualization method it was demonstrated that the enclosed eutectic cells / colonies belong to the same primary austenite grain, and that the pore cavities were formed at the boundaries between the primary austenite grains (the intergranular areas). Further investigations [18], confirmed using EBSD that the shrinkage cavities are located at the boundaries between the primary austenite grain units, as seen in Figure 11. Rough casting surfaces were found to indicate where the shrinkage pores were open and connected to the surrounding atmosphere. Elemental investigation of the internal shrinkage porosity surfaces showed they were strongly oxidized, indicating the early contact of the pore surface with the oxidizing atmosphere while the metal was still hot in the casting cavity. This observation was interpreted to mean that the shrinkage porosity was formed by sucking the gaseous atmosphere from outside the casting into the intergranular area. Consequently the main material transport path during shrinkage porosity formation is in the intergranular zones where a depression is expected to appear, indicating the initiation of SP formation.

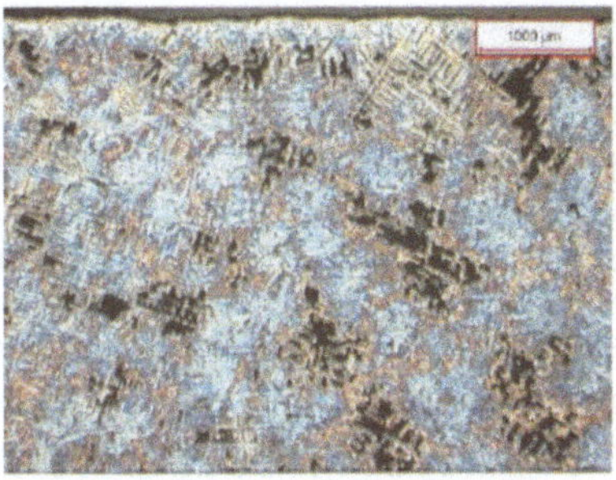

Figure 10. Color-etched microstructure of an LGI casting. Dark areas represent the shrinkage pores enclosing the eutectic cells/colonies.

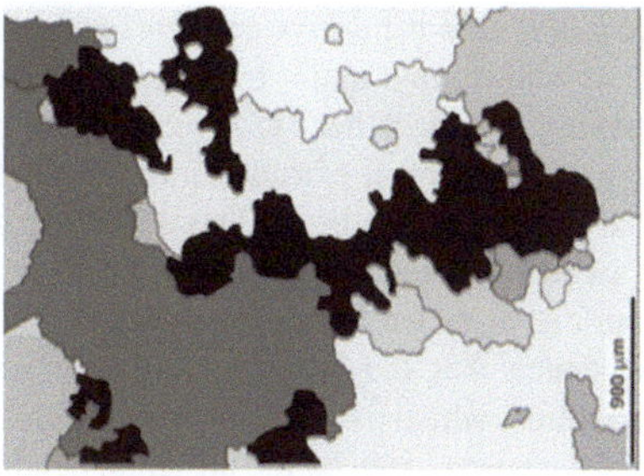

Figure 11. Distribution of shrinkage porosity in relation to the primary austenite grains. Black areas = shrinkage pores. Other colors represent primary austenite grains where the color indicates different crystallographic orientations.

Material transport path and MEP formation

Metal expansion penetration has been observed to form by two different mechanisms [19]. The first case is shown in Figure 12. The casting skin of the critical zone in the last solidifying area bulges outward. It appears that internal pressure forced metal into the metal/mold interface leading to an anomalous metallic structure behind the columnar zone, indicated by a yellow rectangle in the figure. Figure 13 shows the microstructure of this zone, including newly nucleated small primary dendrites embedded in a carbide structure.

Figure 12. Deformed casting surface due to MEP. The yellow rectangle shows the position of an anomalous microstructure.

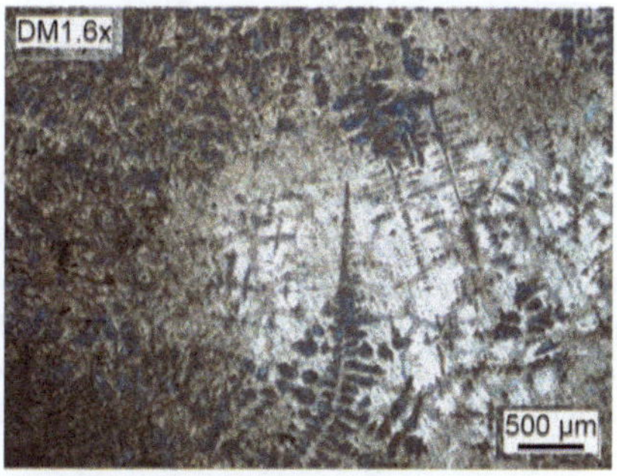

Figure 13. Anomalous microstructure behind the deformed metal/mold interface. Fine primary austenite dendrites are embedded in a carbide phase.

These observations were interpreted as being a result of the interdendritic zone becoming pressurized and the segregated liquid around the primary austenite grains being transported to the extruded zone to form the anomaly. Consequently the main material transport path during MEP formation is in the interdendritic zone (as in the case of SP formation) but the driving force in this case is positive and due to pressurization of the intergranular liquid. In the second case as shown in Figure 14 the liquid metal was squeezed out from the casting cavity occupying the space next to the casting skin, and extruded into the mold sand, as shown in Figure 15. The metal extruded into the sand has a perfect eutectic composition without any primary austenite dendrites. This should be compared to the microstructure above the dashed line (which represents the casting surface) where the microstructure contains a dendritic network as expected from a hypoeutectic alloy composition. This observation was interpreted as evidence that the MEP in this case was formed by squeezing out a perfect eutectic composition which is only expected from the intragranular zone. The intragranular zone means the space between the austenite dendrites which belongs to the same primary austenite grain. Furthermore MEP defects where the liquid metal was forced between the sand grains were found to be predominantly of eutectic composition.

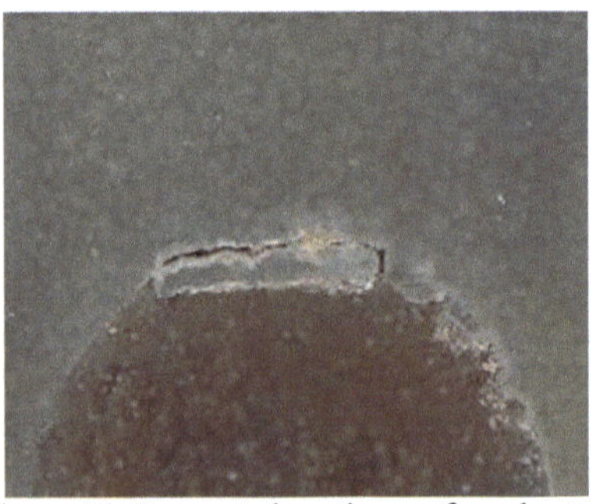

Figure 14. Penetrated casting surface due to MEP.

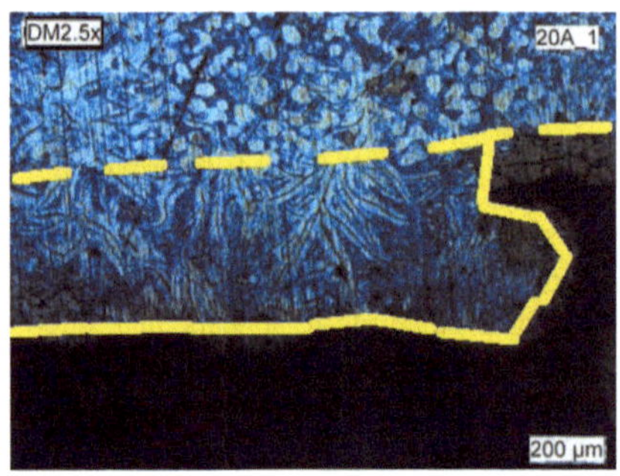

Figure 15. Color-etched extruded phase at the casting/mold interface. The dashed line indicates the original casting/mold interface. The solid line delimits the extruded phase.

Discussion

From the experimental observations it can be summarized that both MEP and SP in LGI form due to material transport across the metal/mold interface between the internal casting cavity and the casting environment. Gaseous atmosphere is sucked in between the primary austenite grains – the intergranular space, during SP formation. Intergranular liquid or eutectic liquid metal is forced outwards, and the casting skin bulges. The mass transport inside the casting, or between the casting and its environment is dependent on the macro solidification structure which is in turn related to the geometrical complexity of the casting. In the case of shape casting the casting cavity is delimited from the mold by a columnar zone which is part of the macro solidification structure and which behaves like a container. In the inner part of the container equiaxed crystals form while outside this zone the columnar zone is in contact with the mold material. In the case of a simple geometrical unit where heat transport away from the casting surface is isotropic the columnar zone has a relatively constant thickness and forms a closed container. Due the growth characteristics the size of the inter-dendritic space in the columnar zone is finer than in the equiaxed zone. The columnar zone is considered impermeable and any material transport between the casting inner zone and the environment is obstructed. This is schematically represented in the upper row of Figure 16. When the metal/mold interface is formed by both concave and convex surfaces heat transport away from the curved surfaces will be non-isotropic and different in the different areas. Convex casting surfaces may transport heat away faster compared to concave surfaces. Consequently even the thickness of the columnar zone will be influenced. A fatal influence on the columnar zone formation with respect to uniform formation of the columnar zone is a combination of basic elemental geometries provoking a variation in heat transport where the mold material becomes heat saturated and instead of being a cooling medium is transformed into an insulating material. Formation of the insulating zones (after heat saturation) retards the nucleation of the columnar zone or delays the solidification time and increases the permeability of the columnar zone. The missing columnar zone or the most permeable part of the macrostructure concentrated in the columnar zone will encourage material transport between the inner zone of the casting (columnar zone) and the casting external (mold) environment. Examples of "fatal" combinations are presented in the middle row of Figure 16, for the case of an internal core supersaturated by heat, mostly found when metal expansion penetration occurs. The lower row of Figure 16 shows the cases when two primitive geometrical parts are placed too close to each other, which also leads to heat saturation and a hot spot between the two cast sections.

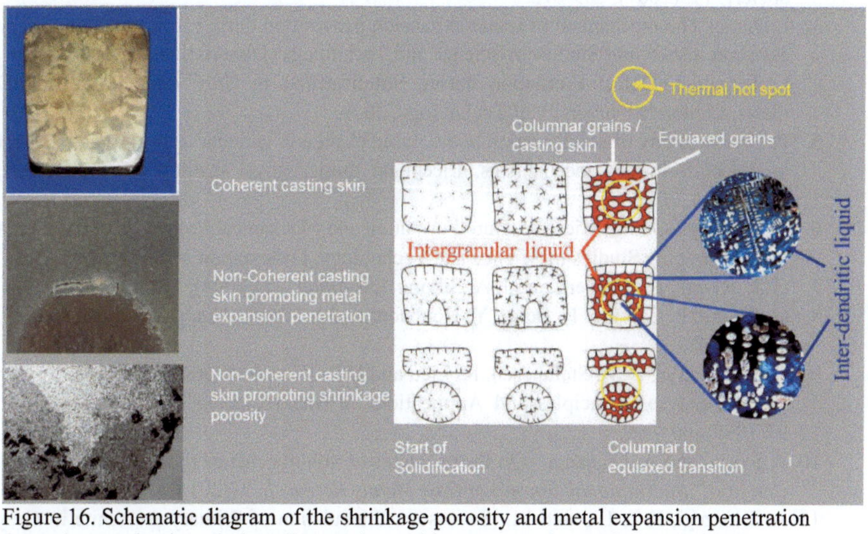

Figure 16. Schematic diagram of the shrinkage porosity and metal expansion penetration formation mechanisms in lamellar cast iron with respect to the casting geometry.

Conclusions

Based on the geometrical complexity of the casting and considering the macrostructural components of the primary phase of LGI it can be concluded that the columnar crystal zone serves as a container and delimits material transport between the mold material and its gaseous environment and the internal mushy zone containing equiaxed crystals and segregated intergranular and intragranular liquid. If the geometrical conditions promote anisotropic heat transport and variable-thickness or missing columnar zones a permeable container of the columnar zone is formed, permitting material transport between the internal and the outer zones. The critical material transport path across the mold/casting surface has been demonstrated to be dependent on the quality of the columnar zone.

Acknowledgement

The present work is performed within a collaboration project in the Casting Innovation Centre, where the collaborating partners are the University of Jönköping, Swerea Swecast AB and the Swedish Foundry industry. The performed work was financed by Vinnova and the Swedish Knowledge Foundation. All collaborating partners and financiers are gratefully acknowledged.

References

1. L. Elmquist, S. Adolfsson and A. Diószegi: "Characterizing Shrinkage Porosity in Gray Cast Iron using Microstructure Investigation". *Transaction of the American Foundrymen Society*, 116(2008), 692-704.
2. A. Diószegi, I.Dugic and I.L. Svensson: "Metal Expansion Penetration on Concave Casting Surfaces of Gray Cast Iron", *Transactions of the American Foundrymen Society*, 115(2007), 609-615.
3. I. Dugic: The mechanism of metal expansion penetration during solidification of grey cast iron. Linköping Studies in Science and Technology. Dissertation No. 1007. 2006.
4. L. Elmquist: Defect Formation during Solidification in Gray Iron Castings. PhD thesis, Chalmers University of Technology. 2009.
5. P. Svidró: Study of solidification and volume change in lamellar cast iron with respect to defect formation mechanisms. Licentiate thesis, Royal Institute of Technology, Stockholm, 2013.
6. A. Diószegi: On the Microstructure Formation and Mechanical Properties in Grey Cast Iron. Linköping Studies in Science and Technology. Dissertation No. 871. May 2004.
7. G. L. Rivera, R. E. Boeri, and J. A. Sikora: *Scr. Mater.*, 50(2004), 331–335.
8. G. Rivera, P. Calvillo, R. Boeri, Y. Houbaert, and J. Sikora: *Mater.Charact.* 59(2008), 1342–1348.
9. K.G. Upadhya, D.M. Stefanescu, K. Lieu and D.P. Yeager: "Computer-Aided Cooling Curve Analyses Principles and Applications in Metal Casting". *AFS Transactions*, 97(1989), 60 – 66.
10. P. Svidró and A. Diószegi: "On the problems of volume change measurements in gray cast iron." *International Journal of Cast Metals Research.* 27(2014). 26-37.
11. A. Diószegi, K.Z. Liu and I.L. Svensson: "Inoculation of Primary Austenite in Grey Cast Iron." *International Journal of Cast Metals Research.* 20(2007), 68-72.
12. A. Diószegi and I. Dugic: "The Mechanism of Metal Expansion Penetration in Gray Cast Iron." Proceedings of the 8th International Symposium on Science & Processing of Cast Iron, SPCI8, October, 2006, Beijing, China.
13. L. Elmquist and A. Diószegi: "Shrinkage Porosity and its Relation to the Solidification Structure of Gray Cast Iron Parts." *International Journal of Cast Metals Research.* 23(2010), 44-50.

14. L. Elmquist and A. Diószegi: "On the Problem of a Migrating Hot Spot. Solidification and Gravity" *Materials Science Forum* 649(2010), 443-448. Trans Tech Publication, Switzerland.
15. A. Diószegi, R. Lora and V. Fourlakidis: "Dynamic Coarsening of Austenite Dendrite in Lamellar Cast Iron Part 1 – Investigation based on interrupted solidification", *Materials Science Forum* Vols. 790-791 (2014), 205-210.
16. V. Fourlakidis, R. Lora and A. Diószegi: "Dynamic Coarsening of Austenite Dendrite in Lamellar Cast Iron Part 2 – The influence of carbon composition", *Materials Science Forum* Vols. 790-791 (2014), 211-216.
17. P. Svidró, L. Elmquist, I. Dugic and A. Diószegi: "Characterisation of primary dendrite morphology in complex shaped lamellar cast iron castings". Proceedings of the 10th International Symposium on Science & Processing of Cast Iron, SPCI10, November, 2014, Mar del Plata, Argentina.
18. L. Elmquist, K. Soivio and A. Diószegi: "Cast iron Solidification Structure and how it is related to Defect Formation", *Materials Science Forum* Vols. 790-791 (2014), 441-446.
19. A. Diószegi, L.Elmquist, J.Orlenius and I.Dugic: "Defect Formation at Casting of Gray Iron Components". *International Journal of Metalcasting* Vol. 3. Issue 4, (2009).

Advances in the Science and Engineering of Casting Solidification
Edited by: Laurentiu Nastac, Baicheng Liu, Hasse Fredriksson, Jacques Lacaze, Chun-Pyo Hong, Adrian V. Catalina, Andreas Buhrig-Polaczek, Charles Monroe, Adrian S. Sabau, Roxana Elena Ligia Ruxanda, Alan Luo, Subhayu Sen, and Attila Diószegi
TMS (The Minerals, Metals & Materials Society), 2015

CHARACTERIZATION OF DIRECTIONALLY SOLIDIFIED GRAY IRON

Elis A. Rivera[1], Tyler Christiansen[1], Amber L. Genau[1], Adrian Catalina[2]

[1]University of Alabama at Birmingham; 1150 10th Ave S Birmingham, AL 35294, USA
[2]Caterpillar; 14009 Old Galena Rd, Mossville, IL 61552, USA

Keywords: Gray Iron, Cast Iron, Eutectic, Directional Solidification, Graphite Spacing

Abstract

Even with decades of study, the complex development of solidification microstructures in cast iron is incompletely understood. Because Fe-C eutectic can produce different morphologies, and even different phases, depending on growth velocities and composition, understanding the conditions under which each forms is important. Directional solidification was used to investigate the effects of alloying additions and solidification velocity on graphite spacing in gray iron. Average and minimum spacing for five compositions, containing varied amounts of Si and Mn, and velocities from 0.5 to 5 μm/s are reported. A critical velocity of around 1 μm/s was observed, above which the graphite structure loses directionality and austenite dendrites appear. A semi-automated MATLAB code was developed for quickly and objectively measuring graphite spacing. The automated results compare favorably with traditional manual measurements, and will allow for more robust measurement of eutectic spacing in systems where the spacing is highly irregular.

Introduction

Gray iron is a type of cast iron with graphitic microstructure. The presence of graphite in the structure gives it a gray color, from which takes its name. It is the most common cast iron and the most widely used cast material based on weight [1]. Gray iron is a common engineering alloy that it is preferred because of its low cost, good machinability, thermal conductivity, and excellent damping capacity. In addition, gray iron experiences less solidification shrinkage and the silicon that it contains promotes good corrosion resistance and increases fluidity when casting [2]. The applications of this material are extensive, including use in internal combustion engine cylinder blocks, pump housings, electrical boxes and decorative castings.

Directional solidification is a well-known process for investigating the behavior of materials under well controlled solidification conditions. Directional solidification on gray iron samples started in the early sixties [3-4] and has shown changes in constitution, morphology and spacing with the variation of growth velocity and composition. Ever since then, more studies by various researchers have been published, in the effort to comprehend the behavior of the complex cast iron. Particularly significant work about lamellar cast iron was carried out by Magnin and Kurz [5-6], who compare the Fe-C behavior to the well-known equation of lamellar spacing given by Jackson and Hunt in 1966 [7]. The Jackson-Hunt relationship shows that lamellar spacing (λ) equals a constant (K) divided by the square root of the growth velocity ($\lambda=K/V^{1/2}$) In Magnin and Kurz's work, they describe the competitive growth of stable and metastable Fe-C eutectic with small amounts of Si, P, Cr, Mn, Ti, Al and S. Their results show that at low solidification rates, gray morphology is purely lamellar eutectic. It is also mentioned that silicon and sulfur content increases the spacing of gray eutectic slightly, while other elements have no effect. For intermediate velocities of solidification (higher than 5 μm/s or 15 μm/s in the presence of sulfur), the spacing of the eutectic displays a higher value than the one predicted. Additional effects on structure, such as spacing increasing upon increasing velocity from 20 to 200 μm/s with the presence of manganese, can be explained by a progressive transition from a lamellar to more fiber-like eutectic.

The purpose of this experiment is to provide benchmark data that will be used to validate new models for predicting the behavior in cast iron. This project collects spacing data in a manner similar to what was done by Magnin and Kurz [5], but for alloys with higher Si and Mn content. The eutectic spacing results will be used to develop and corroborate a quantitative model for predicting growth behavior and spacing under varying solidification conditions and alloying additions.

Experimental Procedure

Gray iron samples were prepared from Sorel pig iron, high purity C, SiC and FeMn melted in an induction furnace at high temperature. The samples that were prepared had different compositions of C, Si and Mn. The alloying elements added were added directly to the melt in the induction furnace. Once the correct chemical composition was achieved individual samples were collected with evacuated glass tubes. A total of five different alloys were created; actual compositions can be seen in Table 1. Carbon and sulfur content was measured using LECO, while other elements were determined using electric arc spectrometer. The table includes the carbon equivalent for each composition, calculated as CE = % C + 0.33% Si + .33% P – 0.027% Mn + 0.4% S. The goal was to make each composition as close as possible to the eutectic composition (4.3 wt% C). The Sorel pig and alloy #1 are slightly hypereutectic, while alloys #2, 3, and 4 are somewhat hypoeutectic.

Directional solidification experiments were carried out in a vertical Bridgman furnace with MoSi heating elements and a water-cooled copper chill block. The maximum temperature of the molten metal in the furnace was 1250°C. Solidification velocities of 0.5μm/s to 5 μm/s were used for the specimens that were processed through the Bridgman furnace. The thermal gradient was measured using two fine gage thermocouples directly into a sample during solidification, and was found to be around 4°C/mm.

Processed samples were approximately 7-9 cm in length and 5mm in diameter. The samples were mounted longitudinally in epoxy resin for polishing and examination. Samples were polished with the grinding equipment with the roughest sandpaper of 60 grit and increasing grit until 1200 grit; lastly the samples were also polished with the one micron diamond paste method. The optical images taken at 50x were used to measure the average graphite spacing, as well as minimum spacing. Maximum spacing is difficult to determine in an irregular eutectic and is not reported. The standard line intercept method was used: drawing a line perpendicular to the direction of the graphite and dividing the length of the line by the number of graphite flakes intercepted. For each sample, a minimum of 25 measurements was taken from a minimum of 5 images. An example of such measurements is shown in Figure 2.

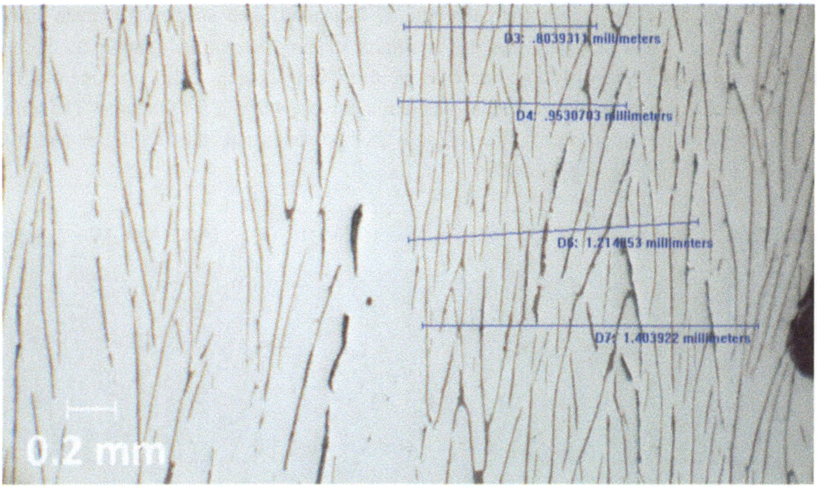

Figure 2: Manual calculation of graphite spacing in a relatively well aligned region for alloy #2 at 1 μm/s.

Table 1: Compositions of the samples taken from the induction furnace given in weight %. Final row indicates carbon equivalent for each composition.

Sample ID	Sorel Pig	#1	#2	#3	#4
C	4.519	3.836	3.420	3.574	3.450
Si	0.211	1.67	2.07	2.01	2.08
Mn	0.0250	0.0744	0.0949	0.593	0.669
P	0.0079	0.0166	0.0148	0.0117	0.0164
S	0.009362	0.007138	0.01001	0.00876	0.01123
Cr	0.0465	0.0488	0.0511	0.0542	0.0542
Mo	0.0575	0.0602	0.0609	0.0618	0.0618
Ni	0.149	0.141	0.135	0.136	0.138
Co	0.0289	0.0332	0.0348	0.0418	0.0398
Cu	0.0652	0.0794	0.0869	0.0956	0.0915
CE	4.59	4.39	4.11	4.23	4.13

Because of the irregular nature of the graphite-austenite eutectic, measuring eutectic spacing in gray iron is not trivial, and the measured spacing can depend highly on the exact placement of lines by the investigator. In order to decrease subjectivity, and increase the number of measurements to improve the statistical averages, a semi-automated method of measurement using MATLAB was developed (outlined in Figure 3). First, regions for measurement are selected and read into MATLAB. For this work, regions with the highest level of alignment were selected. For each region, additional sub-regions are selected by the user. Each sub-region is thresholded, and eighteen radial lines are then drawn on the image at 10° intervals. The number of intersections and line length for each line is measured. The spacing for each line (line length/intersections) is calculated, and the minimum spacing determined (shown as the heavier white line in Fig. 3). Generally, the line with the minimum spacing will be the one that is most closely perpendicular to the flakes and is considered the true spacing for that area. The minimum spacings for all sub-regions are collected and used to determine the minimum and average spacing for that sample or sample region.

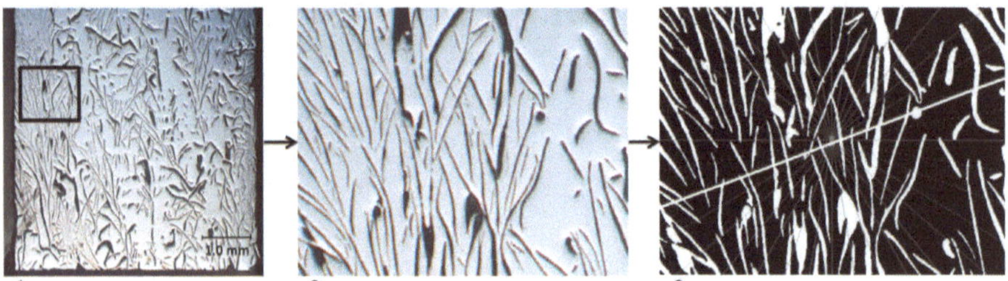

1 2 3

Figure 3: Automated stereology method: (1)Regions for measurement are selected. (2) Image opened by MATLAB code to and sub-region selected. (3) Spacing recorded at angle that produces minimum value (highlighted).

Results and Discussion

An example of the type of microstructures obtained in the directionally solidified samples is shown in Fig. 4. This sample contains a velocity transition in the center. Close up images of the graphite at each velocity are shown in Fig. 5. At the lower velocity (1 μm/s) the graphite structure in the material shows a somewhat aligned morphology in some regions. At the higher velocity (5 μm/s), the graphite structure is much finer and has no directionality or alignment. At both velocities, austenite dendrites are present, which branch and become finer at the higher velocity, as expected. The presence of dendrites in this sample was unexpected, because the composition is hypereutectic, indicating that the primary phase should be graphite. Therefore, it is likely that the critical velocity is below 1.0 μm/s, and the dendrites are a result of destabilization of the eutectic growth front.

Figure 6 shows three examples of reasonably aligned graphite from three different alloy compositions solidified at 0.5 μm/s. No dendrites are visible in these samples. Table 2 shows the average with standard deviation and minimum values of graphite spacing as measured using the manual line intercept method. Care was taken to select regions with relatively well-aligned graphite, and to avoid regions with large, apparently primary graphite. Table 2 also includes information on the average line length used to measure spacing for each sample. This is an indication of alignment, because a longer perpendicular line can be drawn on more highly aligned graphite.

In these samples, the Sorel pig was found to have a significantly larger spacing than any of the others. This may be related to the especially large flakes of graphite found in this alloy (see Fig. 6), although the alignment of the measured graphite is still quite high. In alloy #1, in the region processed at 1 μm/s, a bimodal distribution of graphite was found. Near the edge of the sample, very fine regions of aligned flake were observed. In the center near the dendrites, coarser but still well aligned flake was found. This is illustrated in Figure 5, where two measurements (on the left) have been taken of the coarser graphite, and three measurements (toward the right) have been taken on the finer graphite. The table reflects the spacings for these two regions separately.

There is not currently enough data to say definitively what effect the alloying additions have on spacing under these growth conditions. However, it is notable that adding manganese seems to decrease the alignment of the graphite. In Table 2, the average line spacing for the two samples with higher Mn content is significantly shorter, although the graphite spacing is similar. This is the result of the graphite forming fan-shaped clusters (Fig. 7), rather than growing with the relatively good alignment seen in the samples without Mn.

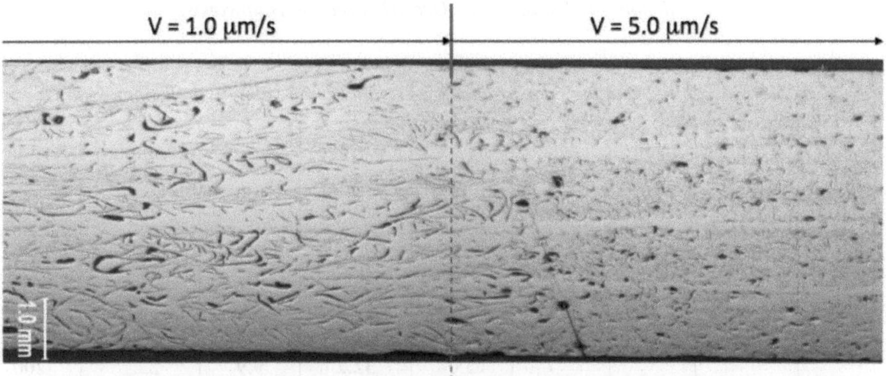

Figure 4: Optical microscope image of unetched gray iron #1 solidified at 1 μm/s and 5 μm/s

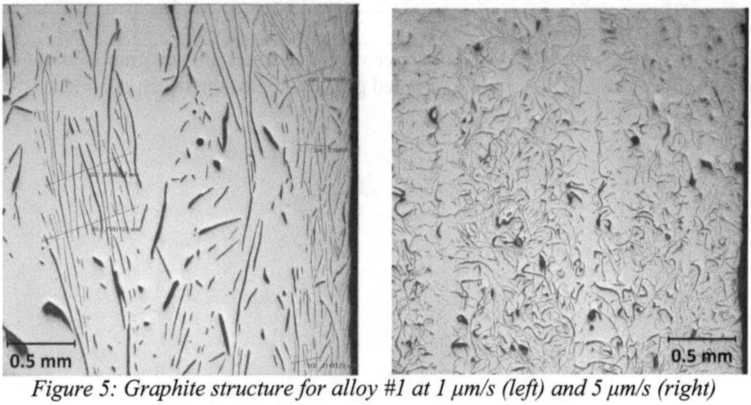

Figure 5: Graphite structure for alloy #1 at 1 μm/s (left) and 5 μm/s (right)

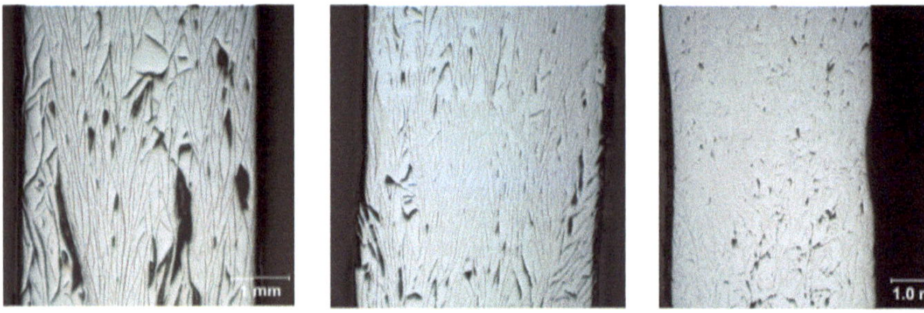

Figure 6: Longitudinal images of Sorel pig (left), gray iron #2 (center), and gray iron #3 (right, all solidified at 0.5 μm/s. Growth direction is upwards.

Table 2: Graphite spacing measurements as a function of velocity and composition.

Alloy	Carbon Equiv (wt%)	Amt Si (wt%)	Amt Mn (wt%)	Velocity (μm/s)	Average Spacing (μm)	STD (μm)	Minimum Spacing (μm)	Avg line length (μm)
Sorel Pig	4.42	0.21	0.03	0.5	70.6	15.4	48.9	1010
1	4.36	1.67	0.07	1.0*	32.8	5.9	22.9	270
					48.2	7.3	36.7	530
				5.0	32.2	9.9	23.1	200
2	4.16	2.07	0.10	0.5	47.2	6.7	34.6	670
3	4.22	2.01	0.60	0.5	45.0	7.3	32.8	450
4	4.17	2.08	0.67	0.5	43.8	7.7	25.4	430

* The graphite in this sample is bimodal: upper values are finer aligned graphite regions near the edge of the sample; lower values are for aligned graphite regions between dendrites.

Figure 7. Fan-shaped clusters of graphite in alloy #3 solidified at 0.5 μm/s.

The graph in Figure 8 shows a comparison of manual and automated gray iron spacing measurements for different regions of one sample. It can be seen that the measurements for auto and manual stereology produce statistically similar results, although the standard deviation values of manual stereology are still somewhat lower. Several iterations were required to achieve good agreement between the two methods, including adding a consideration of shape factor to the code to avoid missing smaller flakes. Continuing refinement of the code is ongoing, in order to further improve the accuracy and precision of the results. Future work will rely increasingly on the automated results.

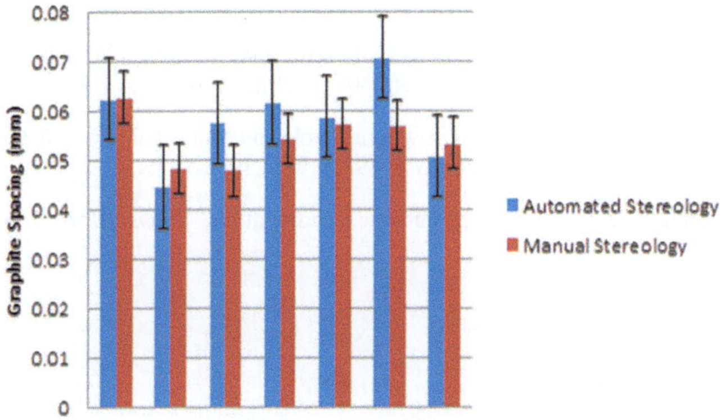

Figure 8: Spacing measurements from seven regions of alloy #2 solidified at 0.5 μm/s

Conclusions and Future Work

In conclusion, gray cast iron shows a difference in graphite spacing that depends on the velocity of growth and composition. For velocities up to 1 μm/s flake graphite shows directional structure and absence of austenite dendrites. Growth velocities higher than 1 μm/s show that flake graphite loses directionality and the presence of austenite dendrite is noticeable. Preliminary spacing results are reported. Mn is shown to have a detrimental effect on the alignment of the graphite. Lastly, spacing measurements between automated and manual stereology methods were completed, showing that the initial MATLAB code that was developed is a significant start to calculate spacing measurements of directionally solidified gray iron. Results showed that measurements of the auto stereology had similar values to the manual stereology results, and this automated method will continue to be refined so that improved spacing measurements on irregular structures can be obtained.

Acknowledgements

The authors would like to acknowledge the University of Alabama at Birmingham, MSE Metals group for all the help into achieving this project. In addition, the authors would like to thank Mr. William Monroe for all the work with the MATLAB code and Mr. Daniel Reeves for the assistance in making the automated spacing measurements. This project was sponsored by Caterpillar Inc., through the Department of Energy supported project DE-EE0005980.

References

1. P. A. Schweitzer, *Metallic Material* (CRC Press, 2003), 72.

2. E. Degarmo, J.T. Black, & A. Ronald, *Materials and Processes in Manufacturing* (Wiley, 2003), 76.

3. K. D. Lakeland, *BCIRA Journal,* 12 (1964), 634-650.

4. R. J. Brigharm, G. R. Purdy, & J. S. Kirkaldy, *Crystal Growth* (H. S. Peiser, ed., Pergamon New York 1967), 161-169.

5. P. Magnin, W. Kurz, "Competitive Growth of Stable and Metastable Fe-C-X: Part 1. Experiments," *Metallurgical Transactions A*, 19A (1988), 1955-1963.

6. P. Magnin, W. Kurz, "Competitive Growth of Stable and Metastable Fe-C-X Eutectics: Part II. Mechanisms," *Metallurgical Transactions A*, 19A (1988),1965-1971.

7. K. A. Jackson, & J.D. Hunt, "Binary Eutectic Solidification," *Trans. AIME*, 236 (1966), 1129-1142.

Advances in the Science and Engineering of Casting Solidification
Edited by: Laurentiu Nastac, Baicheng Liu, Hasse Fredriksson, Jacques Lacaze, Chun-Pyo Hong, Adrian V. Catalina, Andreas Buhrig-Polaczek, Charles Monroe, Adrian S. Sabau, Roxana Elena Ligia Ruxanda, Alan Luo, Subhayu Sen, and Attila Diószegi
TMS (The Minerals, Metals & Materials Society), 2015

AGE-STRENGTHENING OF CAST IRON AND ITS EFFECTS ON MACHINABILITY: REVIEW OF THE LITERATURE

Von L. Richards [1]

[1] Missouri University of Science and Technology, Rolla, MO

Keywords: Cast Iron, Age Strengthening, Machinability

Abstract

This presentation is a review of the research performed over several years to characterize the age-strengthening behavior of graphitic cast iron alloys (gray iron, ductile iron and CG iron.) Nitrogen in metastable solid solution in ferrite is necessary for the age strengthening to occur, similar to quench aging of steels. The activation energy for age strengthening is similar to that for diffusion of nitrogen. Age-strengthening can occur even if the ferrite is present as a phase in pearlite. However, machinability benefits only occur when there is free ferrite in the microstructure.

Introduction

Between 1997 and 2012 AFS supported a number of studies to improve the understanding of the age strengthening in graphitic cast iron. The literature suggests that the process of age strengthening appears to be a precipitation process describable by Avrami-Johnson-Mehl kinetics. Age strengthening was accompanied in many cases by an improvement in machinability which occurred on the same time scale.

Early Research in Cast Iron Age Strengthening

The first published work to suggest the existence of age strengthening was research done in malleable iron by Kempka in 1955.[1] This study showed that annealed malleable iron demonstrated aging behavior somewhat comparable to that of low-carbon steels. The first published literature on the subject [2-5] remained somewhat obscure until interest was rekindled with a study in 1999 from Nicola and Richards [6]. Test bars of cupola melted GCI were poured in this study which was designed to remove the effects of inoculant fade and surface roughness, and average ultimate tensile strength (UTS) was observed to increase by up to 9.9% with 99% statistical confidence. Age strengthening has since been observed to increase average UTS of GCI in the range of 3.3% to 13% in irons from multiple foundries. Early work that found aging in gray cast iron (GCI) was written in 1963 by Ebner, in Germany [2]. Ebner referred to the lower unaged strengths he observed as a "detrimental influence upon tensile strength, derived from melting and pouring" of which he said "declines with increasing storage time." Publications in October 1956[3] and 1969[4] also addressed age strengthening in malleable iron. In 1970, a Russian author published a study of aging effects in GCI [5] Nothing further is evident in the literature until Nicola and Richards[6]. Aging has been observed in cupola melted class 30, 35, and 40 gray irons and in class 30, 35, 40, and 45 induction furnace melted gray irons[6-8]. This is not to say that only these classes of GCI age strengthen, but they are the only GCI with statistically significant results so far. Grade 65-45-12 ductile iron has also displayed age strengthening with >99% confidence of a 4.6% increase in UTS and a 4.1% increase in 0.2%

yield strength[9]. Delaying casting shakeout decreases base GCI strength but does not decrease the percent strength gain from aging[8].

Aging tensile bars in an unmachined condition, or aging after machining, has been determined not to affect the magnitude or rate of strength increase[6]. This finding is significant because it rules out hydrogen as a possible source of strength changes and relates to the discussion of residual stress. Hydrogen diffusivity is so high at room temperature that if aging were caused by diffusion of hydrogen out of the bars, then aging bars in machined condition would significantly increase the rate of aging. This has not been observed.

The Influence of Nitrogen

In the first study by Nicola and Richards [6] aging appeared to be related to the nitrogen content of the irons studied. Based on this observation, the effect of nitrogen on GCI aging became an area of research interest. Nitrogen is now known to be required to produce aging behavior, and is believed to produce the precipitates that cause aging. Several facts support this belief:

- Researchers continue to observe that free nitrogen is required for aging[4-13].
- Nitrogen is known to have significant diffusivity in iron-based alloys at room temperature[14]. This can allow diffusion controlled phase transformations.
- Iron nitride species are known to precipitate in some ferrous alloys at room temperature[14].
- The magnitude of increase of average UTS from age strengthening shows a good linear fit as a function of thermodynamically predicted weight percent Fe_4N, as shown in figure 1 [12].

Differential scanning calorimetry (DSC) results appear to show the metastable $Fe_{16}N_2$ nitride transforming to the stable Fe_4N nitride at ~250°C (482°F)[9]. The DSC results agree with observations by Enrietto[19] of the presence of various nitrides.

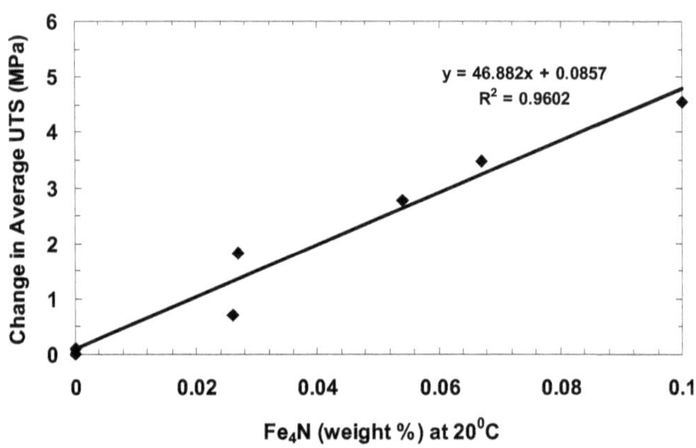

Figure 1: Experimentally measured changes in average UTS from age strengthening as a function of thermodynamically modeled iron-nitride content at room temperature [12]. Linear correlation is a good fit (i.e. R^2=0.96), showing that Fe_4N is an excellent suspect for the precipitate.

For gray iron, foundries can consider an equation presented by Anish for class 30 and 35 gray irons to estimate the increase in strength after 20 days of room temperature aging [12]. This equation, presented as equation 1, applies to irons in which nitrogen and titanium weight percentages are variables, but it assumes that strong nitride-forming elements other than titanium (such as zirconium) are not present in significant amounts.

$$\Delta UTS_{20} = 0.267 + 704.4(\%N) - 99.1(\%Ti) \qquad (1)$$

Kinetics of Nitride-Induced Aging in Cast Iron

Artificial aging studies have been conducted to study GCI aging kinetics [8,13]. The studies have often shown that age strengthening follows standard sigmoidal Avrami-Johnson-Mehl precipitation behavior observed in many metallurgical systems[10]. Figure 2 shows an example of this behavior. Aging slows when done below room temperature and elevated temperatures accelerate aging. Artificial aging at 182°C (360°F) is comparable to experimental times for iron-nitride precipitation observed in steel by Wert at 250°C (482°F)[27]. In the iron aging studies, a small drop in UTS and hardness occurred a few hours prior to the onset of aging at low temperatures such as room temperature and at 100°C (212°F)[15]. In addition to the strength dip, researchers noted that the time required before aging could be observed to increase as the manganese content of the iron increased. A possible explanation for this observation is the occurrence of "interaction solid solution strengthening" such as that found in steel, as discussed in Leslie's text[16] and by Enrietto[21]. These Mn-N interaction regions are not a separate phase. One hypothesis based on internal friction measurements suggested that Mn and N form an ortho-nitride [18]; however, no observation of this has been reported. Instead, it is suggested that manganese lowers the lattice free energy in the ferrite around it, creating preferred interstitial sites for nitrogen. This has been observed to delay the nucleation rate of the nitrides responsible for aging in steel[19] and possibly in cast irons [4,28]. The Mn-N interaction regions are known to inhibit dislocation motion which can be observed as a strength increase. Breakdown of the Mn-N complexes to provide nitrogen for nitride nucleation could explain the slight drop in strength at the start of the aging process. In steel samples, an increase in aging temperature reduced the effect of manganese on precipitation times [21,22]. The only deviations from the artificial aging trend were at 200°C (392°F) and 250°C (482°F), where the apparent dual or sequential precipitation of $Fe_{16}N_2$ and Fe_4N created aging rates that were even higher than those for samples aged at 350°C (662°F). This pattern argues for a two-step precipitation process in GCI like that observed in some nonferrous alloys, a possibility supported by the findings of Richards et al [14] with observation of two kinetic models depending on temperature, each with different activation energy.

Accelerated age strengthening at 182°C (360°F) and 285°C (545°F) in GCI has displayed over-aging[15]. Figure 2 provides an example of such over-aging. This over-aging is likely associated with the growth of some precipitates at the expense of others, which is known to increase the mobility of dislocations and thus to reduce overall strength [27,28].

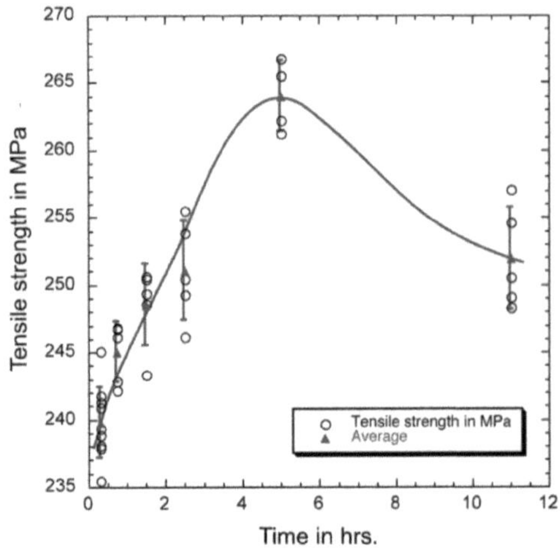

Figure 2: Curve from age strengthening at 285°C (545°F). There is a peak tensile strength at about 5 hours, followed by loss of strength from over-aging [26]. Error bars are one standard deviation in length.

The age strengthening process in GCI follows Avrami-Johnson-Mehl kinetics and can be described by equations 2 and 3. In equation 1, V_f is volume fraction of precipitate, k is a rate constant defined by equation 3, t is time, and n is the time exponent, which is a constant. In equation 3, k_o is the attempt frequency, Q is the activation energy, R is the universal gas constant, and T is absolute temperature.

$$V_f = 1 - \exp\left(-(kt)^n\right) \tag{2}$$

$$k = k_0 \exp\left(\frac{-Q}{RT}\right) \tag{3}$$

The aging process in GCI is not isokinetic [26]. In the iron alloys studied there appears to be a transition in kinetic behavior somewhere between 100°C (212°F) and 182°C (392°F). This transition is apparent in the Arrhenius plot in figure 3. The change in behavior indicates that a different precipitate forms at temperatures above the observed transition. The temperature of transition from either a dual or sequential precipitation process to a single precipitation process is somewhat lower than that observed in steel by Enrietto, who noted a transition to precipitation of only Fe_4N at or above 300°C (572°F)[19]. His study used a high-purity Fe-N alloy, it is conceivable, therefore, that the many elements present in cast iron, both substitutional and interstitial, could affect the precipitation kinetics and account for the different observations.

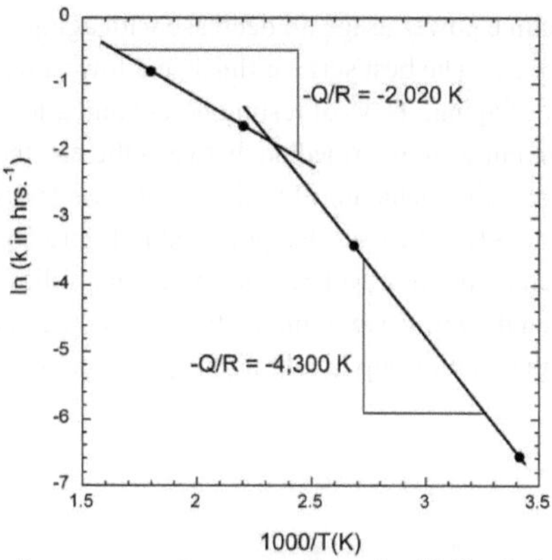

Figure 3: Arrhenius plot for gray cast iron aging kinetics [26]. Aging at 182°C (360°F) and 200°C (392°F) indicates formation of only one precipitate, whereas curves at 20°C (68°F) and 100°C (212°F) indicate the presence of at least two precipitate types and/or morphologies during aging.

Interactions with Substitutional Species

Lekakh, et al showed that manganese can affect the rate of age strengthening decreasing the time to achieve full strength up to a point then slowing the process at higher concentrations as shown in figure 4. Their analysis was as based on Ab intio calculations which suggested an interaction between Mn substitution pairs and N serving as a nucleation site for Fe_4N. However the analysis suggested that at higher Mn levels the α'' $Fe_{16}N_2$ would be stabilized at higher Mn contents thus delaying the Fe_4N formation and consequent age strengthening, This is a case where an interaction between alloy components improves the process up to a point but is deleterious at higher content.

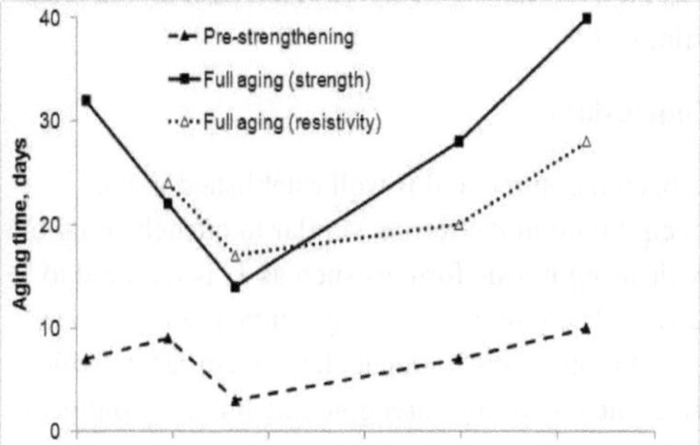

Figure 4 - - Effect of manganese age strengthening time with sufficient nitrogen and 0.015-0.025%S

Machinability Effects

Edington, et al. published the first research attempting to determine the effect of aging on machinability which showed improved machinability with aging [31]. Research showed that tool life increases, while tilt-rate (a possible measure of tool wear rate), surface finish deterioration rate, and machine power usage all decrease with aging. The results of the tool life study may be seen in figure 5a. The best surface finish and lowest required amperage appeared to be on days three and six of aging. Days of testing before and after this showed improvements over the un-aged condition, but were intermediate between the day three and day six behavior and the un-aged behavior. Similar behavior of tool life improvement with aging was observed by Kountanya and Boppana [32]. Their results, provided in figure 5b, showed a decrease in tool wear rate (expressed as increased tool travel per unit wear) that followed a sigmoidal behavior. Teague, et al have discussed the influence of microstructure on the interaction between age-strengthening, explaining the relationship involving the presence of free ferrite [33].

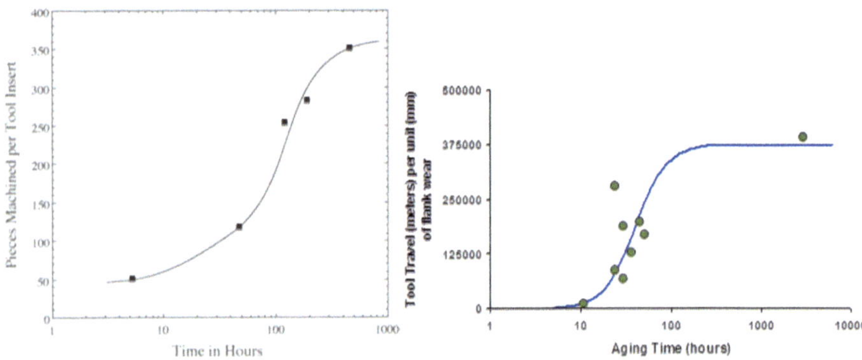

(a) (b)

Figure 5 – (a) Tool life data for foundry D showed dramatic improvements in the time that the tool was considered to be usable for machining [31]. It is noteworthy that the improvements in tool life appear to follow the same behavior as the UTS during aging. (b) Decrease in tool wear rate observed when machining GCI disk castings [32].

Conclusions

Age strengthening in graphitic cast irons has been measured and is well established in the literature. The mechanism appears nitride precipitation in the ferrite, similar to quench aging of steel. Nitrogen in excess of stoichiometry with string nitride formers such as Ti is required to achieve some nitrogen in solid solution in ferrite. The kinetics of the age strengthening process match well with diffusion limited nitride precipitation. Substitutional elements in ferrite which interact with N such as Mn appear to also affect the rate of age strengthening but in a nonlinear way. Machinability of gray cast iron improves on the same time scale as age-strengthening particularly with free ferrite present

References

1. G.E. Kempka, "Embrittlement, Toughening, and Subcritical Thermal Treatment of Malleable Iron," *AFS Transactions,* **63** 675-82 (1955).

2. R. Ebner, "Influence of Ageing, Sampling Site and Machining of Test Specimens Upon Tensile Strength and Hardness of Grey Cast Iron," *BCIRA,* **50** 689-91 (1963).
3. C.A. Beiser, Evans, E. B., in *Modern Casting*, 1956.
4. P.B. Burgess, "Age Hardening Ferritic Malleable," *AFS Transactions,* 172-9 (1969).
5. P.V. Novichkov, "Thermocyclic Aging of Cast Iron at 200 - 280 Degrees C," *Liteinoe Proizvodstvo (in Russian),* 31-5 (1970).
6. W.M. Nicola, Richards, V., "Age Strengthening of Gray Cast Iron, Phase I: Statistical Verification," *AFS Transactions,* 749-55 (1999).
7. W.M. Nicola, Richards, V.L., "Age Strengthening of Gray Cast Iron, Phase II: Nitrogen and Melting Method Effects," *AFS Transactions,* 233-7 (2000).
8. W.M. Nicola, Richards, V., "Age Strengthening of Gray Cast Iron, Phase III: Effect of Aging Temperature," *AFS Transactions,* 1085-95 (2001).
9. V.L. Richards, Van Aken, D.C., Mereau, O.P., Nicola, W.M., "Effects of Room Temperature Aging on Ductile Iron," *AFS Transactions,* (2004).
10. V.L. Richards, Van Aken, D.C., Nicola, W., "Age Strengthening of Gray Cast Iron: Kinetics, Mechanical Property Effects," *AFS Transactions,* (2003).
11. J. Edington, Nicola, W., Richards, V.L., "Age Strengthening of Gray Cast Iron: Nitrogen Effects and Machinability," *AFS Transactions,* (2002).
12. T. Anish, Lekakh, S., Richards, V., "The Effect of Ti and N on Iron Age Strengthening " *AFS Transactions,* **116** (2008).
13. V.L. Richards, Anish, T.V., Lekakh, S., Van Aken, D.C. Nicola, W., "Age Strengthening of Gray Iron - Kinetics Study," *AFS Transactions,* **115** (2007).
14. W.C. Leslie, *The Physical Metallurgy of Steels;* pp. 74-9. CBLS Publisher & Book Distributor, 1981.
15 I. Fall, Genin, J.-M.R., "Mössbauer Spectroscopy Study of the Aging and Tempering of High Nitrogen Quenched Fe-N Alloys: Kinetics of Formation and Fe16N2 Nitride by Interstitual Ordering in Martensite," *Metallurgical and Materials Transactions A,* **27A** (1996).
16 G.R. Booker, Sutton, A. L., Phil, D., "An Investigation of Nitride Precipitates in Pure Iron and Mild Steels," *Journal of the Iron and Steel Institute,* 205-15 (1957).
17. U. Dahmen, Ferguson, P., Westmacott, K. H., "A TEM Study of α''-Fe16N2 and γ'-Fe4N Precipitation in Iron-Nitrogen," *Acta Metallurgica,* **35** [5] 1037-46 (1987).
18. L.J. Dijkstra, "Precipitation Phenomena in the Solid Solutions of Nitrogen and Carbon in Alpha Iron below the Eutectoid Temperature," *Transactions Met. Society of AIME,* **185** 252-60 (1949).
19. J.F. Enrietto, "The Solubility and Precipitation of Nitrides in Alpha-Iron Containing Manganese," *Transactions of the Metallurgical Society of AIME,* **224** 43-8 (1962).
20. K.F. Hale, McLean, D., "Structure of Quench-Aged Iron-Carbon and Iron-Nitrogen Alloys," *Journal of the Iron and Steel Institute of London,* **201** 337-52 (1963).
21. I. Hrivňák, in *Metal Treatment and Drop Forging*, p. 175-81, 1961.
22. I. Hrivňák, in *Metal Treatment and Drop Forging*, p. 233-42, 1961.
23. K.H. Jack, "Binary and Ternary Interstitual Alloys I. The Iron-Nitrogen System: The Structures of Fe_4N and Fe_2N," *Proceedings of the Royal Society of London Series A, Mathematical and Physical Sciences,* **195** [1040] 34-40 (1948).
24. K.H. Jack, "The Occurrence and the Crystal Structure of α"-iron Nitride; A New Type of Interstitial Alloy Formed During Tempering of Nitrogen-Martensite," *Proceedings of the Royal Society of London Series A, Mathematical and Physical Sciences,* **208** [1093] 216-24 (1951).

25. V.A. Phillips, "An Electron Microscope Study of Quench-Aging and Strain-Aging in A Dilute Fe-C-N Alloy," *Transactions of the ASM,* **56** 600-17 (1963).
26. T.V. Anish, "Age Strengthening of Gray Cast Iron: Alloying Effects and Kinetics Study"; Metallurgical Engineering Thesis. University of Missouri-Rolla, Rolla, 2007.
27. C. Wert, "Precipitation Out of Dual Solid Solutions of Carbon and Nitrogen in Alpha-Iron," *Acta Metallurgica,* **2** 361-7 (1954).
28. D.A. Porter, Easterling, K.E., *Phase Transformations in Metals and Alloys,* 2 ed.*;* p. 306. CRC Press, 2004.
29. R.E. Reed-Hill, Abbaschian, R. , *Physical Metallurgy Principles,* 3 ed.*;* p. 533. PWS Publishing Company, 1994.
30. S.N. Lekakh, V. L. Richards, J. Medvedeva, and J. M. Murphy, "Effect of Alloying Elements on Gray Iron Natural Aging. Part 1. Manganese," *AFS Transactions*, 120 (2012).
31. J. Edington, Nicola, W., Richards, V.L., "Age Strengthening of Gray Cast Iron: Nitrogen Effects and Machinability," AFS Transactions, (2002).
32. P. R.K.a.B. Kountanya, "Optimization of machining of automotive components with polycrystalline cubic boron nitride," *Proc. IMechE Part B: Journal Engineering Manufacture*, 222 797-805 (2008).
33. J.A.Teague, V. L. Richards, "Age Strengthening of Cast Irons: Review of Research and Literature," *International Journal of Metalcasting*, 4(2), Spring 2010.

Advances in the Science and Engineering of Casting Solidification
Edited by: Laurentiu Nastac, Baicheng Liu, Hasse Fredriksson, Jacques Lacaze, Chun-Pyo Hong, Adrian V. Catalina, Andreas Buhrig-Polaczek, Charles Monroe, Adrian S. Sabau, Roxana Elena Ligia Ruxanda, Alan Luo, Subhayu Sen, and Attila Diószegi
TMS (The Minerals, Metals & Materials Society), 2015

EXAMINATION OF AUSTENITE SOLIDIFICATION AND SPHEROIDAL GRAPHITE GROWTH IN Ni-Fe-C ALLOYS

J. Qing, V. L. Richards, and D.C. Van Aken

Missouri University of Science & Technology, Rolla, MO

Keywords: Spheroidal Graphite, Multi-Nodular Eutectic Cell, Austenite Engulfment, Quenching Experiment, EBSD, Deep Etching, Staged Graphite Growth.

Abstract

The austenite phase and the spheroidal graphite in a *Ni-Fe-C* alloy at various solidification stages were captured by quenching experiments using spherical ceramic shell molds with inserted thermocouples. Multiple graphite nodules /spheroidal graphite particles inside an eutectic cell were observed using optical microscopy. Individual eutectic cells were differentiated by an intermetallic phase occurring within the interdendritic regions. Electron backscatter diffraction (EBSD) and orientation image mapping (OIM) analyses were performed to investigate the crystallographic orientation and the grain boundary distribution of the austenite grains surrounding individual spheroidal graphite particles. The austenite engulfment process around the spheroidal graphite was revealed by EBSD/OIM analyses. Examination of the graphite nodules extracted by deep etching experiments revealed three stages of growth for the spheroidal graphite particles.

Introduction

A nickel (*Ni*)-carbon *(C)* alloy with a graphite phase shows many similarities to a graphitic *Fe-C* alloy: the graphite morphology varies from flake, compacted, to spheroidal for both alloys; the graphite morphology changes from a spheroidal to a flake by introducing sulfur/oxygen for both alloys; and spheroidal graphite is obtained by either adding magnesium (*Mg*) /cerium (*Ce*) or increasing the cooling rate for both alloys. Nickel has a similar electron structure with iron, and both the iron and the nickel have a primary face centered cubic (FCC) crystal structure during solidification. However, the *Ni-C* alloy does not have the eutectoid reaction which occurs in the *Fe-C* system; thus, *Ni-C* simplifies the metallographic and electron microscopic study. A *Ni-C* alloy has been used previously as a model material for *Fe-C* alloy or by alloying *Fe-C* with nickel to stabilize the austenite in the study of graphite growth. [1-7] A *Ni-Fe-C* alloy with spheroidal graphite was used as a model material to investigate the austenite growth relative the spheroidal graphite growth in ductile iron in this research.

Austenite engulfment of the spheroidal graphite was extensively reported in the *Fe-C-Si* alloy with spheroidal graphite [8-16]. Engulfment/encapsulation is usually seen in the eutectic systems having one or both phases growing as faceted crystals [12], the former of which is thought to be the case for a *Ni-C* alloy or *Fe-C* alloy with spheroidal graphite. Lux, Mollard and Minkoff [12] attributed the austenite engulfment to the difference in growth rates between the austenite and graphite, without coupled growth of the two phases at the solid-liquid interface. To investigate various growth stages, a quenching experiment was usually used to capture the structures. In the iron system, it is easier to recognize the remnant liquid (usually transformed to ledeburite) and austenite phase (usually transformed to pearlite or martensite depending on the solid state cooling rate) after chemical etching. However, it is hard to differentiate the remnant liquid in the

Ni-C alloy from the original solid austenite since the liquid also transforms to the austenite upon quenching. The crystallographic orientation and the grain boundary distribution of the austenite grains will be examined using the electron backscattered diffraction (EBSD) and orientation image mapping (OIM) technique in this study, which makes the identification of the relicts of the liquid phase at quenching in the *Ni-C* alloy possible.

An eutectic cell in the *Fe-C-Si* alloy consists of spheroidal graphite and surrounding austenite shell attached to the austenite dendrite.[17-18] An eutectic cell of the *Fe-C-Si* alloy may include several spheroidal graphite particles.[17] In a model by Ruxanda, Beltran-Sanchez, Massone and Stefanescu[16], multiple graphite nodules were enveloped by a single eutectic cell in the *Fe-C-Si* alloy, which was called the multi-nodular eutectic grain model. A sample preparation technique in which the *Fe-C-Si* alloy experienced direct austempering heat treatment after solidification has revealed tens of spheroidal graphite particles in a single austenite dendrite. [19] Examination of an intermetallic phase at the austenite interdendritic regions will be used to prove the multi-nodular eutectic cell model in this study. EBSD/OIM analyses will also provide evidence to the multi-nodular eutectic cell model.

As for the growth of the spheroidal graphite particle, many theories/models exist but the real mechanism remains unknown. Three well-known theories of spheroidal graphite growth include (i) circumferential growth of curved graphene layers around the graphite nodule [20], and (ii) spiral growth of the graphene layers as helix cones [21], and (iii) spiral growth of a graphene layers as pyramidal cones [22]. Spheroidal graphite growth firstly by a circumferential model and finally by a cone-helix model was reported for a *Ni-C* alloy [6-7].

The objective of this paper is to understand the austenite growth relative to the spheroidal graphite growth in the *Fe-C-Si* alloy using the model *Ni-Fe-C* alloy. Graphite nodules extracted by the deep etching experiment will be used to understand the growth mechanism of spheroidal graphite in the *Ni-Fe-C* alloy.

Experimental Procedure

A spherical mold design was adopted for the quenching experiments, as shown in Fig. 1. It was made by building ceramic shells around a 38 mm diameter ball. A metal inlet and a gas vent were created and a handle was attached. One thermal couple was installed in the center of the mold to monitor the cooling curves.

A nickel alloy containing sixty weight percent (wt.%) nickel was used in this study. The composition of the alloy was measured with an arc spectrometer while the carbon and sulfur was measured with a combustion analyzer, as given in Table 1. This *Ni-Fe-C* alloy is hypereutectic according to the calculated phase diagram using Factsage software, see Fig. 2(a). Evolutions of austenite (γ), ferrite, liquid (L) and graphite phases of this alloy were estimated by the Factsage Scheil Cooling model, as shown in Fig. 2(b). A fully austenitic matrix was predicted.

A fourteen pound charge was melted in an induction furnace under argon protection. The charge consisted of high purity induction iron (0.002% *C* and 0.006% Sulfur (*S*)), pig iron (4.2% *C*, 0.17% silicon (*Si*), and 0.006% *S*), electrolytic nickel (99.9% purity), and pure graphite (99.9% purity). When the liquid metal temperature reached 1515°C it was treated with nickel-magnesium (15% *Mg*). The spherical molds were held on the surface of the liquid metal for ten seconds to preheat before immersed in the liquid metal. Temperature of the metal was controlled to be 1420°C (+/- 10°C) when the molds were immersed. Different growth stages were captured by quenching the metal in the mold in iced brine at different times. One sample was quenched at liquidus temperature at three seconds, and another sample was quenched near the temperature for eutectic thermal arrest after 45 seconds hold in an insulation nest. These two samples were compared with one unquenched sample cooled in the insulation nest. Experimental cooling curves for the three seconds quenched and the unquenched sample are plotted in Fig. 3. The oscillation in cooling curve came from the drastic cooling by iced brine.

Fig. 1. The spherical ceramic shell mold used for the quenching experiments.

Table 1. Chemical compositions (wt.%) of the *Ni-Fe-C* alloy.

Leco C	*Leco S*	*Si*	*Mn*	*Mg*	*Ni*	*Fe*
2.02	0.0025	0.501	0.021	0.17	60.0	balance

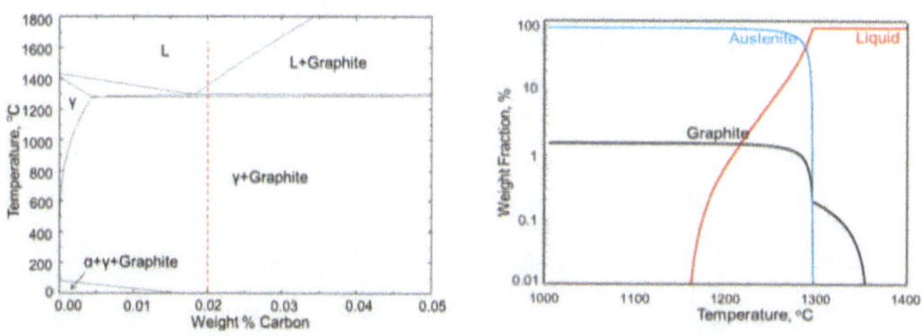

(a) (b)

Fig. 2. Factsage calculations: (a) the phase diagram of *Ni-Fe-C* alloy at a fixed iron content of 37.5 wt.%, and (b) the phase evolutions in the *Ni-Fe-C* alloy with 60 wt.% *Ni* during cooling. The red dashed line in (a) indicates the composition of the *Ni-Fe-C* alloy.

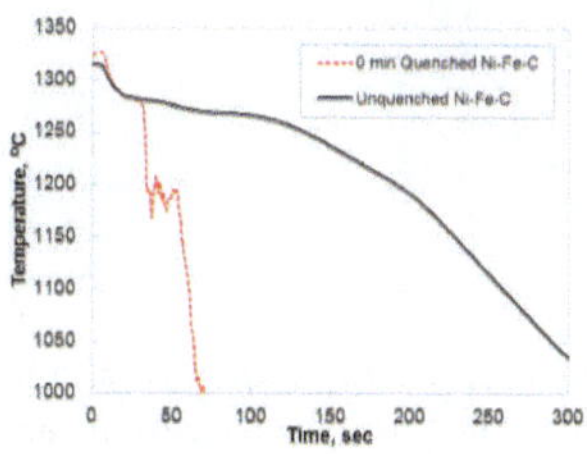

Fig. 3. Measured cooling curves of the three seconds quenched and the unquenched *Ni-Fe-C* castings.

Metallographic specimens were sectioned from the center (close to the tip of thermocouple) of the castings, and they were ground and polished using standard metallography procedures. A second set of specimens was mechanically polished for an extended time then vibratory polished to achieve high surface quality. Electron backscatter diffraction (EBSD) analyses were performed on the second set of specimens using a Helios Nanolab 600 FIB system. In addition, a one gram specimen from the center of each casting was deep etched using boiling concentrated hydrochloric acid. The matrix was removed by this deep etching. The graphite particles were filtered, rinsed and dried, then transferred to the carbon tapes for SEM observations. SEM examination was performed using the Helios Nanolab 600 FIB system.

Experimental Results and Discussions

Multi-Nodular Eutectic Cell

The as-polished microstructures at the center of the castings are shown relative to the quenching sequence in Fig. 4(a-c). Excess magnesium was introduced to ensure high graphite nodularity, and to form an intermetallic phase with iron and nickel in order to enhance the contrast of the austenite interdendritic regions. The intermetallic phases are observed as the dark gray phases occurring in the interdendritic regions, as indicated by the white dotted arrows. Based on the microstructures, distinct increments on graphite particle size and austenite dendrite arm spacing were observed for the quenching sequence. As mentioned earlier, an eutectic cell consists of spheroidal graphite and surrounding austenite shell attached to the austenite dendrite. The eutectic cells were outlined by the intermetallic phase. Multiple nodules inside a single eutectic cell were directly identified on the as-polished microstructures at higher magnifications, as illustrated in Fig. 4(d). This study used an intermetallic phase to verify the multi-nodular eutectic cell model in the *Ni-Fe-C* alloy, and the multi-nodular phenomenon was observed previously in a Fe-C-Si alloy using a color etching metallographic method [18].

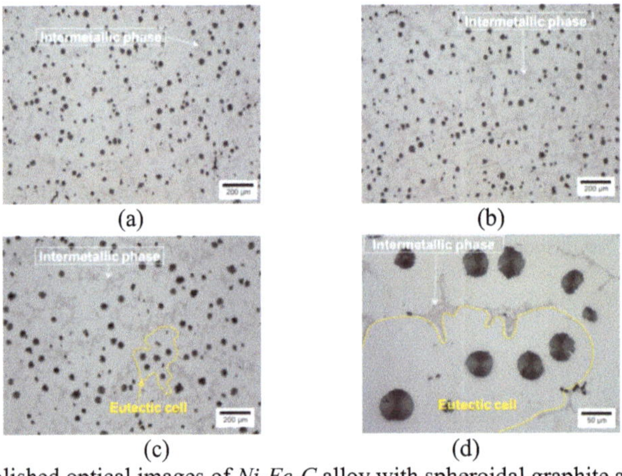

Fig. 4. As-polished optical images of *Ni-Fe-C* alloy with spheroidal graphite and austenitic matrix. The increments on the graphite particle size and the austenite dendrite arm spacing are observed over the three seconds quenched specimen (a), the 45 seconds quenched specimen (b) and the unquenched specimen (c). (d) A magnified microstructure of the unquenched specimen showing the details of the intermetallic phase. The dark gray intermetallic phase outlined the eutectic cells. Eutectic cells outlined by yellowed dotted lines contain multiple graphite nodules.

Austenite Engulfment

Crystallographic orientation image maps from the center of the casting are summarized in Fig. 5. The graphite phase was not included for the crystallographic orientation analyses. Therefore, the graphite particles appear as white blank regions in the maps. The austenite and the intermetallic phase (both characterized as FCC crystal structures) were analyzed for their crystallographic orientations. Different color codes in the map represent different crystallographic orientations, as given in Fig. 5(a). High angle grain boundaries are highlighted with the black bold lines, and low angle boundaries are indicated with the red thin lines in Fig. 5(b)-(e) and the brown thin lines in Fig. 5(f)-(g). It was observed in some cases that a single austenite grain contained multiple graphite nodules. Graphite nodules in contact with one austenite grain, two austenite grains and three austenite grains were also observed. The austenite grain boundaries were typically radially orientated next to a graphite nodule. The original austenite dendrites are recognizable from their shapes. Dendrite arms in the same austenite dendrite had similar crystallographic orientations such that mainly low angle boundaries appeared between the austenite dendrite arms. The intermetallic phase was seen separated from the austenite matrix by the high angle boundaries.

The authors have reported the independent formations of austenite phases and graphite nodules and the later austenite engulfment of the graphite nodules for an *Fe-C-Si* alloy in another paper [23]. Evidence for austenite engulfment was also found in the *Ni-Fe-C* alloy using the EBSD/OIM analyses.[24] It should be noted that in the *Ni-Fe-C* system, the liquid phase also transformed to austenite upon quenching. During the early solidification stages, the graphite particles were surrounded by the liquid phase, and they would be subsequently engulfed by austenite.[23] The single austenite grain observed around a small graphite nodule in the quenched specimen could be a relict of the quenched liquid phase, or it was possibly already a solid austenite shell that engulfed the nodule. Multiple austenite grains around a graphite nodule might indicate the incomplete austenite encapsulation around the graphite nodule, provided high angle boundaries are also present. Among these multiple austenite grains, a small austenite grain connected to the graphite nodule by a narrow "tail" was frequently observed, which might be the relict of remnant liquid phase trapped in the gap between the closing austenite engulfment. A clear austenite dendrite (in green color) was captured in the 45 seconds quenched specimen, as shown in Fig. 5 (d). Inside this austenite dendrite, graphite nodules partially engulfed by the austenite were observed, as indicated by the letter "P", and graphite nodules already fully encapsulated by the austenite were also observed, as indicated by the letter "F". Suggestions of relicts of remnant liquid were also found in the 45 seconds quenched specimen (see Fig. 5 (d-e)). As the solidification proceeded, most of the graphite nodules were fully engulfed by the austenite resulting in mainly a single austenite grain around a graphite nodule, as shown in Fig. 5 (f-g). There also remained few graphite nodules sitting at the junctions of multiple austenite grains in the unquenched specimen, as labelled by the letter "M". This is the reason why typically a single or two austenite grains surround a graphite nodule in the unquenched specimen.

In the past, the matrix around a graphite nodule was thought to be a single grain.[25] Zou et al. reported that the spheroidal graphite in the quenched *Ni-C* alloy was typically surrounded by only one or two austenite grains. They claimed that the austenite around a degenerate graphite particle was poly-crystal for the *Ni-C* alloy, but there was not any grain boundary highlighted in their EBSD orientation maps to support their conclusions.[26] Our study revealed that the number of austenite grains around a spheroidal graphite depended upon the solidification stage. Evidence could be found also in the EBSD orientation maps of a direct austempered *Fe-C-Si* alloy [19], in which further growth of the austenite dendrite during the austempering heat treatment led to further austenite encapsulation of the graphite nodules. Therefore, a single austenite grain was basically observed around the spheroidal graphite in their austempered *Fe-C-Si* alloy.[19] No direct correlation was found between the number of austenite grains around a graphite particle and the morphology of graphite particle, but this will merit further exploration by EBSD/OIM.

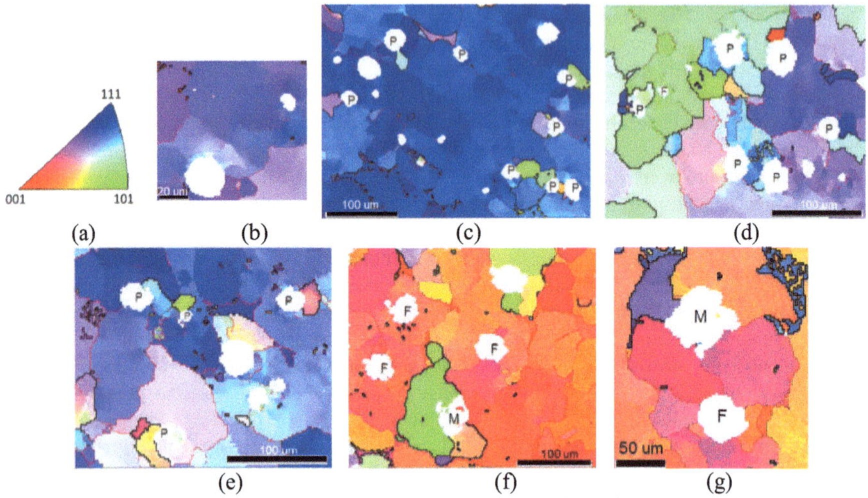

Fig. 5. (a) Color codes for the EBSD orientation map. Orientation maps of the three seconds quenched specimen (b-c), the 45 seconds quenched specimen (d-e), and the unquenched specimen (f-g). P: graphite nodule partially engulfed by the encroaching austenite; F: graphite nodule fully engulfed by an austenite grain; M: graphite particle surrounded by multiple austenite grains in the unquenched specimen.

Three Stages of Growth for Spheroidal Graphite

Columnar features were frequently observed within the larger sized graphite nodules but smooth surfaces were observed for the smaller sized nodules in the microstructures (see Fig. 6). Examinations of the graphite nodules (see Fig. 7) extracted by deep etching experiment revealed (1) curved graphene layers wrapping around the smaller sized graphite (~20 μm diameter) which formed growth steps on the surface, and (2) columns of graphite crystals radially oriented in the medium sized graphite (~35 μm diameter), and (3) radially oriented, faceted pyramidal graphite crystals separated by "valleys" in the large sized graphite (~55 μm diameter). The growth steps on the smooth surface of the smaller sized nodule indicate a circumferential growth model [20], while the growths of columns and pyramids seems to follow a cone-helix growth model [21] and a conical pyramidal growth model [22], respectively. The growth of spheroidal graphite therefore can be divided into three stages, which follow different growth models, initially a circumferential growth model, followed by a cone-helix growth model, and finally a conical pyramidal growth model, as schematically shown in Fig 8. As for the transition from the columnar crystals to the faceted pyramidal crystals, it was thought caused by a decrease in undercooling due to the releasing of latent heat.[22] Statistical measurements will be performed to determine the graphite feature size associated with this transition.

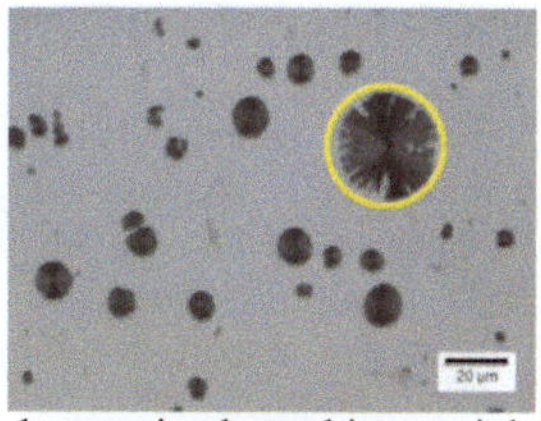

Fig. 6. Columnar features in a larger sized graphite particle indicated by the yellow circle.

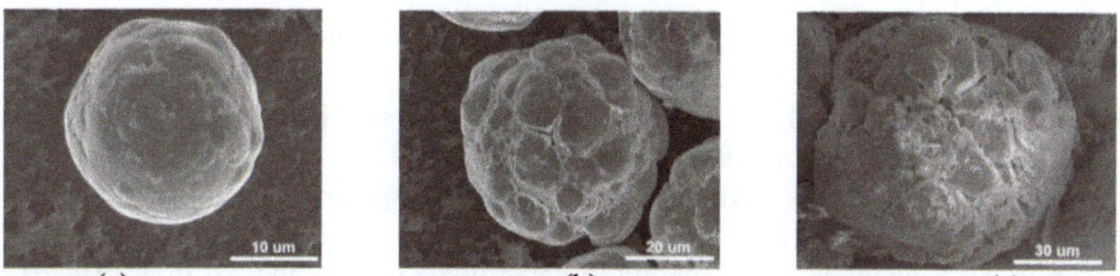

(a) (b) (c)

Fig. 7. Secondary electron images of graphite particles removed by the deep etching with diameters of 20 μm (a), 33 μm (b) and 55 μm (c). Images (a) and (b) were of graphite particles extracted from the three seconds quenched specimen, and (c) is from the unquenched specimen.

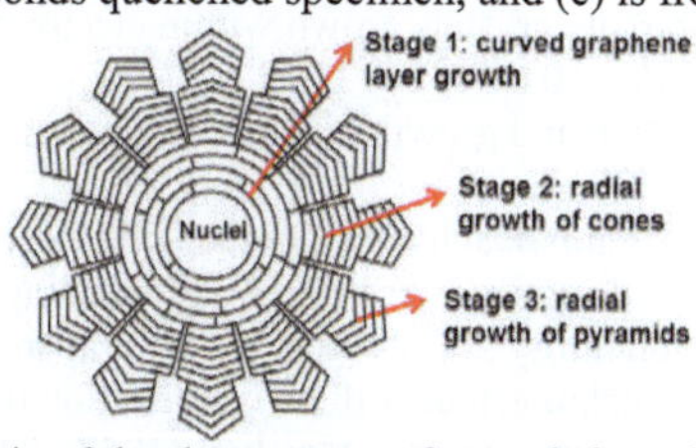

Fig. 8. Schematic of the three stages of growth for spheroidal graphite.

CONCLUSIONS

The solidification of a *Ni-Fe-C* alloy with spheroidal graphite was experimentally studied by quenching methods using spherical ceramic molds in this study. Multiple graphite nodules were observed in a single eutectic cell for the studied *Ni-Fe-C* alloy. The austenite grain boundary distributions revealed by EBSD/OIM analyses showed evidence for austenite engulfment around the spheroidal graphite. Relicts of the remnant liquid phase trapped between the closing engulfment of the austenite grain in the quenched specimens provided evidence of the ongoing austenite engulfment process. The austenite grain around the graphite nodule in the unquenched specimen is mainly a single grain because of the completion of austenite engulfment. Examinations of the deep etched graphite particles supported the model that the spheroidal graphite growth occurs in three stages: (1) firstly by circumferential growth, (2) followed by a cone-helix growth, (3) and finally by conical pyramidal growth. Transition from a columnar crystal to a pyramidal crystal is a result of the recalenscense. Element segregations should be used to differentiate the original austenite and remnant liquid phase combined with the EBSD analyses in future work. The mechanism underlying the transitions of the graphite growth models needs further exploration. The graphite feature size associated with transition between growth models should be statistically measured.

ACKNOWLEDGEMENTS

The authors would like to gratefully thank Dr. Simon Lekakh for his help on experimental design and technical discussions. Mingzhi Xu, Marc Harris and Joseph Kramer are acknowledged for their help with sample preparations. Clarrisa Wisner and Jessica Terbush are acknowledged for their guidance on HRSEM and EBSD operations. Thanks to ArcelorMittal for donation of the electrolyte nickel, and Hickman, Williams & Company for donation of *NiMg* alloy.

REFERENCES

1. A. Hellawell, M.P. Wilkinson, "Directional Freezing of Eutectic Alloys Containing Graphite", *J. Iron Steel Insts.*, (203)1965, 834–835.
2. R.J. Brigham, G.R. Purdy, J.S. Kirkaldy, "Unidirectional solidification of Fe–C, Ni–C and Fe–C–Si Eutectics", *J. Crys Growth*, 20–24 (1967), 161–169.
3. M.J. Hunter, G.A. Chadwick, "Nucleation and growth of spheroidal graphite alloys", *J. Iron Steel Insts.*, 210 (1972), 707–717.
4. S. Banerjee, "A Review of the Formation of Spheroidal Graphite in Cast Iron", *The British Foundaryman*, (1965), 344–352.
5. H. Morrogh, W.J. Williams, "Graphite Formation in Cast Irons and in Nickel–Carbon and Cobalt–Carbon Alloys", *J. Iron Steel Insts.,* 155 (1947), 321–369.
6. A. Shaahin, K. Haamun, "Graphite crystals grown within electromagnetically levitated metallic droplets". *Acta Materialia*, 60 (20) (2012), 7123–7131.
7. A. Shaahin, A. Reza, "Nucleation and growth kinetics of graphene layers from a molten phase". *Carbon*, 51(2013), 110–123.
8. D. M. Stefanescu, and D. K. Brandyopadhyay, "On the solidification kinetics of spheroidal graphite cast iron", *Physical Metallurgy of cast iron IV*, (Proceedings of the Fourth International Symposium on The Physical Metallurgy of Cast Iron, Tokyo, Japan,1990), 15-26.
9. E.Scheil und L. Hutter, Untersuchungen über die Kristallisation des Gusseisens mit Kugelgraphit. Archiv fur das Eisenhuttenwesen, Mai/Juni (1953), 237-246.
10. H. Hecht and J. C. Margerie, *Mem. Sci. Rev. Metallurg.*, 68(5) (1971), 325-338.
11. Solidification of Eutectic Alloys: Cast Iron, Casting, ASM Handbook, Vol. 15, (ASM International, 2008), 317–329.
12. B. Lux, F. Mollard, and I. Minkoff, "On the formation of envelopes around graphite in cast iron", *The Metallurgy of Cast Iron,* (Proceedings of the second international symposium on the metallurgy of cast iron, Geneva, Switzerland, Georgi Publishing, 1974), 371-400.
13. R. Boeri, F. Weinberg, "Microsegregation in Ductile Iron", *AFS Transactions*, 97 (1989), 197-184.
14. G. Rivera, R. Boeri, and J. Sikora, "Research Advances in Ductile Iron Solidification", *AFS Transactions*, 111 (2003), 979-989.
15. C. Van de Velde, "A New Approach to the solidification of Ductile Iron", *Modern Casting AFS*, (1998) 143.
16. R. Ruxanda, L. Beltran-Sanches, and J. Massone, and D. M. Stefanecu, "On the Eutectoid Solidification of Spheroidal graphite Iron: An Experimental and Mathematical Modeling Approach", *AFS Transactions*, 109 (2001), 1037-1048.
17. D. M. Stefanescu, "Modeling of Cast Iron Solidification— The Defining Moments", *Metallurgical and Materials Transactions A*, 38A (2007), 1433.
18. G. Rivera, R. Boeri, and J. Sikora, "Counting Eutectic Grains in S.G. Cast Iron", *J. Adv. Mater. Res.*, 4 (1997), 169-174.

19. G. Rivera, P.R. Calvillo, R. Boeri, Y. Houbaert, J. Sikora, "Examination of the solidification macrostructure of spheroidal and flake graphite cast irons using DAAS and ESBD", *Materials Characterization*, 59 (2008), 1342 –1348.
20. J.P. Sadocha, J.E. Gruzleski, "The Mechanism of Graphite Spheroid Formation in Pure Fe–C–Si alloys", *The Metallurgy of Cast Iron,* (Proceedings of the second international symposium on the metallurgy of cast iron, Geneva, Switzerland, Georgi Publishing,1974), 443–456.
21. D.D Double, A. Hellawell, "Cone-Helix Growth Forms of Graphite", *Acta Metallurgica*, 22 (1974), 481-487.
22. I. Minkoff, "The Sphetrulitic Growth of Graphite", *Mat. Res. Soc. Symp. Proc.*, 34 (1985), 37-45.
23. J. Qing, V. L. Richards, D. C. Van Aken, "Examination of Nodular Graphite Formation and Austenite Solidification in Ductile Iron", *AFS Transactions,* (2015) (paper submitted).
24. J. Qing, V. L. Richards, "A Review: Atomic level analysis of graphite growth morphology", *AFS Transactions*, 122 (2014) 265-272.
25. K. R. Olen and R. W. Heine, "A Revision of the Fe-C-Si System", *AFS Cast Met. Res. J.*, 2 (1968), 28-43.
26. Y. Zou, Y. Kato and H. Nakae, "Influence of Melting Conditions on Graphite Morphology in Ni–C Alloy and Grain Number of Matrix", *ISIJ International*, 52(2012), 441–446.

Advances in the Science and Engineering of Casting Solidification
Edited by: Laurentiu Nastac, Baicheng Liu, Hasse Fredriksson, Jacques Lacaze, Chun-Pyo Hong, Adrian V. Catalina, Andreas Buhrig-Polaczek, Charles Monroe, Adrian S. Sabau, Roxana Elena Ligia Ruxanda, Alan Luo, Subhayu Sen, and Attila Diószegi
TMS (The Minerals, Metals & Materials Society), 2015

PRODUCTION OF SELECTED KEY DUCTILE IRON CASTINGS USED IN LARGE-SCALE WINDMILLS

Yung-Ning Pan[1*], Hsuan-Te Lin[1], Chi-Chia Lin[1] and Re-Mo Chang[2]

[1]Department of Mechanical Engineering, National Taiwan University, Taipei, Taiwan
[2] Metal Industries Research and Development Centre, Kaohsiung, Taiwan,
*Corresponding author: Yung-Ning Pan, panyn@ntu.edu.tw

Keywords: Heavy-section ductile iron, Tensile property, Impact value, Alloy design, Microstructure

Abstract

Both the optimal alloy design and microstructures that conform to the mechanical properties requirements of selected key components used in large-scale windmills have been established in this study. The target specifications in this study are EN-GJS-350-22U-LT, EN-GJS-350-22U-LT and EN-GJS-700-2U. In order to meet the impact requirement of spec. EN-GJS-350-22U-LT, the Si content should be kept below 1.97%, and also the maximum pearlite content shouldn't exceed 7.8%. On the other hand, Si content below 2.15% and pearlite content below 12.5% were registered for specification EN-GJS-400-18U-LT. On the other hand, the optimal alloy designs that can comply with specification EN-GJS-700-2U include 0.25%Mn+0.6%Cu+0.05%Sn, 0.25%Mn+0.8%Cu+0.01%Sn and 0.45%Mn+0.6%Cu+0.01%Sn. Furthermore, based upon the experimental results, multiple regression analyses have been performed to correlate the mechanical properties with chemical compositions and microstructures. The derived regression equations can be used to attain the optimal alloy design for castings with target specifications. Furthermore, by employing these regression equations, the mechanical properties can be predicted based upon the chemical compositions and microstructures of cast irons.

Introduction

With the wind energy having increased substantially and steadily for the past decade, and also much more wind development is anticipated in the coming years, especially in the area of large scale windmills, both of onshore and offshore [1, 2], the establishment of the production technologies for key components of large utility-scale wind turbines becomes very vital to cope with the current development of wind power. Those components include hub, rotor shaft, main frame, gear box, etc., with the materials being mainly of ductile cast irons.

The production of sound heavy section ductile iron castings is not an easy task. In authors' previous study [3], appropriate casting conditions had been suggested and are given in Table 1, which took into account the following issues that are generally encountered in the production of heavy section ductile cast irons: (1) The formations and their elimination of the detrimental structure features that are frequently observed in heavy-section ductile iron castings, including various forms of degenerate graphite, carbon flotation on the cope surface, alloying segregation, etc. [4-10]. (2) Obtaining a relatively high nodule count is a prerequist in order to achieve a sound heavy section ductile iron microstructure. Accordingly, both SiC pre-inoculation and Fe-Si late inoculation were suggested, other than the regular ladle inoculation. (3) Regarding the sub-zero impact toughness, the control of Si content is essential, because an increase in Si content will raise the ductile-to-brittle transition temperature, and hence the impact toughness suffers [11, 12]. Accordingly, the Si content should be kept below some 2.0%.

The current work is a long-term on-going research project aiming to establish casting technologies for major components used in large windmills that can meet specific specifications. The optimal alloy design and microstructure characteristics of heavy section ductile cast irons that can comply with spec. EN-GJS-400-18U-LT had been studied and reported previously [3], and this paper deals with the casting conditions required to comply with the specifications of both EN-GJS-350-22U-LT and EN-GJS-700-2U (Table I).

Table I. Specifications for various ductile iron castings for large-scale windmills

Specification	Casting thickness [mm]	Ave. impact value [J]	Individual impact value [J]	Tensile strength [MPa]	Yield strength [MPa]	El. [%]
EN-GJS-350-22U-LT	60<t≦200	≧10(-40°C)	≧7	≧320	≧200	≧15
EN-GJS-400-18U-LT	60<t≦200	≧10(-20°C)	≧7	≧370	≧220	≧12
EN-GJS-700-2U	30<t≦60	--	--	≧700	≧420	≧2

Experimental Procedures

Alloy Design

To assess the effect of alloy compositions (C, Si, Mn and Cu) on the microstructure and mechanical properties (tensile properties and impact value) of heavy section ductile cast irons that can meet both specifications of EN-GJS-350-22U-LT and EN-GJS-400-18U-LT, the Si content was set at around 2.0% with minor variations of some 0.3%, and Mn, Cu and Ni contents were varied to attain various pearlite percentages in the matrix. The alloy design reported in Table II is the final chemical analyses of all heats. On the other hand, for heat series B, pearlite-forming elements, such as Mn, Cu and Sn, were added, either individually or in combination, to achieve the required matrix pearlite content in order to fulfill the relatively high strength requirement of specification EN-GJS-700-2U. The alloy design of Heat series B, together with the results of tensile property tests are presented in Table III.

Table II. Results of chemical analyses of all heats (%)

Heat No.	C	Si	C.E.	Mn	P	S	Mg	Ni	Cu	Pb	Fe
A1	3.62	1.76	4.21	0.18	0.028	0.009	0.059	0.29	0.016	0.012	Bal.
A2	3.51	1.77	4.10	0.18	0.028	0.009	0.050	0.29	0.016	0.012	Bal.
A3	3.47	1.68	4.03	0.17	0.029	0.010	0.049	0.30	0.017	0.012	Bal.
A4	3.60	1.56	4.12	0.21	0.042	0.011	0.039	0.38	0.171	0.014	Bal.
A5	3.51	2.25	4.26	0.20	0.037	0.009	0.047	0.38	0.165	0.014	Bal.
A6	3.53	1.83	4.14	0.20	0.042	0.010	0.039	0.38	0.163	0.014	Bal.
A7	3.47	2.16	4.19	0.27	0.024	0.009	0.052	0.30	0.300	0.014	Bal.
A8	3.42	2.11	4.12	0.27	0.027	0.009	0.053	0.29	0.310	0.014	Bal.
A9	3.58	2.22	4.32	0.21	0.033	0.011	0.049	0.01	0.197	0.014	Bal.
A10	3.50	1.90	4.13	0.20	0.028	0.012	0.048	0.30	0.197	0.014	Bal.
A11	3.61	2.08	4.30	0.14	0.028	0.011	0.049	0.20	0.013	0.007	Bal.
A12	3.51	2.10	4.21	0.14	0.029	0.011	0.053	0.20	0.013	0.007	Bal.
A13	3.55	2.28	4.31	0.23	0.031	0.010	0.057	0.04	0.024	0.006	Bal.

Table III. The mechanical property test results for 25mm Y-block specimens of Heat series B

Heat No.	Mn (wt%)	Cu (wt%)	Sn (wt%)	Tensile strength (Mpa)	Yield strength (Mpa)	Elongation (%)	Conform to GJS700
B1	0.25	0.6	0.01	621	412	9.1	×
B2	0.25	0.6	0.03	641	432	5.2	×
B3	0.25	0.6	0.05	712	564	4.3	○
B4	0.25	0.8	0.01	710	523	6.3	○
B5	0.25	1.0	0.01	792	592	5.1	○
B6	0.45	0.6	0.01	766	464	6.8	○
B7	0.45	0.6	0.03	792	487	5.4	○
B8	0.45	0.6	0.05	782	528	4.7	○

Melt Preparation

Heats were prepared in a 10-ton capacity low-frequency induction furnace using charge materials that consist of low sulfur pig iron, steel scrap, Fe-Si alloy (for Si content adjustment) and SiC (for pre-inoculation). The nodularization treatment was conducted at about 1480°C by adding 1.3% of an Mg-FeSi alloy containing minor amounts of RE, Ca and Al using the sandwich method. Ladle inoculation was performed by stirring the inoculant (0.5%) into the melt stream while the melt was transferred to a pouring ladle. The inoculant employed is a proprietary Fe-Si alloy that contains Ba, Ca and Al. Late inoculation was performed by adding 0.1% of another proprietary Fe-Si alloy into the melt stream during the mold filling process. Furthermore, the specimens used for analysis were obtained from 70mm-thick Y-blocks attached to the actual castings (cast-on specimens). On the other hand, for Heat series B, a total of 8 heats were prepared in a 600kg-capacity induction furnace by employing the same foundry practice as in heat series A, except that pearlite-formers (Mn, Cu and Sn) were added and varied in the alloy design, as mentioned above. In addition, the specimens used for microstructure analyses and tensile property tests were obtained from separately cast 25mm-thick Y-blocks.

Tensile Property Test

The specimens taken from the Y-block castings were machined as per ASTM A536-84 to produce test pieces having a 12.5-mm (0.5-in) gauge diameter and a 50.8-mm (2-in) gauge length [13]. Each reported datum point is the average of 4 test pieces. Tensile tests were performed with Shimadzu UH-20A machine as per ASTM E8/E8M-09.

V-notched Impact Test

V-notched impact test specimens in accordance with ASTM A327-91 were machined also from the same cast-on Y-block castings [14]. For sub-zero temperature impact tests (both -20±2°C and -40±2°C), the specimens were immersed in a solution with appropriate ratios of alcohol and dry ice to achieve the target test temperature. Four test specimens were conducted for each experimental condition and the average of these four tests was reported.

Results and Discussion

Table IV lists the results of mechanical property tests and also the microstructure analyses for cast-on 70mm-thick Y-blocks of Heat series A. The effect of matrix pearlite percentage on the tensile strength and elongation is shown in Fig. 1. As expected, both the tensile strength and yield strength increase, while the elongation decreases, as the pearlite percentage increases. However, it has to be noted that a certain degree of variation in property values can be observed at a fixed amount of pearlite. Detailed analysis of the data indicates that in addition to the influence of alloying elements, e.g., Mn and Cu, nodule count also plays an important role in determining the pearlite (or ferrite) content in the matrix. Therefore, the evaluation of tensile properties of cast irons should consider both the alloying element contents and microstructures, and this will be discussed later.

Table IV. Results of mechanical property tests and microstructure analyses for 70mm-thick cast-on specimens of all heats

Heat No.	T.S. [MPa]	Y.S. [MPa]	El. [%]	BHN	$I_{-20}{}^{o}C$ [J]	$I_{-40}{}^{o}C$ [J]	Nodu. [%]	N.C. [#/mm^2]	Pearlite [%]
A1	370	229	28.0	127	19.1	16.0	90	253	~0
A2	368	216	26.5	126	18.0	16.8	89	188	~0
A3	364	221	29.0	126	21.2	16.2	83	218	~0
A4	380	244	27.0	132	18.8	16.1	88	332	~0
A5	376	242	27.0	137	18.1	16.8	88	295	~0
A6	393	252	27.0	134	15.3	8.8	89	175	~0
A7	665	361	9.8	220	4.9	4.1	91	261	84
A8	646	357	8.1	216	4.5	4.0	89	260	84
A9	444	297	23.0	159	12.4	8.5	90	482	13
A10	530	299	16.0	183	6.1	5.7	93	460	13
A11	397	251	27.0	143	11.2	5.9	91	215	12
A12	399	265	26.0	141	10.4	6.7	90	208	13
A13	381	236	29.0	133	13.7	9.0	82	206	~0

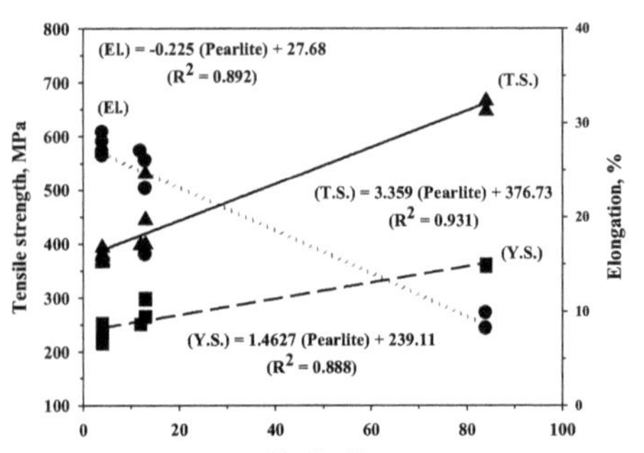

Figure 1. The effect of matrix pearlite percentage on the tensile strength, yield strength and elongation.

The effect of Si content on the impact value (at both -20°C and -40°C) is shown in Fig. 2(a). It is apparent that both the impact values of $I_{-20}{}^{o}C$ and $I_{-40}{}^{o}C$ decrease with increasing Si content in a linear form. It is clear from Fig. 2(a) that in order to meet the impact requirement of specifications EN-GJS-400-18U-LT and EN-GJS-350-22U-LT, the Si content should be less than 2.15% and 1.97%, respectively. On the other hand, the effect of pearlite percentage on the impact values (both at -20°C and -40°C) is shown in Fig. 2(b). Note that the impact values of both $I_{-20}{}^{o}C$ and $I_{-40}{}^{o}C$ decreases with increasing pearlite percentage in a curvilinear form, with an abrupt reduction in impact value at lower side of pearlite content. This implies that to produce a primarily ferritic matrix structure and also the control of pearlite content are essential for achieving high sub-zero impact values. The maximum pearlite content should be kept below 12.5% in order to meet the impact requirement for specification EN-GJS-400-18U-LT ($I_{-20^{\circ}C} \geqq 10J$), and 7.8% for specification EN-GJS-350-22U-LT ($I_{-40^{\circ}C} \geqq 10J$).

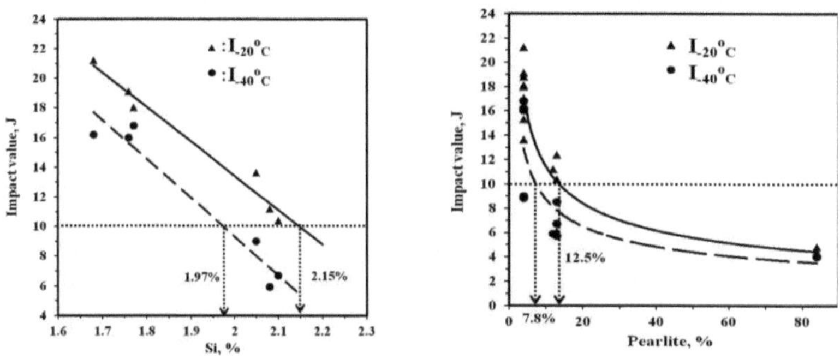

(a) (b)

Figure 2. The dependence of impact values at both -20°C and -40°C on (a) the Si content, and (b) the pearlite percentage.

Furthermore, some variations in data points in both Figs. 2(a) and 2(b) may indicate that other metallurgical variables, in addition to Si content and pearlite percentage, may affect the cast iron impact property, i.e., nodule count. Therefore, multiple regression analyses were performed to correlate the alloy compositions and microstructures with tensile properties and impact values both at ambient and sub-zero temperatures of heavy section ductile cast irons. The analyses results are summarized in Table IV, where regression equations for tensile properties (tensile strength and elongation) and impact value as functions of metallurgical variables, including Si, Mn, Cu, %pearlite and nodule count were derived. The correlations among impact values at both -20°C and -40°C, tensile strength and elongation are depicted in Fig. 3. This figure can be used to determine what tensile properties shall be if the required impact values are to be met. For example, from the illustration depicted in Fig. 3(b), one can conclude that the tensile strength should be kept below 383MPa, whereas the elongation should be higher than some 27.0%, if the $L_{40^{\circ}C}$ needs to exceed 10J.

(a) (b)

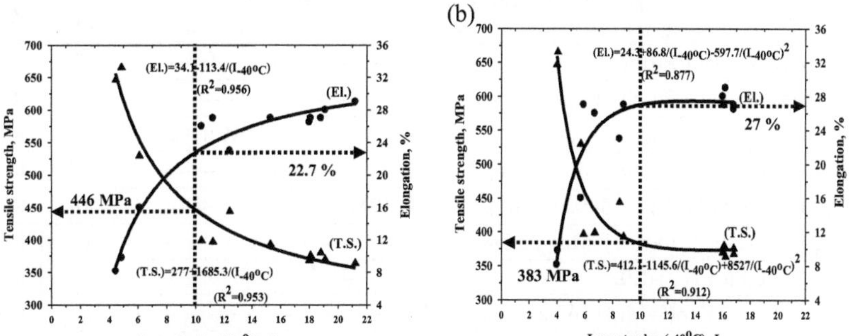

Figure 3. Correlations among impact values, tensile strength and elongation (a) $L_{20^{\circ}C}$, and (b) $L_{40^{\circ}C}$.

On the other hand, for heat series B, pearlite-forming elements, such as Mn, Cu and Sn, were added, either individually or in combination, to increase the matrix pearlite content and hence the strength of the alloys. The alloy design of Heat series B, together with the results of tensile property tests are presented in Table III and also shown in Figs. 4 & 5. Regarding the results of Table III, even though the Heats B3-B8 can all meet the specification EN-GJS-700-2U, the optimal compositions of Mn, Cu and Sn contents are 0.25%Mn + 0.6%Cu + 0.05%Sn (B3), 0.25%Mn + 0.8%Cu + 0.01%Sn (B4), and 0.45%Mn + 0.6%Cu + 0.01%Sn (B6), as shown in Fig. 5.

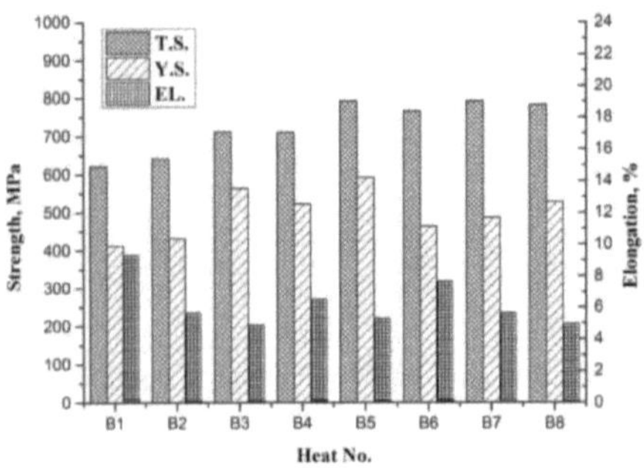

Figure 4. The tensile properties of Heats series B.

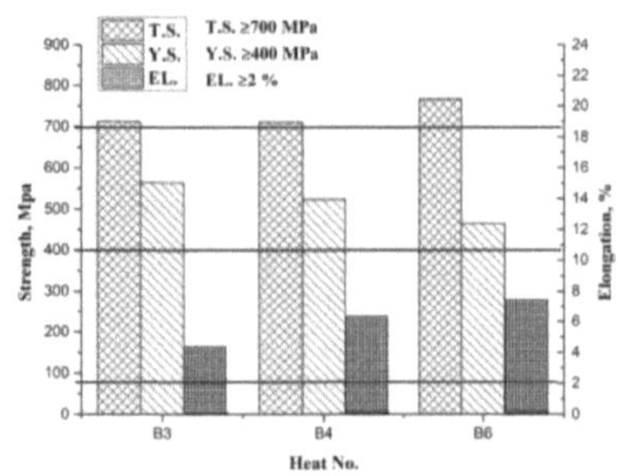

Figure 5. The tensile properties of Heats B3, B4 and B6 in series B that meet the specification EN-GJS-700-2U.

The optimal alloy design and pearlite content for conforming to the designated specifications can be calculated based upon the regression equations listed in Table V. The results are presented in Table VI, together with the predicted mechanical properties. Furthermore, Representative microstructures of castings that conform to specifications EN-GJS-350-22U-LT and EN-GJS-700-2U are shown in Figs. 6(a) and 6(b), respectively.

Table V. Regression equations derived for tensile properties and impact value as functions of metallurgical variables, including %Si, %Mn, %Cu, %pearlite and nodule count (N.C.)

(TS, MPa) = -0.14(N.C.)+2.96(%Pearlite)+3.1(%Si)+261.9(%Mn) +45.6(%Cu)+348.8 --------(1)	R^2=0.99
(TS, MPa)=3.31(%Pearlite)+367 --------(2)	R^2=0.98
(YS, MPa) = 1.46(%Pearlite)+239.1 --------(3)	R^2=0.89
(El, %) = 0.005(N.C.)-0.2(%Pearlite)+1.5(%Si)-18.4(%Mn)-4.7(%Cu)+27.7 ---(4)	R^2=0.98
(El, %) = -6.9(%Si)-117.9(%Mn)-13.2(%Cu)+61.3 --------(5)	R^2=0.98
(I_{-20} °C, J) = 0.004(N.C.)-0.13(%Pearlite)-6.6(%Si)+28.2 --------(6)	R^2=0.86
(I_{-20} °C, J) = -1.3(%Si)-88.4(%Mn)-22.3(%Cu)+38.5 --------(7)	R^2=0.83
(I_{-40} °C, J) = 0.85(I_{-20} °C)-1.02 --------(8)	R^2=0.87
(Pearlite, %) = 257.6(%Mn)+162.2(%Cu)-45.4 --------(9)	R^2=0.75

Table VI.The optimal alloy design and pearlite percentage for meeting specific specifications, together with the predicted mechanical properties

Specification	Mechanical property	Optimal alloy design and pearlite content	Predicted mechanical property
EN-GJS-350-22U-LT (60≦t<200)	T.S.≧320MPa Y.S≧200MPa El.≧15% $I_{-40°C}$≧10J	Si: 1.95±0.05% Mn: (0.16~0.19) % Cu: Trace (≦0.015%) Pearlite: 7.5±1%	T.S.=(367~383)MPa Y.S.=(235~242)MPa El.=(25~29)% $I_{-40°C}$=(15.2~17.4)J
EN-GJS-400-18U-LT (60≦t<200)	T.S.≧370MPa Y.S.≧220MPa El.≧12% $I_{-20°C}$≧10J	Si: 2.05±0.05% Mn: (0.21~0.23) % Cu: Trace (≦0.015%) Pearlite: 12.0±1%	T.S.=(403~419)MPa Y.S.=(250~258)MPa El.= (20~22)% $I_{-20°C}$=(15.2~17.0)J
EN-GJS-700-2U (30≦t<60)	T.S.≧700 MPa Y.S≧420 MPa El.≧1%	0.25Mn+0.8Cu+0.01Sn 0.25Mn+0.6Cu+0.05Sn 0.45Mn+0.6Cu+0.01Sn	T.S.= (710~766)Mpa Y.S= (464~523)MPa El.= (4~6)%

(a) 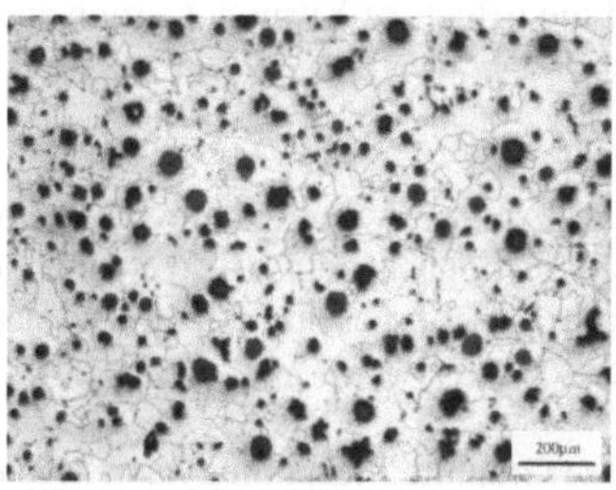

(b) 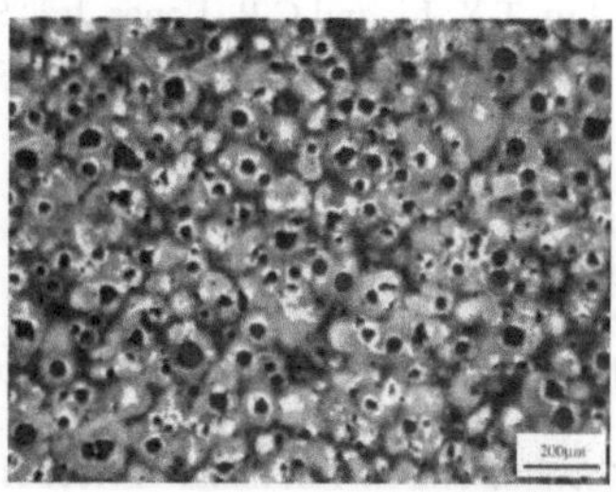

Figure 6. Representative microstructures of castings conforming to various specifications: (a) EN-GJS-350-22U-LT and (b) EN-GJS-700-2U.

Conclusions

Based upon the results obtained in this study, the following conclusions can be drawn:

(1) The optimal alloy design for achieving the desired microstructures and mechanical properties to meet the specifications of both EN-GJS-400-18U-LT and EN-GJS-350-22U-LT had been established.

(2) In order to meet the impact requirement of specification EN-GJS-350-22U-LT, the Si content should be kept below 1.97%, and also the maximum pearlite content shouldn't exceed 7.8%. On the other hand, Si content below 2.15% and pearlite content below 12.5% were registered for specification EN-GJS-400-18U-LT.

(3) The optimal alloy designs for complying with specification EN-GJS-700-2U include 0.25%Mn + 0.6%Cu + 0.05%Sn, 0.25%Mn + 0.8%Cu + 0.01%Sn and 0.45%Mn + 0.6%Cu + 0.01%Sn.

(4) The derived regression equations can be used to attain the optimal alloy design for castings with target specifications. By employing these regression equations, the mechanical properties can also be predicted based upon the chemical compositions and microstructures of cast irons.

Acknowledgement

This work was supported by the Ministry of Science and Technology of Taiwan, Grant No.: MOST 103-3113-E-002-004. The authors would like to thank YING CHIEN Foundry Industry Co., Ltd. for the collaboration in melting and casting, data collection and analysis.

References

1. World in Transition: Towards Sustainable Energy Systems, German Advisory Council on Global Change, (2003).

2. World Energy Outlook, International Energy Agency, (2007).

3. E.N. Pan, C.C. Lin, and R.M. Chang, "The Optimal Alloy Design and Process Conditions for Heavy Section Ductile Cast Irons With High Sub-Zero Impact Toughness," *AFS Trans.,* 120 (2012), 289-296.

4. J.F. Wallace and J.C. Sawyer, "Effects and Neutralization of Trace Elements in Gray and Ductile Irons," *AFS Trans.*, 76 (1968), 368-404.

5. P.C. Liu, T.X. Li, and C.R. Loper, Jr., "Study of the Effects of Yttrium, Cerium and Rare Earths on the Graphite Morphology in Heavy Section Ductile Iron," *AFS Trans.*, 97 (1989), 11-16.

6. P. Larrañaga, I. Asenjo, and J. Sertucha, "Effect of Antimony and Cerium on the formation of Chunky Graphite during Solidification of Heavy-Section Castings of Near-Eutectic Spheroidal Graphite Irons," Metallurgical and Materials Transactions, Part A, Physical Metallurgy and Materials Science, 40A (2009), 654-661.

7. L. Zhenhua and L. Yanxiang, "Effect of Sb on the Graphite Morphology and Mechanical Properties in Heavy Section ductile Iron," Materials Science Forum, 475-479 (2005), 2769-2772.

8. I. Riposan, M. Chisamera, and S. Stan, "Performance of Heavy Ductile Iron Castings for Windmills," China Foundry, 7(2) (2010), 163-170.

9. P.C. Liu and C.R. Loper, Jr., "The Role of Sb in Heavy-Section Ductile Cast Iron," *AFS Trans.*, 98 (1990), 753-757.

10. E.N. Pan, C.N. Lin, and H.S. Chiou, "Effects of Pb and Solidification Conditions on the Graphite Structure of Heavy Section Ductile Cast Irons," *AFS Trans.*, 103 (1995), 265-273.

11. *Ductile Iron Handbook*, American Foundrymen's Society, Inc., Des Plaines, IL (1992)

12. C. Labrecque, P.M. Cabanne, and E.C. Muratore, "Ductile Iron Characteristics and Impact Strength at Low Temperature," *AFS Trans.*, 118 (2010), 233-242.

13. ASTM designation A536-84: Standard Specification for Ductile Iron Castings, (2009).

14. ASTM designation A327-91: Standard Test Methods for Impact Testing of Cast Irons, (2006).

Advances in the Science and Engineering of Casting Solidification
Edited by: Laurentiu Nastac, Baicheng Liu, Hasse Fredriksson, Jacques Lacaze, Chun-Pyo Hong, Adrian V. Catalina, Andreas Buhrig-Polaczek, Charles Monroe, Adrian S. Sabau, Roxana Elena Ligia Ruxanda, Alan Luo, Subhayu Sen, and Attila Diószegi
TMS (The Minerals, Metals & Materials Society), 2015

AN OVERVIEW OF ISOTHERMAL COARSENING IN HYPOEUTECTIC LAMELLAR CAST IRON

Juan Carlos Hernando[1], Attila Diószegi[1]

[1] School of Engineering Jönköping University, Box 1026, 551-11, Jönköping, Sweden
Corresponding author e-mail: Juan-Carlos.Hernando@jth.hj.se

Keywords: Lamellar Cast Iron, Coarsening, Ripening, Dendrite Morphology, Solidification

Abstract

A complete qualitative characterization of the isothermal coarsening process in hypoeutectic lamellar cast iron is presented for the first time in this work. Interrupted solidification experiments were used to study the evolution of the dendritic austenite network under long term isothermal conditions. Cylindrical samples were re-melted and isothermally coarsened for times from 2 minutes to 6 days at 1175°C after dendritic coherence was reached. Micrographs from horizontal and vertical sections of the coarsened samples are presented. Complete fragmentation of the dendrite network and further rearrangement of the solid phase are reported as new behaviors in the coarsening process in lamellar cast iron. A linear increase in secondary dendrite arm spacing in agreement with the literature is observed in the first several samples confirming qualitative observations. A new model is proposed which describes the entire coarsening process observed in this investigation.

Introduction

The solidification process of a cast component is the single most important factor which influences the final properties achieved in service. Our ability to improve the performance of real castings depends on further development and new tools to control and predict this solidification process. The primary solidification and its primary austenite phase have so far received little attention in cast iron investigations. However, the metallic network of primary austenite that forms during this part of the solidification is to a large extent responsible for the final properties, and also influences the subsequent events in the solidification process such as graphite solidification and shrinkage porosity. The growth mechanisms of the primary austenite in hypoeutectic lamellar iron includes a nucleation stage followed by dendritic growth [1]. Dendrites constitute the principal growth morphology of a casting during the early stage of the solidification [2]. Parallel with the beginning of dendritic growth, coarsening appears due to the free energy associated with the dendritic interfaces, and continues during the remaining stages of the solidification process [3]. Regardless of the early appearance of coarsening during solidification, its main influence happens after dendrite coherence is attained, when coarsening becomes the principal form of growth [4]. The key driving force of coarsening is the interfacial curvature dependence of the chemical potential [4, 5, 6]. A continuous modification of the interface morphology takes place during coarsening, with transport of atoms from regions of high curvature, through the liquid matrix, to regions of low curvature, thereby generating a reduction of the total interface area [7]. A major requirement for coarsening is therefore the presence of interdendritic liquid which allows a rapid redistribution of the solid [3]. In the classic process of coarsening, the smallest dendritic arms re-melt and disappear, feeding at the same time the adjacent larger arms, which continue to grow and increase their surface area [3]. At the same time, in the teardrop-shaped dendrite arms, material is transported from the base to the tip, creating a new independent spheroidal entity and producing a fragmentation of the dendrite network after long coarsening times [2]. Traditionally, secondary dendrite arm spacing (SDAS) has been used as a parameter to study the size scale and coarsening in metal alloys and has been reported to be dependent on the time during which solid and liquid coexist during solidification [8, 9, 10]. Several authors express the relation between the SDAS and coarsening through the total solidification time [2, 3]. The goal of this work is to study the morphological variations of the primary austenite phase under isothermal conditions in a hypoeutectic lamellar cast iron. Numerous works have presented coarsening studies in real metallic alloys [8, 9, 10, 11], and more recently other works studied dynamic coarsening in lamellar cast iron solidification [7, 12], but our lack of knowledge about isothermal coarsening of primary austenite in lamellar cast iron after long coarsening times requires an experimental study that catalogues the different events taking place as a function of time.

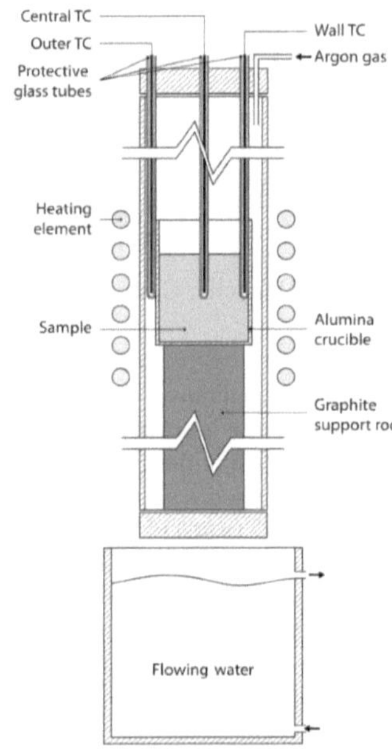

Figure 1. Experimental equipment [13].

Material

A hypoeutectic lamellar cast iron was produced under industrial foundry conditions from a single melt. Cylindrical samples of ϕ50 x 100 mm were produced in sand molds with the composition shown in Table I. From this base material, samples of 400g were machined and re-melted to obtain a final specimen of ϕ42 x 42 mm after the experimental treatment.

Table I: Average Chemical Composition of the Experimental Alloy

Element	C	Si	Mn	P	S	Cr	Mo	CE
Wt. %	3.34	1.78	0.58	0.034	0.086	0.149	0.225	3.94

Experimental Equipment

The main component of the experimental equipment is a vertical chamber programmable resistance furnace, extensively described in Lora et al. [13] and which was used for the re-melting experiments. The samples were placed in an alumina crucible and inserted into the furnace through the base. A graphite support rod held the sample exactly in the middle of the heating elements of the furnace. The bottom of the furnace was closed with a sliding shutter that facilitates a quick release of the sample into a quenching media. The top of the furnace was sealed with a water-cooled lid that includes an argon gas inlet to ensure a neutral atmosphere inside the chamber. The top lid is prepared in order to use three thermocouples (TCs), one at the center of the sample (Central TC), one at the inner wall of the crucible (Wall TC) and one at the outer crucible wall (Outer TC), see Figure 1. The TCs were type S (Pt / Pt+10%Rh) mounted in an alumina sheath which insulates the TC wires, leaving only the welded joint accessible. The TCs were placed inside a protective glass tube to prevent direct contact with the melt. The TCs were connected to a commercial data acquisition system that records all the data measured by the TC at a sampling rate of 75 Hz. The quenching equipment consisted of a bath of water below the chamber of the furnace. The bottom shutter of the furnace could be released so that the crucible containing the sample dropped directly into a water tank below. A pump ensured continuous circulation and turbulent flow during the quenching.

Experimental Procedure

The experimental work started with the acquisition of the natural cooling curve (NCC), considered as the solidification reference curve of the cast iron samples. A sample was re-melted inside the alumina crucible by a heating cycle of 90 minutes from room temperature to 1450 °C and a holding time of 30 minutes at that temperature. At this point the furnace was switched off and the solidification occurred under natural cooling to room temperature. During this stage, the temperature was recorded by three TCs (Central TC, Wall TC and Outer TC) to determine the coherence point which was identified as the maximum temperature difference during the primary solidification between Central TC and Wall TC [14] (only the Central TC measurement is represented as NCC in Figure 2). Based on the data obtained from NCC, a new isothermal coarsening temperature profile (ICTP) was designed to fulfill three conditions: a) the isothermal stage should detach smoothly from the NCC after the coherence point was reached, b) the holding temperature should be high enough to avoid the formation of eutectic cells and c) the holding temperature should be stable during the whole experiment. The target temperature to study the isothermal coarsening was 1175 °C. For the isothermal coarsening experiments, after 17.2 minutes under natural cooling and immediately after the coherence point was reached, the furnace was re-started and a programmed thermal cycle guaranteed a smooth transition between the natural cooling and the isothermal holding temperature. This temperature was stable during the times used to study the isothermal coarsening of the primary austenite. The

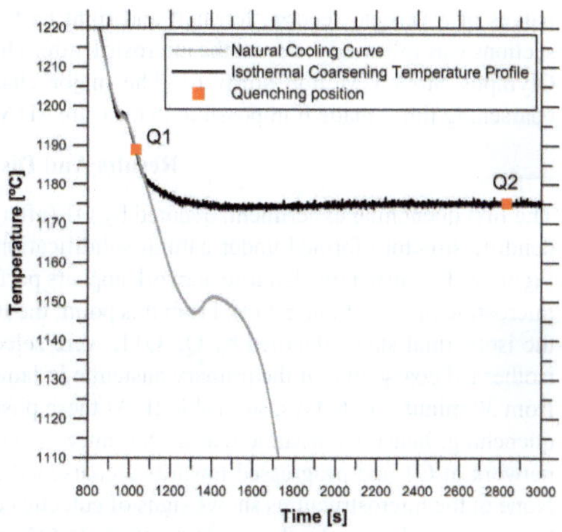

Figure 2. Natural cooling curve and isothermal coarsening temperature profile including the two first quenching positions.

last stage of the experimental procedure was to perform the quenching experiments. After the specimen had been maintained on the ICTP for the precise coarsening time, the shutter of the furnace was opened quickly and the specimen was quenched into the turbulent flowing water, remaining in the quenching media for 5 minutes.

Microstructure Investigation

The quenched samples were first sectioned perpendicularly to the longitudinal axis at the geometrical center using wet cutting. The samples were ground, polished and then etched with a picric acid solution used at 108 °C containing 200 ml of distilled water, 160g KOH, 40g NaOH and 40g picric acid. Micrographs capturing the whole cross-section of the specimen were taken for both halves of the specimen, bottom half and top half. An average of 200 micrographs were assembled into one single picture of the whole horizontal section. Once the horizontal micrographs of the specimens were successfully obtained, the vertical micrographs were prepared in the same manner. A cut parallel to the longitudinal axis at the center of each half was done producing two halves of a vertical section, left half and right half. The same procedure as with the horizontal sections was followed to reveal the microstructure. The SDAS was measured using the commercial Olympus Stream Motion software. The major changes in the austenite morphology at long coarsening times made it impossible to measure SDAS in all micrographs.

Results And Discussion

The first quenching experiment, denoted by Q1 (after 2 min), was chosen to investigate the initial dendrite structure formed under natural solidification conditions just before the ICTP starts, see Figure 2. The austenitic dendrite network appears perfectly developed and coherent throughout the microstructure, see Figure 4 (a). From this point, the ICTP begins and consecutive positions along the isothermal stage, denoted by Q2-Q11, were selected to complete the characterization of the isothermal coarsening of the primary austenite in lamellar cast iron from short to very long times, from 30 minutes to 6 days, see Table II. At these positions no solidification should occur prior to quenching, hence we assume that all the microstructures started from the same initial austenite network at Q1 and progressed towards a coarsened austenite network at constant solid fraction. None of the microstructures shows signs of eutectic colonies, validating the isothermal coarsening temperature chosen. In the microstructure at Q2 (after 30 min) we can begin to distinguish coarsening of the dendrite network. Despite the fact that the network is completely coherent we can see that the dendrites are thicker compared to Q1. A reduction in the surface area due to the isothermal coarsening can be assumed. The three next microstructures Q3 (1,5h), Q4 (3 h) and Q5 (6h) can be considered as showing the transition between a coherent austenite network and an incoherent coarsened dispersed primary phase. In Q3 the equiaxed zone starts to show an incoherent behavior in the dendrite network and with increasing time that incoherent zone expands towards the columnar zone as can be observed in Q4 and Q5. At the end of this series of samples, in Q5, the equiaxed zone is totally incoherent and the columnar zone seems to adopt the same behavior as the time increases. Although it is now very difficult, we can find some dendrites suitable to quantify the SDAS. It is easy to observe that the austenite units follow the classic theory of coarsening, adopting a more spherical shape, decreasing the total interfacial area and separating from the primary arm as time elapses.

Table II: Quenching Positions, Solidification Time, Isothermal Coarsening Time and SDAS

Sample	Total Solidification Time (s)	Coarsening Time (s)	Approximate Isothermal Coarsening Time	SDAS (μm)
Q1	1032	132	2 min (dynamic coarsening)	38
Q2	2832	1932	30 min	85
Q3	6432	5532	1,5 h	126
Q4	11832	10932	3 h	173
Q5	22632	21732	6 h	190
Q6	44232	43332	12 h	283
Q7	87432	86532	1day	-
Q8	173832	172932	2 days	-
Q9	260232	259332	3 days	-
Q10	346632	345732	4 days	-
Q11	519432	518532	6 days	-

In Q6 (12h) the remaining dendrite network starts to fragment, reducing the number of secondary dendrites drastically and making very hard to measure SDAS. The fragmented and now independent austenite adopts more a spheroidal morphology than at the previous quenching time. We can distinguish a small zone free of austenite close to the center of the sample. The microstructure Q7 (24 h) is totally incoherent throughout, fragmentation of the dendrite network is complete, and the austenite in the sample continues to coarsen as predicted by the classic theories, adopting an almost perfect spherical morphology. SDAS cannot be measured in this and subsequent samples. In this position (Q7) a larger area around the middle of the sample shows a lack of austenite. The remaining microstructures; Q8 (2 days), Q9 (3 days), Q10 (4 days) and Q11 (6 days) show total fragmentation of the dendrite network, which allows the movement of incoherent particles of austenite through the liquid, and starting a new ripening process. Some dispersed spheroidal units of austenite start to cluster in large linear structures to minimize the perimeter and form stable agglomerates of austenite in a coalescence process. This can be observed in the columnar zone of the samples.

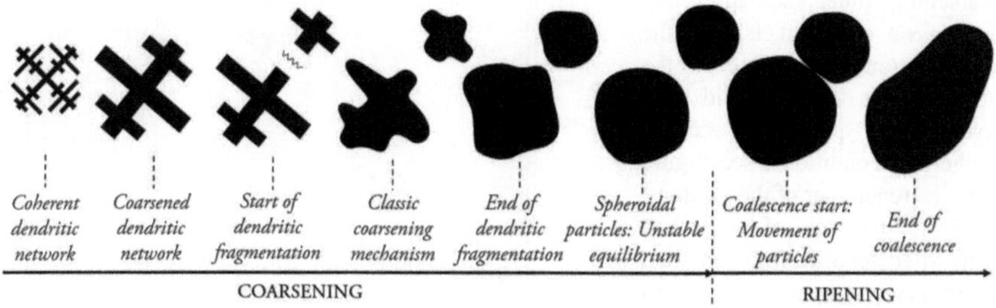

Figure 3. Representation of the isothermal coarsening process in a hypoeutectic lamellar cast iron.

The remaining independent particles of austenite maintain a spheroidal shape. In these microstructures there is a greater fraction of liquid than solid present and in the horizontal sections of the sample we can see a zone completely free from austenite. From the vertical sections we can clearly observe how the austenite has migrated towards the top part of the sample during the ripening process, see Figure 6. In the last samples (Q10 and Q11) we find fully austenitic domains with almost no presence of liquid phase present at the top of the sample. Figure 3 shows schematically how the austenite dendrites coarsen, the dendritic network fragments and the austenite units spheroidise and eventually coalesce after long isothermal coarsening times.

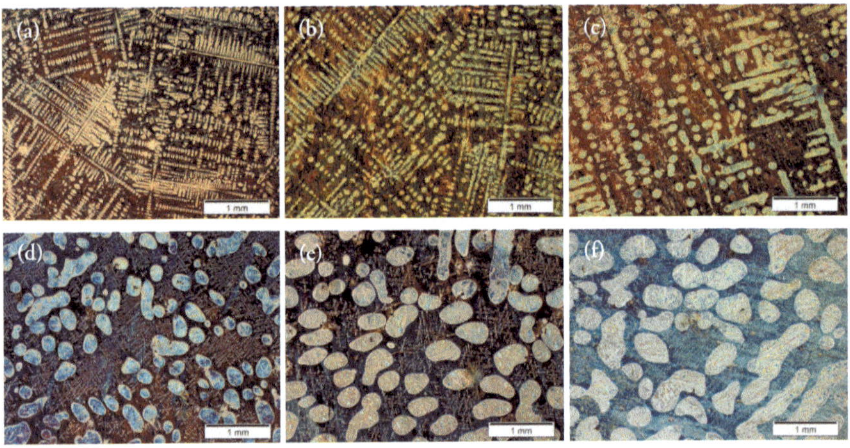

Figure 4. Micrographs from horizontal section: (a) Q1: after 2 min, (b) Q2: 30min, (c) Q4: 90 min, (d) Q7: 24 h, (e) Q8: 2 days, (f) Q9: 3 days of isothermal coarsening.

Measurements made on those samples where SDAS could be measured, Q1-Q6, confirm that SDAS increases proportional to the cube root of the isothermal coarsening time, indicating that the classic exponent 1/3 for the solidification time found in the literature is also valid for coarsening processes under isothermal conditions, see Figure 5. In the remainder of the samples, Q7-Q11, the lack of secondary arms after dendrite fragmentation made quantification of the SDAS impossible.

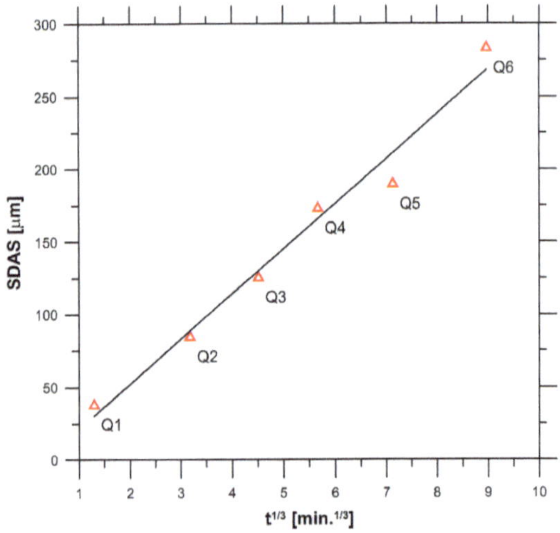

Figure 5. SDAS as function of the cube root of the isothermal coarsening time.

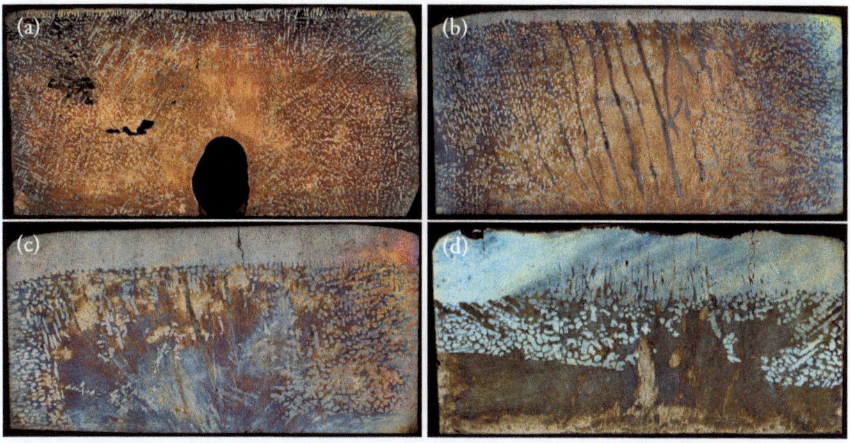

Figure 6. Complete micrographs of vertical top section: (a) Q6: after 12 h, (b) Q7: 1 day, (c) Q10: 4 days, (d) Q11: 6 days of isothermal coarsening. Top part of the figure represents the top part of the sample. Diameter of the sample ϕ42 mm.

Summary

1. A characterization of isothermal coarsening of the primary austenite in lamellar cast iron has been achieved through an experimental method combining interrupted solidification experiments, color etching, image analysis and SDAS measurement.
2. Observation of the microstructures allows the description of a new model for the coarsening mechanism in lamellar cast iron, see Figure 3.
3. The classic mechanism of coarsening, traditionally referred to as Ostwald ripening, occurs during normal isothermal coarsening times. The reduction of the dendrite surface area accompanied by progressive dendrite fragmentation leads the primary austenite into an unstable state where the dendrite network is completely broken and the primary austenite is present as a solid dispersed phase with a spherical morphology. From this point, after long coarsening times, a ripening mechanism commences and movement of incoherent particles of austenite takes place through the liquid, followed by a coalescence process which results in a linear clustering in some zones of the sample, creating large austenite units.
4. These results confirm that SDAS is a good parameter to characterize isothermal coarsening of the primary austenite, but only while the dendritic structure is still coherent and retains a discernible dendrite network.
5. The relation between SDAS and time of coexistence of liquid phase and solid phase, demonstrated for the solidification process, is proved to be valid in case of isothermal processes and the 1/3 exponent remains constant.
6. Fragmentation of the dendritic structure over time shows that SDAS is not applicable for long isothermal coarsening times and new parameters should be used to characterize the coarsening processes.

References

1. P. Beeley, *Foundry Technology*, (Oxford, UK, Butterworth-Heinemann, 2001), 51-63.
2. D.M. Stefanescu, *Science and Engineering of Casting Solidification*, (New York, NY, Kluwer Academic/Plenum Publishers, 2002), 174-185.
3. M.E. Glicksman and P.W. Voorhees, "Ostwald Ripening and Relaxation in Dendritic Structures," *Metallurgical Transactions A,* 15 (1984), 995-1001.
4. J. Lacaze and G. Lesoult, "Birth, Growth and Ripening of Solidification Structures," *Acta Stereologica,* 5 (2) (1986), 331-336.
5. I.M. Lifshitz and V.V. Slyozov, "The Kinetics of Precipitation From Supersaturated Solid Solutions," *Journal of Physics and Chemistry of Solids,* 19 (1961), 35-50.
6. P.W. Voorhees, "The Theory of Ostwald Ripening," *Journal of Physics and Chemistry of Solids,* 38 (1985), 231-252.
7. R. Lora and A. Diószegi, "Dynamic Coarsening of 3.3C-1.9Si Gray Iron," *Metallurgical and Materials Transactions A,* 43A (2012), 5165-5172.
8. P.W. Wilson, T.Z. Kattamis and Y. Shiohara, "Coarsening during solidification of aluminium-copper alloys," *Journal of Material Science,* 23 (1988), 2882-2892.
9. S.P. Marsh and M.E. Glicksman, "Overview of Geometric Effects on Coarsening of Mushy Zones," *Metallurgical and Materials Transactions A,* 27 (1996), 557-567.
10. D. Kammer and P.W. Voorhees, "The morphological evolution of dendritic microstructures during coarsening," *Acta Materiala,* 54 (2006), 1549-1558.
11. R. Mendoza, J. Alkemper and P.W. Voorhees, "The Morphological Evolution of Dendritic Microstructures during Coarsening," *Metallurgical and Materials Transactions A,* 34 (2003), 481-489.
12. A. Diószegi, R. Lora and V. Fourlakidis, "Dynamic Coarsening of Austenite Dendrite in Lamellar Cast Iron Part 1 – Investigation based on interrupted solidification," *Materials Science Forum,* 790-791 (2014), 205-210.
13. R. Lora, A. Diószegi and L. Elmquist, "Solidification Study of Gray Cast Iron in a Resistance Furnace," *Key Engineering Materials,* 457 (2011), 108-113.
14. G. Chai et al., "Dendrite Coherency during Equiaxed Solidification in Binary Aluminum Alloys," *Metallurgical and Materials Transactions A,* 26 A (1995), 965-970.

Advances in the Science and Engineering of

CASTING SOLIDIFICATION

An MPMD Symposium Honoring Doru Michael Stefanescu

Cast Iron II

Session Chair:

Adrian V. Catalina

Advances in the Science and Engineering of Casting Solidification
Edited by: Laurentiu Nastac, Baicheng Liu, Hasse Fredriksson, Jacques Lacaze, Chun-Pyo Hong, Adrian V. Catalina, Andreas Buhrig-Polaczek, Charles Monroe, Adrian S. Sabau, Roxana Elena Ligia Ruxanda, Alan Luo, Subhayu Sen, and Attila Diószegi
TMS (The Minerals, Metals & Materials Society), 2015

EXPERIMENTAL STUDIES OF GRAY CAST IRON SOLIDIFICATION WITH LINEAR VARIABLE DIFFERENTIAL TRANSFORMER

Abel Tadesse[1], Hasse Fredriksson[1]

[1]KTH (Royal Institute of Technology); Brinellvägen 23; Stockholm; SE-100 44; Sweden

Keywords: Gray cast iron, LVDT, Inoculant, Eutectic cells, Flake graphite

Abstract

Expansion during the solidification of gray cast iron was studied by the help of Linear Variable Differential Transformer (LVDT). The chemical composition of the samples was altered by adding two types of inoculant; Superseed® (50% Si, 1% Sr and 0.5% Al) and SMZ® (69% Si, 1.9% Ca, 0.7% Ba, 5% Zr, 4.5% Mn and 1.3% Al). During the solidification, the melt shows hardly no shrinkage in the primary austenite formation region, but the eutectic region shows higher expansion. The expansion during the eutectic growth increase, when the inoculant weight percentage escalates. At the same time, the eutectic cells get smaller and increases in cells number. The micrograph reveals undercooled and interdendritic graphite transformed to homogenized flake graphite. The inoculation process reduces the solidification rate along with different stable oxide and sulfide nuclei's created prior to the solidification, as a result eutectic cell gets more sites to grow. The change in micrograph and solidification rate was believed to modify the mechanical property of the cast.

Introduction

Gray cast iron has high application in automotive industries due to its properties. Its easy castability and machinability makes gray iron one among the few metals used for engine blocks in automobile industry [1]. The properties of gray cast can change by changing different parameters. The chemical composition and solidification rate are the factors responsible for property changes. Depending upon the constituents in the melt, the solidifying cast undergo variable range of expansion and contraction in different phase region. The solidifying cast hardly shows any freezing shrinkage, because the liquid shrinkage and contraction during the formation of austenite is compensated by the growth of graphite during the eutectic formation [2].

By keeping this fact in mind different types and amount of strong de-oxidizing and de-sulfurizing inoculant were added in the liquid melt in order to homogenize and facilitate the eutectic formation [3, 4, 5]. The principle behind inoculation was to form stable oxide and sulfide prior to solidification, and create enough nucleation sites for the graphite growth in the eutectic region [4]. During solidification the growth of austenite will occur first and eutectic formation will come after. The precipitation of graphite in stable eutectic structure can be favored by the addition of inoculant [6]. This inoculant create heterogeneities within the oxides and sulfides to enhance the nucleation of graphite on and around the oxide-sulfide particles.

Experimental Procedure

One base material was used in all experiments. A total of eighteen experiments was conducted with the help of Linear Variable Differential Transducers (LVDT sensors). Among them, four were without inoculant, and the rest of the experiments involves inoculation. Each condition was repeated at least once during the experimental work. The expansion and contraction of solidifying melt were recorded with the help of LVDT sensors (accuracy <±0.1%). Four sensors have been used in each experiment at four different locations (see Figure 1a and b). This will help us to retrieve complete data from representative sides of the sample. The data from the LVDT sensors combined with the temperature data logger have been used to study the effects of inoculation on the cast during solidification.

Molding and Sensor Positioning

Dry Silica sand was used for mold making and water glass (Na_2SiO_3) as a binder (the weight of the binder was determined by considering the amount of sand [7, 8]). Equal amount of sand and binder have been used in each experiment and the mold were kept in heating furnace for the same curing time. The pattern was coated with zircon during molding in order to remove it easily from the cavity. The sand and water glass were thoroughly mixed in a sand mixing machine. Adequate ramming was applied during molding. After the removal of the pattern from the sand mold, the mold cavity was coated with refractory material. Finally, the sand mold placed in a heating furnace for curing.
Quartz rods were used to connect the sensor tip with the melt in the mold cavity. Quartz tubes used to separate the quartz rod from holes in the sand mold, and also create smooth motion for the rods. The tips of the quartz rods placed inside the mold cavity was bent in order to stick with the solidifying shell in the walls and follow the movement of the solid part. The other end of the quartz rods was connected to the sensor tip with the help of connectors.
The mold cavity has a height of 7 cm, length of 5 cm and thickness of 1 cm. All the four sensors have been located exactly at mid distance of the mold cavity as shown in the Figure 1a and b. By the help of rigid metallic frame the sensors were fastened in a stationary location.

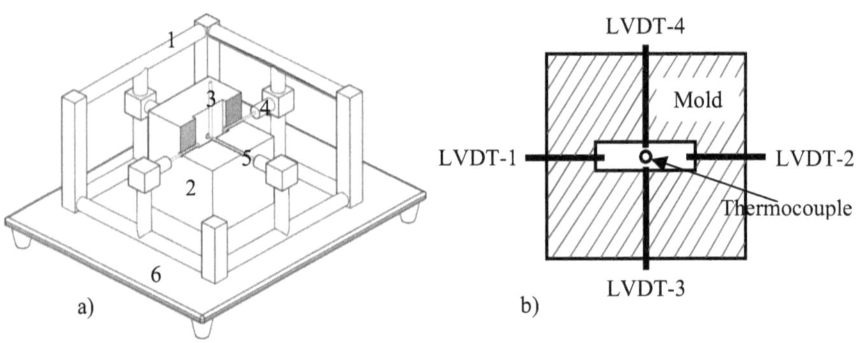

Figure 1. (a) 3D half sectioned image showing the arrangement of the sand mold with sensor positioning and temperature logger; 1- metallic frame, 2- sand mold, 3- temperature logger, 4- LVDT sensor, 5- quartz rod and 6- Stand. (b) Half sectioned top view for mold, thermocouple and LVDT sensors; the longitudinal displacement measurement (D2) is the sum of LVDT-1 and 2, the transversal displacement measurement (D1) is the sum of LVDT-3 and 4, the thermocouple located at the center of the mold cavity.

Melting and alloys

The melting process was done in a high-frequency induction furnace. During melting argon gas was supplied as a shielding gas in order to protect the melt from excessive oxidation. The temperature during melting was monitored by a temperature data logger. When the temperature reaches to 1400ºC, inoculates (see Table 1) were introduced inside the crucible and kept the melt temperature constant approximately for 1 minute. The magnetic field created by the induction coil will create a stirring effect on the melt, as a result the inoculant homogenized throughout the melt. Among eighteen expansion experiments, seven were performed by adding a different percentage of SS (0.14, 0.4 and 1 wt. %), the other seven was performed by adding a different percentage of SMZ (0.14, 0.4 and 1 wt. %) and the last four was without inoculant (see Table 2). The inoculant composition is summarized in table 1.

Table 1. Superseed (SS) and SMZ inoculants chemical composition in (wt. %).

Inoculant Type	Si	Ca	Ba	Zr	Mn	Sr	Al
SS	46-50	≤0.1	-	-	-	0.6-1	≤0,5
SMZ	62-69	0.6-1.9	0.3-0.7	3-5	2.8-4.5	-	0.55-1.3

Table 2. Composition of Alloys with Weight Percentage (wt. %) and amount of Inoculant.

Alloy	Ino.%	C	Si	S	Mn	Mo	Cr	N	Sr	Ti	Zr	Ceq
BGI	0	3.16	1.78	0.054	0.57	0.27	0.28	0.007	0.0	0.01	0.0	3.73
GIS	0.14	3.15	1.84	0.054	0.57	0.27	0.28	0.007	0.001	0.01	0.0	3.75
GIZ	0.14	3.15	1.86	0.054	0.57	0.27	0.28	0.007	0.0	0.01	0.004	3.76
GIS	0.4	3.15	1.95	0.054	0.57	0.26	0.279	0.007	0.002	0.01	0.0	3.78
GIZ	0.4	3.15	2.02	0.054	0.58	0.26	0.279	0.007	0.0	0.01	0.011	3.8
GIS	1	3.13	2.22	0.054	0.56	0.26	0.277	0.0069	0.006	0.01	0.0	3.84
GIZ	1	3.13	2.37	0.054	0.59	0.26	0.277	0.0069	0.0	0.01	0.03	3.87

BGI; Base Gray Iron, GIS,Z; Gray Iron with SS and SMZ respectively, Ino.%; weight percent of Inoculant, Ceq; carbon equivalent.

Microstructural studies

All the experimental samples undergo optical microscopy analysis and electron microscopy examination (SEM S-3700N). Backscattered electrons (BSE) were used to analyze the different types of oxide formed prior to solidification and eutectic growth. Stead reagent was used for etching the samples to reveal the eutectic cells [9, 10]. The solution was made by mixing 25ml HCl (32% aqueous solution), 10g Cupric chloride, 40g Magnesium chloride and 1000ml of ethanol (99%). Then by using the electrolytic technique the samples were etched in the stead solution for about 60sec. After that it undergoes microanalysis for the eutectic cell count.

Results

Solidification analysis

In almost all the samples, the primary austenite formation region shows hardly any shrinkage. When the eutectic formation starts, the expansion of the side's increases until the solidification ends. Increasing the inoculant percentage will influence the expansion in the longitudinal and transversal directions during the solidification. A higher expansion in the longitudinal direction was measured for samples with high percentage of inoculants, whereas the expansion occurred in the transversal direction decreases (see Figure 2). Physically it is shown that, when a change in slope (dT/dt) occurs from temperature curve it corresponds with change in expansion or contraction from the LVDT sensors. These emphasize the phase change occur within this gap and corresponds to those points. Sample with no inoculation shows higher expansion in transversal direction but the opposite sides have lower expansion during eutectic formation.

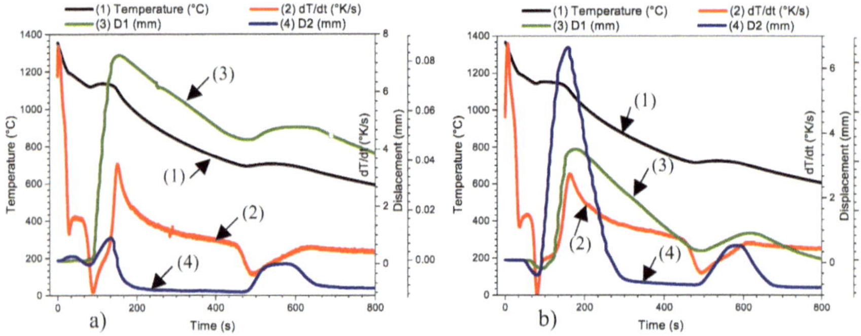

Figure 2. Temperature, displacement and slope of cooling curves vs time; (a) sample with no inoculation, (b) samples with 1% SMZ addition.

According to the results, the expansion starts when the eutectic reaction occurs and continues until it reach a maximum close to the end of the solidification process. Then, contraction starts and become negative after some time. This is normal contraction of the material after solidification. The total expansion during and after solidification was evaluated as well as the temperature when the expansion changed to contraction (see Table 3).

Table 3. Summarized result of the experiments during and after solidification.

Sample type	Ino. %	End of Solidification (°C)	Maximum expansion during solidification		Maximum contraction after solidification	
			D1(mm)	D2(mm)	D1(mm)	D2(mm)
BGI	0	1101.7	0.0824	0.0089	0.0776	-0.0109
GIS	0.14	1102.5	0.058	0.0122	0.0575	-0.0071
GIS	0.4	1100	0.0162	0.0285	0.0145	-0.0066
GIS	1	1102.2	0.013	0.0415	0.0058	-0.0092
GIZ	0.14	1100.1	0.0469	0.012	0.0423	-0.012
GIZ	0.4	1100.9	0.0411	0.0642	0.0302	-0.0077
GIZ	1	1103.5	0.045	0.0862	0.0278	-0.0075

Microstructural analysis

The experimental result shows that the addition of the inoculant improves the microstructure of the samples by changing the solidification rate and gives the graphite more time to grow longer in the size and distributed more or less uniformly throughout the structure (see Figure 3). Figure 3a and e clearly shows that the undercooled graphite and interdendritic graphite dominant, whereas the samples with inoculant increases the formation of longer flake graphite and it also provides uniform distribution throughout the sample. The chemical analysis in SEM indicates that, the elements present in the inoculant forms stable oxides and sulfides prior to the solidification and create enough nucleation sites for the graphite to nucleate (see Figure 4 & 6). The stead solution electrolysis also shows that the eutectic cell numbers increase with increasing inoculant percentage. The samples with no inoculant shows bigger eutectic cells with fewer in number. The principle behind the Stead solution electrolysis was; the reagent will react with the grain boundaries of the eutectic cells and exposed the outline of the cell boundaries by attacking the phosphorous micro segregated areas. Samples with high inoculates percentage shows higher phosphorous segregation, the eutectic cell size reduced and much higher numbers (see Figure 4). An increase in addition of inoculant results in more nuclei creation with more vacant places for growth the eutectic cell and refine the cells, as a result the flake graphite growth is higher.

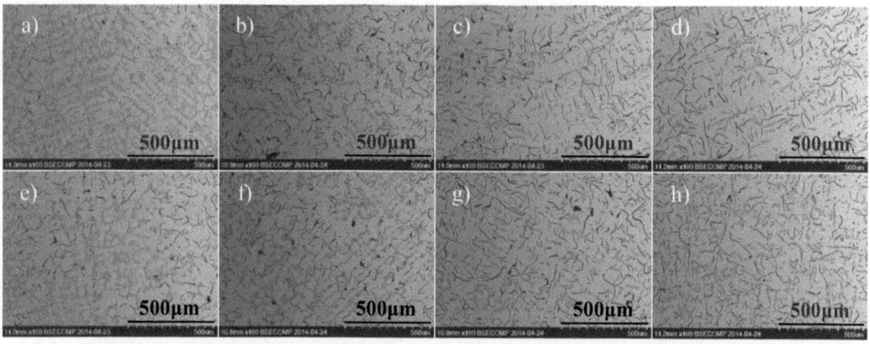

Figure 3. Microstructural evolution upon increasing inoculant percentage from left to right. The first row represent samples with SS inoculant (0, 0.14, 0.4 and 1 wt. % addition respectively). The second row represent samples with SMZ inoculant (0, 0.14, 0.4 and 1 wt. % addition respectively).

The micrograph in Figure 3g & h shows the flake graphite produced by using SMZ, these flakes are larger in size and many in number.

Discussion

The addition of the inoculant into the melt changes the composition, at the same time the microstructure gets improved by reducing the undercooled and interdendritic graphite. It promotes more flake graphite growth and uniformly distribute throughout the structure. These effects were analyzed by measuring the statistical distribution of the flakes at randomly selected locations (see Figure 5a and b). The statistical result shows that the number of the flake graphite within the length of 24µm and above increase drastically upon an increase in inoculant percentage (see Figure 5a). Analyzing two random positions in a sample also resemble larger differences in sample with no inoculant and samples with less amount of inoculants (see Figure 5b). Whereas the sample with higher inoculant content shows closeness within the recorded data and the median of the data series of randomly selected positions. The length of the flake which lays between 25-75% of the recorded data show an increase as we go to the right of the graphs. According to these results, more number of longer flakes are observed as the inoculant percentage escalates. This change also observed clearly in the micrographs.

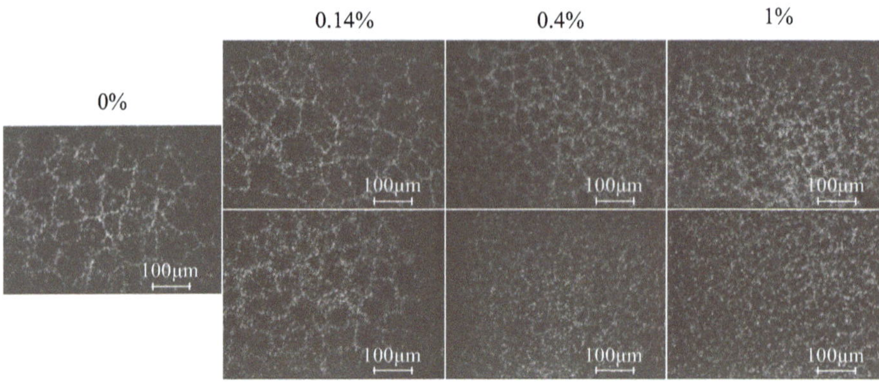

Figure 4. Eutectic cells reveled by using stead solution and electrolysis. The first row from the second column represent samples with SS inoculant (0.14, 0.4 and 1 wt. % addition respectively). The second row form second column represent samples with SMZ inoculant (0.14, 0.4 and 1 wt. % addition respectively).

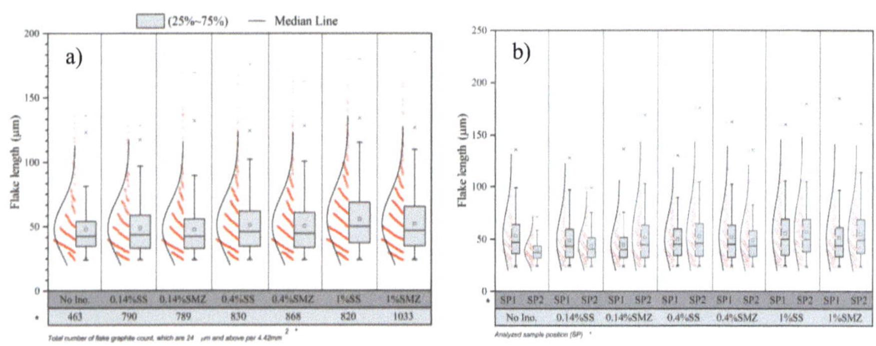

Figure 5. (a) Flake length data distribution vs the total number of flake count per 4.42 mm^2 for each experiments; the red circles represents the measured data points, the gray rectangle represents the measured points that fall between 25-75% of the total data, the line inside the gray rectangle represents the median of the measurment. (b) Flake length data distribution for two randomly selected sample positions (SP) for samples with and without inoculant; the amount of inoculant increases to the right, the minimum flake length take for this statistical distribution is 24 μm, the gray rectangles represent the measured points that fall between 25-75% of the total data, the line inside the gray rectangle represents the median of the measurement.

The addition of different types of inoculant will changes the solidification behavior of the metals. The above change was as a result of the chemical composition of the original melt and nature the constituents of the inoculant like; Zr, Sr, Ca, Ba, Mn and Al are strong oxidizer or sulfurized (see

Figure 6). Thus, this elements will create a stable oxide and sulfides prior to the solidification according with their affinity level. The oxides and sulfides size are small and act as a nuclei to facilitate the growth of the austenite and eutectic cells. According to the elements affinity level, those having high affinity towards oxygen will create oxide prior to solidification began. The same way elements having higher affinity for sulfur will create sulfide prior to the solidification. These oxides and sulfides will promotes flake graphite on and around them. It will also provide sufficient space for the growth of the eutectic cells.

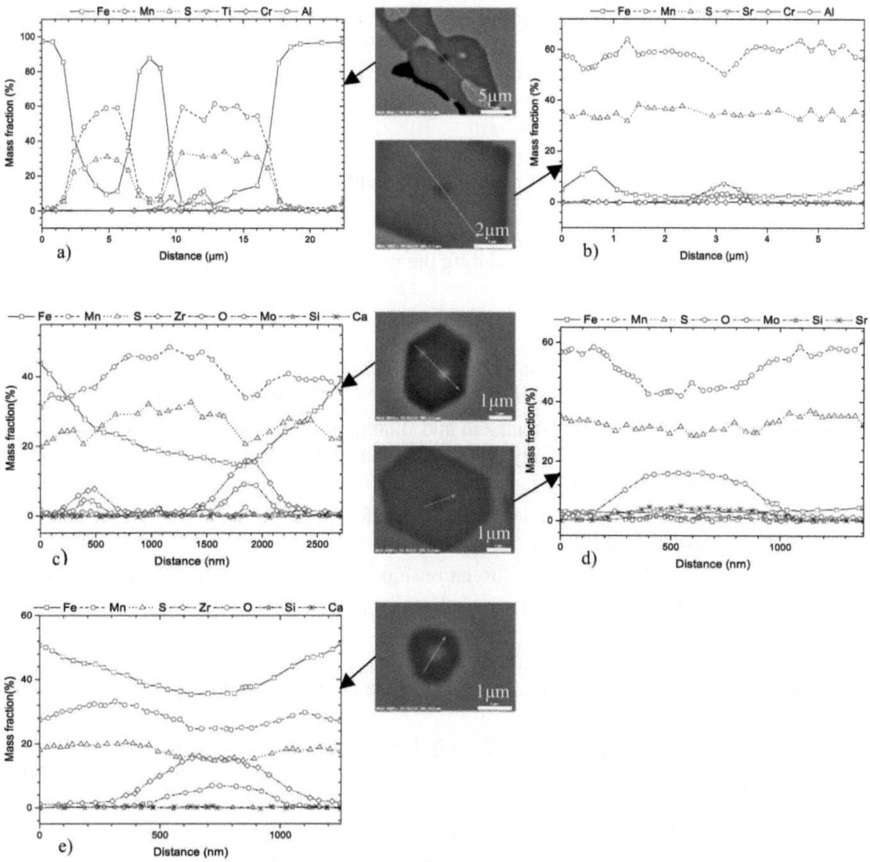

Figure 6. BSE line analysis; (a) Sample with no Inoculant: MnS, Al_2O_3, Cr and Ti. (b) Sample with 0.4% SS: MnS, SrS, Cr and Al. (c) Sample with 0.4% SMZ: MnS, ZrO_2, CaO, MoO_2 and Si. (d) Sample with 1% SS: MnS, MoO_2, SrS and SrO. (e) Sample with 1% SMZ: MnS, ZrO_2, CaO, CaS.

Conclusion

During the solidification of gray cast iron, the metal shows hardly any shrinkage during the austenite formation. The eutectic formation region shows a higher expansion upon increasing the inoculant percentage. The flake graphite gets modified with inoculant addition by reducing the undercooled and interdendritic graphite. Reducing the solidification rate and modifying the morphology of the flake by adding a different amount of inoculants, one can reduce the total volumetric shrinkage during solidification. Lowering the volumetric shrinkage will reduce the occurrence of internal stresses, as a result internal and surface cracks can be minimized. The eutectic cells drastically increase with an increase in inoculants. The addition of the inoculant creates stable nuclei that act as a nucleation site for the flake growth in the eutectic cells. But the chemical composition of the inoculant must take into consideration during the inoculation process.

Acknowledgment

The authors would like to acknowledge ELKEM AS, FOUNDRY TECHNOLOGY PRODUCTS R&D for financial and material supports during the work.

References

[1] M. Behnam et al., "Effect of cooling rate on microstructure and mechanical properties of gray cast iron," *Material Science and Engineering A,* 528 (2010), 583-588.

[2] R. Hummer, "A study of the shrinkage and dilatation during solidifcation of nodular cast iron-its relation to the morphology in crystallization," *Material Research Society Symposia Proceedings,* 34 (1985), 213-222.

[3] H. Nakae, "Influence of inoculation on solidification in cast iron," *International Journal of Cast Metals Research,* 21 (2008), 7-10.

[4] H. Muhmond and H. Fredriksson, "Relationship between inoculants and the morphologies of MnS and graphite in gray cast iron," *Metallurgical and Materials Transactions B,* 44B (2013), 283-298.

[5] L. Elmquist et al., "Inoculation and its effect on primary solidification structure of hypoeutectic gray cast iron," *International Journal of Cast Metals Research,* 23 (2010), 124-129.

[6] R. Castillo and T. Baker, "The relationship between tensile strength and microstructure in flake graphite cast iron," *Materials Research Society Symposia Proceedings,* 34 (1985), 487-495.

[7] K. Granat et al., "The influence of microwave heating and waterglass kind on the properties of molding sands," *Archives of Foundary Engineering,* 8 (2008), 119-122.

[8] M. Stachowicz et al., "Influence of water glass grade and quantity on residual strength of microwave-hardened molding sands," *Archives of Foundry Engineering,* 11 (2011), 93-98.

[9] R. D. Burkhart, "Method for electrolytic eaching of gray irons with stead's reagent". Iowa Patent 3,996,119, 2 Feb 1976.

[10] E. Fras et al., "Eutectic cell and Nodule count in cast iron," *ISIJ International,* 47 (2007), 269-276.

Advances in the Science and Engineering of Casting Solidification
Edited by: Laurentiu Nastac, Baicheng Liu, Hasse Fredriksson, Jacques Lacaze, Chun-Pyo Hong, Adrian V. Catalina, Andreas Buhrig-Polaczek, Charles Monroe, Adrian S. Sabau, Roxana Elena Ligia Ruxanda, Alan Luo, Subhayu Sen, and Attila Diószegi
TMS (The Minerals, Metals & Materials Society), 2015

UNDERCOOLING, COOLING CURVES AND NODULE COUNT FOR HYPO-, HYPER- AND EUTECTIC THIN-WALLED DUCTILE IRON CASTINGS

Wojciech Kapturkiewicz and Andriy Burbelko

AGH University of Science and Technology, 23 Reymonta Str., 30-059 Krakow, Poland

Keywords: modeling, ductile iron, solidification, thin walled casting

Abstract

Solidification model and numerical calculations are presented describing the solidification of a thin wall ductile iron with hypo-, hyper- and eutectic composition. The principal assumptions of the kinetic nature of growth, depending on undercooling in respect of the equilibrium lines, have been adopted, disregarding the diffusion processes, which was justified by the rapid course of the crystallization process in a thin-walled casting. This kinetic model was operating in a correct mode when it was completed with adjusted calculations of the carbon amount diffusing through the austenite film around the graphite nodules. The applied model of diffusion determined jointly with the kinetic model of the growth of graphite and austenite resulted in high-speed calculation program. Quite interesting are the results showing distinct differences in the kinetics of solidification and final structure of the cast iron with the same degree of eutectic saturation, but different content of C and Si.

Introduction

Nodular graphite cast iron, also known as ductile iron, ductile cast iron, nodular graphite iron and spheroidal graphite iron (SGI), has major applications in critical engineering parts due to its excellent properties and castabillity. To predict the mechanical properties of nodular graphite iron it is necessary to simulate the refinement and volume fraction of the individual structural constituents present in this cast iron.

Most of the computer modeling programs described in literature is devoted to eutectic transformation [1 - 8] under the pre-assumed stationary conditions of carbon diffusion in austenite. In [6,7] a physical model of solidification of the nodular graphite cast iron which quantitatively accounts for the formation of non-eutectic austenite during cooling and solidification of hypereutectic as well as hypoeutectic cast iron has been presented. In investigations described in [8], process modeling techniques have been applied to describe the multiple phase changes occurring during solidification and subsequent cooling of near-eutectic nodular graphite cast iron, based on the internal state variable approach.

According to [9] at the eutectic temperature, austenite dendrites and graphite spheroids nucleate independently in the liquid. This mechanism has been confirmed in the experiments [10,11] and modeling [12, 13] by both for hypo-eutectic and eutectic, as well as hyper-eutectic SGI. The non-stationary diffusion in nodular graphite iron casting has been presented in [14].

In [15, 16] some formulae have been introduced for uni-nodular models assuming that a basic unit of solidification is formed by a graphite nodule and austenite shell covering this nodule, whereas multi-nodular models assume that each unit of solidification is formed by a grain of dendritic austenite containing several graphite spheroids.

In [17], a simple kinetic, non-diffusional model of the SGI solidification was presented, using knowledge available so far for near-eutectic composition, confronted with experiments [18,19] regarding both the cooling curves and graphite nodule count in a real thin-walled casting. In the present study, this model and the results were extended to include the ductile iron of hypo- and hypereutectic composition. A model of the formation of eutectic grains was also modified taking into account the nucleation and growth of primary graphite crystals in the liquid.

In calculations, the equations and physicochemical parameters of the alloy as well as the boundary conditions of the process according to previous work [17] were used, taking into account the following analysis regarding the partition coefficient for carbon.

Process Model Description

The aim of the present study was to develop a rapid and flexible simulation programme to reproduce the solidification process of cast iron with nodular graphite, occurring at the micro level in a thin-walled casting. The model assumes a flat shape of the wall with unidirectional heat flow. The conditions of cooling occurring in a casting of this type allow the following assumptions to be adopted:

— heat flux from the casting to a mould is described by Chvorinov dependence, taking into account the temperature of the casting surface, the initial mould temperature and the heat diffusivity coefficient [$Ws^{1/2}/(m^2K)$] of the moulding material;

— the diffusion mechanism of the grain or dendrites growth is ignored. It is assumed that it has a kinetic character, and the growth rate of the solidified phase is directly dependent on undercooling relative to the equilibrium liquidus temperature of the austenite and graphite. The equilibrium temperature depends on current chemical composition of the liquid phase, allowing for changes in the content of C and Si in the liquid. At this stage of the study, the effect of phosphorus has been neglected;

— change in the composition of the liquid phase is calculated using partition coefficients and basing on the mass balance of components. A constant value of the partition coefficient of silicon was adopted. Based on the analysis of test results [20] it has been concluded that the value of the partition coefficient of carbon k_C is strictly related with the silicon content in alloy. Using the available data [20], a functional relationship between k_C and silicon content in the alloy was determined;

— below the equilibrium temperature of graphite, crystals of free graphite nucleate and grow in the liquid. Graphite nucleation is interrupted after reaching the maximum undercooling relative to the equilibrium temperature of graphite. In case of exceeding this maximum, in a further stage of the crystallization process, graphite can continue its nucleation until the next maximum of this undercooling is reached. If actual temperature falls below the equilibrium temperature of the austenite, the crystals of graphite are coated with a film of austenite to form spherical grains of the graphite eutectic;

— further growth of the grains of the graphite eutectic depends on the undercooling relative to the equilibrium temperature of the austenite with carbon diffusing through the austenite film. A stable process of diffusion is assumed with a dependence on the current limit of carbon content on the graphite / austenite and austenite / liquid interface and on the thickness of the austenite film, where the driving force is the difference between the values of the equilibrium concentration of carbon in austenite in contact with the liquid phase and graphite (according to the revised model [5]);

— in both hypoeutectic alloys, in the case of temperature drop below the liquidus line of austenite, and in hypereutectic alloys, in the case of temperature drop below the liquidus line, the crystallization of off-eutectic austenite is admitted. The nucleation of dendrites of the off-eutectic austenite is interrupted upon exceeding the maximum local undercooling of the alloy;

According to [1, 6], balance on the graphite-liquid interface has the following form:

$$\rho_{gr}\left(C_{gr/\gamma}-C_{\gamma/gr}\right)\frac{dr_{gr}}{d\tau}=D_{\gamma}\rho_{\gamma}\left.\frac{dC_{\gamma}}{dr}\right|_{r^{gr}} \tag{1}$$

where $C_{gr/\gamma}$, $C_{\gamma/gr}$ – carbon content at the graphite/austenite and austenite/graphite interface, respectively,

ρ_{gr}, ρ_{γ} – the density of graphite and austenite, respectively,

r_{gr} – radius of the graphite nodule,

D_{γ} – carbon diffusion coefficient in austenite.

In our opinion, the concentration on both sides of the interface ($C_{gr/\gamma}$, $C_{\gamma/gr}$ - left side of the equation) should be linked to the corresponding densities:

$$\left(C_{gr/\gamma}\rho_{gr}-C_{\gamma/gr}\rho_{\gamma}\right)\frac{dr_{gr}}{d\tau}=D_{\gamma}\rho_{\gamma}\left.\frac{dC_{\gamma}}{dr}\right|_{r^{gr}} \tag{2}$$

As a result, the differential growth of the radius of graphite (surrounded by a layer of austenite) has slightly different form as compared to equation (14) [6])

$$\Delta r_{gr}=\Delta\tau D_{\gamma}\frac{r_{\gamma}}{\left(r_{\gamma}-r_{gr}\right)\cdot r_{gr}}\cdot\frac{C_{\gamma/L}-C_{\gamma/gr}}{\frac{\rho_{gr}}{\rho_{\gamma}}-C_{\gamma/gr}}, \tag{3}$$

and this equation we have used in the calculations.

The addition of Si to Fe-C alloy increases carbon activity in the liquid and solid solution. This results in a shift of both the liquidus and solidus lines of austenite in the direction of lower carbon content and change in the equilibrium distribution of carbon between the liquid and austenite. Based on the available data [20, Fig. 4 (d) - (g)], within the range of the Si content from 0 to 5.2 wt%, this dependence can be described with a linear equation:

$$k_C = 0.49 - 0.03 * C_{Si} \tag{4}$$

where C_{Si} - Si concentration in weight% .

Results and Discussion

In calculations, the equations and physicochemical parameters of the alloy as well as the boundary conditions of the process according to previous work [17] were used, taking into account the following analysis regarding the partition coefficient for carbon.

The diffusion growth mechanism on which the earlier studies are based, in the case of high-speed process of casting thin-walled parts may not satisfy, due to the high diffusion resistance and a short process duration, the conditions of crystal growth (as demonstrated by own attempts at modeling). Neither does it seem to satisfy the adopted assumptions of a "fast and flexible" program.

The results of simulation computations were carried out for a flat, 3 mm thick, casting cooling in a ceramic mold. The results are qualitatively similar to the experiments known and described in the literature. A quantitative comparison with the experiments will be provided in further stage of the studies.

The aim of the calculations was to plot for a flat, 3 mm thick, casting the cooling curves, determine the graphite nodule count, the graphite nodule size, and the amount of crystallized off-eutectic austenite. Another aim was to predict the occurrence risk of hard spots (cementite eutectic) in casting depending on the saturation coefficient S_c and Si content in the Fe-C alloy.

Figure 1a shows a sample graph plotted for the central part of casting, including a cooling curve with the course of the equilibrium liquidus temperature of austenite T_γ, the equilibrium temperature of graphite T_{gr}, the equilibrium temperature of eutectic transformation Te, and the equilibrium temperature of metastable eutectic T_{mst}.

Fig. 1b shows for these curves the course of undercooling in respect of ΔT_γ, ΔT_{gr} and ΔT_{mst}. Similar curves for a fragment near the casting surface (a layer 0.15 mm thick) - Figs. 2 a and b - indicate small differences compared to the course of curves plotted for the center of the casting. Very characteristic is the course of the undercooling curves ΔT_{gr} (Figs. 1b and 2b), indicating that, practically, the ΔT_{gr} value remains negative over the whole process, which means that there is no threat that the hard spots may occur in casting. A minor threat does occur at the surface of the casting (Figs. 2a and 2b), when the minimum temperature appears on the cooling curve within the time interval of 2 - 2.5 s.

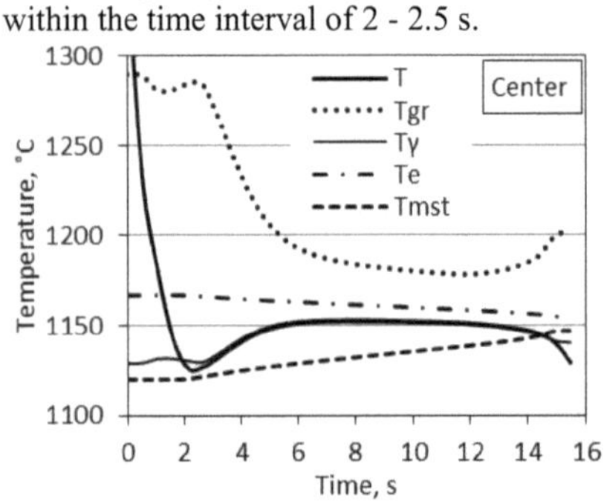

a)

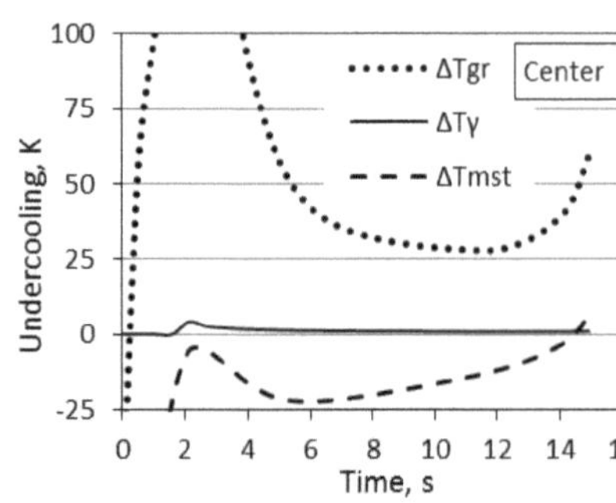

b)

Fig. 1. Cooling curve T, liquidus equilibrium temperature of austenite T_γ, equilibrium liquidus temperature of graphite T_{gr}, temperature of stable eutectic T_e and temperature of metastable eutectic T_{mst} (a); undercooling ΔT_{gr}, ΔT_γ and ΔT_{mst} for center of 3 mm plate casting till end of solidification (b).

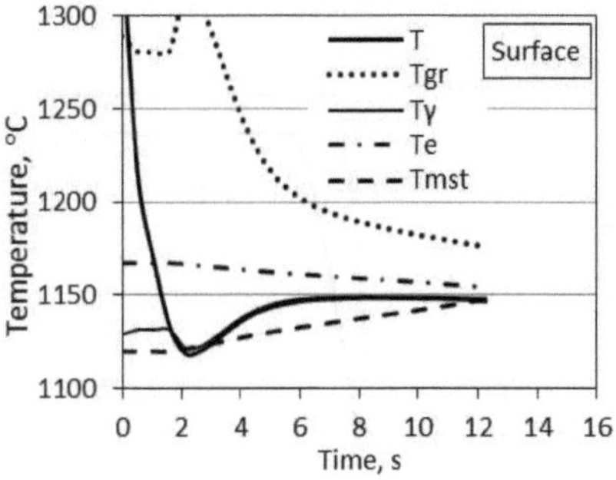

a)

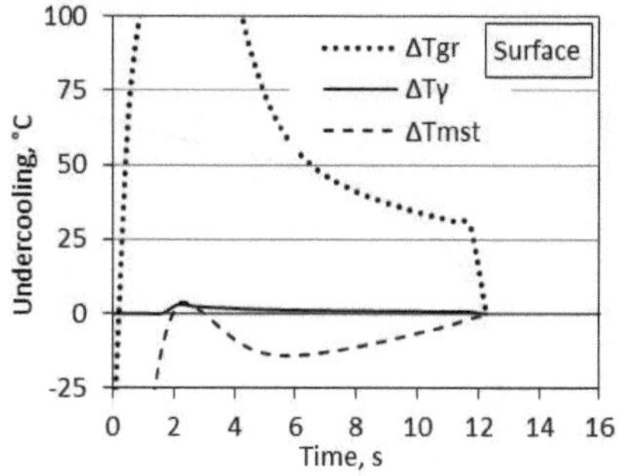

b)

Fig. 2. Cooling curve T, liquidus equilibrium temperature of austenite T_γ, equilibrium liquidus temperature of graphite T_{gr}, temperature of stable eutectic T_e and temperature of metastable eutectic T_{mst} (a); undercooling ΔT_{gr}, ΔT_γ and ΔT_{mst} near the surface (an element 0.15 mm thick) of 3 mm plate casting till the end of solidification (b).

The graph of the temperature field in the cross-section of the casting - Fig. 3 - shows a small temperature gradient in the thickness of the thin-walled casting, which is causing the above-mentioned small differences visible between Figs. 1a, 2a and 1b and 2b, respectively. It is typical that in the range of recalescence (curve 1 in Fig. 3), the temperature in the entire cross-section of the casting is lower than in further stages of the casting cooling (curves 2 - 4) with the exception of the temperature at the end of solidification (curve 5).

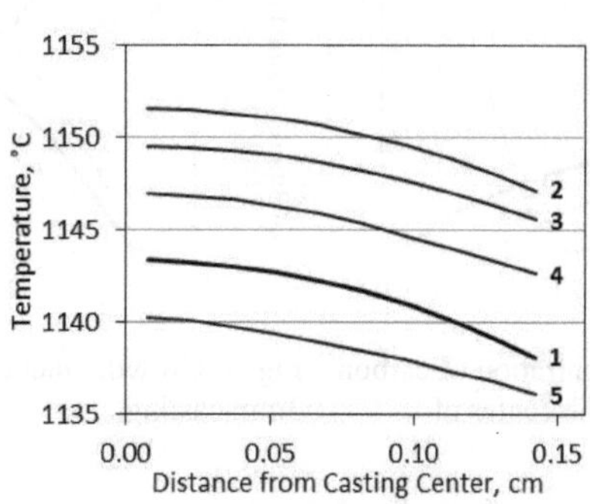

Fig. 3. Temperature distribution in the section of the 3 mm plate casting (cooling curve Fig.1 and Fig. 2). Number on the curves: (1) - 4.3 s, (2) – 6.4 s, (3) – 12.8 s, (4) – 13.9 s, (5) – 14.9 s of process time.

Figure 4 shows the kinetics of graphite growth in the center of the casting and in a fragment near the surface (a layer 0.15 mm thick). Similarly, Fig. 5 shows the growth kinetics of off-eutectic austenite (austenite dendrites) in the alloy with a near-eutectic composition (actually slightly hypereutectic; S_c = 1.003) for both casting center and the subsurface layer. The result confirms the literature data stating the presence of primary austenite in alloy with a eutectic or hypereutectic composition.

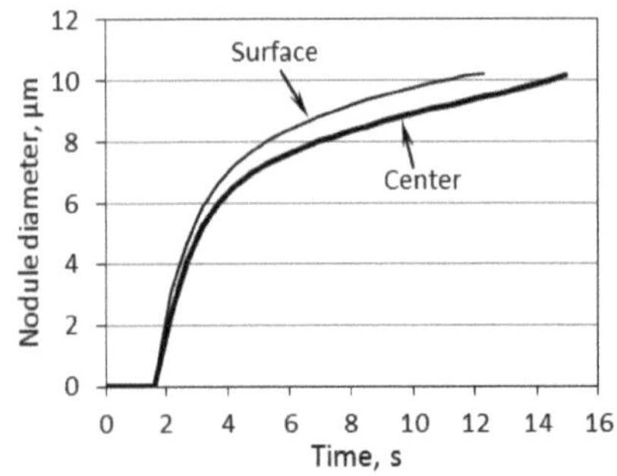

Fig. 4. Kinetics of graphite nodules growth in the center of casting and in its fragment near the surface (an element 0.15 mm thick).

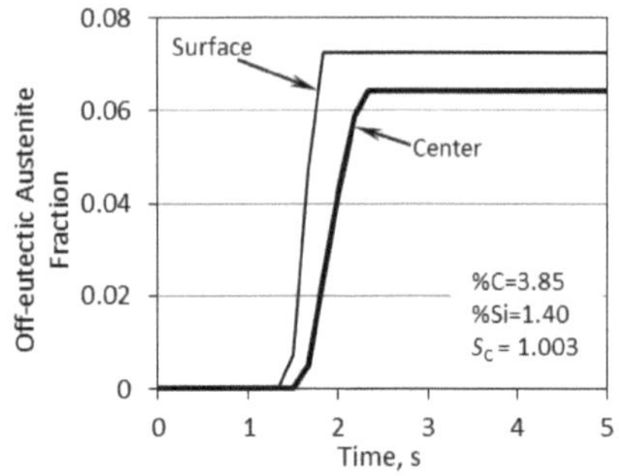

Fig. 5. Kinetics of off-eutectic austenite growth for the near eutectic composition in the center of casting and near the surface (an element 0.15 mm thick).

In a similar way, changes in the concentration of carbon and silicon in the liquid (Fig. 6), and the growth kinetics of solidified phase (Fig. 7) were depicted.

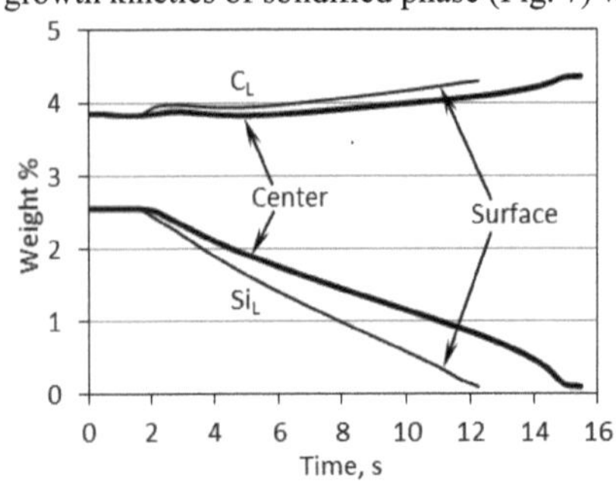

Fig. 6. Changes in the concentration of carbon and silicon in the liquid in the center of casting and near the surface.

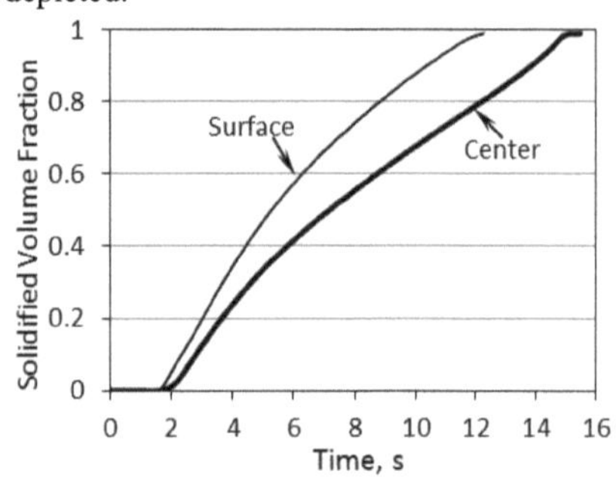

Fig. 7. Growth kinetics of solidified phase in 3mm casting.

Figures 8 a, b show the effect of saturation degree on the graphite nodule count and on the size of graphite nodules. Figs. 9a and 9b show the effect of saturation degree on the amount of the crystallized off-eutectic austenite (a), and on the metastable undercooling (b).

Figures 10 a, b and 11 a, b show the effect of the Si content in the alloy on the aforementioned phenomena. The results relate to the center of the casting.

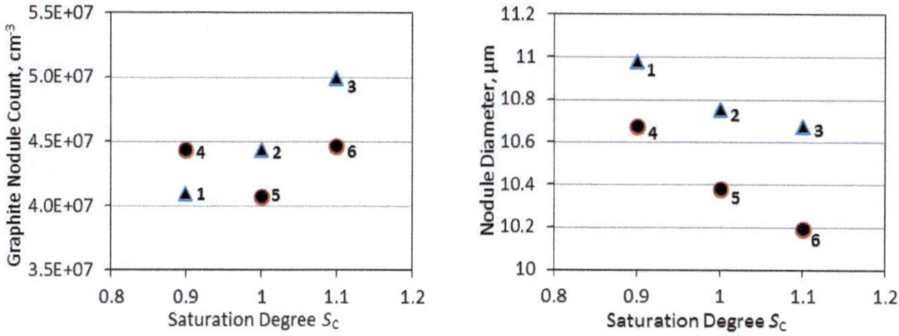

a) b)

Fig. 8. Effect of saturation degree on graphite nodule count (a) and on size of graphite nodules (b). 1 – 3.85%C, 0%Si; 2 – 4.26%C, 0%Si; 3 – 4.7%C, 0%Si; 4 – 3.46%C, 1.4%Si; 5 – 3.6%C, 2.2%Si; 6 – 3.85%C, 2.55%Si. The same connotations near points in Fig. 8 relate to Fig. 9.

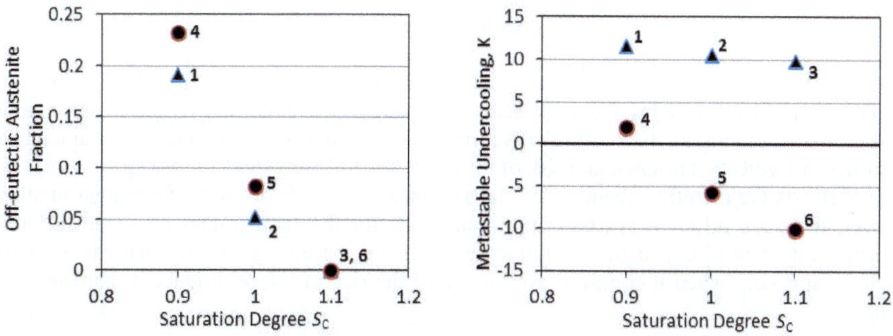

a) b)

Fig. 9. Effect of saturation degree on the amount of the crystallized off-eutectic austenite (a), and on the metastable undercooling (b).

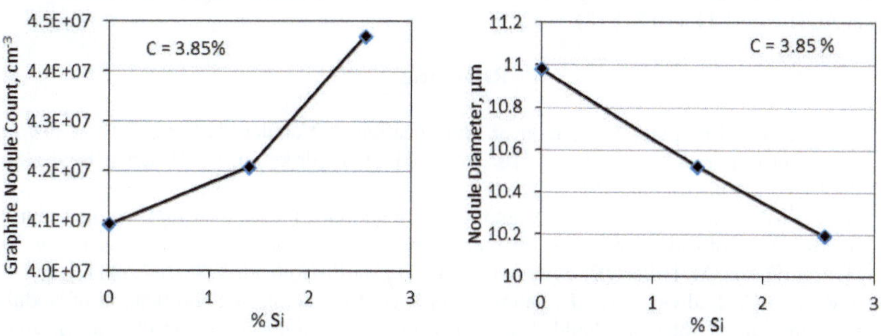

a) b)

Fig. 10. Effect of the Si content in the alloy on the graphite nodule count (a) and on the size of graphite nodules (b).

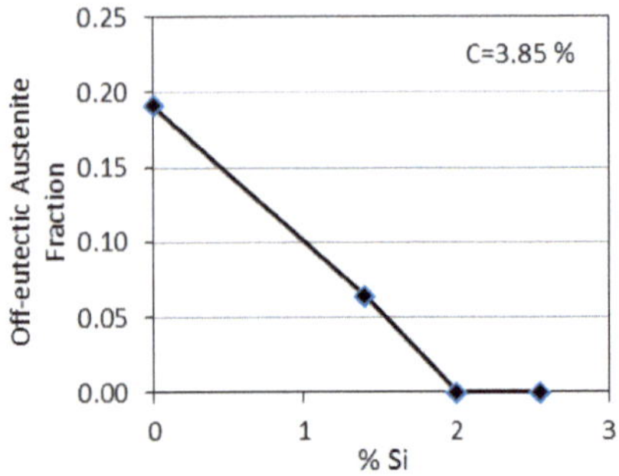

a)

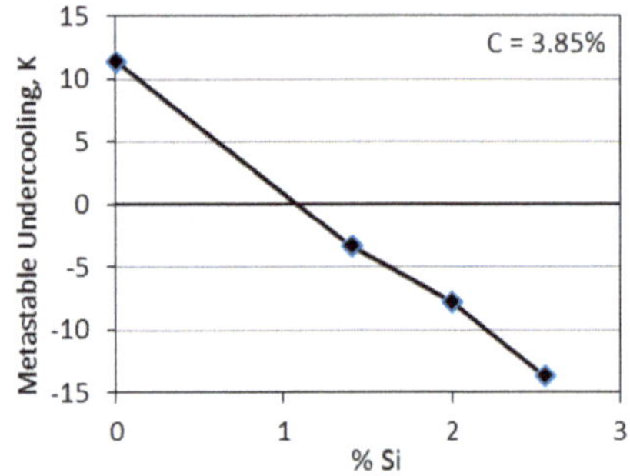

b)

Fig. 11. Effect of the Si content in the alloy on the amount of the crystallized off-eutectic austenite (a), and on the metastable undercooling (b).

Conclusions

1. The developed model, based on kinetic nucleation and growth of crystals, allows obtaining quickly such results describing the solidification of ductile iron under the conditions of cooling of a thin-walled casting as the cooling curve, nodule count and nodule diameter, the amount of crystallized off-eutectic austenite and risk of the occurrence of hard spots in casting.
2. These effects are directly dependent on the saturation degree S_c or on the Si content in alloy. However, there are different crystallization conditions for the same values of S_c but different proportional content of C and Si. Saturation degree is not the only parameter that characterizes the alloy, especially when it comes to assessment of the risk of the occurrence of hard spots in casting.

Acknowledgements

The present study was supported by the Polish Ministry of Science and Higher Education. Project AGH No 11.11.170.318 (8).

References

1. K. Su et al., "Computer Simulation of Solidification of Nodular Cast Iron," *Proc. 3rd Int. Symp. on Physical Metallurgy of Cast Iron,* ed. H. Fredriksson, M. Hillert (Elsevier Sci. Publ. Co., North-Holland, 1985), 181-189.
2 H. Fredriksson, I. Svensson: "Computer Simulation of the Structure formed during Solidification of Cast Iron," *Proc. 3rd Int. Symp. on Physical Metallurgy of Cast Iron,* ed. H. Fredriksson, M. Hillert (Elsevier Sci. Publ. Co., North-Holland, 1985), 273-284.
3. S.M. Yoo, A. Ludwig, and P.R. Sahm, "Numerical Simulation of Solidification of Nodular Cast Iron in Permanent Moulds," *Solidification Processing*, ed. J. Beech and H. Jones (Ranmoor House, Univ. of Sheffield, 1997), 494-497.
4. S. Chang, D. Shangguan, and D.M. Stefanescu, "Modeling of the Liquid/solid and Eutectoid Phase Transformation in Spheroidal Cast Iron," *Metall. Trans. A*, 23A (1992), 1333-1346.

5. T. Skaland, O. Grong and T. Grong, "A Model for the Graphite Formation in Ductile Cast-Iron. Solid-State Transformation Reactions," *Metal. Trans. A*, 24A (1993), 2347-2353.
6. G. Lesoult, M. Castro, and J. Lacaze, "Solidification of Spheroidal Graphite Cast Irons – I. Physical Modelling," *Acta Mater.,* 46 (3) (1998), 983-995.
7. J. Lacaze, M. Castro and G. Lesoult, "Solidification of Spheroidal Graphite Cast Irons – II. Numerical Modeling," *Acta Mater.,* 46 (3) (1998), 997-1010.
8. M.I. Onsoien et al., "A process Model for the Microstructure Evolution in Ductile Cast Iron. Part I. The Model," and "Part II: Application of the Model," *Metall. Mater. Trans. A*, 30A (1999), 1053-1068, and 1069-1078.
9. D.M. Stefanescu and D.K. Bandyopadhyay, "On the Solidification Kinetics of Spheroidal Graphite Cast Iron," *Proc. 3rd Int. Symp. on "Metallurgy of Cast Iron"*, ed. G. Ohira, T. Kusakawa, and E. Niyama (MRS, Tokyo, Japan, 1990), 15-26.
10. D.K. Banerjee and D.M. Stefanescu, "Structural Transitions and Solidification Kinetics of SG Cast Iron During Directional Solidification Experiments," *AFS Trans.,* 99(1991), 747-759.
11. G.L. Rivera, R. Boeri and J. Sikora, "Revealing the Solidification Structure of Nodular Iron," *Int. J. Cast Metals Res.,* 8 (1995), 1-5.
12. A. Burbelko et al., "Simulation of the Ductile Iron Solidification Using a Cellular Automaton," *Key Eng. Mat.*, 457 (2011), 330-336.
13. K.M. Pedersen, J.H. Hattel, N. Tiedje, "Numerical Modelling of Thin-walled Hypereutectic Ductile Cast Iron Parts," *Acta Materialia*, 54 (2006), 5103-5114.
14. E. Fraś et al, "Modeling of Graphitization Kinetics in Nodular Cast Iron Casting," *Modeling of Casting, Welding and Advanced Solidification Processes – IX*, ed. P.R. Sahm et al., (Aachen, Shaker, 2000), 885-892.
15. D.J. Celentano et al., "Computational Simulation of Microstructure Evolution During Solidification of Ductile Iron," *Int. J. Cast Met. Res.* Vol. 21 (6) (2008), 416-426.
16. P.M. Dardati et al. "Analysis of Ductile Cast Iron Solidification: Numerical Simulation and Experimental Validation," *Int. J. Cast Met. Res.,* 22 (5) (2009), 390-400.
17. W. Kapturkiewicz, A. Burbelko, M. Górny, "Undercooling, cooling curves and nodule count for near-eutectic thin-walled ductile iron castings," *ISIJ Int.,* 54 (2) (2014), 288-293.
18. E. Fras et al., "Eutectic cell and nodule count in cast irons," *Int. J. Cast Met. Res.* Vol. 20, 2007, 233-239.
19. K.M. Pedersen, N. Tiedje, "Undercooling and nodule count in thin walled ductile iron castings," *Int. J. Cast Met. Res.*, 20 (3) (2007), 145-150.
20. D.M. Stefanescu, "Thermodynamic Properties of Iron-Base Alloys", *Metals Handbook*, vol.15, Casting, D.M. Stefanescu et al. eds., (Metals Park, OH, 1998), 61-70.

Advances in the Science and Engineering of Casting Solidification
Edited by: Laurentiu Nastac, Baicheng Liu, Hasse Fredriksson, Jacques Lacaze, Chun-Pyo Hong, Adrian V. Catalina, Andreas Buhrig-Polaczek, Charles Monroe, Adrian S. Sabau, Roxana Elena Ligia Ruxanda, Alan Luo, Subhayu Sen, and Attila Diószegi
TMS (The Minerals, Metals & Materials Society), 2015

GRAPHITE GROWTH MORPHOLOGIES IN HIGH Al CAST IRON

H. M. Muhmond, H. Fredriksson

Royal Institute of Technology (KTH), Stockholm, Sweden

Keywords: Graphite morphology, high Al cast iron, SiAl alloy, and micro-segregation.

Abstract

High Si and Al cast iron has been investigated experimentally and thermodynamically. Alloys were prepared in the laboratory with low to high Al and Si concentrations. Experiments were performed using Mg treatment of the melt. In another series of experiments, the effect of Ca, Sr and Ce on graphite morphology was investigated. In the last experiments, the effect of Ca and Ce in the absence of Mg was studied. The samples were analyzed using field emission-scanning electron microscope (FE-SEM) and energy dispersive x-ray spectroscopy (EDX). In alloys with high Al and Si, the formation of chunky graphite is crucial for control. The effects of Al, Si and inoculants were analyzed with the help of thermodynamics. It was found that the activity of oxygen changed due to the high concentrations of Al and Si, which influenced the nucleation of MgO and other oxides in the melt. The oxygen level in the melt determines the graphite morphology.

Introduction

High aluminum cast iron is useful due to its oxidation resistance, high impact strength and better wear resistance at room temperature and higher temperatures. This family of cast iron Fe-C-Al is not very new. It has been researched by Löhberg et. al (1938, 1969) [1, 2], Aleksandrov et. al [3] and Bobro [4] (1964) in Russia. Several authors [5, 6, 7, 8, 9, 10] have concluded that Al is a stronger graphitizer than Si. Aluminum segregates negatively and reduces the solubility of C in the austenite, which increases the graphitization. Al stabilizes pearlite during the eutectoid reaction.

According to Carlberg et al. [6], the eutectic temperature of the alloy increases by more than 20 °C for every percent increase in the Al content. These authors [3] performed unidirectional solidification experiments and found that with increasing Al content, the solidification front progresses from smooth to an irregular front, where the two phases do not co-operate with each other, but as the Si content increases, the front does not change. These authors [3,7] showed that the coupled zone decreases at high Al contents, which explains the lack of co-operation between the two phase regions. Several authors [3, 7] found that Al promotes coarse flake graphite in the absence of Mg treatment. The solidification interval also increases as the Al content increases [3]. Fe-C-Al alloys have a high consumption rate of Mg.

Experimental methods

Cast iron alloys with variable Si and Al compositions were prepared in the laboratory. The compositions and the detail about the additives is provided in Table 1. The S content was below 0.005 % in all of the samples except SiAl-2. All samples were melted in an argon atmosphere using a high frequency induction furnace. The crucibles were placed inside a graphite susceptor protected by an alumina tube. The temperature was monitored using an S-type thermocouple, placed inside a thin-walled alumina tube with one end closed, which was also used to stir the melt while mixing the alloy and treating the melt with different additives, such as FSM (FeSiMg: 5 % Mg, 0.3 to 0.7 % Ca, 47 % Si and balance Fe), and commercially available inoculants which we named as CCA1 (0.75-1.25 % Al, 0.75-1.25 % Ca, 1.5-2 % Ce, 70-76 % Si and Fe balance) and CCA2 (same composition as CCA1 but CCA2 surface is coated with about <1 % O/S). All of the additions to the melt were performed between 1425 °C and 1400 °C.

Table 1. Experiments and composition of the samples. (C.E= %C + %Si/3 + 0.125 % Al)

Sample	Final Composition					FSM	Other additives
	C	Si	Al	Fe	C.E	%	
SiAl-1	3.60	2.30	3.25	90.43	4.77	----	----
SiAl-2	3.60	2.30	3.25	90.43	4.77	----	0.05 % S
SiAl-3	3.22	5.05	----	91.62	4.90	2.50	----
SiAl-4	3.22	4.56	0.49	91.61	4.80	2.50	----
SiAl-5	3.22	4.56	1.95	90.15	4.98	2.50	----
SiAl-6	3.22	4.56	3.41	88.69	5.17	2.50	----
SiAl-7	3.70	2.61	3.41	90.15	5.00	2.50	----
SiAl-8	3.70	1.15	3.90	91.13	4.57	2.50	----
SiAl-9	3.22	4.56	3.41	88.69	5.17	2.50	1 % Ca
SiAl-10	3.20	4.31	3.40	88.02	5.06	2.00	0.5 % Mg 0.5 % Ca
SiAl-11	3.20	4.31	3.40	88.02	5.06	2.00	1 % Sr 1 % Ca
SiAl-12	3.49	3.87	3.15	89.02	5.17	2.00	1 % CCA2 0.5 % Mg
SiAl-13	3.49	4.36	3.15	88.47	5.34	----	3 % CCA1 0.2 % Ca
SiAl-14	3.34	3.48	3.38	89.80	4.92	----	5 % CCA1

In the samples SiAl-3 to 8, the aluminum was added after the addition of Mg in an attempt to let the Mg dissolve effectively in a slag-free melt. The samples were allowed to cool slowly (~50 °C/min) in an argon atmosphere. The samples were cross sectioned perpendicularly. The final polishing was performed with 1 micron diamond suspension. All samples were etched using 4 % Nital. Some samples were deep etched using 5 % brome-methanol solution for approximately 15 minutes at room temperature to analyze the morphology of the graphite. This type of etching was performed in the presence of a rotating magnet in the solution at room temperature, which enhances the dissolution rate of the desired phase. Finally, color etching was performed using Klemm I for approximately 2 min at approximately 30 °C.

Results

The untreated SiAl sample (SiAl-1) resulted in graphite flakes, uniformly distributed in the matrix. By increasing the sulfur content to 0.05 %, thick graphite flakes appeared in the microstructure. In both samples, the graphite flakes were heavily branched as shown in the deep etched sample in Fig. 1 (SiAl-1, 2). In the following description, we have defined compacted graphite non nodular with thick sections and shorter lengths, vermicular graphite as flake like thin and long graphite with irregular growth and chunky graphite are those which have rounded ends and grow in branches to form a complex large network. The Mg-treated samples without Al but with high Si (SiAl-3, Fig. 1) resulted in a compacted and a small fraction of vermicular graphite morphology with low nodularity. After a slight increase in the Al content (0.49 %), the fraction of nodules reduced (SiAl-4, Fig. 1). After increasing the Al content to 1.95 % (SiAl-5), chunky graphite was mostly observed with a small fraction of vermicular graphite and compacted type of graphite. From the deep etched samples, one can see that after certain growth, the primary nodules growth continued with a complex morphology as shown in Fig. 1 (SiAl-5). The nodules nucleated first, and then, the vermicular graphite formed. The chunky graphite formed at the end of the eutectic reaction. At higher level of both Al and Si (SiAl-6, Fig. 1), three types of primary graphite were formed. There was star-like graphite, nodular graphite with branches growing tangentially from it and normal nodular graphite. Fine, chunky graphite was formed during the eutectic reaction in this sample.

At higher Al contents but with lower Si contents (SiAl-7, Fig. 1), primary nodular graphite and a large number of branched nodules from the primary graphite formed. The eutectic graphite was once again of the chunky type, compacted graphite and vermicular graphite. In the lower part of the sample between the dendrite arms, only chunky graphite was present, which precipitated in a network as shown in the deep etched sample in Fig. 1 (SiAl-7). In another sample with a high Al content but a lower Si content (SiAl-8), large compacted graphite formed with chunky and vermicular graphite. A small fraction of nodular graphite was also present, which seems to have formed prior to the compacted graphite. The chunky and vermicular graphite are formed during the last stages of the eutectic solidification. The brighter dendritic areas are due to the etching effect of the Al rich austenite.

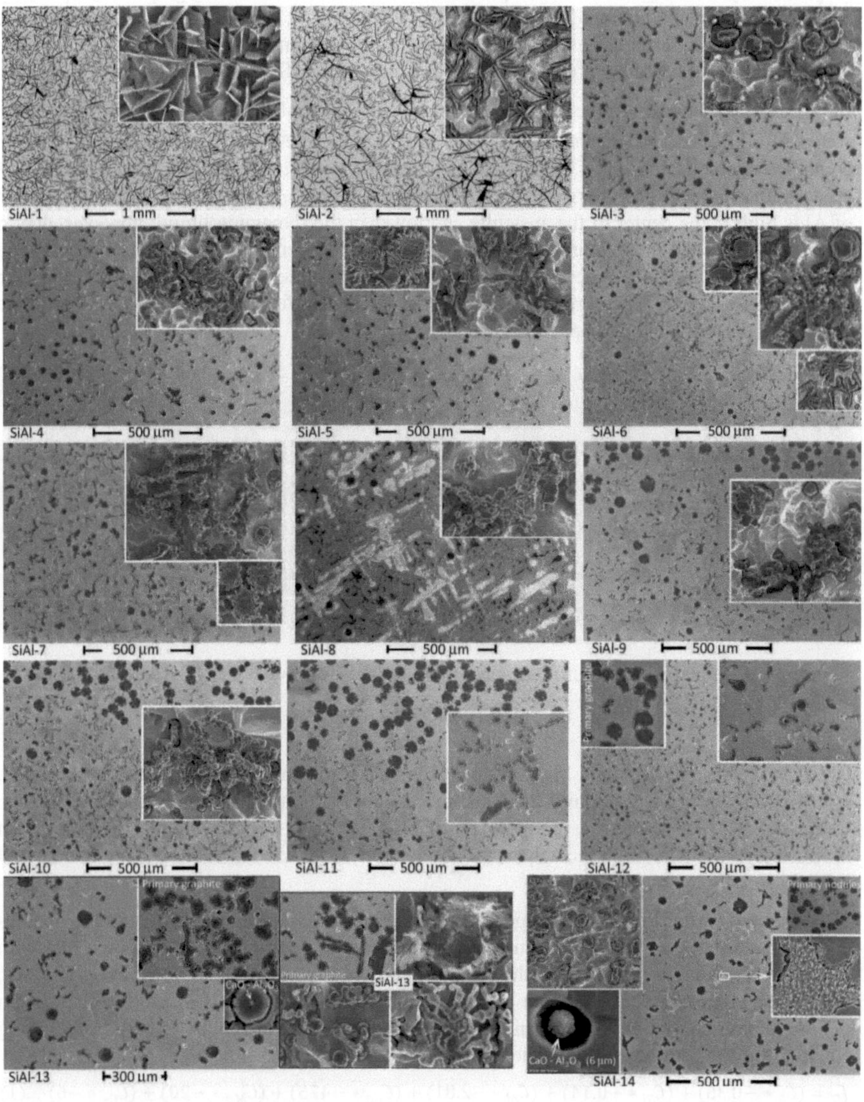

Figure 1. SEM images of all samples.

In another series of experiments, the samples were treated with pure Ca together with FSM treatment. The result was primary graphite nodules, but the chunky vermicular graphite could not be eliminated (Fig. 1, SiAl-9). By adding a small amount of pure Mg together with FSM, a large fraction of fine chunky graphite resulted, as shown in Fig. 1 (SiAl-10). Sr did not create any differences since it is difficult to add it to the melt in pure form because it is a highly volatile substance (Fig. 1, SiAl-11). Differences were observed in the graphite structure after the addition of the CCA2 inoculant, which contains Ca and Ce (Fig. 1, SiAl-12). However, fine vermicular graphite could still be observed, but the nodule count was better than in the previous samples. At 3 % CCA1 inoculants addition, there were mostly nodular and compacted type of graphite, only some regions in the sample contained flake like and chunky graphite morphology. The nodularity was increased by adding the CCA1 inoculant, which also contains Ca and Ce but no oxygen. Different type of carbides were formed in sample SiAl-14, which might have affected the nodule count and nodularity, as shown in Fig. 1 (SiAl-14).

EDX analysis of the small inclusions in the matrices of samples 10 and 14 showed that there were only AlN particles in the matrix with small traces of Al_2O_3 (Fig. 2). Some particles contained a small amount of Mg, which must be an oxide. The color etching of SiAl-14 revealed the segregation pattern of Al and Si as shown in Fig. 3. It was observed that the content of Al is high in the center of the austenite and lower in the remaining liquid; however, due to the high Al content, the silicon segregates positively in the remaining liquid. The silicon-rich areas are shown in blue, while the Al-rich areas appear brighter; the yellowish areas show the intermediate regions. These results were confirmed by performing EDX analysis of some selected spots with different appearances.

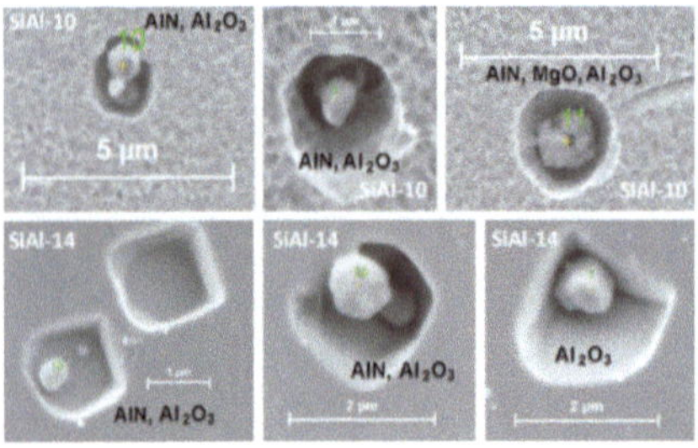

Figure 2: EDX analysis of micro particles

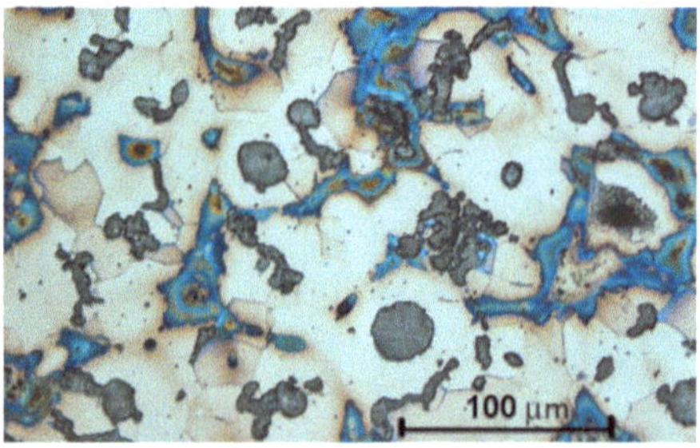

Figure 3: Color etching of sample SiAl-14

Discussion

It was very difficult to obtain higher nodularity and nodule count in any of the samples. Al, which is known to be a graphite flake stabilizer, causes the nodularization to be ineffective. Using the Mg treatment, the graphite morphology tended towards chunky graphite with increasing Al content. With the help of thermodynamics, we will try to understand this type of behavior.

It was shown earlier that a low oxygen content is very important to the formation of nodules and MgO particles are needed to nucleate graphite nodules. In Fig. 4, it is shown that with increasing Al content in the alloy, the equilibrium between oxygen and deoxidizers (Si, Mg, Ca, Ce etc.) trends towards a higher concentration of oxygen, which is due to the decrease in the interaction coefficient of oxygen with the mentioned deoxidizers by using the following equation for the interaction co-efficient of oxygen in the melt.

$$f_O = (C_C * -0.35) + (C_{Si} * -0.14) + (C_{Al} * -2.81) + (C_{Ca} * -475) + (C_{Mg} * -20) + (C_{Ce} * -6) \quad (1)$$

In eq. (1), interaction co-efficient is largely influenced by the concentration of Al in the melt. This change in activity of oxygen decreases with increasing Si content in the melt compared to Al, as shown in Fig. 4, for the formation of MgO and Ce_2O_3. When both the Al and Si contents are high in the alloy (as it is in our case), then the oxygen level in the melt will remain high, then either one have to increase the amount of inoculants or chose the one which have strong de-oxidizers.

The surface energy required for the formation of stable nuclei was considered a variable with a higher value at high temperatures and a lower value at low temperatures. During cooling and solidification, elements such as O, S, Mg, Ca, Sr and Ce would be consumed by forming compounds with oxides. An assumed range was defined for Mg, Ca, Sr and Ce after the addition but before the solidification starts. This range of compositions was based on the consumption behavior of Mg, Ca, Sr and Ce. These are only rough values which provides a base for the calculation. After knowing this, then one can enter the real values and make a customized calculation. However when it comes to the consumption, we categorized Ca is the fastest consuming element followed by Ce, Sr and Mg.

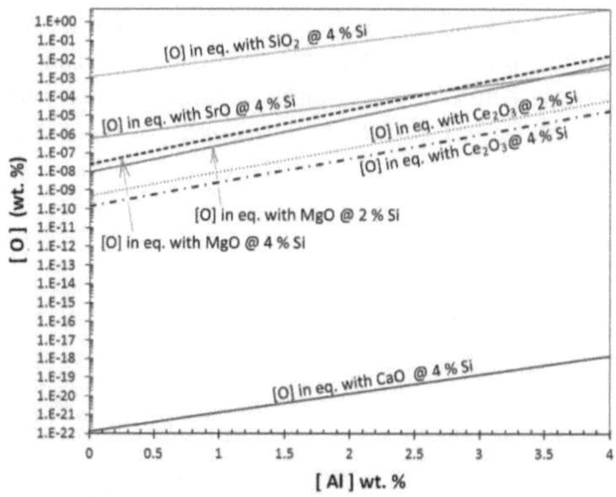

Figure 4: The effect of Al content in the melt on the equilibrium concentration of oxygen with Mg, Ce, Ca, Sr and Si. The oxygen concentration in equilibrium with MgO and Ce2O3 is shown at 4 % and 2 % Si.

The interaction coefficients depend on the composition of the remaining melt. During cooling, the interaction coefficient will remain constant and independent of the temperature, but during solidification due to the micro segregations of C, Si, Al, and other elements, the values will change according to equation 1. The micro segregation of the elements during solidification is provided in Figs. 5 and 6 and calculated using Scheil's equation and the Lever rule. During solidification (from approximately 1250 °C to 1150 °C) the remaining liquid will be enriched with O, Mg, Ca, Ce, Si and C, but the Al content will decrease in the melt. After analyzing the results from color etching, it was concluded that the partition coefficient of Al should be >1 and for Si should be <1. These changes in the composition were considered in calculating the required oxygen content for nucleation of different oxide particles in Fig. 5. Because there is a large Al content in the melt, the oxygen content in the melt should be calculated by considering an equilibrium with Al_2O_3 (Fig. 2). The nucleation of a stable Al_2O_3 particle requires a much higher oxygen content in the melt. The oxygen level required for nucleating MgO is much higher (Fig. 5) compared to alloys with low Si and Al contents.

In Fig. 6, the available oxygen concentration in the melt is shown by the equilibrium of oxygen with Al_2O_3. After adding FSM or inoculants, the oxygen concentration will change according to the type of oxides are nucleated. The nucleation line of MgO is above the available oxygen concentration until approximately 40 % of the melt is solidified, before which it will not be possible to nucleation MgO particles. Similarly, in the absence of Mg, Ce and Ca, SrO and Al_2O_3 could be nucleated when more than 50 % melt is solidified. After treating the melt with Ce, the oxygen level will drop further by forming Ce_2O_3 particles in the melt at the beginning of solidification. Later, due to enrichment of O in the melt, MgO could be formed at an approximately 0.7 fraction of solid SrO and Al_2O_3 at an approximately 0.8 solid fraction.

At a normal nitrogen level in the melt (approximately 0.002 %), it is easy to form AlN particles due to the very low equilibrium value of N with Al. The formation energy of AlN is -137 KJ/mol at 1450 °C, but the interaction coefficients are such that the nucleation of AlN is easy compared to the formation of Al_2O_3. From the EDX analysis, it was found that all of the analyzed particles were AlN, and Al_2O_3 was present at trace amounts. The oxygen level required to nucleate CaO particles ranges from 1E-14 % at 1450 °C to 4E-26 % at 1150 °C. At the current oxygen level, Ca can be completely consumed by forming CaO particles prior to solidification. The chunky graphite was almost eliminated in the last sample by the formation of Ce_2O_3 and CaO in the melt due to the low level of oxygen. In the earlier samples where Mg treatment was performed, the nodularity was not good, and the morphology of most of the graphite was of the chunky type. As a result of the high oxygen level in the melt due to the decreased activity of oxygen from the Al, the nucleation of certain oxides (SrO, MgO) would become difficult as shown by Fig. 5. It was found that with an increasing Si content in the melt, the equilibrium value of MgO with O is not strongly influenced, as shown in Fig. 4; however, the combination of both high Si and Al affects the activity of the oxygen, which makes the nucleation of MgO in the melt difficult.

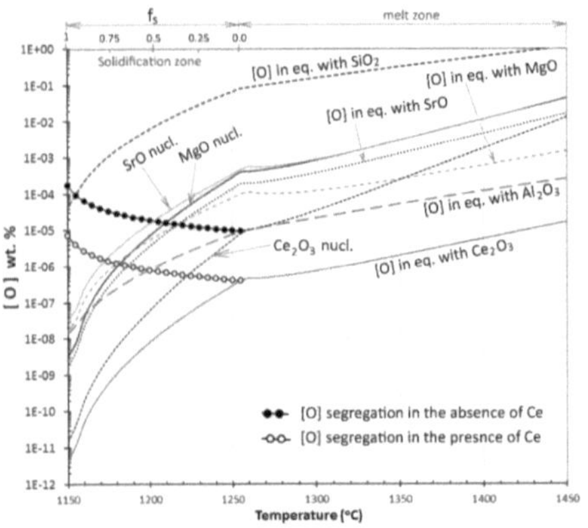

Figure 5: Nucleation of oxide particles and the initial level of oxygen prior to solidification.

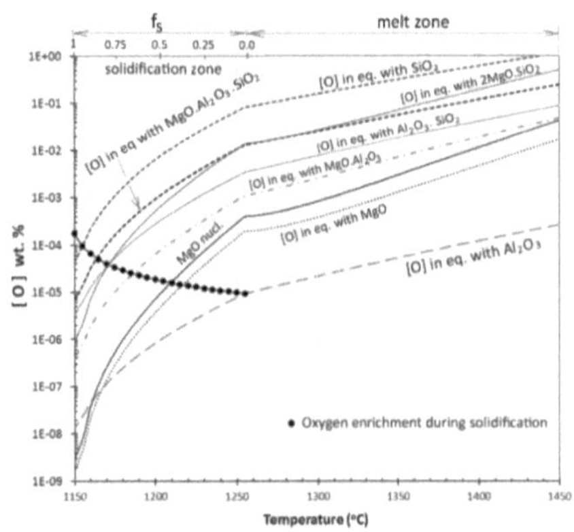

Figure 6: Equilibrium of oxygen with complex oxides, homogeneous oxides and available oxygen concentration.

The possibility of forming complex oxides, such as $MgO.Al_2O_3.SiO_2$, $Al_2O_3.SiO_2$, $MgO.SiO_2$ etc., was analyzed in Fig. 6. According to the available oxygen concentration in the melt, which is in equilibrium with Al_2O_3, it is not possible to form complex oxides prior to solidification; however, during solidification, oxygen enrichment in the remaining liquid can increase the oxygen concentration to a limit where MgO and/or complex oxides of $MgO.Al_2O_3$ could form (Fig. 6). As a result, the oxygen concentration will remain at relatively high values, which can have a strong effect on the graphite morphology. In the case of only Mg treatment, MgO can nucleate during the last stages of solidification, but from the start of solidification until the MgO is formed, vermicular and/or chunky graphite form.

Conclusion

Magnesium treatment of high Si and Al cast iron alloys is ineffective due to the difficulty of formation of MgO. To increase the nodularity and reduce the formation of chunky graphite in

SiAl alloys, one must use strong deoxidizers, such as Ca, Ce or La. The other solution would be to reduce the Al content to a level where MgO can effectively nucleate.

Appendix

The initial concentrations of the added elements will not remain constant during cooling. We assumed the reduction in the elements during cooling to be approximately 10 times. The activities of MgO, SiO_2, Al_2O_3 and complex oxides are equal to 1.

Table 2. Energy of formation of oxides and nitrides at standard state of wt. %, surface tension of particles and molar volumes.

Oxides	*ΔG°*	*Ref*	*Surface tension*	*Molar volumes*	*Initial concentrations*
	(J/mol)		1450 to 1150 °C	m^3/mol	wt. %
Al_2O_3	-1131799.8-10.46T*logT+257.2T	[12]	1 to 0.5	26E-5	Mg = 0.01
SiO_2	4.1868(142000-55T)	[12]	1 to 0.5	23E-5	Ca = 0.001
MgO	-4.1868*(181600+7.37T*logT-75.7T)	[13]	0.75 to 0.35	11E-5	Ce = 0.001
CaO	-1284906.4+214.55T	[14]	0.75 to 0.2	17E-5	Sr = 0.001
Ce_2O_3	-4.1868*(425621-66T)	[15]	0.75 to 0.2	43E-5	N = 0.002
SrO	-565000+100T	[16]	0.75 to 0.2	22E-5	
AlN	4.1868*(-77700+26.08T)	[12]	1 to 0.5	13E-5	
$2MgO.SiO_2$	-2106300+702.9T	[20]	---	---	
$MgO.Al_2O_3$	-1944430+610.2T	[20]	---	---	
$MgO.Al_2O_3.SiO_2$	-6473290+2129.2T	[20]	---	---	
$Al_2O_3.SiO_2$	-4774154+1592T	[20]	---	----	

Table 3. Interaction parameters of elements j= O, S, Ca, Mg, Ce, La, Al, Si and Sr upon elements i= C, Si, P, S, Mo, Al, O and N, for the SiAl alloy. Values marked with * are assumed based on their formation energies in comparison to other similar element for which the interaction parameter is known. During solidification, due to micro segregation of C, Si and Al, these values will change. Values provided below are from 1450 to 1250 °C. The concentrations of C, Si and Al used in these calculations are 3.5, 4.56 and 3.16 respectively.

i	e_O^i	e_{Ca}^i	e_{Mg}^i	e_{Ce}^i	e_{Al}^i	e_{Si}^i	e_{Sr}^i	e_N^i
C	-0.35 [12]	-0.097 [18]	-0.24 [20]	-0.037 [18]	0.091 [20]	0.24 [12]	-0.24 *	0.5
Si	-0.14 [12]	-0.066 [18]	-0.088 [20]	-0.2 *	0.006 [20]	0.32 [12]	-0.088 *	0.093
P	-0.07 [12]	-3.1 [18]				0.086[12]		
S	-0.133 [12]	-110 [18]	-2 *	-40 [15]	0.049 [12]	0.057 [12]	-0.5 *	
Mo	0.003 [12]							
Al	-2.81 [11]	-0.072 [19]	-0.51 [18]	-0.52 [18]	0.048 [12]	0.063 [12]		0.13
O	-0.2 [11]	-600 [19]	-430 [18]	-12.1 [18]	-1.98 [20]	-0.25 [12]	-200 *	
N	0.057 [11]		-0.5 *		-0.005 [12]			
Ca	-475 [17]	-0.002 [17]						
Ce	-6 *			0.006 [18]		0.093 [12]		
La	-2 *							
Mg	-20 *							0.5
f (j)	2.302E-5	5.87E-3	4.97E-4	2.091E-3	3.2	318.9	3.58E-2	286.2

Supersaturation of oxygen (C_O) required to nucleate a particle M_xN_y and he micro-segregation of element in the melt, was calculated by using the model in our previous article [23].

References

[1] K. Löhberg and W. Schmidt, "Die Eisenecke des Systems Eisen-Aluminum-Kohlenstoff," *Arch. Eisenhuttenwessen,* vol. 11, no. 12, pp. 607-614, 1938.

[2] K. Löhberg and A. Ueberchauer, "Die stabilen Phasengleichgewicht in der Eisenecke des Systems Eisen-Kohlenstoff-Aluminium," *Geissereiforschung,* vol. 21, no. 4, pp. 163-169, 1969.

[3] N. N. Aleksandrov and N. I. Klochnev, Technology and Properties of Heat-Resistant Cast Irons, Moscow: Mashinostroenie, 1964.

[4] Y. G. Bobro, "Aluminum Cast Irons," Charkovskogo University, 1964.

[5] W. J. Keep, M. C. F and L. D. Vorce, "Influence of Al on Iron," *American Associated for the advancement of Science,* pp. 220-231, 1988.

[6] T. Carlberg and H. Fredriksson, "Influence of Si and Al on the solidification of cast iron," in *Solidification of Casting of Metals*, London, 1979.

[7] J. Fargues and J. C. Margerie, *Fonderie-Fondeur D'Aujourd'hui,* vol. 42, pp. 23-35, 1985.

[8] S. K. Shaha, S. Dyuti, M. M. Haque and M. A. Maleque, *J. Applied Sci.,* vol. 10, no. 12, pp. 1196-1199, 2010.

[9] A. Shayesteh-Zeeraati, H. Naser-Zoshki, K.-R. A. R and M. R. Yousef-Sani, *Proc. IMechE Part L, J. Materials: Design and Applications,* vol. 224, pp. 117-122, 2010.

[10] N. Janakiev, *Giesserei-Praxis,* vol. 16, pp. 295-298, 1971.

[11] A. Ghosh and G. V. R. Murthy, *Transaction ISIJ,* vol. 26, pp. 629-637, 1986.

[12] J. F. Elliott, M. Gleiser and V. Ramakrishna, Thermochemistry for Steelmaking, vol. 2, London: Addison-Wesley, 1963, pp. 547-567.

[13] O. Kubaschewski and C. B. Alcock, Metallurgical Thermochemistry, 5 ed., England: Pergamon Press ltd., 1979, pp. 274-384.

[14] H. Li and G. Selvaduray, "Elligham Diagram Web Project," San Jose State Univeristy, 09 05 2014. [Online]. Available: http://www.engr.sjsu.edu/ellingham/ellingham_tool_p1.php. [Accessed 2014].

[15] A. Vahed and D. A. R. Kay, "Thermodynamics of rare earths in steel making," *Met. Trans. B,,* vol. 7B, pp. 375-383, 1976.

[16] A. Brandes and G. B. Brook, Smithells Metals Reference Book, 7 ed., Butterworth-Heinemann ltd, 1992.

[17] Q. Han, X. Zhang, D. Chen and P. Wang, "The calcium-phosphorous and the simultaneous calcium- oxygen and calcium-sulfur equilibria in liquid iron," *Met. Trans B,* vol. 19B, pp. 617-622, 1988.

[18] Q. Han, "Thermodynamic behavior of rare earth and alkaline earth elements in molten iron and nickel," in *Proceedings of the sixth international iron and steel congress*, Nagoya, 1990.

[19] T. Kimura and H. Suito, "Calcium de-oxidation equilibrium in liquid iron," *Met. Mat. Trans. B,* vol. 25B, pp. 33-42, 1994.

[20] J. Zhou-hua, L. I. Shuang-jiang and L. I. Yang, "Thermodynamic calculation of inclusion formation in Mg-Al-Si-O system of 430 stainless steel melts," *Journal of iron and steel research,* vol. 18, no. 2, pp. 14-17, 2011.

[21] H. Fredriksson and U. Åkerlind, Materials Processing During Casting, England: John wileys, 2005, pp. 280-291.

[22] Q. Han, X. Zhang, D. Chen and P. Wang, "The calcium-phosphorous and the simultaneous calcium- oxygen and calcium-sulfur equilibria in liquid iron," *Met. Trans B,* vol. 19B, pp. 617-622, 1988.

[23] H. M. Muhmond and H. Fredriksson, "Relationship Between Inoculants and the Morphologies of MnS and Graphite in Gray Cast Iron," *Metall. Mater. Trans. B,* vol. 44B, pp. 283-298, 2012.

Advances in the Science and Engineering of Casting Solidification
Edited by: Laurentiu Nastac, Baicheng Liu, Hasse Fredriksson, Jacques Lacaze, Chun-Pyo Hong, Adrian V. Catalina, Andreas Buhrig-Polaczek, Charles Monroe, Adrian S. Sabau, Roxana Elena Ligia Ruxanda, Alan Luo, Subhayu Sen, and Attila Diószegi
TMS (The Minerals, Metals & Materials Society), 2015

Thermophysical Properties of Thin Walled Compacted Graphite Iron Castings

Marcin Górny, Janusz Lelito, Magdalena Kawalec

AGH University of Science and Technology; Reymonta 23, 30-059 Krakow, Poland

Keywords: compacted graphite iron; thermal conductivity; thermal diffusivity; specific heat; graphite morphology; thin walled iron castings

Abstract

The thermal conductivity, diffusivity and specific heat were investigated in thin walled compacted graphite iron castings. The research was conducted for thin-walled iron castings with a 3-mm wall thickness. This study addresses the effect of cooling rate and of titanium additions on the exhibited microstructure and thermophysical parameters of thin-walled compacted graphite iron (TWCI) castings as determined by changing the molding media (silica sand and insulating sand LDASC), and Ferro Titanium. The tested material represents the occurrence of graphite in the shape of nodules, flakes (C and D types) and compacted graphite with a different shape factor and percent nodularity. Thermophysical parameters have been evaluated by the laser flash technique in a temperature range of 22-600°C. The results show that the cooling rates together with the titanium content largely influence the microstructure, graphite morphology and finally thermophysical properties of thin walled castings.

1. Introduction

Compacted Graphite Iron (CGI) also known as Vermicular Graphite Iron is a modern engineering alloy with attractive features that enable its use in the automotive industry. Good thermophysical properties of CGI are of high importance, especially in thin wall castings which are simultaneously, thermally and mechanically loaded, such as cylinder blocks, heads and brake systems [1-4].

Thermophysical properties of thin walled compacted graphite iron castings are strongly influenced by graphite fraction, its morphology, eutectic grains, and metallic matrix [5-9]. Ferrite metallic matrix has higher thermal conductivity then a pearlitic one. The thermal conductivity of graphite is strongly anisotropic and along the hexagonal planes, the conductivity is very high. At the room temperature, thermal conductivity of graphite can be as high as 500 $Wm^{-1}K^{-1}$ [5]. In the case of white cast iron, the carbon present in the form of cementite reduces thermal conductivity (about 8 $Wm^{-1}K^{-1}$).

Compacted graphite cast iron may have a complex microstructure, especially in a thin sections. This is due to the fact that the process of obtaining thin-walled castings is not simple, because it is associated with a wide range of cooling rates at the beginning of graphite eutectic solidification [10,11]. With increasing cooling rates in thin-walled CGI castings, thermal undercooling increases and graphite gradually becomes nodular, resulting in an increased nodule count and lower compact graphite ratio. Therefore, the production of thin-walled compacted iron castings is more difficult than that of thicker section iron [12]. The formation of compacted graphite in thin wall castings is thus a very difficult process with only a narrow margin of residual Mg: too much Mg will give an excess of nodules, whereas too little Mg will lead to the formation of gray iron flake structures [13]. Therefore, a typical compacted graphite cast iron has a mixed structure where 5%-30% of the graphite has a spheroidal structure. The presence of the flake graphite, which has the biggest impact on the thermophysical properties is unacceptable.

Surface-active chemical elements (oxygen, sulfur) in the liquid metal significantly affect the morphology of crystallizing graphite and thus for cast iron thermal and mechanical properties. These surface-active chemical elements may also be in the mold and then there are the conditions for the creation of degenerated graphite zone near the surface layer of cast iron, which is characterized by different properties relative to the base material. The graphite degradation in the surface layer is the most critical for thin wall castings, where it could become more than 10% of the total section. It affects also castings of thicker walls, due to the long solidification time providing an extended metal/mould interaction time.
The literature provides limited data [14] on the relation of high cooling rate, structure and thermophysical properties of thin wall compacted graphite cast iron. In this work we consider the occurrence of graphite in the shape of nodules, flakes (types C and D) and compacted graphite with a different shape factor and percent nodularity to the thermophysical properties at ambient and elevated temperatures in thin wall castings with a wall thickness of 3 mm.

2. Experimental

The experimental melts were produced in an electric induction furnace of intermediate frequency and a 15 kg capacity crucible. The furnace charge consisted of Sorelmetal, technically pure silica, Fe-Mn, and steel scrap. Melting was carried out at 1490°C, with the liquid metal held at this temperature for 2 min. This was followed by vermicularization and inoculation operations using a bell method. For the vermicularization process Fe-Si-Mg (6% Mg), as well as Fe-Ti (alloys II-IV) were used to obtain differential titanium concentrations. The Fe-Si alloy (75% Si, 0.75-1.25% Ca, 0.75-1.25% Ba, 0.75-1.25% Al) in an amount of 0.6 wt.% was used for inoculation. The chemical composition of the investigated iron is given in Table I. The pouring temperature was approximately 1400°C. Experimental castings applied in investigations were test blocks, according to the ASTM A 536-84 Standard, with plate section sizes of g=3 mm in the working part. The sand molds (alloys I, II, III, VI) were made using conventional green molding sand consisting of silica sand, bentonite (7 wt.%), a water/bentonite ratio of 0.4 and a granularity of 100-200 μm.
In addition, sand molds (alloys IV, V, VII) were made using Low-Density Alumina-Silicate Ceramic (LDASC, composition 25-45% Al_2O_3 and 55-75% SiO_2) whose heat transfer properties are drastically reduced (approx. by 4 fold) when compared with those of silica sands. Moreover, the sand is characterized by a low density 0.35-0.45 $g{\cdot}cm^{-3}$. The molds were instrumented with Pt/PtRh10 thermocouples of 0.1 mm diameter and with their tips located in the geometrical center of each mold cavity. An Agilent 34970A electronic module was employed for temperature recording.

Table I. Results of chemical composition

	Chemical composition (wt. %)										
Alloy	C	Si	Mn	P	S	Cr	Ni	Cu	Al	Ti	Mg
I	3.63	2.47	0.03	0.026	0.017	0.03	0.00	0.050	0.010	0.009	0.010
II	3.66	2.55	0.04	0.027	0.020	0.03	0.01	0.040	0.010	0.070	0.005
III	3.60	2.57	0.05	0.023	0.018	0.04	0.040	0.060	0.021	0.133	0.021
IV	3.60	2.57	0.05	0.023	0.018	0.04	0.040	0.060	0.021	0.133	0.021
V	3.62	2.43	0.04	0.030	0.014	0.02	0.030	0.010	0.020	0.008	0.014
VI	3.62	2.43	0.04	0.030	0.014	0.02	0.030	0.010	0.020	0.008	0.014
VII	3.60	2.58	0.06	0.028	0.022	0.03	0.030	0.010	0.011	0.009	0.013

A metallographic characterization was made using a Leica MEF 4M microscope and a QWin v3.5 quantitative analyzer at various magnifications in order to determine the graphite and matrix morphologies. At least five areas in the central part of the sample were used for these measurements.

In order to relate the microstructure to the thermal conductivity of cast iron (alloys from I to VII, Table I), the thermal diffusivity of cast iron samples was measured using a LFA 427 device using pulsar laser method (Laser Flash Analysis). The laser flash measurements were performed in a continuously flowing argon atmosphere to prevent oxidation using a computational model of the "Cape-Lehmann + pulse correction". At each temperature three measurements for statistical purposes were performed. The thermal conductivity was calculated by using the obtained data together with a bulk density and specific heat:

$$k(T)=a(T)\cdot c_p(T)\cdot \rho(T) \tag{1}$$

where:

$a(T)$ - thermal diffusivity [m^2s^{-1}]; $c_p(T)$ - specific heat [J $kg^{-1}K^{-1}$]; $\rho(T)$ - material density [kg m^{-3}]

The specific heat was determined using the comparative method with the use of the pulsar laser having a reference-Inconel 600 sample (with a known coefficient of thermal expansion). The measurement of the thermal diffusivity of the reference and cast iron sample was performed. On the basis of these measurements, the specific heat value was determined using the formula:

$$c_p^{sample} = \frac{T_\infty^{ref}}{T_\infty^{sample}} \cdot \frac{Q^{sample}}{Q^{ref}} \cdot \frac{V^{sample}}{V^{ref}} \cdot \frac{\rho^{ref} \cdot D^{ref}}{\rho^{sample} \cdot D^{sample}} \cdot \frac{d_{hole}^{2,sample}}{d_{hole}^{2,ref}} \cdot c_p^{ref} \tag{2}$$

where:

c_p - specific heat of sample/reference [J $kg^{-1}K^{-1}$]; T - temperature of sample/reference [K]; Q - energy absorbed by sample/reference [J]; V - amplitude of signal amplification of sample/reference; ρ - density of the sample/reference [kg m^{-3}]; D - thickness of the material [m]; d - hole diameter of measurement sample/reference [m].

Materials density is a tested temperature range (22-600°C) were determined with DIL 402C Netzsch dilatometer. The measurements were performed at a temperature rate of 5 K/min.

The thermophysical parameters were assessed at a room temperature of 22°C rising to approximately 600°C. Plate shaped samples were cut out from the centre of the working portion of each casting for the thermophyisical parameter measurements. The specimens had dimensions of 10 x 10 x 2 mm.

3. Results and Discussion

3.1 Thermal Analysis

Figure 1 shows the cooling curves obtained from an experimental investigation for thin walled castings.

On the basis of a thermal analysis, the cooling rates near the equilibrium graphite eutectic solidification temperature were determined. Molding materials with a low ability to absorb heat (e.g., LDASC sand) have a significant effect in reducing the cooling rates. Attention should be paid to when LDASC sand is used as casting molds ($a \approx 0.025$ J°C^{-1} cm^{-2} $s^{-1/2}$) instead of silica sand ($a \approx 0.10$ J°C^{-1} cm^{-2} $s^{-1/2}$), the cooling rates at the beginning of the solidification decrease from about 29.1 Ks^{-1} to 5.8 Ks^{-1}.

The influence of graphite morphology on the cooling curves is shown in the initial stage of the crystallization process (Fig. 1b). The undercooling that initializes solidification of eutectic in cast iron V is higher than in cast iron IV with a compacted graphite eutectic. It should be noted that, to obtain alloys IV and V the same amount of inoculant and vermicularizator were used. It is to ensured the same nucleation potential of the graphite from the point of view of the inoculation. Also in the case of thin-walled castings obtained in the mould based on quartz, a higher value of the maximum undercooling at the beginning of the eutectic crystallization for a higher fraction of nodular graphite was recorded (Fig. 1a). In the case of the preparation of the cast iron with compacted graphite using the by a controlled amount of inoculant method, undercooling that initializes graphite eutectic solidification can be of the same order as in the case of ductile cast iron [15] or even lower [16-17]. Graphite eutectic growth in cast iron with flake graphite (cast iron VII, Fig. 1b) requires less driving force and thus eutectic undercooling is lower. It shows on the curve that temperature of eutectic recalescence is much higher compared to those in SGI and CGI. In this way, thermal analysis provides the opportunity for the foundry process control – from the view point of nucleation potential, chilling tendency and graphite morphology.

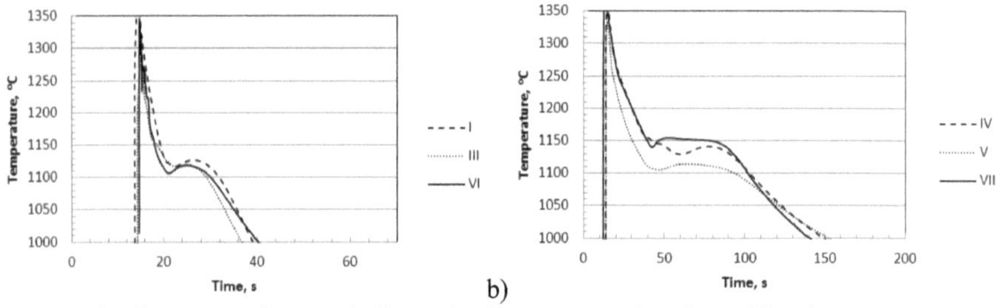

Figure 1. Cooling curves for tested alloys: a) Alloys I, III and VI from silica foundry molds, b) Alloys IV, V and VII from LDASC foundry molds.

3.2. Microstructure

The results of metallographic analysis are shown in the Table II. Figures 2 show the exhibited microstructures found in thin walled castings from alloys I-VII.

Metallographic examinations revealed a significant effect of Ti additions on compacted graphite in thin walled castings. However, the addition of Ti requires the addition of extra magnesium to avoid any risks for graphite flake formation (see Fig. 2b). Thus, it is common to set a limit of 20% nodularity in order to meet CGI specifications. In the case of thin-walled castings, the natural tendency of CGI is to solidify with a higher nodularity in thin outer walls (<4-5 mm) with up to 50% nodularity. Studies show that in TWCI castings with a wall thickness of 3 mm, additions of 0.13% Ti result, in reductions in the graphite nodule fraction from 97% for base iron, down to 34%.

Metallographic evaluations clearly show that the use of LDASC insulation sand significantly reduces cooling rates (Fig. 1). In turn, this causes a significant increase in the compacted graphite fraction exhibited in the TWCI castings. When LDASC sands are employed in castings with wall thicknesses of 3 mm, the compacted graphite fraction exceeds 65% (Fig. 2e). In the case of LDASC molding sand and with the addition of titanium at the level of 0.13% in castings with wall thicknesses of 3 mm the compacted graphite fraction exceeds 80% (Fig. 2d). Thus, it satisfies the ISO Standard [20] concerning a minimum volume fraction of compacted graphite.

Table II. Results of metallographic examinations

Alloy	Graphite type	Graphite fraction %	Graphite nodule fraction %	Ferrite fraction, f_f %	The maximum length of the graphite particles*, μm	Shape factor** S	Interparticle distance*** λ, μm
I	Nodular	15	97	37	10.84	0.81	65.9
II	Flake of D type	20	-	70	15.12	<0.2	3.3
III	Nodular+compacted	13	34	30	58.29	0.67	72.5
IV	Nodular+compacted	12	15	75	75.63	0.55	72.4
V	Nodular+compacted	13	35	75	91.42	0.66	83.6
VI	Nodular+compacted	14	73	67	49.11	0.76	67.8
VII	Flake of C type	20	-	57	151.50	0.45	33.3

* The maximum length of graphite particles were determined based on the average length of the five largest graphite particles are in the measured field and marked with no less than three places of the polished section

** Determination of the degree of compactness was based on the ratio of (S) defined as [18]:

$$S = \frac{4A}{p^2} \quad (3)$$

where:

A – surface area occupied by graphite particles,

p – circumference of the graphite particles.

*** Interparticle distances were determined based on Fullman equation [19]

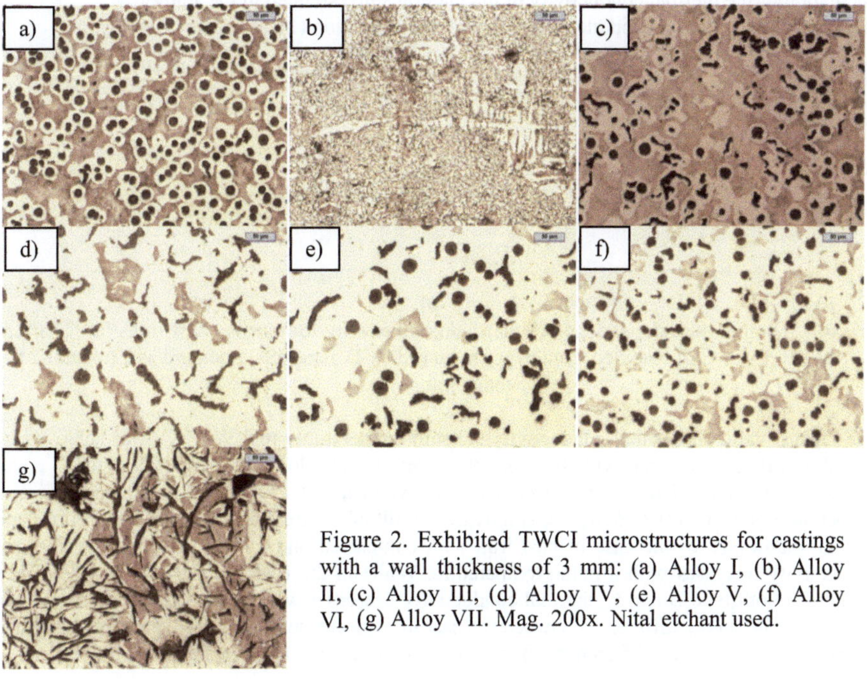

Figure 2. Exhibited TWCI microstructures for castings with a wall thickness of 3 mm: (a) Alloy I, (b) Alloy II, (c) Alloy III, (d) Alloy IV, (e) Alloy V, (f) Alloy VI, (g) Alloy VII. Mag. 200x. Nital etchant used.

3.3. Thermophysical properties

The results of the influence of temperature on the thermal diffusivity and specific heat are shown in Fig. 3 for different of cast iron types. Figure 3a shows significant differences in the thermal diffusivity at the temperature 22°C and it becomes more similar at elevated temperatures. At a temperature of 22°C, a difference of more than $6.4x10^{-6}$ m^2s^{-1} exists between the cast iron with flake graphite type C (alloy VII) and ductile iron (alloy I).

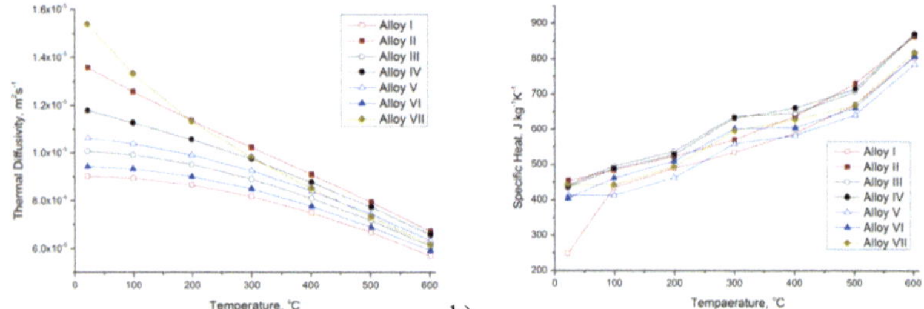

a) b)

Figure 3. Thermal diffusivity (a) and specific heat (b) for various cast iron as a function of temperature.

Figure 3b shows that the value of the specific heat increases from 400 $Jkg^{-1}K^{-1}$ at the temperature of 22°C up to 800 $Jkg^{-1}K^{-11}$ at the temperature of 600°C. The difference of specific heat values between side points are similar throughout the temperature range and are equal about to 100 $Jkg^{-1}K^{-1}$. The results of cast iron thermal diffusivity and specific heat were then used in the calculation of the thermal conductivity of the materials investigated in the study.

The results of the thermal conductivity, calculated by equation (1), are depicted in Fig. 4a.

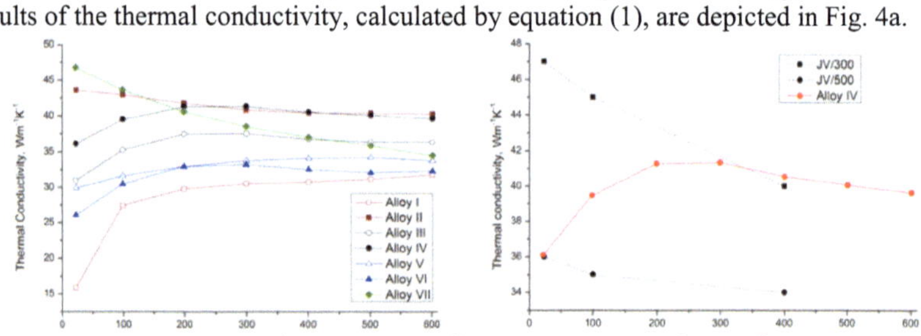

a) b)

Figure 4. Temperature dependence of thermal conductivity for various cast irons (a), Thermal conductivity of CGI obtained from experiment (Alloy IV) and ISO Standard as a function of temperature (b).

The highest value of thermal conductivity at room temperature, 46.8 $Wm^{-1}K^{-1}$ was achieved by the alloy VII for gray iron with flake graphite type C. The lowest value, 15.9 $Wm^{-1}K^{-1}$, was attained by the alloy I for ductile iron. Thus, a divergence of approximately 30 $Wm^{-1}K^{-1}$ at a temperature of 22°C was established, simply as a result of an alteration of cooling conditions and titanium additions. In the case of cast iron with compacted and nodular graphite, the thermal conductivity increases with a rising temperature. While in the case of cast iron in which flake graphite is both type D and C, thermal conductivity decreases with a rising temperature. As can be seen in Fig. 4a the difference between the cast irons decreases at elevated temperatures. Such a tendency of smaller differences between grey, ductile and compacted graphite iron at increasing temperatures, has been reported in work done [7] for castings with typical wall thickness.

It is worth noting the effect of graphite morphology on the thermophysical properties of the thin walled casting. In the case of cast iron with flake graphite of type C higher thermal conductivity in a temperature ranging from 22 to 100°C was obtained in comparison to graphite of type D. From a temperature of 200°C, a higher thermal conductivity was obtained in gray cast iron with a flake graphite of type D. This change may be associated with a significant reduction in the distance between the graphite particles (precipitate) (Table II), and also with the increase of ferrite fraction in the microstructure. According to the [16] conductivity for pearlitic grades over the same temperature range, there is approximately twenty percent less in comparison to ferritic grades. It should also be noted that the cast iron with flake graphite of type D shows much smaller differences in thermal conductivity with a temperature change in the range tested (Fig. 4a).

The effect of the percentage of compacted graphite on the thermal conductivity at different temperatures is also shown in Fig. 4a. It is clear from this figure, that increasing the percentage of compacted graphite causes an increase in thermal conductivity. The presence of graphite nodules within the microstructure negatively affects the thermal conductivity of compacted graphite cast iron. It is worth noting that the thin-walled castings of alloy IV are characterized by high thermal conductivity at elevated temperatures (with an effect from 200°C), which reaches the level of those obtained for the cast iron with flake graphite. As in the case of grey cast iron with flake graphite type D, compacted graphite cast iron (Alloy IV) shows little change in the thermal conductivity in the investigated temperature range, especially compared to ductile iron. This feature should be attributed primarily to short and thick compacted graphite with rounded edges (when compared to normal type A flake eutectic graphite) which are randomly oriented and can be connected to their nearest neighbours within an eutectic cell (as in grey cast iron) [21]. Figure 4b showed values of thermal conductivity for cast iron with compacted graphite in accordance with a ISO Standard [20] compared to the value of thermal conductivity of cast iron obtained for alloy IV. It can be concluded, that the obtained values of thermal conductivity in thin-walled castings with a wall thickness of 3 mm are within the limits prescribed for CGI. Furthermore, it can be added that the thin-walled compacted graphite cast iron can characterize good mechanical properties [11] and at the same time high thermophysical properties that predispose them to work in variable thermal loaded condition. It is also a very important fact, that tool life is increased with higher thermal conductivity.

Summing up, TWCI provides an opportunity to meet the performance requirements of the next generation of engineered components, particularly castings exposed to variable thermal loaded condition (cylinder blocks and heads).The current casting market is driving the need for stronger cast irons with lower total assembly weight than grey cast iron parts, but with improved machinability, thermal-fatigue resistance, damping capacity, casting mould yield, and castability compared to ductile iron parts. From the point of view of economics and ecology, TWCI can compete with aluminium alloy, providing unique mechanical and thermal properties and ensuring new opportunities to satisfy the requirements of lightweight thin-walled castings.

Conclusions

1. The thin walled castings made of cast iron with compacted graphite have high thermal conductivity at an elevated temperature (from 200°C), which achieve the level of those obtained for the cast iron with flake graphite.
2. The thin walled castings made of cast iron with compacted graphite similar to cast iron with type D flake graphite shows little change in the value of thermal conductivity in the investigated temperature range of (22-600°C) in comparison, especially to ductile iron.

3. The thin walled castings made of cast iron with compacted graphite have good mechanical properties [11] and simultaneously high thermophysical properties, which predispose them to work in variable thermal loaded conditions by minimizing the accumulation of thermally induced stress.

4. The existence of spheroidal graphite within the microstructure negatively affects the thermal conductivity of compacted graphite cast iron in thin sections. It is manifested in reduction thermal conductivity as well as an increased sensitivity to change in variable thermal loaded conditions.

Acknowledgements

The authors acknowledge financial support from Polish National Science Centre (through grant No. 2013/09/B/ST8/00210)

References

1. W. Guesser, T. Schroeder, S. Dawson, "Production Experience with Compacted Graphite Iron Automotive Components", *AFS Trans,* 109 (2001), 1–11.
2. S. Dawson, "Controlling the production of compacted graphite iron". *Mod Cast,* 88 (1998), 38–41.
3. S. Dawson, "Compacted graphite iron – A material solution for modern diesel engine cylinder blocks and heads," *China foundry,* 6 (3) (2009), 241-246.
4. D. Holmgren, R. Kallbom, I. Svensson, "Influences of graphite growth direction on the thermal conductivity of cast iron," *Metall. Mater. Trans. A*, 38(2) (2007), 268–275.
5. J. Helsing, G. Grimvall, "Thermal conductivity of cast iron: Models and analysis of experiments," *J. Appl. Phys*, 70 (1991), 1198-1206.
6. Rukadikar, MC& Reddy, GP, "Influence of chemical composition and microstructure on thermal conductivity of alloyed pearlitic flake graphite cast irons ", J. Mater. Sci. , vol. 21, 12, 1986, p.4403-4410.
7. D. Holmgren, "Review of thermal conductivity of cast iron ", *Int. J. Cast Metal Res.*, 18, (6) (2005), 331-345.
8. R. Monroe, C. Bates, "Some thermal and mechanical properties of compacted graphite iron," *AFS Trans.* 90 (1982), 615-624.
9. L. Hecht, R. Dinwiddie, H. Wang, "The effect of graphite flake morphology on the thermal diffusivity of gray cast irons used for automotive brake discs, " *J. Mater. Sci.*, 34 (19) (1999), 4775-4781.
10. M. Gorny, M. Kawalec, "Effects of Titanium Addition on Microstructure and Mechanical Properties of Thin-Walled Compacted Graphite Iron Castings," *J Mater Eng Perform*, 22 (5) (2013), 1519-1524.
11. M. Gorny, M. Kawalec, G. Sikora, H. Lopez, "Effect of Cooling Rate and Titanium Additions on Microstructure of Thin-Walled Compacted Graphite Iron Castings," *ISIJ Int.*, 54 (10) (2014), 2288–2293.
12. J. Zhou, "Colour Metallography of Cast Iron," *China Foundry*, 8(1) (2011),154–165.
13. I. Riposan, M. Chisamera, R. Kelley, M. Barstow, R.L. Naro, "Magnesium-Sulfur Relationships in Ductile and Compacted Graphite Cast Irons as Influenced by Late Sulfur Additions," *AFS Trans.*, 2003, 111, p 869–883.
14. O. Erfan, O. Elmabrouk, "Influence of Section Thickness on the Thermal Conductivity of Compacted Graphite Cast Iron at Elevated Temperatures," *Int. J.Eng.Innovative Technology*, 3 (7) (2014),29-33.
15. L. Jinhai, Y. Litao, L. Guolu, L. Changqi, L. Yinguo Y. Zhaoyu, "Influence of fading on characteristics of analysis curve of compacted graphite iron," *China Foundry*, 3 (2011), 295-299.
16. ASM Specialty Handbook: Cast Irons. ASM Int. Handbook Committee, Ed. J.R. Davis, 2006.
17. D. Stefanescu, F. Martinez, I.Chen, "Solidification Behavior of Hypoeutectic and Eutectic Compacted Graphite Cast Irons. Chilling tendency and eutectic cells, *AFS Trans*, 91 (1983), 205-216.
18. S. Charoenvilaisiri, D. Stefanescu, R. Ruxanda, T. Piwonka, "Thin Wall Compacted Graphite Iron Castings ,"*AFS Trans.*, 110 (2002), 1113-1130.
19. R.L. Fullman, Measurement of particle sizes in opaque bodies, *Trans.AIME*, (1953) 447-452.
20. ISO 16112:2006, Compacted graphite cast irons - Classification, I.S.O. Standardization, 2006.
21. M. Konig, "Literature review of microstructure formation in compacted graphite iron, " *Int J Cast Metal Res* 23 (3) (2010), 185-192.

Advances in the Science and Engineering of Casting Solidification
Edited by: Laurentiu Nastac, Baicheng Liu, Hasse Fredriksson, Jacques Lacaze, Chun-Pyo Hong, Adrian V. Catalina, Andreas Buhrig-Polaczek, Charles Monroe, Adrian S. Sabau, Roxana Elena Ligia Ruxanda, Alan Luo, Subhayu Sen, and Attila Diószegi
TMS (The Minerals, Metals & Materials Society), 2015

UNDERSTANDING CAST IRON MATERIALS AND COMPONENTS: A NEVER ENDING STORY

Ingvar L Svensson and Jakob Olofsson

Jönköping University, School of Engineering,
Materials and Manufacturing
P.O. Box 1026, SE-551 10 Jönköping, Sweden

Keywords: cast iron, solidification, microstructure, mechanical properties, castings

Abstract

How can an in principal binary alloy of iron and carbon show so many fascinating phenomena and still today give surprises to users, foundrymen and researchers? This paper points out some critical steps in the understanding of the whole chain, from the melt to a cast iron product in service. The understanding of the material is gradually improved, assisted by the advances of other fields, e.g. analyzing methods and computational techniques. The heart in cast iron is the graphite, which is a highly difficult phase to understand but gives the material its unique properties. The linkage between understanding and modelling is necessary to calculate/simulate the processes occurring, where the precipitation, nucleation and growth of the different phases are the keys. Proper nucleation and growth models have been introduced to predict e.g.

- primary precipitation of austenite and graphite
- eutectic growth of different morphologies of graphite or cementite and austenite
- solid state transformation of austenite into ferrite and pearlite in both grey and ductile irons

and now gives realistic microstructures and solidification curves for most practical cases. The microstructure formation models gives input to shrinkage and volume calculations to predict porosities, and to predictions of mechanical properties. By linking microstructure formation models, characterization models for mechanical properties and Finite Element Analysis (FEA) it is today possible to use local properties in simulations of the behavior of cast iron components.

Many phenomena in cast iron, however, still remain unexplained. As one student labelled one of his experimental files on ductile iron, cast iron materials and simulations are indeed *a never ending story,* with a bright future in industrial applications

Introduction

The development of materials and products is focusing on environmental and sustainability aspects. Cast irons are in this perspective a group of materials with many advantages; e.g. recyclable without quality loss, low cost, good and unique properties. In other words, a material for the future. Cast irons have a complex development of microstructure, and sometimes gives unexpected results and properties due to a lack of understanding of the phenomena occurring during solidification and production.

The manufacturing process gives cast iron components properties that are dependent on component design, metallurgy and casting method. Factors as local wall thickness influences the coarseness and type of microstructure developing, and the material will have local properties depending on the local metallurgical and thermal history [1]. Simulations of the stress/strain behavior of cast products are typically performed using constant material properties throughout the castings, but the local properties can vary essentially in the volume of the casting. This makes it difficult to optimize the component with good accuracy. However, using modern simulation tools to predict local solidification and microstructure, it is today possible to calculate the local material properties for all parts of the casting, and with high accuracy determine its stress and deformation behavior.

Cast irons are today modern materials which continue to be developed, and have the potential to fulfill even more customer needs in the future. We are now in a paradigm shift in knowledge about the cast iron material and products. The contribution to the knowledge of cast iron have been made by hundreds of researchers, where we would like to especially bring out researchers as Doru Stefanescu, J. Sikora, M. Hillert, H. Fredriksson, E. Piwowarsky and C. R. Looper. Still there are many secrets to be revealed for the next generation researchers. Cast irons are thus indeed – *a never ending story.*

Cast iron research for sustainable products

Cast iron may be an environmentally good material, depending on e.g. how the melting energy is generated. The carbon dioxide outlet must to be reduced and the fossil fuel must decrease to a low level to give a sustainable society. In e.g. Sweden, where most of the energy comes from water-power, nuclear power and wind power, the carbon dioxide outlet is approximate 113 kg per 1000 kg castings, while in some countries using fossil energy the outlet of CO_2 is 1700 kg CO_2 /1000 kg castings [2]. Important steps in the lifecycle of a cast iron product can be described by *the casting circle* (Figure 1). The starting point is the melt and the alloy preparation, going all the way to product behavior in service and product recycling. To maintain sustainable growth in the knowledge of cast iron the following should be highlighted. Observations and documentation of cast iron phenomena should be treated in a way that it in the end can be formulated as a mathematical model to simulate the observed phenomena. If a good model can be derived, the research findings are "brought to life", providing researchers and industries with meaningful results which can be expanded and coupled to other models (Figure 2). Much research in the cast iron area today starts from scratch every time. This leads to a slow expansion of the knowledge front. Mathematical models can be used as a fundament for the next researcher, leading to a more rapid expansion of the knowledge front.

The casting circle shows two main areas; process and metallurgy on the left side, and material behavior in a product on the right. In order to obtain a good industrial product development process, and a high performance of the product in service, the transfer of knowledge and collaboration between these two areas have to be strong [3]. In the following sections, some specific areas which require further exploitation in order to strengthen this collaboration are pointed out.

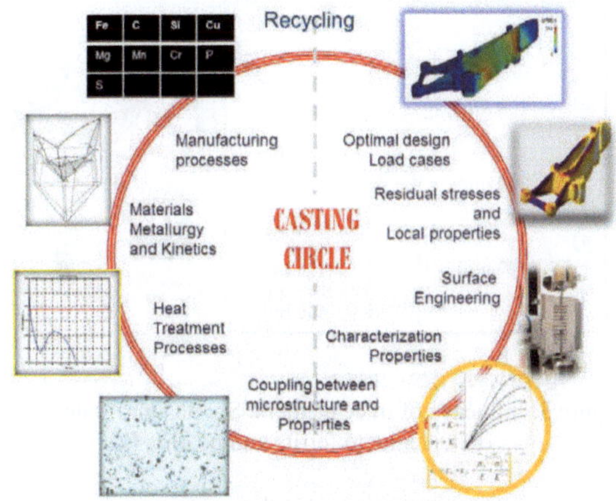

Figure 1. The casting circle, showing important steps in the lifecycle of a cast iron product.

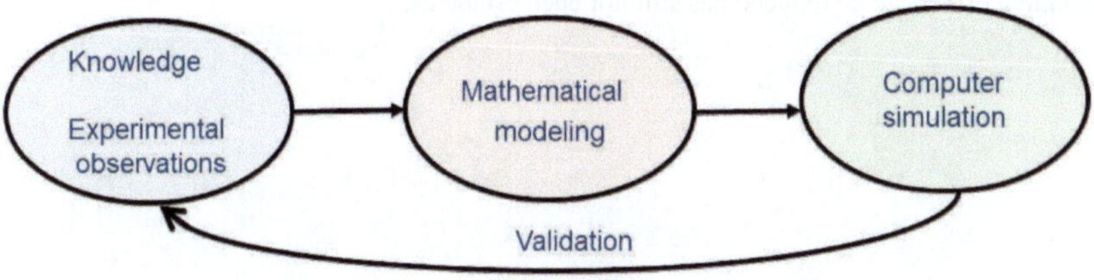

Figure 2. Coupling of Experimental observations – mathematical modeling– computer simulation

Solidification and microstructure development in cast iron

Preparation of the melt

The production of the melt and the composition of the batch, pig iron, steel scrap and recycled returns from the society will influence the quality of the melt. For cast iron production this means that the influence of trace elements (on the ppm level) on the nucleation and growth of phases during solidification as well as the solid state transformation needs to be established. These trace elements have been noted to have a significant effect on e.g. pearlite formation in heavy castings, Figure 3, but it is still an open question how these affects can be modelled and handled in an industrial context [4].

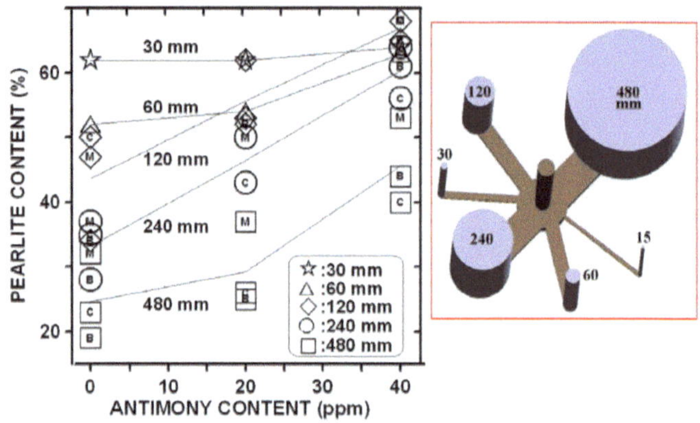

Figure 3. Influence of trace element, Sb, on solid state reaction of austenite to ferrite and pearlite [4]

<u>Nucleation</u>

The *nucleation* of phases in the *melt* is not yet well understood, as well as the interior or the melt. The number of nucleation sites, natural or artificially added are highly dependent on the experimental setup. Especially the nucleation of graphite is still an unresolved issue, though several theories have been proposed. In most commercial cast iron alloys artificial nucleation agents are added. The fading of the nucleation agent is a well-documented phenomena, but how the fading effect can be reduced has still not been established.

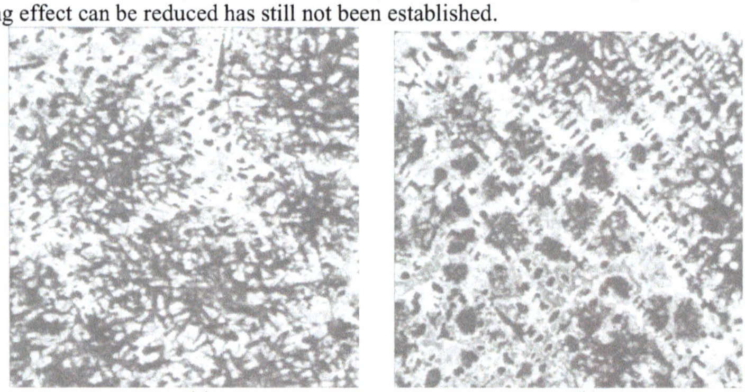

(a) (b)

Figure 3. Grey iron without (a) and with (b) a secondary eutectic population [5].

The sequence of nucleation of primary and eutectics are still to be revealed. Figure 3a shows a eutectic cell size population, and Figure 3b shows a secondary population with small cells. This can also be seen very clearly in some ductile iron materials, and has been shown to be important contributions to metal expansion penetration and shrinkage formation [5].

The *nucleation of pores* is not well understood. This is a complex course of events involving volume changes during solidification, gases, feeding, clean melts and particles which nucleates the pores [6-7].

At *solid state* the *nucleation* problems are obvious, and many times it can be described with a CCT-diagram [8], see figure 4. One important factor in the growth of ferrite is the time between passing the equilibrium temperature and the temperature when the first nuclei will start to grow. In steel this is well documented and to a certain degree simulated, but in cast irons further work is needed.

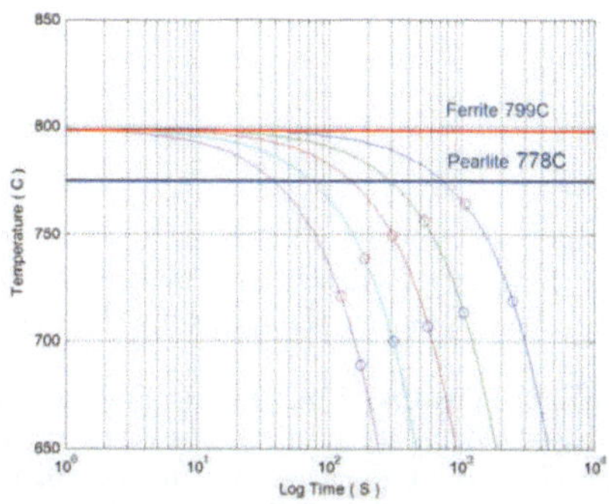

Figure 4. CCT-diagram of a Ductile iron, EN-GJS-500-7, 3.83 % C 2.26% Si and 0.40 Mn [8]

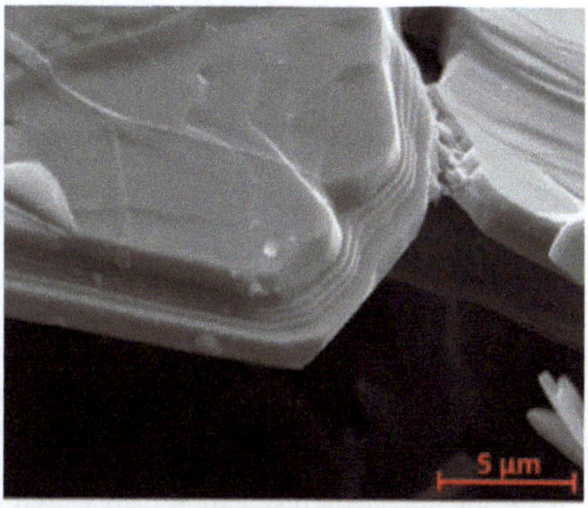

Figure 5. A graphite nuclei which have been formed on the basal plane growing ledges at the tip of a flake [9]

Growth kinetics

Primary *austenite* dendrite solidification has been studied under long time but knowledge about nucleation and growth in cast iron of the austenite is limited. The growth of *graphite* is a key subject, both when it grows in the melt and in contact with austenite. Different elements will control the way the carbon will attach to the graphite nuclei. New hope of increased knowledge regarding the growth of graphite is the large interest in graphene [9], which opens up a new scenario to create new microstructures.

The *ferrite growth* is not fully understood, especially when alloying elements are used. How is e.g. the ferrite growth reduced by copper? Several mechanisms have been proposed [10], but the answer is still not clear. The *pearlite growth* influences the mechanical properties by changing the pearlite lamellar spacing. Here there is a clear coupling to CCT- diagrams, which shows the start temperature of pearlite growth, see Figure 4.

The understanding and modelling of growth of the *porosities* in cast iron is still under discussion. Here the volume changes are qualitatively understood but models to provide reasonable simulation results still needs to be established.

Microstructure and mechanical properties

Deformation mechanisms and characterization of mechanical behavior

It is well known by that the microstructure is very important to the mechanical properties. Commonly, properties as elastic modulus, ultimate tensile strength, yield strength and ultimate elongation have been reported in the literature. For computational structural analyses of cast iron components these properties are not enough, but the entire shape of the tensile test curve is required.

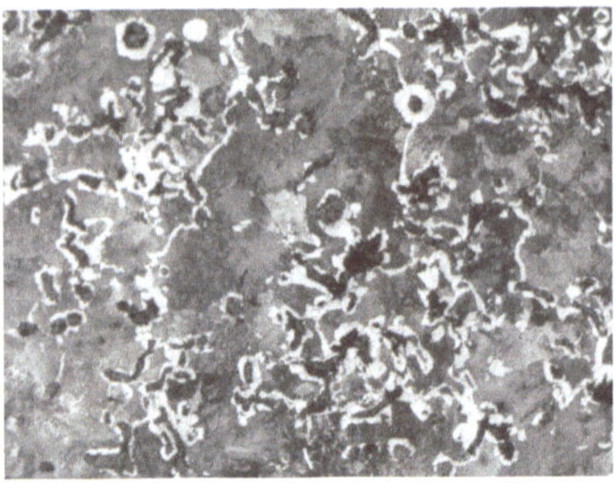

Figure 6a. CGI with a complex micostructure to follow the deformation procedure

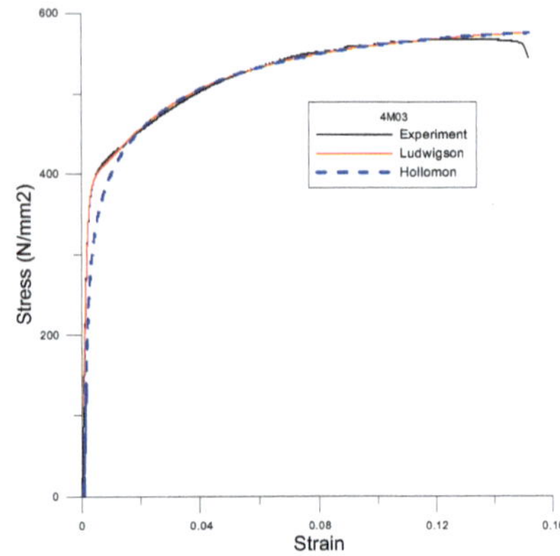

Figure 6b. Evaluated and simulated with the Ludwigson and Hollomon models [11-13]

The plastic deformation of a material is depending on the dislocation movements. For one phase or simple shaped structured materials, physically based models are available. For cast iron, with unsymmetrical phases in 3-D, Figure 6a, it is not yet possible to characterize the deformation behavior fully with dislocation theory. With the use of empirical curve fitting methods, e.g. the Hollomon [11] and the Ludwigson [12] equations, we have started to get an understanding for how the tensile test curve [13] can be interpreted in terms of microstructure.

Component behavior depending on local microstructure

During the solidification of a cast iron component, all of the previously outlined phenomena act together and form a highly complex interaction between e.g. casting process parameters, component geometry and local solidification conditions. This leads to local variations in microstructure throughout a component, and thus local variations in material behavior. Though it is of great importance to study and model all the outlined phenomena separately, it is thus also of significant scientific and industrial importance to put the derived knowledge into a larger perspective and see how they contribute to "the big picture", the performance and behavior of the component. The development of casting process simulation software have here contributed by enabling calculations of several steps of the casting circle (Figure 1), where simulations can be performed using mathematical models for e.g. nucleation, microstructure evolution during solidification and the effect of heat treatment to predict the local variations in final microstructure. However, the effect of these local variations in microstructure on the behavior and performance of a cast iron component in service is rarely reported in the literature. Recently the current authors proposed a simulation strategy denoted *the closed chain of simulations for cast components* [14]. Here casting process simulations and local microstructure formation and material characterization are incorporated into a Finite Element Method (FEM) analysis of the mechanical behavior of the component. In the FEM simulation each element of the FEM mesh has its own microstructure-based mechanical behavior. This enables consideration of local variations in mechanical behavior throughout a cast iron component, rather than homogeneous material behavior. It has been shown that these local variations are important contributions to the mechanical behaviour of the component, which a homogenous FEM simulation fails to describe [15]. Calculation of a stress distribution at a certain load level is shown in Figure 7. Using the local properties gives a more realistic performance of a component at load. The stress distribution is highly dependent on the previous plastic deformation in the component. As the deformation behaviour is depending on the type of microstructure and fineness, the deformation will not always depending on the stresses. It has been previously shown [14] that the local variations in mechanical behaviour are important contributions to the mechanical behaviour of the component, which a homogenous FEM simulation fails to describe.This redistribution is also found in cast aluminium, where it has been physically observed using Digital Image Correlation (DIC) [16].

These results highlights the importance and relevance of taking an integrated product development perspective on cast iron components. Cast iron solidification is more than an issue for material scientists and metallurgists, it is also of great importance for product development and designers. This also highlights the importance of establishing accurate models of cast iron phenomena in order to increase the predictive power of structural analyses of cast iron components. Structural analyses are never better than the material data used in the simulation. To establish accurate material models which produce accurate local material data for structural analyses will therefore be a continued significant challenge for cast iron scientists in the future.

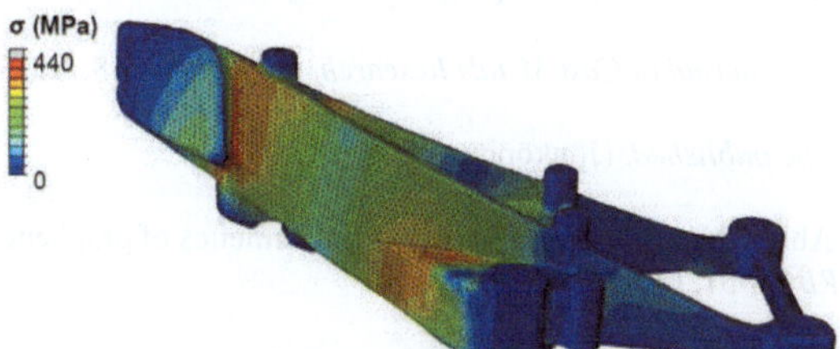

Figure 7. Simulated stress distribution using microstructure-based mechanical behaviour through the derived Ludwigson equation characterisation models

Conclusions

The solidification of cast iron components is a highly complex interaction between multiple phenomena from the nucleation of different phases in the liquid to growth of them in at solidification and at solid state transformation. To produce a casting of industrial relevance is not only about the solidification and microstructure, but predicting and obtaining the right final mechanical behavior is of large importance for successful product development and manufacturing processes. We propose that by adopting an integrated product development perspective on cast iron solidification, solidification modelling and numerical simulation, cast iron will continue to have a bright future in the 21st century. Thanks to the great amount of knowledge developed by smart and hardworking scientists in the previous decades, *cast iron will remain – a never ending story.*

Acknowledgement

The main author thanks Professor D. M. Stefanescu who has been a valuable discussion partner and friend during the development of the different steps in the casting circle. Thanks also to Professor H.Fredriksson and Professor M. Hillert who provided fundamental knowledge in the Physical metallurgy and casting area, during my master, research studies and following research.

References

1. D.M. Stefanescu, "Modeling of cast iron solidification - The defining moments." Symposium on Solidification Modeling and Microstructure Formation *2006 TMS Annual Meeting*. (7 2007). 1433 – 1447.

2. Swiss Centre for life cycle inventories www.ecoinvent.org

3. J. Olofsson, "Simulation of Microstructure-based Mechanical Behaviour of Cast Components", *Ph.D. thesis*, (Jönköping University, School of Engineering, 2014).

4. M. Wessén, I.L. Svensson, and R. Aagaard, *International Journal of Cast Metals Research,* **16** (1-3), (2003) pp. 119-124.

5. I. Dugic, "An Investigation of the Effect of Inoculants on the Metal Expansion Penetration in Grey Iron". *Journal of Cast Metals Research, (*1999), Volume 11, number 5, pp 333-338.

6. I. Dugic, *Dissertation* No. 1007, (Linköping University 2006).

7. D.M. Stefanescu, *Int. Journal of Cast Metals Research*, (2005), Vol. 18, No. 3, 129-143

8. I.L. Svensson, *To be published,* (Jönköping University, 2014).

9. S. Amini and R. Abbaschian, "Nucleation and growth kinetics of graphene layers from a molten phase", *CARBON* 51, (2013), 110 – 123

10. M. Wessén, "On the Mechanisms of Structure Formation in Nodular Cast iron", *Ph.D. thesis*, (Dept. of Casting of Metals, Royal Institute of Technology, Stockholm, 1997).

11. H. Hollomon, *Trans. AIME* 162, (1945), 268

12. D. C. Ludwigson, "Modified stress-strain relation for FCC metals and alloys", *Metal. Trans.* 2, (1971), 2825-2828.

13. I.L. Svensson and J. Olofsson, "On microstructure-based mechanical behaviour of a ductile iron component". *.Proceedings of "X:th Int. Symposium on Physical Metallurgy of Cast Iron",* 2014, Mar del Plata, Argentina, Nov. 10-14.

14. J. Olofsson and I.L. Svensson, "Incorporating predicted local mechanical behaviour of cast components into finite element simulations", *Materials & Design*, (2012), no. 34:494-500.

15. J. Olofsson and I.L. Svensson, "The effects of local variations in mechanical behaviour – numerical investigation of a ductile iron component", *Materials & Design*, (2013), no. 43:264-271.

16. J. Olofsson, I. L. Svensson, P. Lava and D. Debruyne, "Characterisation and investigation of local variations in mechanical behaviour n cast aluminium using gradient solidification, digital image correlation and finite element simulation", *Materials & Design*, (2014), no. 56:755-762.

Advances in the Science and Engineering of Casting Solidification
Edited by: Laurentiu Nastac, Baicheng Liu, Hasse Fredriksson, Jacques Lacaze, Chun-Pyo Hong, Adrian V. Catalina, Andreas Buhrig-Polaczek, Charles Monroe, Adrian S. Sabau, Roxana Elena Ligia Ruxanda, Alan Luo, Subhayu Sen, and Attila Diószegi
TMS (The Minerals, Metals & Materials Society), 2015

UNDERSTANDING SUPERFINE GRAPHITE IRON SOLIDIFICATION THROUGH INTERRUPTED SOLIDIFICATION EXPERIMENTS

G. Alonso[1], D.M. Stefanescu[2], P. Larrañaga[1], E. De la Fuente[1], E. Aguado[1], R. Suarez[1,3]

[1]Área Ingeniería, I+D y Procesos Metalúrgicos, IK4-Azterlan, Durango, Spain
[2]The Ohio State University, Columbus, OH and The University of Alabama, Tuscaloosa, AL
[3]Veigalán Estudio 2010, Durango, Spain

Keywords: Gray Iron, Graphite Morphology, Interrupted Solidification, Titanium Additions, Oxygen Content

Abstract

The tensile strength of near-eutectic gray iron can be increased from 230-300 to 300-345MPa, without a significant increase in hardness, through 0.3-0.4%Ti addition to low sulfur (<0.01%S) iron. This is due to the combination of higher primary austenite/eutectic ratio and the precipitation of superfine-interdendritic-graphite (SIG), characterized by a fine (10-20μm) and highly branched fibrous structure.

To reveal the influence of the %Ti on graphite shape evolution during solidification and its relationship to the solid fraction, quenching experiments at successive solidification stages were carried out on hypoeutectic alloys with 0.18% and 0.32%Ti. The graphite shape factors were measured, and their evolution as a function of the titanium content and the solid fraction was analyzed. SEM was used to evaluate the change in graphite shape during early solidification, as well as its nucleation and growth. The correlation between the oxygen in the melt and SIG formation was also explored. It was concluded that nucleation of graphite in SIG irons occurs on graphite substrates at the austenite/liquid interface because of carbon supersaturation.

Background

The standard graphite (Gr) shapes in cast iron are lamellar graphite (LG), compacted/vermicular graphite (CG) and nodular/spheroidal graphite (SG). LG can be further classified as a function of the distribution and size of the graphite lamellae into uniformly distributed type-A (LAG), and interdendritic type-D (LDG) or type-E. Other intermediate shapes have been identified, such as coral graphite and superfine interdendritic graphite (SIG). Coral graphite is a highly branched fibrous type of graphite that is different from either LDG or CG. It has been obtained in pure Fe-C-Si alloys containing very small amounts of sulfur, typically around 0.001%S, at high cooling rates [1]. The SIG is short (10-20 μm) and stubby, exhibiting rounder edges than type-D, similar to the coral graphite. It is obtained in low-S and moderate-Ti additions (*e.g.* <0.01%S, 0.3-0.4%Ti) irons [2]. The combination of this graphite with high amounts of fine primary austenite [3] produces high strength in high carbon equivalent LG iron combined with low hardness (~ 250 HB). Taking full advantage of the excellent static mechanical properties of this iron requires an understanding of its solidification, including SIG nucleation and growth.

Titanium is well-known to be a deoxidizer and structure refiner in steel. For the case of cast iron it was argued that Ti additions nucleate small equiaxed dendrites [4, 5]. Okada [6] suggested that Ti additions resulting in the formation of TiC produce low carbon regions at the solid/liquid (S/L) interface, favoring formation of type-D graphite. Titanium additions increase the type-A-to-D graphite transition temperature and thus favor type-D graphite formation at smaller undercooling [7]. Yet, no explanation for this behavior was provided.

According to Wilford and Wilson [8], Ti reacts with N producing TiN or Ti(CN) that affects the solidification of primary austenite. The excess Ti will then react with S. Formation of TiS decreases the available S for MnS formation and increases undercooling which is responsible for type-D graphite formation. No direct nucleation influence of Ti on graphite was found [9,10,11]. Thus, the effect of Ti on Gr shape may not be related to Gr nucleation, but to Gr growth. It is well accepted that the main parameters affecting Gr shape are the level of surface active elements, such as sulfur and oxygen, and the cooling rate (solidification velocity, V).
The effect of sulfur on graphite shape depends on its content. It was found that for low S irons (0.007%S) an abrupt decrease in the lamellar spacing occurs at some critical solidification velocity [12]. Such a transition was not found in higher S irons (0.047-0.45%). These findings corroborate those of other researchers [13,14,15], albeit for some slightly different S levels or solidification rates as shown in the example in Figure 1. The sudden transition in low-S irons seems to indicate a type-A-to-D transition.

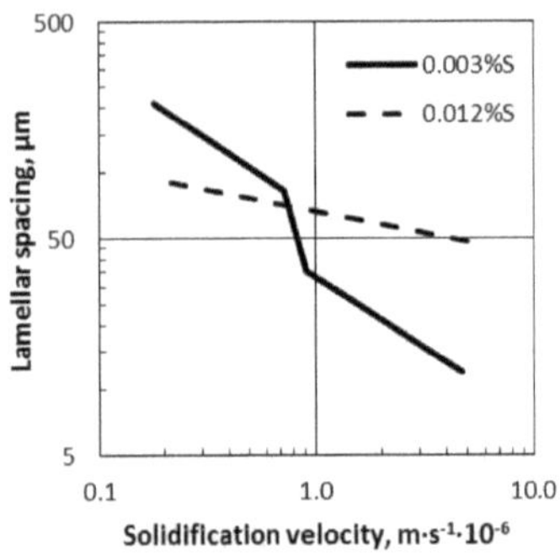

Figure 1. Effect of sulfur and solidification velo-city on the graphite spacing in directionally solidified irons; based on data from Ref. [15].

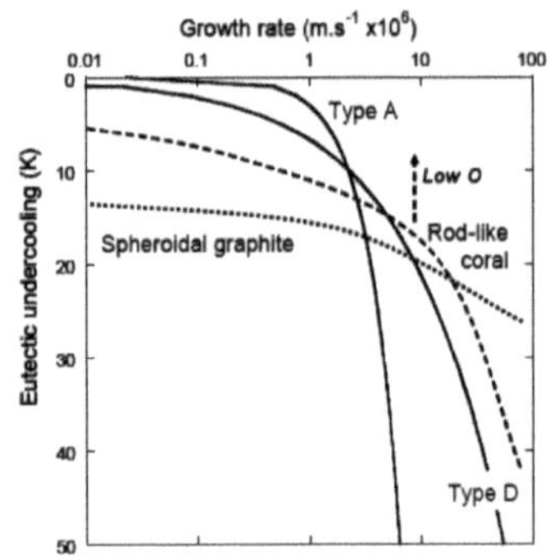

Figure 2. Effect of growth velocity and eutectic undercooling on transitions between various graphite shapes in pure Fe-C-Si alloys [16].

Another parameter that affects graphite morphology is the cooling rate (solidification velocity). Figure 2 shows the transition between various graphite shapes as a function of growth velocity and undercooling with respect to the stable temperature. This semi-quantitative graph was produced by Lacaze *et al.* [16] using data from various sources [17, 18, 19, 20, 21, 22]. Working under the assumption that the structure evolving from solidification is the one that grows at the lower undercooling, it is concluded that type-A Gr forms at low solidification velocities. As the velocity increases, a transition to type-D will occur. Note that the abrupt decrease in Gr spacing in Figure 1, and the sudden increase in eutectic undercooling for type-A in Figure 2, take place at $V \approx 1\mu m/s$, where the type-A-to-D graphite transition occurs. This can be construed as an argument for a change in solidification mechanism upon the type A-to-D transition. Decreasing the O level can decrease the undercooling for coral growth to such an extent that there is no window for type D to appear in very pure Fe-C-Si alloys, as illustrated with the arrow in Figure 2 [16]. On the contrary, addition of 0.02% S to pure Fe-C-Si alloys leads to the disappearance of the A-to-coral transition, which is replaced by the A-to-D transition [22].
It is now widely accepted that LG grows in the crystallographic A-directions with the {0001} basal planes parallel to the plane of the lamellae [23]. It appears that the growth starts with the formation of graphene sheets. Graphene is a 2-dimensional, crystalline allotrope of carbon, in

which carbon atoms are densely packed in a regular hexagonal pattern. To produce LG, the sheets will have to grow also in the *C*-direction and produce multilayer sheets. The thickening of the graphene sheets could occur through spiral growth.

The goal of this paper is to attempt further understanding of the role of Ti in refining the graphite from LAG to LDG to SIG. A method widely used to study nucleation and growth phenomena is interrupted solidification experiments. It provides a glimpse at early solidification. Such experiments were conducted by the authors to understand early solidification of CG iron [24]. Interrupted solidification experiments conducted on irons with 0.13 to 0.22% Mg showed that graphite formation starts with small SG, which develops tails at about 0.4 fraction solid (f_S). At 0.55 to 0.69 f_S, depending on the %Mg, CG is formed. It is thus apparent that austenite dendrites affect graphite growth and therefore graphite shape. Similar experiments were undertaken in this work.

Experimental Strategy

Melting and Casting. Two heats (*Quench I* and *II*) were produced in a 100 kg medium frequency induction furnace (250 Hz, 100 kw). The charge for each heat consisted of 37 kg of ductile iron returns and 13 kg of high purity pig iron. Ferro-manganese (74.8%Mn, 2.13%Si, 5.49%C, 0.082%P and 17.3%Fe) and carburizer additions were made as needed. After superheating to 1500°C, the iron was transferred into the pouring ladle. The heats had two different levels of titanium, 0.18% and 0.32%Ti. The Ti was added as FeTi (72.3%Ti, 4.5%Al). From each heat, six standard thermal analysis cups were poured and the cooling curves were recorded. Inoculation was made directly in the cups through the addition of 0.2% of a commercial inoculant having the following chemical analysis: 68.1%Si, 0.89%Al, 1.65%Ca, 0.45%Bi, 0.38%Ba and 0.37%RE. The solidification of the iron in the cup was interrupted by quenching in brine at increasing times, as shown in the example in Figure 3. After cooling to room temperature, the cups were sectioned and prepared for metallographic examination.

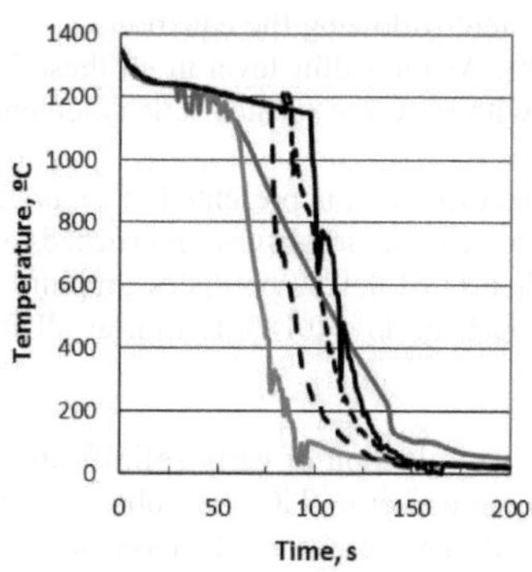

Figure 3. Example of cooling curves obtained during the quenching experiments showing the time and temperature of quenching (*Quench II*)

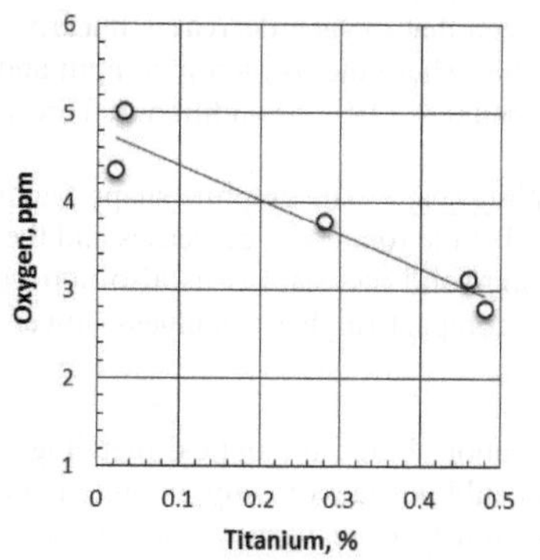

Figure 4. Effect of Ti in the melt on the oxygen content

Two additional heats (*Oxygen I* and *II*) were produced in the same furnace to evaluate the effect of Ti on the oxygen content of the melt. The charge materials were the same as for the previous heat. Ti additions were made in the furnace and an oxygen probe was used to measure the temperature and the oxygen level in the melt. The chemical analysis of all experimental irons is reported in Table I.

Table I. Chemical Analysis (wt%) of the Quenched Irons.

Heat	Furnace additions	C	Si	Mn	P	S	Ti	ppm O
Quench I	Ti	2.94	1.92	0.58	0.022	0.011	0.18	ND
Quench II	Ti	3.10	1.85	0.53	0.021	0.011	0.32	ND
Oxygen I	none	3.47	1.99	0.62	0.021	0.009	0.021	4.35
	Ti	3.43	2.00	0.62	0.020	0.010	0.28	3.77
	Ti	3.41	1.97	0.59	0.022	0.011	0.48	2.78
Oxygen II	none	3.43	1.94	0.60	0.019	0.007	0.032	5.01
	Ti	3.42	1.95	0.60	0.018	0.010	0.46	3.13

Note: reported oxygen values were adjusted for a temperature of 1500°C

Characterization. The fraction of the area occupied by the solid after different solidification times was measured through quantitative metallography techniques on color and Nital etched samples, as described in detail in Ref.[25]. A grid was superimposed on the micrographs. The intersection of the grid with the liquid fraction (iron carbides) was marked by dots. The ratio between the number of dots and the total number of the intersections on the grid represents the area fraction of liquid, f_L. The solid fraction is then calculated simply as $f_S = 1 - f_L$.
The graphite shape was evaluated through quantitative metallography with the software Image J on unetched samples. The measured graphite shape parameters included the aspect ratio ($AR = major\ axis/minor\ axis$) and the roundness ($Rnd = 4\ area/(\pi \cdot major\_axis^2)$). After Nital etching the fraction solid was evaluated on photographs, and a record of graphite shape factor - fraction solid was made.

Experimental results

Effect of oxygen. The effect of Ti on the oxygen content of the melt is summarized in Figure 4. It is seen that oxygen decreases linearly with the Ti content following the equation O = -3.88·Ti + 4.798, where the oxygen is in ppm and the Ti in wt%. As the sulfur level in all these irons is less than 0.011%S, the addition of Ti results in an iron with very low surface active elements.

Graphite shape. The graphite shape factor-solid fraction correlation is presented in Figure 5. It is seen that the roundness decreases and the aspect ratio increases as solidification proceeds and the fraction solid increases. It is also noticed that the higher Ti of 0.32% produces graphite that is more compact (higher roundness, lower aspect ratio) than the low 0.18%Ti iron at all fraction solids.

Nucleation. Extensive optical metallography and SEM examination of early solidification samples could not identify any graphite nuclei. While a high number of TiC were observed in samples from *Quench II*, they seldom acted as nuclei. No MnS were observed. It was also noticed that the SIG aggregates nucleate at the austenite (martensite)/liquid (γ/L) interface and then grow into the liquid (carbides) (Figure 6a,b). Parallel graphite plates emerging from the γ-phase are seen on Figure 6c, which suggests a cooperative γ - Gr growth into the liquid. The orientation of the Gr in the proximity of the austenite, seems to favor a direction parallel to the γ/L interface (Figure 6d,e).
Since graphite nuclei for SIG were not identified in this research or in previous research (*e.g.* Ref. [11]), it is reasonable to assume that SIG nucleation occurs on graphite substrates produced at the austenite/liquid interface in regions enriched in carbon because of carbon rejection by the austenite. These Gr aggregates then grow into the liquid as spherical graphite/austenite grains

(Figure 7a). At higher fraction solid, the Gr aggregate incorporates the austenite dendrites while continuing to grow in contact with the liquid (Figure 7b).

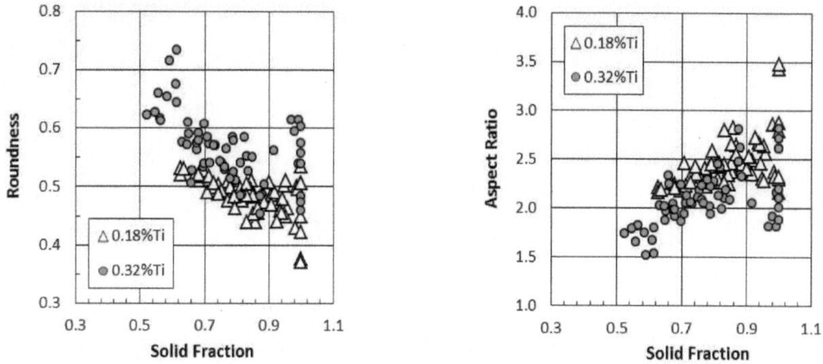

Figure 5. Correlation between the graphite shape factors and the fraction solid

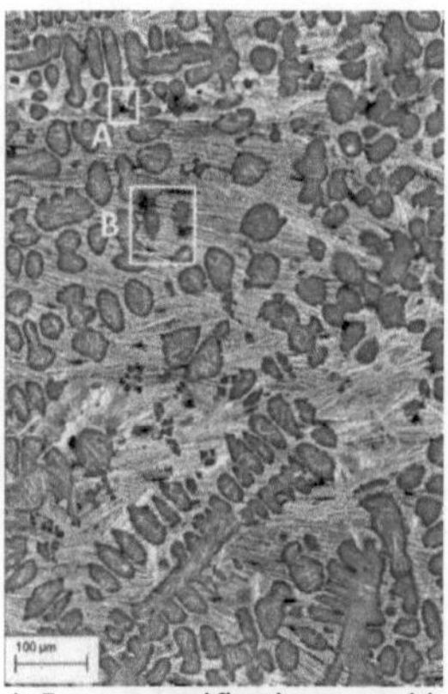

a) Low magnification; graphite always in contact with austenite

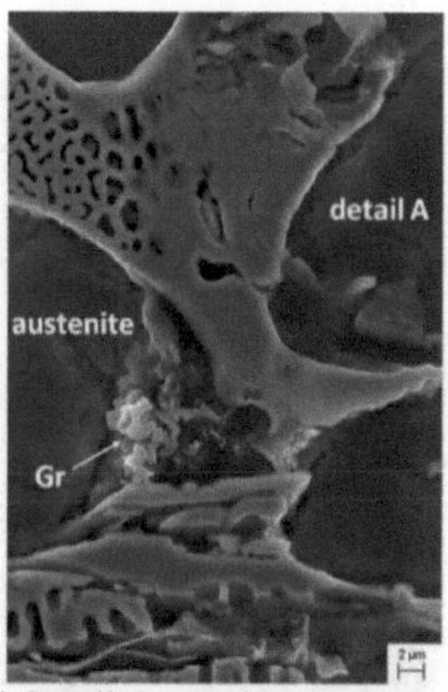

b) Detail A; graphite nucleation and growth at austenite/liquid interface

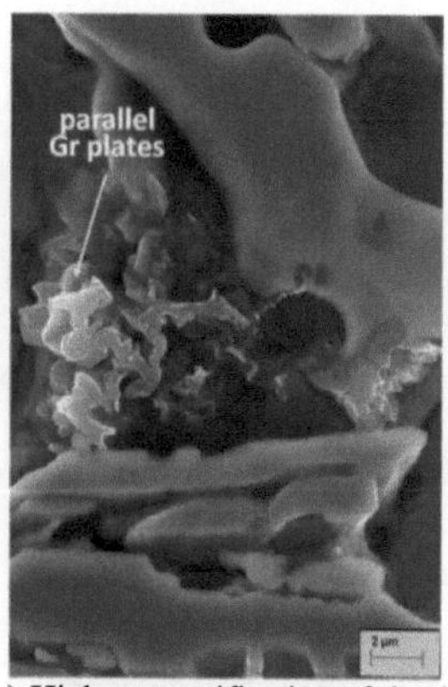

c) Higher magnification of detail A; growth of parallel graphite plates

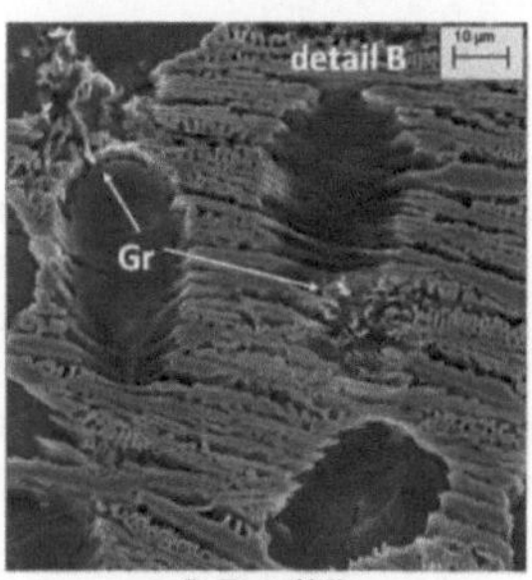

d) Detail B

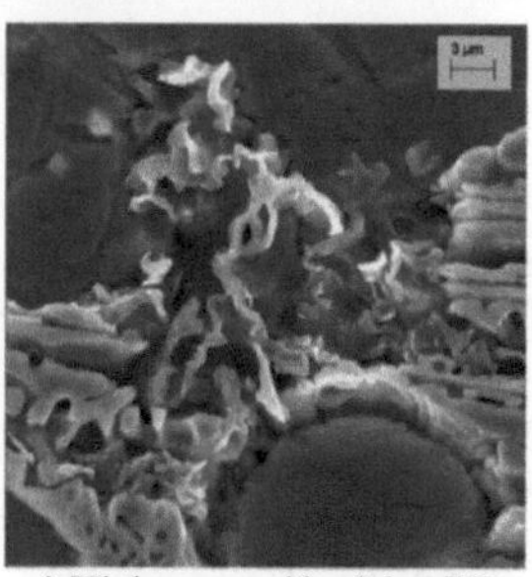

e) Higher magnif. of detail B

Figure 6. Early solidification (f_S = 0.49) SEM images for the 0.18%Ti iron (*Quench I*)

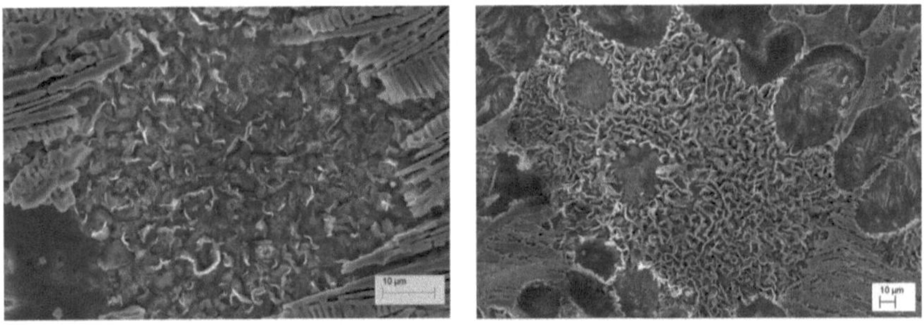

a) Early solidification; spherical graphite/ austenite aggregate

b) Late solidification; Gr/austenite aggregate growing into the liquid

Figure 7. Growth of Gr/austenite aggregate; SEM image; 0.18%Ti iron (Quench I)

As solidification proceeds the narrow graphite plates widen, while growing cooperatively with the austenite (see parallel plates on Figure 8) into the liquid.

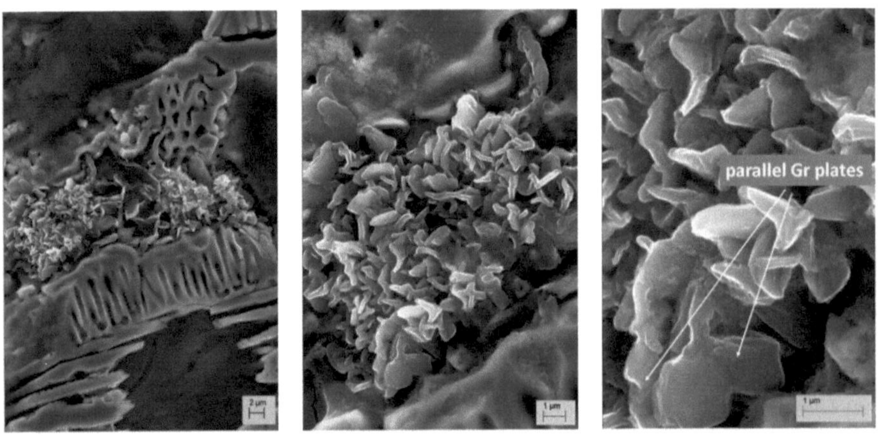

Figure 8. Solidification SEM images ($f_S = 0.57$) of the 0.18%Ti iron (Quench I); increased magnification from left to right

Discussion

To understand the effect of Ti on graphite shape we will use an undercooling (ΔT) – solidification velocity (V) diagram similar to that advocated by Park and Verhoeven [22] to explain the effect of S and Te on the A-to-D transition, and by Lacaze *et al.* [16] to explain formation of chunky graphite (see Figure 2). At low solidification velocity LAG grows at the lowest undercooling and is therefore the stable shape. As V increases, the undercooling becomes favorable to LDG at velocities higher than the transition velocity $V_{A \to D}$. Coral graphite becomes the stable shape at $V_{D \to coral}$ (Figure 9a).

Scanning Auger microscopy revealed that surface active elements such as S and Te segregate to the Fe/Gr interface [22]. Sulfur forms a 0.24 nm thick monolayer. Elements that adsorb at the Fe/Gr interface are likely to increase the undercooling required for growth. In addition, as different surface active elements induce different fault densities in the Gr, they are expected to affect the undercooling in various degrees. According to Park and Verhoeven [22], the adsorption of S on the graphite is probably responsible for the larger number of stalking faults, which causes the frequent twinning observed in LDG. If the S content of the iron is decreased the ΔT-V curves will move up. As the effect of S is stronger for LDG than for LAG (less faults in LAG) the graph shows only the LDG curve moving up. The $A \rightarrow D$ transition is displaced to lower velocities and thus type-D graphite is more stable. Also, as both D and coral graphite have larger Fe/Gr interface than type-A, the effect of S is expected to be stronger for coral than for type-A graphite. Accordingly, decreasing the S content will move the coral kinetic curve upward with respect to that for LAG (Figure 9a).

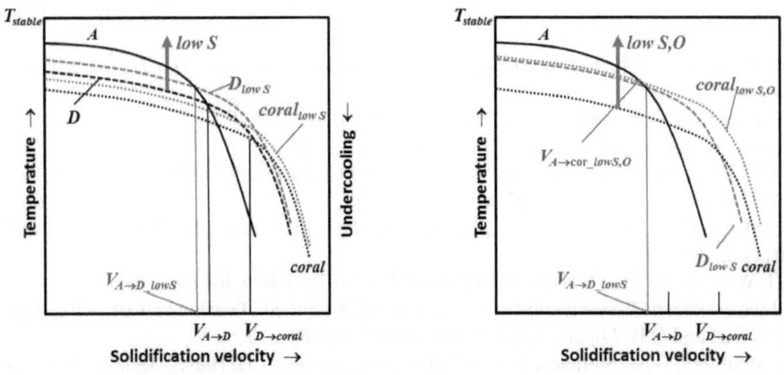

Figure 9. Effect of sulfur and oxygen on graphite shape transitions

The effect of oxygen is less clear. One Auger work [26] found both S and O on the Gr/Fe interfaces in iron castings, and it was argued that the two elements formed there during solidification. A second Auger study [22] concluded that O also segregates to Fe/Gr interfaces in type-A but not type-D eutectic. While S segregates to Fe/Gr interfaces at the solid/liquid front, O segregates to these interfaces during cool down. This last study also found that there is no significant effect of O on the A-to-D transition, but that increased oxygen activity decreases the $V_{A\rightarrow coral}$. However, this last conclusion is inconsistent with the general understanding that purer iron favors coral graphite formation. Assuming then that lower O decreases the undercooling, the difference between the LAG-to-coral transition moves upward and may even occur at lower undercooling than the LDG-to-coral Gr transition (Figure 9b).

To produce SIG iron both low sulfur and some Ti are required. As confirmed in this work, Ti is a strong deoxidizer. The low-S + Ti combination would have the effects of favoring type-D graphite occurrence at lower cooling rates than usual, with the possibility of some coral graphite formation. This is consistent with our experimental results.

The growth mechanism of LDG and SIG appears to be different than that of LAG. LAG nucleates in the liquid, grows as spherical Gr/austenite aggregates, and then incorporates dendrites (if they exist). LDG and SIG nucleate at the dendrite/liquid interface, grow as spherical Gr/austenite aggregates until they make contact with other dendrites, after which they grow around the dendrites.

Conclusions

Interrupted solidification experiments were conducted to shed light on the solidification of SIG iron. The low S of the base metal and the low O resulting from Ti deoxidation contribute to the nucleation and growth of very fine interdendritic graphite that nucleates at the austenite/liquid interface, apparently on graphite substrates. The transitions form type-A-to-D-to-coral graphite can be explained qualitatively through solidification velocity-undercooling diagrams.

References

[1] B. Lux, *Giesserei Forschung*, 19 (1967) 141

[2] P. Larrañaga, J. Sertucha, A. Loizaga, R. Suárez, D.M. Stefanescu, *AFS Trans*. 120 (2012) 347-353

[3] P. Larrañaga, J. Sertucha, A. Loizaga, R. Suárez, D.M. Stefanescu, *AFS Trans*. 120 (2012) 337-346

[4] P. Basutkar, S Yew and C. Loper, *Trans. AFS* 77 (1969) 321-28

[5] G. Ruff and J.F. Wallace, *Trans. AFS* 84 (1976) 705-728

[6] A. Okada and H. Miyake, The Unknown World of Cast Iron (in Japanese), *KANSAI Univ. Press* (1996) 141

[7] H. Nakae and K. Fujimoto, in *Key Engineering Materials*, Trans Tech Publ. Switzerland 457 (2011) 25-30

[8] K.B. Wilford and F.G. Wilson, *British Foundryman* 78 (1985) 301 and 364

[9] B. Lux, in *Recent Research on Cast Iron*, H.D. Merchant ed., Gordon and Breach (l968) 241

[10] G.X. Sun and C.R. Loper, *Trans. AFS* 91 (1983) 639-646

[11] E. Moumeni, D.M. Stefanescu, N.S. Tiedje, P. Larrañaga, J.H. Hattel, *Metall. Mater. Trans. A*, 44A, 11 (2013) 5134-5146

[12] G. Ohira, T.Sato, Y. Sayama, in *The Metallurgy of Cast Iron*, St. Saphorin, Switzerland, Georgi Publ. Co. (1974) 295

[13] H. Nieswaag, A.J. Zuithoff, in: *The Metallurgy of Cast Iron*, B. Lux, I. Minkoff, F. Mollard editors, Georgi Pub. Co., St. Saphorin, Switzerland (1975) 327-351

[14] T. Fujikawa, K. Nakamura, H. Sumimoto, S. Kiguchi, M Hatate, *Int. J. Cast Metals Res*. 11 (5) 313

[15] H. Nakae, H. Sin, *Int. J. Cast Metals Res*. 11 (5) 345

[16] J. Lacaze, L. Magnusson Åberg, J. Sertucha, in 2013 *Keith Millis Symposium on Ductile Iron*, American Foundry Soc., Nashville, TN (2013) 232-240

[17] B. Lux, W. Kurz, ISI publication 110 (1967) 193-203

[18] E. Lundbäck, I.L. Svensson, J.T. Thorgrimsson, in *Solidification 87*, Inst. of Metals (1987) 113-117

[19] B. Dhindaw and J.D. Verhoeven, *Metall. Trans. A*, 11A (1980) 1049-57

[20] J.C. Ruth, M. Turpin, Mém. Sci. Rev. Métall. 66 (1969) 633-640

[21] R.J. Brigham, G.R. Purdy, J.S. Kirkaldy, *Canadian Metall. Quaterly* 3 (1964) 239-243

[22] J.S. Park and J.D. Verhoeven, *Metall. Mater. Trans. A,* 27A (1996) 2740

[23] B. Lux, I. Minkoff, F. Mollard, E. Thury, in: *The Metallurgy of Cast Iron*, B. Lux, I. Minkoff, F. Mollard editors, Georgi Pub. Co., St. Saphorin, Switzerland (1975) 495-505

[24] G. Alonso, D.M. Stefanescu , P. Larrañaga, R. Suarez, in *Science and Engineering of Cast Iron 10*, Mar del Plata, Argentina (2014) to be published

[25] G. Alonso, D.M. Stefanescu, P. Larrañaga, J. Sertucha, R. Suárez, *AFS Trans*. 120 (2012) 329-335

[26] W.C. Johnson and H.B. Smartt, *Metall. Trans. A* 8A (1977) 553-65

Advances in the Science and Engineering of

CASTING SOLIDIFICATION

An MPMD Symposium Honoring Doru Michael Stefanescu

Novel Casting and Molding Processes

Session Chair:

Laurentiu Nastac

Advances in the Science and Engineering of Casting Solidification
Edited by: Laurentiu Nastac, Baicheng Liu, Hasse Fredriksson, Jacques Lacaze, Chun-Pyo Hong, Adrian V. Catalina, Andreas Buhrig-Polaczek, Charles Monroe, Adrian S. Sabau, Roxana Elena Ligia Ruxanda, Alan Luo, Subhayu Sen, and Attila Diószegi
TMS (The Minerals, Metals & Materials Society), 2015

COUNTER GRAVITY CASTING

John Campbell

University of Birmingham, UK.

Keywords: Casting Defects; Oxides; Bifilms; Fatigue

Abstract

The pouring of metals under gravity is generally an extremely damaging process, creating defects in the casting which limit properties and performance, but the damage can be limited to some extent by extreme care. In contrast, counter-gravity casting has the potential to fill moulds without the production of any defects. Historic problems which have impaired success are analysed and solutions presented. Counter-gravity is recommended as the process for the future.

Introduction

Why not pour castings using gravity? Gravity costs nothing, never suffers a power failure, and is ubiquitously available. However, there are a few problems: these include (i) the effect of metal velocity; and (ii) the effect of air entrainment. We shall consider these in order below.

Metal velocity

If we can accept that the recently discovered limit on the safe velocity of liquid metal entering a mold is approximately 0.5 m/s (and there is much evidence that this seems correct) then how far does the melt have to fall under the acceleration due to gravity to reach and exceed this speed? This is easily and quickly calculated and found to be close to 10 mm.

Unfortunately, this is a ridiculously small distance. Any fall of liquid metal is likely to be more than this, if not very much more! By the time the melt has fallen 200 mm, the height of the down sprue in a small mold, its velocity is close to 2 m/s, and after a fall of 1 m for a fairly large mold the velocity is nearer 5 m/s.

When it is realised that the damage done to a melt is proportional to its kinetic energy, proportional to the square of the velocity, we see that we have increased the potential for damage, measured in terms of additional square meters of bifilm cracks, by substantial margins.

Thus it is disappointing to note how gravity accelerates metals to speeds we do not want, far in excess of speeds which can be extremely damaging to melts.

Air entrainment

Practically all filling systems for molds use a conical pouring basin, sometimes called a trumpet. Especially when filled from the nozzle in the base of a bottom-poured ladle, the trumpet acts as a pump, entraining air into the metal stream. The consequence of this is a mixture of at least 50%, if not 80%, of air in the falling metal. The energy involved, especially when falling a meter or more, ensures that the metal entering the mold is in the form of an air/metal emulsion. Whereas some of the air can escape by the buoyancy of its bubbles, the abundant acreage of oxide films remain as permanent damage, constituting a dense population of crack-like features throughout the casting.

One of the most common examples of seriously damage metal is provided by the practice of casting special steels into ingot molds. The arrangement has a central trumpet and downsprue, leading to a series of radial runners to ingots places around the sprue. The metal in the ingots is seen to 'boil' with the mass of rising air bubbles, and the consequent damage to the metal is seen in the poor fatigue lives of bearings, and low toughness and possibly low hardenability.

The population of bifilms acts to reduce mechanical properties, particularly elongation, by initiating multitudes of mini-failures which grow into the so-called ductile dimples seen on the fracture surface. As unbonded interfaces in the liquid metal hydrogen or other gases can diffuse in, expanding the bifilm into a bubble, now constituting micro-porosity. Similarly the bifilms can be expanded by reducing the pressure of their environment, thus growing shrinkage porosity, possibly starting as micro-porosity, but with the potential to grow into macro-porosity if the lack of feeding is significant. The bifilms, sometimes as macroscopic films covering large areas of the casting, can be pulled apart by contractional forces, thus creating hot tearing. When bridging wall-to-wall in a casting they are extremely efficienct at creating leak paths. Thus bifilms form the starting point of all the major common casting defects which tend to control our lives in the foundry.

There is now plenty of evidence that all the above casting problems can be eliminated if bifilms are eliminated. However, when pouring under gravity, this is easier said than done.

Gravity Pouring Solutions

The arrival of Contact Pouring on the foundry scene is one of the most important and most welcome developments of recent years. The elimination of the trumpet eliminates the major source of air entrainment, with the result that the casting benefits by a reduction in bifilm problems by at least an order of magnitude. It is a major benefit, not to be underestimated.

The remainder of the filling system also has a major effect on metal quality. Most filling systems, especially those formed from pre-formed refractory tubing are usually grossly oversized, with the result that the air inside such systems has plenty of opportunity to mix with the high velocity metal, once again creating generous quantities of bifilms, to give copious problems in the casting. The use of a completely sand-molded filling system, designed to have its cross section areas at every point along its length of only the size required for the metal (so there is no room for air) is a further step forward.

Having said all of this, the fact that a filling system necessary starts out containing 100% air and requires to be converted to 100% metal means that the transition period, the priming period, must be a mixture. The priming air is a problem in even the best gravity pouring systems. The provision of run-offs to take away some of the damaged priming metal is normally never adequate when it is considered that the priming time is often 10 % of the total fill time; flow-off traps are usually never sufficiently large, and if made sufficiently large, would seriously impact the economics of casting production. Such problems tend to be less serious for castings of 500 to 1000 kg or more. For castings of 1 kg to a few hundred kg the problem is serious. However, for very small castings, smaller than 1kg, the filling system channels are so small that surface tension assists the melt to fill the whole channel, pushing the air ahead like a piston. There is no priming problem here providing channels are correctly sized as required by the naturally pressurised system.

The usual problem is that whereas most foundries have the incentive to adopt a good filling system design for a heavy, expensive casting, the foundry has hundreds of small and medium sizes castings for which the computation of a naturally pressurised system and its implementation as a piece of patternwork is too costly for a casting only worth a few dollars. This situation is completely solved by the counter gravity approach described below.

Efforts to address the priming problem in medium and heavy castings continue with new designs of filling systems which can better deal with the priming issue. These developments

continue to this day. Perhaps there is a tolerably successful future for gravity pouring; it remains to be fully demonstrated at this time.
In the meantime, prior to counter gravity being installed as part of the foundry's next 5-year plan, the adoption of contact pouring, plus the adoption of naturally pressurised filling system designs, will be a good short term low cost interim solution to boost customer goodwill in addition to company profits.

Counter Gravity Casting

The single overwhelming benefit of the counter gravity filling of molds is the control over filling velocity. This velocity is completely under the control of the founder. It need never exceed the theoretical 0.5 m/s, and so can fill the mold cavity without creating a single defect. Production managers have been known to recoil at such statements of what appears to be slow speeds, on the ground that they require to meet production speeds. However, in principle, a 0.5 m high mold could be filled in 1 second, and a massive 2 m high mold in 4 seconds. These mold filling times are of course fanciful, but demonstrate a point: the filling speed for the mold can be set to whatever is necessary, within reasonable limits. Thus a 7 second fill time for an automated greensand plant is usually not a problem; the maximum speed is controlled only by the velocity in the gate. If the initial ingate speed is above about 1.2 m/s the fountaining effect will create bifilms by the tumbling and splashing of the melt. However, when the gate is covered with a sufficient depth of metal, the speed can be ramped up without problem. Such control, varying the speed of fill, is unthinkable for the gravity filling scenario.
For those foundries with the hundreds of small and medium sized castings for which expensive naturally pressurised filling systems cannot reasonably be provided, the counter gravity approach offers a complete solution. The filling system design now consists of a hole drilled in bottom of the mold to let in the metal. It does require a sealing device such as a steel slide gate to prevent the loss of metal if the mold is lifted clear of the casting unit immediately after filling, which would be useful to maximise productivity, and would be recommended for additional technical reasons as explained below.
It is not recommended to allow the mold to sit on the casting station until the casting has solidified, keeping on the pressure in an effort to get the casting to feed to soundness. This method is subject to the great danger of convection, in which hot metal convects up part of the riser tube, remelting parts of the casting prior to cooling and convecting down adjacent areas of the riser tube. This circulation of thermal energy in continuous turbulent streams from the hot furnace below can prevent the casting from ever becoming solid. It can, of course, be countered by such devices as providing a narrow slot gate which can freeze quickly before convective effects build up. However, even this is relatively slow and therefore likely to have the production manager exploding with frustration.
The advantage of the ability to control the inward velocity of metal into the mold at any point during the filling process should not be underestimated. For instance, with a gravity filled mold if the initial fill is found to be too slow, resulting in an incompletely filled casting, the design of the filling system will require to be revised by the casting engineer, and a new pattern made, all often taking several days. The counter gravity system, with its simple drilled hole filling design, is far more easily addressed. If the first mold does not fill, the casting engineer can simply increase the fill speed at the push of a button, so that with no pattern changes and no delay, the next mold can be filled immediately, and will normally be an instant success.
A further benefit of counter gravity filling is predicted associated with its special action on the oxide covering the rising meniscus. In contrast to gravity pouring in which the oxide is entrained into the melt by splashes and waves, during the uphill motion of counter gravity

filling the oxide splits, moving to one side, becoming the skin of the casting. The movement of the oxide across the top meniscus takes time, in which the oxide thickens. Thus it provides the casting with a thick oxide skin. The thickness of this outer layer assists to bridge between sand grains and other imperfections, improving the surface finish of the casting, and contrasting with gravity pouring in which the mold is hammered by fast moving metal, and so penetrated to different degrees. Even grey iron castings I have made 'shine' when filled nicely uphill. It is also to be expected that the thick, home grown 'ceramic skin' will act as a useful barrier against mold reactions, thus reducing such phenomena as metal/mold reactions.

Counter Gravity Methods

The most common way to encourage the melt to rise against gravity is by the so-called 'low pressure' casting system. In this process the melt is held in a furnace which can be pressurised by a gas, forcing the melt up a riser tube and into the mold. This technique requires the furnace to be refilled from time to time, and is therefore a process interrupted by refilling pours from a ladle carried by fork lift truck.

Electromagnetic pumps have been proven to work well to drive liquid Al alloy uphill, even though they do appear to suffer from some limited kind of outgassing in the form of champagne type bubbles rising through the metal. Centrifugal pumps are also known to be practical to displace Al alloys and some lower melting point metals under some degree of control into molds. Both of these have immense flow velocities in their working volumes which is a cause for concern – although any real damaging effect has yet to be proven.

The author has some experience with use of small, low cost pneumatic pumps for filling sand molds. This experience, while admittedly only limited, gave excellent results. The use of pneumatic pumps for volume production has yet to be demonstrated. Their small volume would allow pressurisation with argon if necessary. The use of a protective gas would almost certainly yield a reasonable lifetime of the pump body by limiting oxidation. A pump might last a week at the rate of one casting per minute, after which the low cost SiC refractories of the body would be discarded and replaced, retaining the complex head plate assembly which contained the mechanical actuation for the stoppers.

For the counter gravity production of relatively large steel castings melting is typically carried out in a separate furnace, such as a large arc furnace which can melt economically large pieces of cheap scrap. After any appropriate steelmaking treatments, the melt is tapped into a ladle which is then transferred and lowered into a pit. A pressure tight lid is secured over the pit and the melt transferred into molds which travel overhead.

Adoption of the No-Pour Principle

Some low pressure permanent mold foundries casting Al alloy wheels do not pour their liquid metal under gravity by the expedient of putting their charge of ingots and foundry returns directly into the pressurised furnace. This is a mistake. The primary oxide skins on the charge materials are serious defects which can be transferred into the casting. However, the provision of filtration may prevent the worst aspects of this problem, although the occasional blockage of the filter by such large films has to introduce variability into the filling time, with uncertain results for the saleability of the casting.

For aluminum alloys it is possible to create an ideal system in which the liquid metal is never once subjected to pouring and can be delivered into the mold practically free from defects. Such a system might include the use of a dry hearth furnace (to eliminate primary oxide skins on charge materials) connected directly to a same-level launder or holding furnace, possibly via a filtration system (to reduce the bifilm population already in the charge materials), and finally connected (once again all at the same level) to a pump which can deliver metal uphill into a mold. The absence of pouring in the foundry and the use of an upward fill system is a council of perfection, and not difficult to implement. The Cosworth Casting system invented over thirty years ago by the author is still going strong, still producing complex cylinder block castings at about every 45 seconds. It is a pity few have so far managed to combine this principle with a good melting and non-pouring melt handling system. Eventually such foundries will have the potential for castings with immense properties, such as possibly infinite fatigue life as one instance.

For steels the concept of a no-pouring foundry is more of a challenge. Here the implementation of counter gravity has been the pressurisation of a ladle of liquid metal sat in a pit; the ladle having been filled by tilting the furnace and pouring (tapping) into the ladle in the usual manner. The pouring action is known to any witness to be horrendously turbulent; the fall involved is often 2 or 3 m, creating velocities of 6 or 7 m/s, so that huge damage is done to liquid steels by pouring from furnaces.

Fortunately, some of this damage floats out soon after tapping is complete, but not all of the entrained inclusions float clear. Others remain to degrade properties.

This threat to properties is almost certainly less serious than that introduced during the casting process, particularly during the casting of bottom-filled ingot moulds. The defects created at this stage are currently well known in the early failure of windmill bearings by fatigue, damaging the commercial prospects of wind power. I grieve over every helicopter crash I learn about; perhaps this tragedy was unnecessary. There are multitudes of other examples of mysterious apparently brittle failure of steels which are most probably initiated by bifilms resulting from pouring from furnaces.

The usual process of pouring from the melting furnace into the ladle which is subsequently lowered into the pit for pressurised transfer into the mold could be avoided if melting was carried out in an induction furnace, and if the body of the furnace itself (subsequent to any treatment) could be pressurised so as to fill moulds directly. An initial design study indicates that such a furnace might be capable of melting and casting approximately 5 tons, even though smaller furnaces in the range of 0.5 to 1.0 tons would be much more easily converted for counter gravity production directly from the furnace.

The Depressurisation Bubble Danger

One relatively unknown danger of many counter gravity systems has plagued the aluminum low pressure permanent molded wheel production used world-wide. This is the creation of bubbles from the linings of pressurised furnaces when the pressure is released after a casting is made. If air is the pressurising gas, air is forced into the permeable furnace lining during the pressurisation of the furnace used to displace the melt up the riser tube and into the mold. The air can only fill the refractory lining from the outside because the liquid metal protects the inside of the lining, but the permeation period usually lasts several minutes. However, after solidification, when the pressure is released, allowing the melt to fall back down the riser tube, the gas contained under pressure in the lining now escapes from both front and back of the lining. Those bubbles escaping from the front rise through the melt, creating bubble trails as massive bifilm defects. After some time making castings the melt becomes so viscous with films that it cannot be cast. In the meantime, horribly brittle and leaky castings complete with surface imperfections on the beauty face have been produced.

The use of nitrogen for the pressurisation of such furnaces is of no help, probably as a result of the formation of nitride bifilms and nitride bubble trails. The employment of argon is a solution of sorts, but the creation of disturbance by bubbles, even if composed mainly of an inert gas, is not expected to be helpful for good melt quality. In addition of course, this partial solution is expensive.

The problem is reduced if no high pressurisation step is used (this is not necessary anyway with good quality metal). It is further reduced if the pressure is not totally released, maintaining the metal level near the top of the riser tube between castings. This elimination of the fall of the metal down the riser tube has additional major benefits by (i) reducing the creation of oxide skins on the inside of the riser tube; while the tube remains full of metal it remains clean; and (ii) the 'whoosh' of metal into the base of the furnace stirs sediments back into suspension, reducing the quality of metal entering the following casting.

The bubbling problem is eliminated if the refractory container for the liquid metal is impermeable to gases. For this reason crucibles of hot isostatically pressed (HIPped) SiC are known to be free from this phenomenon.

An alternative solution is the use of a pump; electromagnetic or centrifugal, although concerns have been noted above. The use of small pressurised volumes associated with pneumatic pumps is a potential solution because the pressurising crucible feature of the pump can be easily obtained in a HIPped condition to avoid any degradation of quality of the melt in the pump by air bubble formation.

Conclusions

Although counter gravity is strongly recommended as the ultimate process for the production of defect-free cast products, until such plants can be produced and installed there are interim solutions for upgrading our current gravity pouring processes.

1. All gravity poured castings would benefit from the elimination of all types of pouring basins, but particularly the elimination of the conical basin, and the consequent adoption of the contact pour technique.
2. Large gravity poured castings, particularly ingots for subsequent forging or rolling, would additionally benefit from the adoption of a sand-molded naturally pressurised filling system.
3. Counter gravity filling of molds can ensure the filling process itself does not introduce defects in the casting. However, to achieve optimum results, prior processes to eliminate the oxide skins on the charge, and to eliminate the oxide bifilm populations already present in the charge, require to be implemented.
4. The full implementation of melt cleaning and turbulence-free mold filling promises to revolutionize both the casting industry, metallurgy, and engineering by the supply of metals resistant to such failure modes as fracture, fatigue and various modes of pitting and grain boundary corrosion.

Further reading by the author

1. "Complete Casting Handbook" Elsevier 2011.
2. "Quality Castings – a personal history of the Cosworth Casting Process" under consideration for publication by AFS 2015.
3. "Stop Pouring; Start Casting" The AFS Hoyte Memorial Lecture, AFS Congress Proceedings 2012; and AFS International J. Metalcasting 2012 vol 6 (No 3) pp 7-18
4. "The origin of Griffith cracks" Trans. Metall&Mater B 2011, vol 42B, (No 12) pp 1091-1097

Advances in the Science and Engineering of Casting Solidification
Edited by: Laurentiu Nastac, Baicheng Liu, Hasse Fredriksson, Jacques Lacaze, Chun-Pyo Hong, Adrian V. Catalina, Andreas Buhrig-Polaczek, Charles Monroe, Adrian S. Sabau, Roxana Elena Ligia Ruxanda, Alan Luo, Subhayu Sen, and Attila Diószegi
TMS (The Minerals, Metals & Materials Society), 2015

A Numerical Model for Predicting the Gas Evolution in Silica Sand (Furan Binder) Mold Castings

Laurentiu Nastac[1], Shian Jia[1], Mihaela N. Nastac[2] and Robert Wood[2]

[1]The University of Alabama, Tuscaloosa, AL, 35487, USA, email: lnastac@eng.ua.edu

[2]ExOne, 127 Industry Boulevard, North Huntingdon, PA 15642, USA

Keywords: Silica Sand Mold Castings, Furan Binder Pyrolysis, 3D Mold Printing Technology, Numerical Modeling

Abstract

Modeling of gas evolution during sand-mold castings is one of the most important technical and environmental issues facing the metal casting industry. The current effort funded by ExOne focused on developing the capability of numerically predicting the gas evolution for the furan binder-silica sand system. Specifically, the decomposition of furan was experimentally analyzed and then predicted based upon the work developed in the current project. This methodology can be easily implemented into existing commercial casting codes. A parametric experimental and theoretical study was also performed for A356 cylinders (D100xH200m) and bars (H50mmxW50mmxL250mm) cast into silica sand molds (furan binder) of 50 mm mold wall thickness to investigate the effects of superheat and heating/cooling conditions of the mold on the gas evolution. Such information would enable more technically and environmentally friendly decisions to be made concerning the process design used to make a given casting.

Introduction

Casting is an important process for economical manufacture of metallic parts. The most widely used mold material for this process is sand, which is held together by one of several available binders. Upon proper curing, these binders hold the sand particles into a rigid shape suitable for producing the desirable part shape. In a similar fashion, interior passageways are created by use of sand cores, which are inserted into the mold cavity. These cores also rely upon binders to rigidly hold the sand particles into the desired shape.

By design, these binders chemically break down due to heat (*i.e.*, the binders go through pyrolysis) from the molten metal. This loosens much of the surrounding sand, which allows for easy removal of the solidified casting. Pyrolysis of the binders also results in the creation of a wide variety of byproducts – some of which are hazardous air pollutants (HAPs).

One experimental study [1] completed at The University of Alabama in 1997 showed that 39 separate compounds were released during pyrolysis of a common resin-coated sand. This study concluded that the specific compounds that were released depended upon the initial chemical makeup of the binder and the temperature history of the resin during exposure to heat. Prediction of the types and amounts of each compound was found to be too complex for hand calculation. Assessment of the resulting compounds in a foundry setting results in further complications due to the wide variations in temperatures throughout a sand mold. Consequently, foundry personnel have not attempted to quantify the relationship between processing conditions (e.g., metal pouring temperature, sand thickness, binder-to-sand ratio, sand-to-metal ratio, and use of metal delivery and feeding aids) and the generation of HAPs during production of castings. However, such information would enable more environmentally friendly decisions to be made concerning the process design used to make a given casting.

Modeling of emissions has been identified as one of the top environmental issues facing the cast metal industry [2]. The kinetics of the pyrolysis reactions for both phenolic novolac and phenolic urethane cold box resins and the kinetic parameters needed for modeling are described in [3]. Pyrolysis of these resins will result in the formation of a complex mixture of fixed gases (methane and carbon monoxide), light gases, volatile compounds, and semi-volatile compounds. The hazardous air pollutants (HAPs) created by pyrolysis are mostly semi-volatile compounds, having a mass fraction of just about 0.0007 wt.% for a phenolic novolac resin at 980°C (see Table I in Ref. [3]) and about 0.0013 wt.% for a phenolic urethane cold box resin at 900°C (see Table VII in Ref. [3]). Some of the light gases and volatile liquids are also listed as HAPs (see Table I in Ref. [3]). The pyrolysis of furan binder is determined in the current study.

Model Description

The main goal of this work was to develop a computer model to predict emissions from furan binder during sand castings. The methodology for gas-emission modeling is presented below. This methodology can be implemented into existing commercial casting codes.

Time scale analysis. A time-scale analysis was performed to determine the relevant phenomena in emission modeling. From the time-scale analysis it was conclude that (a) gas convection and molecular diffusion time-scales are of the same order of magnitude and they must be accounted for in emission modeling and (b) sand-mold heat penetration (heat diffusive time) is about one order of magnitude higher than gas convection and diffusion. In this case, high thermal diffusivity sand molds (such as zircon molds) will enhance the emissions of HAPs while low thermal diffusivity molds (silica or alumina insulators) should result in decreased generation of HAPs. Therefore, the methodology for gas emission modeling should include the following four aspects:

i) Heat transfer through the mold;
ii) Kinetics of the pyrolysis reactions;
iii) Generated gases, which are convected through a porous mold; the gas flow is driven by a low pressure gradient (D'Arcy's flow) [4]; the gas pressure is proportional to the number of moles of gas as described by classical thermodynamics;
iv) Molecular diffusion.

The mathematical representation of the above approach in 3D Cartesian coordinates (*x*, *y*, and *z*) is as follows:

Energy Transfer in the mold. The temperature field evolution in the mold can be computed with

$$\frac{\partial}{\partial t}\left(\rho_m c_p T\right)=\frac{\partial}{\partial x}\left(k\frac{\partial T}{\partial x}\right)+\frac{\partial}{\partial y}\left(k\frac{\partial T}{\partial y}\right)+\frac{\partial}{\partial z}\left(k\frac{\partial T}{\partial z}\right)-L\left\{\frac{\partial}{\partial t}\left(\rho_m g_g\right)\right\} \tag{1}$$

where t is time, T is the temperature, k is the mold thermal conductivity, ρ_m is the mold density, c_p is the mold specific heat, L is the latent heat of vaporization/condensation, and g_g is the gas mass fraction. The last term on the right hand side of Eq. 1 accounts for the binder vaporization and/or condensation.

Gas Pressure. Gas pressure (P) is calculated with the empirically derived equation of state for a thermally perfect gas from classical thermodynamics,

$$P\,v = \frac{\overline{C}_g}{M} R\,T \tag{2}$$

where T is the gas absolute temperature (K), v is the specific volume (m^3/kg), $\overline{C}_g$ is the average gas concentration within the elemental (computational cell) volume (kg gas/kg sand), M is the average atomic weight of gases (kg/mol), and R is the universal gas constant (R = 8.314 J/kg-mol/K). The specific pore volume in the sand, v, is calculated as

$$v = \frac{1}{\rho_m} f_p \tag{3}$$

where f_p is the pore fraction of the sand.

Kinetics of Pyrolysis Reactions. This parameter can be approximated by the following first-order kinetics chemical reaction [2]:

$$\frac{\partial W_g}{\partial t} = k_r\left(W_{max} - W_g\right) \quad \text{with} \quad k_r = A\,e^{-\Delta H / RT} \tag{4}$$

where k_r is the first-order reaction constant (s^{-1}), A is a constant called the frequency factor (s^{-1}), ΔH is the activation energy, and W_{max} is the maximum amount of component that can be produced at the operating temperature (in kg gas/kg sand). A previous study [3] shows the values of the kinetic parameters for shell resins. Similar parameters can be determined for furan binder based on the current experimental effort.
Thus, the amount of gas produced at any time, W_g (kg gas/kg sand) is updated each time step as:

$$W_g^{\,n} = W_g^{\,o} + k_r^{\,o}\left(W_{max} - W_g^{\,o}\right) \tag{5}$$

where n and o denote the new and old values, respectively. Note that the maximum value of W_g is equal to W_{max}.

Species Transport Equation. The conservation of species can be written as:

$$\frac{\partial}{\partial t}\left(\rho\,\overline{C}_g\right) + \frac{\partial}{\partial x}\left(\rho u_g \overline{C}_g\right) + \frac{\partial}{\partial y}\left(\rho v_g \overline{C}_g\right) + \frac{\partial}{\partial z}\left(\rho w_g \overline{C}_g\right) = \frac{\partial}{\partial x}\left(\rho D_g \frac{\partial}{\partial x}\left(\overline{C}_g\right)\right) + \frac{\partial}{\partial y}\left(\rho D_g \frac{\partial}{\partial y}\left(\overline{C}_g\right)\right) + \frac{\partial}{\partial z}\left(\rho D_g \frac{\partial}{\partial z}\left(\overline{C}_g\right)\right) \tag{6}$$

where ρ is the gas density, D_g is the gas molecular diffusivity, $\overline{C}_g$ is the average gas concentration within the elemental volume, and u_g, v_g and w_g are the velocity components in the x, y, and z directions, respectively.

The average gas concentration within the elemental volume, $\overline{C}_g$ (kg gas/kg sand) is updated at the beginning of each time-step calculation as follows:

$$\overline{C}_g^{\,n} = \overline{C}_g^{\,o} + k_r^{\,o}\left(W_{max} - W_g^{\,o}\right) \tag{7}$$

Velocity components. Velocity components (u_g, v_g, and w_g in the x, y, and z directions, respectively) are calculated from D'Arcy's flow as follows:

$$u_g = k_D \frac{\partial P}{\partial x} \quad v_g = k_D \frac{\partial P}{\partial y} \quad \text{and} \quad w_g = k_D \frac{\partial P}{\partial y} \tag{8}$$

where k_D is the permeability coefficient.

Algorithm. The flowchart for the emission model is presented in Figure 1, where T is the new temperature distribution, T_{cr} is the minimum temperature for binder breakdown, c_g and w_g are the new gas concentration distribution and the new amount of gas produced at time t, respectively, δt_{min} is the current time step used in calculations based on gas molecular diffusivity and gas velocity, and IC and BC stand for initial and boundary conditions, respectively. The plotted variables, c_g and w_g, are defined as follows:

$$w_g = W_g / W_{max} \quad \text{and} \quad c_g = \overline{C}_g / W_{max} \tag{9}$$

The model output consists of saving c_g, w_g, and T at any chosen time for all cells. Also, the final values of c_g, w_g, and T of all cells are saved at the end of computations on the computer disk.

Results and Discussion

To demonstrate the gas emissions from silica sand molds, 50mmx50mmx250mm bar and D100mmxH200mm cylindrical geometries were used (see Figure. 1). The mold wall thickness was 50 mm for both geometries. The pouring temperatures of the A356 alloy were 710°C (superheat of 100°C) and 810°C (superheat of 200°C). The ambient temperature was 25°C. The temperature evolution in middle of the bar and cylindrical castings and in the bar and cylindrical molds (at 25 mm and 10 mm from the metal-mold interface) are shown in Figure 3. Thermo-physical properties of the mold materials are shown in Table 1. Temperature-averaged gas properties are shown in Table 2. Boundary conditions for gas concentration are zero-gradient at bottom of the mold and zero-value at the left, right, and top of the mold. The loss of ignition (LOI) contents were also measured in the bar and cylindrical molds at locations near the thermocouple areas and at the metal/mold interface (see Tables 3 and 4). As it can be seen in Tables 3 and 4, higher mold temperatures provided higher values of the LOI content.

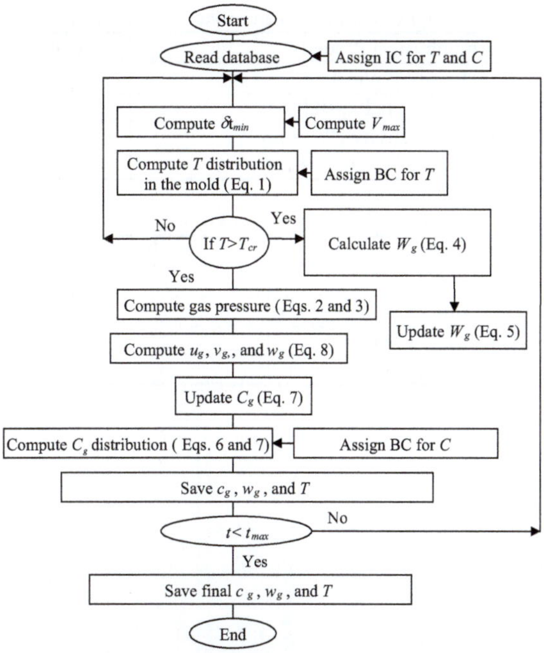

Figure 1. Flowchart of the emission model.

Figure 1. Tested ExOne molds and the cast bar and cylindrical geometries.

Table 1. Thermophysical Properties of the Tested Silica Sand Molds

Thermophysical Property	Symbol	Value
Density [kg/m^3]	ρ_m	1400
Specific heat [J/kg•K]	c_p	1150
Thermal conductivity [W/m•K]	k	0.70
Permeability coefficient [N•s/m^4]	k_D	1.0e-07

Table 2. Average Gas Properties [5]

Parameter	Symbol	Value
Density [kg/m^3]	ρ	0.5
Dynamic viscosity [N•s/m^2]	η	2 x10^{-5}
Thermal conductivity [W/m•K]	k	0.05
Molecular diffusivity [m^2/s]	D_g	5 x10^{-5}

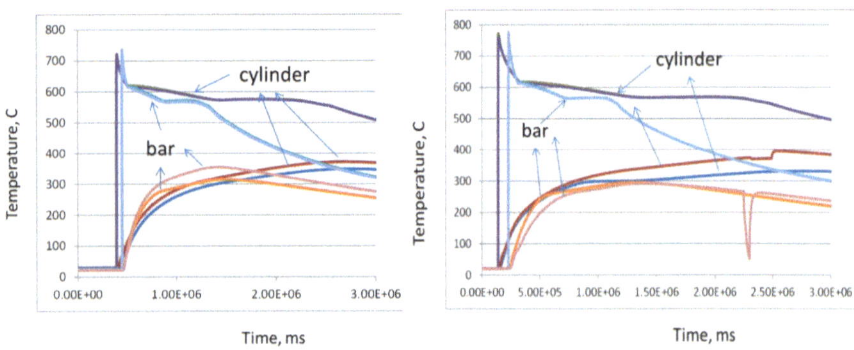

Figure 2 The temperature evolution for the tested molds (bars and cylinders): (a) superheat of 100°C and (b) superheat of 200°C.

Table 3. LOI Results for the Bar and Cylindrical Geometries (100°C Superheat)

Distance from the metal/mold interface	Bar (bottom, front)	Cylinder (front, left)
25 mm	1.32; 1.39	1.0; 1.3
10 mm	1.19; 1.07	0.86; 0.92
all	1,52; 1.26	1.0; 1.23
corner	1.33	1.34

Table 4. LOI Results for the Bar and Cylindrical Geometries (200°C Superheat)

Distance from the metal/mold interface	Bar (bottom, front)	Cylinder (front, left)
25 mm	1.30; 1.38	1.43; 1.35
10 mm	1.12; 1.06	1.04; 0.97
all	1.28; 1.27	1.27; 1.26

In Figure 3, the test sample weight loss as a function of temperature as determined by TGA. The test sample size was 50 mg, the heating rate was 20°C/min and the tested temperature range was 25-900°C. As it can be seen from Figure 3, the pyrolysis of the furan binder is mostly completed at a temperature of about 530°C.

Figures 4 and 5 shows the predicted distribution of temperature, rate of HAPs generated, and HAPs concentration in the cylindrical (after 1000s) and bar (after 300s) geometries, respectively. The gas pressure distribution patterns for both casting geometries can also be determined based on Fig. 4c and 5c and Eq. 2. Thus, these types of results are indeed very useful to estimate the potential outgassing of the mold.

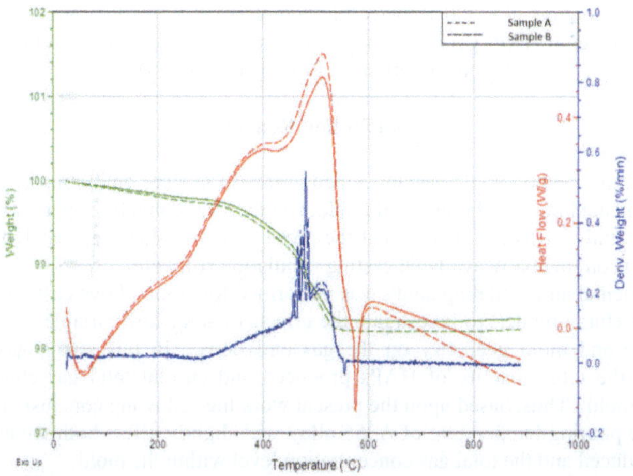

Figure 3 Furan binder pyrolysis determined by TGA for two samples taken at different times.

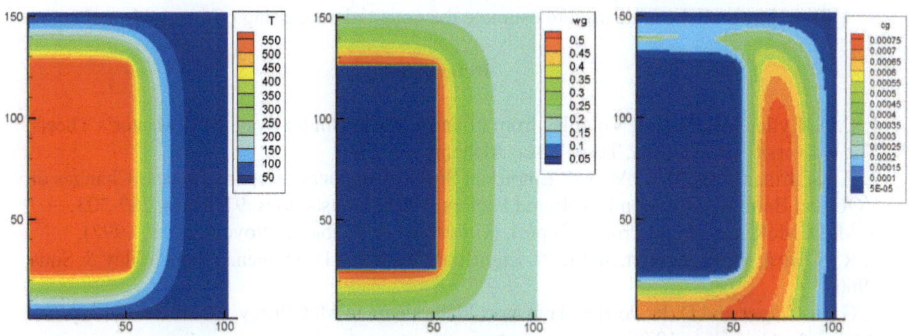

(a) (b) (c)

Figure 4 Distribution after 1000 seconds (cylindrical geometry): (a) Temperature (T in deg. C), (b) Current local rate of HAPs generated (w_g is defined by Eq. 9), and (c) Average local HAPs concentration (c_g is defined by Eq. 9).

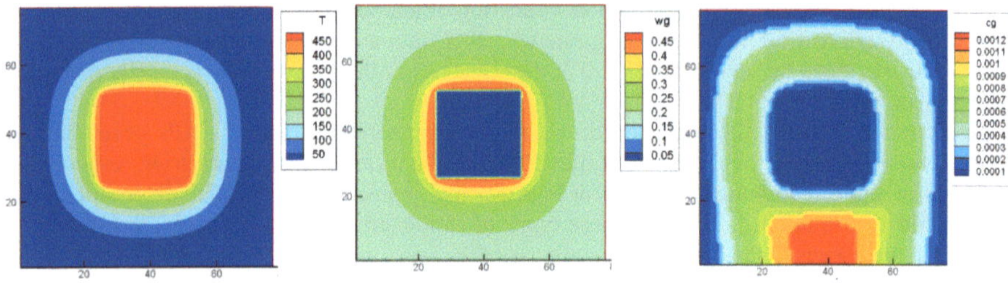

(a) (b) (c)

Figure 5 Distribution after 300 seconds (bar geometry): (a) Temperature (T in deg. C), (b) Current local rate of HAPs generated (w_g is defined by Eq. 9),and (c) Average local HAPs concentration (c_g is defined by Eq. 9).

Concluding Remarks

1. A 3D modeling methodology was developed to investigate gas emissions in sand castings. It was implemented into a 2D numerical model and then validated by experiments performed in the current study. It is anticipated that the present methodology can easily be implemented into any 3D commercially available casting simulation software.
2. An experimental and modeling study was performed for A356 alloys cast into Exone's silica sand molds (furan binder) to investigate the effects of some key variables, including pouring temperature and mold geometry on the gas emissions. In this study, gas emissions are defined as the total quantity of HAPs produced and current total gas concentration level within the mold. Thus, based upon the present work the following conclusions were reached:
   (i) The pouring temperature of A356 alloy will slightly affect both the amount of HAPs produced and the total gas concentration level within the mold.
   (ii) A very important parameter is the volume ratio of the mold to the casting. A larger volume ratio (bar geometry) will decrease the amount of HAPs produced but will increase the total gas concentration level within the mold.
   (iii) The model can be used to estimate the potential outgassing of the molds and cores.

References

1. C. A. Lytle, "Analysis of Volatiles from Foundry Resin Binder Systems", (Master's Thesis, The University of Alabama, Tuscaloosa, Alabama, 1997).
2. P. Kauffmann, and R. C. Voigt, "Empirical Study of Impact of Casting Process Changes on VOC and Benzene Emission Levels and Factors," AFS Transactions, 97(1993), 297-303.
3. M. D. McKinley, Final Project Report, (University of Alabama, November 13, 1997).
4. R. B. Bird, W. E. Stewart, and E. N. Lightfoot, Transport Phenomena, (John Wiley & Sons, 1960).
5. G. H. Geiger and D. R. Poirier, Transport Phenomena in Metallurgy, (Addison-Wesley Publishing Company, 1978).

Advances in the Science and Engineering of

CASTING SOLIDIFICATION

An MPMD Symposium Honoring Doru Michael Stefanescu

Solidification Processing V

Session Chair:

Afina Lupulescu

Advances in the Science and Engineering of Casting Solidification
Edited by: Laurentiu Nastac, Baicheng Liu, Hasse Fredriksson, Jacques Lacaze, Chun-Pyo Hong, Adrian V. Catalina, Andreas Buhrig-Polaczek, Charles Monroe, Adrian S. Sabau, Roxana Elena Ligia Ruxanda, Alan Luo, Subhayu Sen, and Attila Diószegi
TMS (The Minerals, Metals & Materials Society), 2015

EVALUATION OF THE CASTING/CHILL INTERFACE THERMAL BEHAVIOUR DURING A319 ALLOY SAND CASTING PROCESS

Farzaneh Farhang Mehr, Steve Cockcroft, Carl Reilly, and Daan Maijer

Department of Materials Engineering, University of British Columbia;
6350 Stores Rd.; Vancouver, BC V6T 1Z4, Canada

Keywords: Interfacial Heat Transfer Coefficient, A319 Alloy Sand Casting, Chill, Air-gap, Mathematical Modelling

Abstract

One approach to improve fatigue resistance of A319 alloy cast components is to reduce the size of microstructural discontinuities by increasing cooling rates during solidification. A combination of experimental and numerical methods were used to explore the efficacy of using a water-cooled chill to increase cooling rates during solidification in A319 alloy cast in a chilled-wedge format. A solid and a water-cooled chill were designed and installed in a bonded sand mould package instrumented with thermocouples to measure cooling rates and two Linear Variable Differential Transducers to measure the displacement of the casting/chill interface. A mathematical model was developed to evaluate the casting/chill interface heat transfer behaviour in the casting produced with the two chill configurations. The experimental results show that for the casting format studied, the water-cooled chill is able to significantly increase cooling rates during solidification up to 50mm from the chill and the macro-scale gap that develops at the interface is significantly larger for the case of the solid chill compared with the water-cooled chill.

Introduction

One current trend in the automotive industry is to produce small, lightweight, high-power gas and diesel engines. This is being driven by consumer demand to improve fuel economy, without sacrificing performance, and by increasingly stringent CAFE (Corporate Average Fleet Economy) requirements in North America and in Europe [1-3]. A consequence of this trend is the need for an improvement in the thermal and mechanical fatigue durability of the engine blocks, particularly in the engine block main bearing bulkheads. In the absence of altering alloy chemistry, one practical way to improve fatigue performance is to reduce the length scales of the microstructural features arising from solidification. This in turn can be achieved by increasing the cooling rate during solidification, providing cold-shuts and a columnar structure can be avoided [4,5].

In terms of aluminium engine block manufacturing technology, the precision sand casting process is a favourable technology owing to its ability to produce parts with a high degree of design complexity and with relatively good surface quality [6-8]. The utilization of a well-designed metallic chill offers the potential to increase the local cooling rate during solidification of the cast metal, thereby improving fatigue performance in a given alloy system [9,10]. The challenges with this approach include avoiding the occurrence of cold shuts and the formation of a columnar structure adjacent to the chill, while still reducing the length scales of the microstructure to a distance from the chill sufficient to improve component fatigue performance. In industry, excessive cooling adjacent to the chill is avoided by preheating the chills to temperatures up to 200°C. In the context of using a water-cooled chill this would require that the water be switched on after the mould filling process is complete and a prescribed time has elapsed sufficient to avoid the formation of a columnar structure adjacent to the chill.

In addition to the thermal capacity of the chill, the evolution of the casting/chill interface thermal resistance plays a critical role in governing the ability of the chill to extract heat. In general, the contact between the mould and the casting changes with time, owing to the interplay between thermal contraction of the casting and thermal expansion of the chill/mould [11,12]. As a result, the resistance to heat flow that develops at the interface –expressed mathematically as $1/h$, where h is the heat transfer coefficient (W m^{-2} K^{-1})– has a significant influence on the solidification rate of the metal [13,14].

The present work is focused on determining quantitatively the efficacy of a water-cooled chill in increasing the cooling rate relative to a solid chill in a standard wedge chill mould fabricated in precision sand. As part of this work differences in the chill/casting interface behaviour will also be examined.

Experimental

In order to reproduce the broad range of solidification rates a sand cast engine block experiences during solidification, a standard wedge-shaped A319 casting (Figure 1) was cast in a bonded sand mould (designed and provided by "Nemak Canada Ltd."). The wedge-mould is designed to allow solidification horizontally from the narrow end of the wedge, where the chill is located, toward the thicker end, which is attached to a large riser. Both a solid chill configuration and a water-cooled chill configuration were tested. In this research, A319 alloy was used, as it represents the dominant alloy used for the commercial production of engine blocks and cylinder heads in North America. The alloy composition is presented in Table 1. Also, commercially pure copper (99.9%) was used in the production of the chills (Figure 2.a).

Table 1. Chemical Composition of A319 Alloy

Alloying Element	Si	Fe	Cu	Mn	Mg	Zn	Ti	Sr	Al
Balance (Wt%)	8.10-8.75	0.27-0.59	2.60-2.95	(0.7·Fe)-0.44	0.31-0.4	0.4-0.8	0.12-0.16	0.004 max	Bal.

The sand mould package and chills were instrumented with seven thermocouples (type K, Super OMEGACLAD XL) to obtain temperature data from within the casting and chills. Four thermocouples were placed at mid height in the mould cavity to measure temperatures at different distances from the chill (at 5, 10, 30, and 50 mm) – see Figure 1. These measurements enabled the investigation of the variation in cooling rate with distance from the chill. To examine the thermal behaviour of the chills, three holes were drilled into the chills to facilitate placement of thermocouples, 2mm from the interface. The interface air-gap variation was measured with two Linear Variable Differential Transducers (LVDT, Schaevitz[1] – HR 050, and HR 100). One

[1] Schaevitz is a trademark for Measurement Specialties Corporation.

was placed in contact with the chill in proximity to the interface with the cavity and the other placed such that it just protruded into the cavity (Figure 2.b) and would become solidified into the casting.

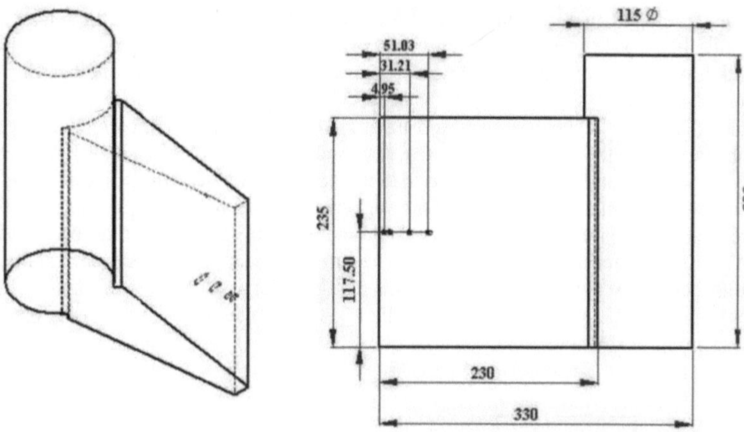

Figure 1: Cast wedge and riser configurations (all dimensions are in mm).

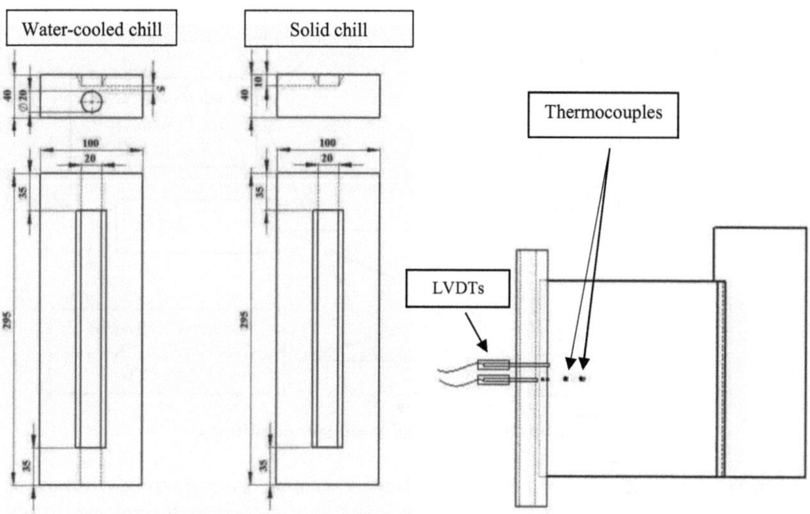

(a) (b)

Figure 2: a) Configuration of the chills (all dimensions are in mm), b) placement of LVDT in casting and chill

The liquid metal was poured into the mould cavity at 740±5°C, and the water flow rate in the chill cooling channel was set to 50 L min^{-1} and measured with an inline water flow meter. The temperature of the main water supply was fairly constant at ~10°C, and water was switched on 10 seconds after the completion of pouring process.

Model Development

In the current work, a mathematical based heat transfer model of the casting, the chills and the mould was developed in the commercial finite element program ABAQUS[1] version 16.3. A thermal-only analysis was completed. Previous work [9] with thermocouples at various heights revealed that only a small vertical temperature gradient was developed in the casting. Thus, it was assumed to a first approximation that the calculation domain can be limited to a two-dimensional (2D) slice of the casting at mid-height. As a final step in reducing the computational size of the problem, the thermal field was assumed to be symmetric about a vertical plane bisecting the casting and hence only one-half of the geometry was modelled (Figure 3). The model domain, including identification of the various boundary conditions, is shown in Figure 3. Details related to the material properties used and methodology for accounting for the latent heat of solidification are presented elsewhere [9]. The same boundary conditions were applied to the solid chill model, except at the cooling channel boundary where a constant heat transfer coefficient of 8.7×10^{-3} (W m^{-2} K^{-1}) was applied based on the flow conditions existing in the channel assuming steady state flow.

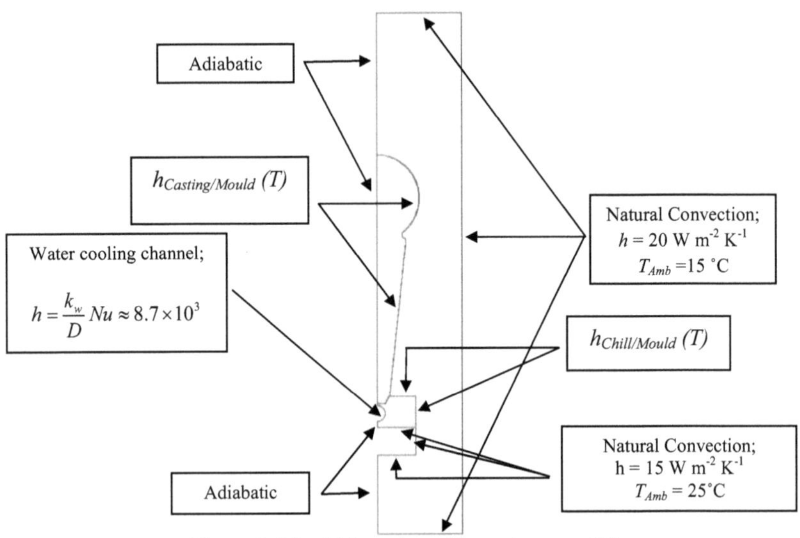

Figure 3: Mould domain and boundary conditions

All of the surfaces normal to the plane of the analysis domain are assumed to be adiabatic as is the symmetry plane. Three sets of temperature dependent heat transfer coefficients were used to describe the casting/sand mould boundary, casting/chill interface and chill/sand mould boundary. Since the modelling does not simulate the mould filling process, the initial temperature of the melt was set to the temperature recorded by the thermocouples at the end of the pouring process.

---

[1] ABAQUS is a trademark of Dassault Systems

The model was used to estimate the heat transfer coefficient by aligning the model predictions to the thermocouple measurements using a trial-and-error methodology whereby changes were made to the temperature dependent interfacial heat transfer coefficient (IHTC) between the casting and the chill until satisfactory agreement was achieved between the model predictions and experimental measurements. A maximum temperature difference of 20°C was assumed to be representative of satisfactory agreement.

Results and Discussion

Comparisons between the model predictions and measurements are shown in Figure 4 (a) and (b). As can be seen, the model is able to reproduce the thermal behaviour obtained from the thermocouples within the castings and chills to within approximately 20°C for most of the locations examined. Thus, the estimates obtained for behaviour of the interfacial heat transfer coefficient may also be expected to be reasonable.

For the case of the water-cooled chill, the experimental data and model data show an initial increase in chill temperature to a maximum of ~100°C, in the period when the water cooling is delayed. The temperature then drops to ~30°C, where it remains largely unchanged, once the water is introduced to the cooling channel. The introduction of the water allows a large temperature difference between the water-cooled chill and the casting to be maintained thus sustaining a large driving force for heat transfer. For the case of the solid chill, the temperature in chill rises to a peak temperature of approximately 300°C, over 1000s, thus there is a substantial reduction in the driving force for heat transfer. As a consequence of this, the solidification time at depths of 30 and 50mm is greatly increased in the solid chill relative to the water-cooled chill (note the difference in time scale in the two plots). At 50mm the average cooling rate over the solidification temperature in the solid chill is ~0.1°C/s, whereas in the water-cooled chill it was ~0.6°C/s.

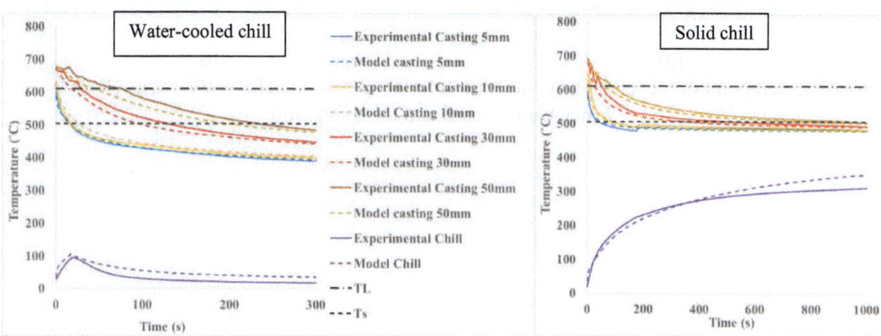

(a) (b)

Figure 4: comparison between the experimental data and model predictions

Figure 5 (a) and (b) show the temperature dependence of the casting/chill IHTCs derived with the model for the two cases using the trial and error fitting approach –i.e. this represents the variation in IHTC input to the model that resulted in a reasonable fit to the data. Vertical lines have been added showing the liquidus and solidus temperatures. The IHTC for both cases starts at a high value 3000 W m^{-2} K^{-1} and remains constant until the liquidus is reached (stage 1). A relatively rapid drop is observed between the liquidus and solidus, to a value of 2100 W m^{-2} K^{-1} for the water-cooled chill and a value of 1100 W m^{-2} K^{-1} for the solid chill (stage 2). Below the solidus temperature (stage 3) the heat transfer coefficient continues to drop rapidly with temperature until between 400 and 300°C for the water-cooled chill, and 470°C for the solid chill at which point the rate of decrease with temperature moderates.

In stage 1, good physical contact between the casting and the chill is achieved as the molten metal fills at least a portion of the asperities present on the chill surface due to the metals high fluidity. Complete contact with the chill is not generally achieved due to surface tension effects. In stage 2, the liquid to solid transformation starts at the interface and contact between the casting and the chill begins to degrade, as the solidification structure at the interface develops shrinkage occurs and a gap begins to form. In stage 3, solidification proceeds into the casting resulting in further shrinkage and the formation of a larger gap. Note: that there is a more rapid drop in the IHTC observed in stage 2 in the solid chill than in the water-cooled chill.

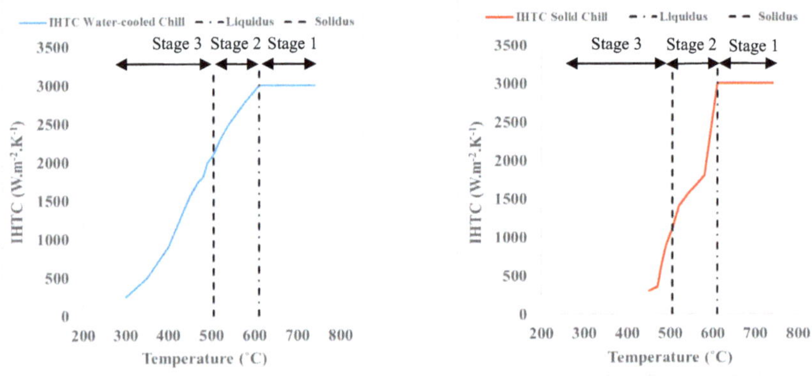

(a) (b)

Figure 5: Temperature dependent casting/chill interfacial heat transfer coefficients (IHTC) – (a) Water-Cooled chill and (b) Solid chill.

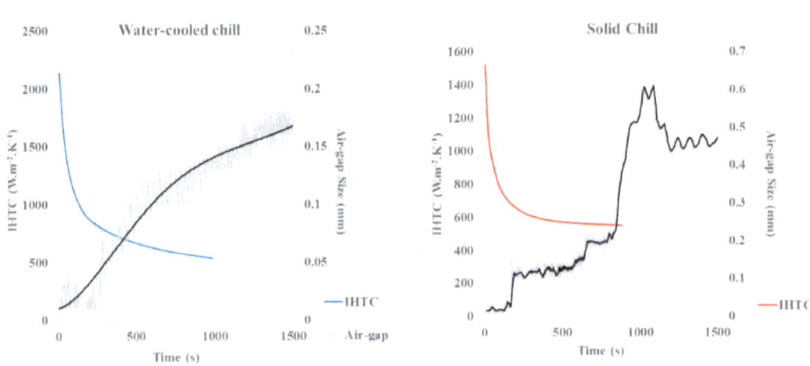

(a) (b)

Figure 6: IHTC predictions and air-gap size measurements – (a) Water-Cooled chill and (b) Solid chill.

Figure 6 shows the IHTC and air-gap size variations with time, for the first 1000 seconds in the two castings (note the difference and the primary and secondary y axis scales in the two plots). The light grey lines are the raw data and the black lines are a moving average of the raw data, taken to remove some of the noise. This data may be used to understand the variations in IHTC with temperature for the two configurations shown in Figure 6. Initially, the IHTC drops significantly at the onset of the formation of the air gap for both cases. The comparatively high insulating nature of air and loss of contact causes a rapid drop in the interfacial heat transfer coefficient/increase in the thermal resistant. The IHTC then continues to decrease, but at a lower rate, since the resistance does not change rapidly with increasing growth of the air-gap. Note that in Figure 6, the gap develops more rapidly in the solid chill compared with the water-cooled chill, which is consistent with a greater rate of drop in the IHTC in stage 2 in the solid chill compared with the water-cooled chill –see Figure 5. Further work is planned to develop a coupled thermal stress model for analysis of the results and to explain the difference in behaviour of the gap in the water-cooled chill casting compared with the solid chill casting.

Conclusions

The present work focused on the quantitative determination of the casting/chill interface thermal behaviour in A319 bonded sand casting cooled with both a solid copper chill and water-cooled copper chill. The results show that water-cooling was effective in maintaining a high driving force for heat transfer compared with the solid chill, helping to achieve high cooling rates deeper into the wedge casting. Furthermore, a smaller air gap formed at the casting/chill interface with the water-cooled chill compared with the solid chill. The main conclusions from this work are summarised as follows:

1. Water-cooling was found to significantly increase the cooling rate during solidification at a distance of 50mm for the chill, in a wedge format casting, from ~0.1°C/s, for the solid chill, to ~0.6°C/s, for the water-cooled chill.
2. Characterisation of the variation in interfacial heat transfer under the casting conditions examined in the study with a 2D transient heat conduction model revealed differences in behaviour of the thermal resistance at the interface between the solid chill and the water-cooled chill. Notably, the heat transfer coefficient dropped more rapidly in the solid chill case relative to the water-cooled chill. The LVDT measurements indicated that this was due to a more rapid formation of an air gap at the casting/chill interface in the case of the solid chill.

Acknowledgment

The writers would like to acknowledge "Nemak Canada", and in particular, Dr. Robert MacKay for providing the sand moulds and A319 alloy and the NCE-Auto21 for providing partial support for this work.

References

[1] M.J. Caton, J.W. Jones, H. Mayer, S. Stanzl-Tschegg, J.E. Allison, Demonstration of an endurance limit in cast 319 aluminium, Metallurgical and Materials Transactions A. 34 (2003) 33-41.

[2] T.J. Smith, H. Sehitoglu, E. Fleury, H.J. Maier, J. Allison, Modelling high-temperature stress-strain behaviour of cast aluminium alloys, Metallurgical and Materials Transactions A. 30 (1999) 133-146.
[3] P. Bastani, J. Heywood, C. Hope, US CAFE Standards: Potential for Meeting Light-Duty Vehicle Fuel Economy Targets, 2016–2025, (2012).
[4] S. Shabestari, M. Malekan, Thermal analysis study of the effect of the cooling rate on the microstructure and solidification parameters of 319 aluminium alloy, Can. Metall. Q. 44 (2005) 305-312.
[5] L. Ananthanarayanan, F. Samuel, J. Gruzleski, The thermal analysis studies on the effect of cooling rate on the microstructure of the 319 aluminium alloy, AFS Transactions. 100 (1992) 383-391.
[6] R.W. Grenkowitz, M.J. Braskich, A.D. Ackerman, Gravity precision sand casting of aluminium and equivalent metals, Gravity precision sand casting of aluminium and equivalent metals. (1997).
[7] R.I. Mackay, D. Cusinato, J.H. Sokolowski, Chemistry optimisation to improve casting durability of engine blocks, International Journal of Cast Metals Research, (2010) Vol. 23, pp. 137-149.
[8] D.M. Stefanescu, ASM Handbook: Casting, Asm Intl, 1992.
[9] F. Farhang Mehr, Quantitative assessment of the effect of copper chills on casting/chill interface behaviour and the microstructure of sand cast A319 alloy, M.A.Sc Thesis, Department of Materials Engineering, University of British Columbia, Vancouver, BC, Canada, 2012.
[10] L. Heysler, F. Feikus, M. Otte, Alloy and casting process optimization for engine block application, AFS Transactions. 50 (2001) 1-9.
[11] T. Loulou, E. Artyukhin, J. Bardon, Estimation of thermal contact resistance during the first stages of metal solidification process: I—experiment principle and modelisation, Int. J. Heat Mass Transfer. 42 (1999) 2119-2127.
[12] C. Hallam, W. Griffiths, A model of the interfacial heat-transfer coefficient for the aluminium gravity die-casting process, Metallurgical and Materials Transactions B. 35 (2004) 721-733.
[13] T.S.P. Kumar, K.N. Prabhu, Heat flux transients at the casting/chill interface during solidification of aluminium base alloys, Metallurgical and Materials Transactions B. 22 (1991) 717-727.
[14] M.A. Gafur, M.N. Haque, K.N. Prabhu, Effect of chill thickness and superheat on casting/chill interfacial heat transfer during solidification of commercially pure aluminium, J. Mater. Process. Technol. 133 (2003) 257-265.

Advances in the Science and Engineering of Casting Solidification
Edited by: Laurentiu Nastac, Baicheng Liu, Hasse Fredriksson, Jacques Lacaze, Chun-Pyo Hong, Adrian V. Catalina, Andreas Buhrig-Polaczek, Charles Monroe, Adrian S. Sabau, Roxana Elena Ligia Ruxanda, Alan Luo, Subhayu Sen, and Attila Diószegi
TMS (The Minerals, Metals & Materials Society), 2015

DEPENDENCE OF HARDNESS ON MICROSTRUCTURE OF A DIRECTIONALLY SOLIDIFIED Sn-40wt.%Bi-0.7wt.%Cu ALLOY

Bismarck Luiz Silva [1] and José Eduardo Spinelli [1]

[1] *Department of Materials Engineering, Federal University of São Carlos – UFSCar, Washington Luis Rd, km 235, 13565-905, São Carlos, São Paulo, Brazil*

Keywords: Sn-Bi-Cu alloys; microstructure; hardness.

Abstract

Continuous efforts in replacing Sn-Pb eutectic solder have been attempted. Sn-Bi alloys are candidates as lead-free solders for low temperature soldering due to its adequate cost and general consistency. Additions of Cu within these alloys can improve the ductility, mechanical strength and wettability. Nevertheless, a deeper study considering the characterization of the as-soldered ternary Sn-40wt%Bi-0.7wt%Cu microstructures remains still to be performed. Thus, the present experiments were carried out by directional solidification (DS) of the Sn-40Bi-0.7Cu with a view to establish experimental interrelations involving solidification thermal parameters (growth rate - V_L and cooling rate - $\dot{T}_L$) with some microstructure features (λ_2). The experimental relations obtained for this alloy have been compared with those already determined for the binary Sn-0.7wt.%Cu and Sn-40wt.%Bi alloys. The microstructures from the bottom to the top along the Sn-Bi-Cu alloy were entirely dendritic. Further, a Hall-Petch equation is proposed relating HV to λ_2 for the Sn-40Bi-0.7Cu alloy.

INTRODUCTION

Due to the inherent toxicity of lead (Pb), environmental regulations around the world have been targeted to eliminate the usage of Pb-bearing solders in electronic assemblies. This has prompted the development of ''Pb-free'' solders, that's why that the research activities in this field have been developed [1-3].

There are strict performance requirements for solder alloys used in microelectronics. In general, the solder alloy must meet the expected levels of reliability and mechanical properties (as the hardness), as well as electrical and mechanical performance. Within this framework, Sn-Bi [2,4,5,6] and Sn-Cu eutectic [7-9] arise as promising Pb-free alternative due characteristics as low melting point, low cost, low dissolution of Cu substrate and feasible for soldering processes.

According to Zu and co-authors [6] the *liquidus* temperature (T_L) and eutectic temperature (T_E) for hypoeutectic Sn-40wt%Bi alloy are 169.5°C and 136.5°C, respectively, whereas Osório *et al.* [4] cite T_L=172°C and T_E=136°C. This latter work showed that the microstructure of the Sn-40wt%Bi is constituted of Sn-rich dendritic matrix and a eutectic mixture in the interdendritic regions. However, some studies [6,10] showed that microstructure for this binary is composed by Sn-rich matrix with Bi precipitates embedded inside this structure and eutectic mixture with alternate-layered lamellar structure.

Meanwhile, the Sn-Cu eutectic reaction is between a faceted intermetallic Cu_6Sn_5 phase and a non-faceted Sn-rich phase and occurs at 227°C. During the eutectic reaction, the Cu_6Sn_5 phase grows as rod-like embedded in a continuous Sn-rich matrix [9,11]. Further, recent studies [12,13] of the directional solidification under unsteady-state conditions showed that eutectic Sn-Cu alloy can exhibit both dendritic structure and cell (eutectic colonies) structures, depending of the cooling rates, with primary Cu_6Sn_5 intermetallic particles dispersed in a Sn-rich matrix. Others works [7,8] also have described similar structures for Sn-0.7wt%Cu solder.

Despite presenting some advantages such as low melting point, the hypoeutectic Sn-40wt%Bi yet need to improve others important properties for application in the electronics industry as UTS-Ultimate Tensile Strength, YS-Yield Tensile Strength, Ductility, Hardness, Wettability, Electrical and Thermal conductivity, Corrosion and Fatigue Resistance. Under this context, few references of literature [14-17] indicate that additions of Cu on binary Sn-Bi solders may promote significant enhancements in the properties listed above, emerging as an alloy of high potential for soldering operations. Typical microstructures for ternary Sn-40wt%Bi-0.1wt%Cu solder were reported for Takao et al [17]. The microstructures are quite similar those obtained for Sn-40wt%Bi solders, but with a presence of primary Cu_6Sn_5 intermetallic particles both in the Sn-Bi eutectics and Sn-rich phase.

Therefore, this study focuses on the influences of the thermal parameters, growth rate – V_L and cooling rate – $\dot{T}_L$, in the evolution of secondary dendritic spacing (λ_2) in the Sn-40wt%Bi-0.7wt%Cu solder solidified under unsteady-state conditions, which in turn was correlated with Vickers hardness (HV) values along of the Sn-Bi-Cu alloy ingot. A Hall-Petch relation between HV and λ_2 was discussed.

EXPERIMENTAL PROCEDURE

The Sn-40wt%Bi-0.7wt%Cu solder alloy was prepared using 99.94% pure tin, 99.98% pure bismuth and 99.94% pure copper. Heat is directionally extracted only through a water-cooled bottom made of low carbon steel (SAE 1020), promoting vertical upward directional solidification. The casting assembly used in the solidification experiment is reported in previous works [18-20]. Continuous temperature measurements in the casting were performed during solidification by fine type J thermocouples (0.2 mm diameter wire sheathed in 1.0mm outside diameter stainless steel tubes) placed along the casting length.

The cylindrical ingot was sectioned along its vertical axis so that longitudinal and transverse samples could be obtained. Selected transverse (perpendicular to the growth direction) and longitudinal sections of the directionally solidified Sn-Bi-Cu casting at different positions from the metal/mold interface were polished and etched (solution of 2mL HCl, 10mL $FeCl_3$ and 100mL H_2O) for metallography. An optical image processing system Olympus, GX51 (Olympus Co., Japan) and Field Emission Gun (FEG) - Scanning Electron Microscope (SEM) Philips (XL30 FEG) were used to acquire the images.

The secondary dendritic arm spacing (λ_2) was measured on longitudinal section of the casting through of intercept method [21]. At least 35 measurements were performed for each selected position. X-ray diffraction (XRD) measurements were carried out with a view to determining the phases forming the Sn-40wt%Bi-0.7wt%Cu as-cast microstructure. XRD patterns were obtained with a 2-theta range from 20° to 90°, Cu-Kα radiation with a wavelength, λ, of 0.15406 nm.

Before hardness tests the specimen surfaces were polished with fine sandpaper to remove any machining marks, polished and etched (same etchant solution cited earlier). Hardness Vickers tests were performed (according to the ASTM E 384-11 standard) on the longitudinal sections of the samples by using a test load of 500g and a dwell time of 15s. The evaluated positions were 5.0mm, 15.0mm, 20.0mm, 25.0mm, 40.0mm, 50.0mm and 70.0mm from cooled surface of the casting. A Shimadzu HMV-G 20ST model hardness measuring tester was used. The average of at least 5 measurements on each sample was adopted as the hardness value of a representative position.

RESULTS AND DISCUSSIONS

Figure 1 shows the macrostructure of the Sn-40wt%Bi-0.7wt%Cu solder casting and typical longitudinal microstructures obtained after metallographic examination. Columnar grains prevailed along the directionally solidified casting, which means that vertically aligned grains have grown from the bottom of the casting. The dendritic growth prevailed along entire casting. It can be seen a gradual increase of secondary dendritic arm (λ_2) due to decreased of cooling rates ($\dot{T}_L$) and growth rate (V_L) in the progress of directional solidification.

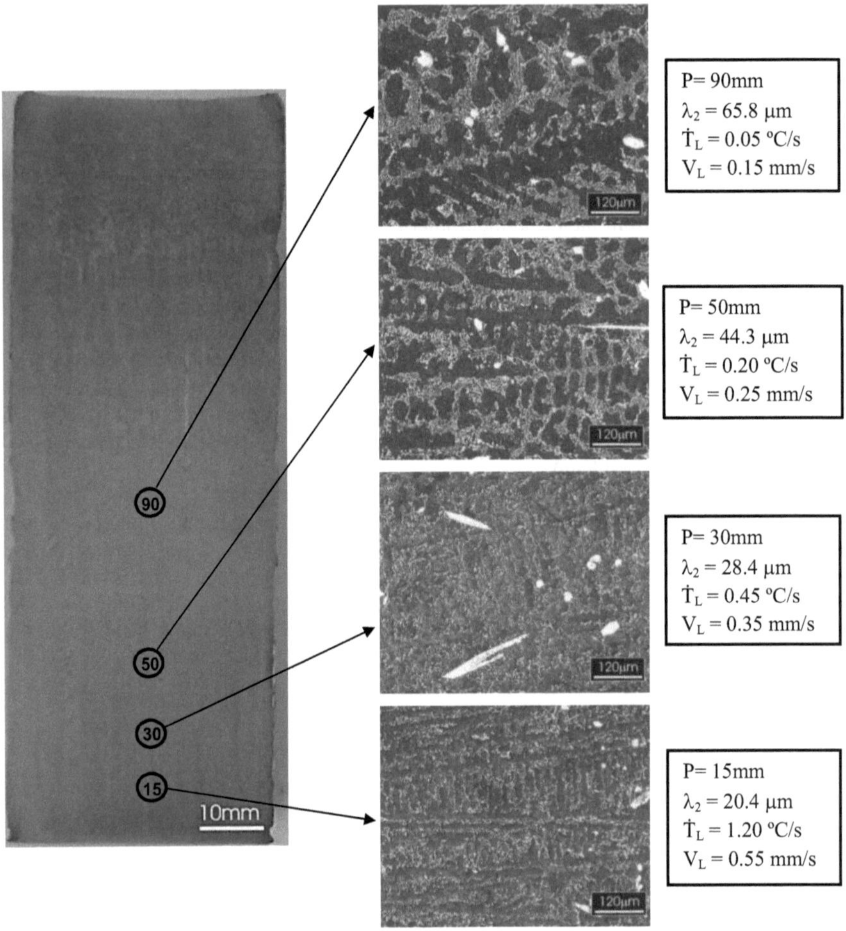

Figure 1. Macrostructure and longitudinal microstructures of the Sn-40wt%Bi-0.7wt%Cu alloy casting. "P" is the position from the metal/mold interface. The black arrows indicate the microstructures (longitudinal section) correspondents for each position observed.

Microstructures of the transverse section show the constituent phases of ternary Sn-Bi-Cu (Figure 2a). It was found that the as-cast microstructure was arranged by Sn-rich dendrites surrounded with a eutectic mixture (Bi-rich and Sn-rich phases). Further, the Sn-rich dendrites were found to be decorated with Bi precipitates in their own core. The primary Cu_6Sn_5 intermetallic particles (indicated by gray arrow) are non-homogeneously distributed along the microstructure. This result is also according with predicted by Takao et al [17]. Figure 2b shows the X-ray diffractograms for the ternary Sn-Bi-Cu solder and the presence of peaks associated with the Cu_6Sn_5 and Cu_3Sn intermetallics compounds (IMCs), Sn-rich and Bi-rich phases in the examined positions from the metal/mold interface. Although the Cu_3Sn particles could not be identified by microscopy, characteristic X-ray peaks were identified. Also, the

X-ray spectra show that the Sn-rich and the Bi-rich phases corresponding peaks' intensities are increased and decreased, respectively, as cooling rate is decreased.

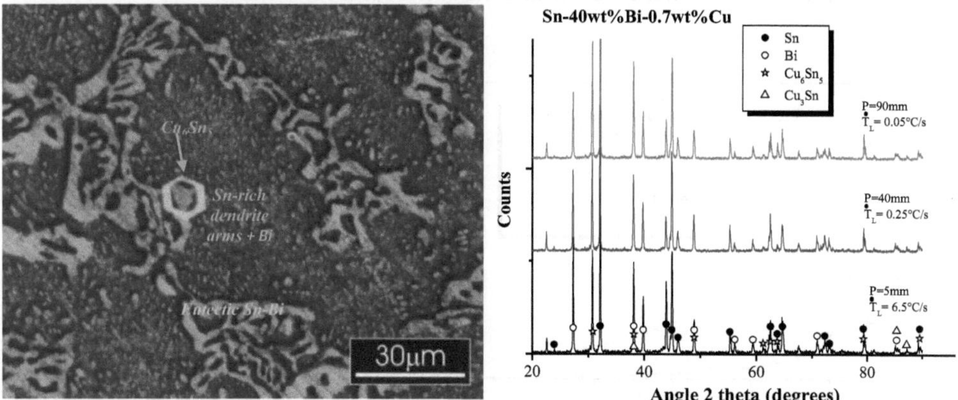

(a) (b)

Figure 2. (a) Microstructure emphasizing the phases formed in the Sn-40wt%Bi-0.7wt%Cu solder alloy; (b) X-ray diffraction (XRD) patterns of Sn-40wt%Bi-0.7wt%Cu solder samples for different positions along the directionally solidified casting length.

From the thermal readings during directional solidification is possible to obtain a power function, position from the metal/mold interface *vs* time of the *liquidus* isotherm (T_L=165.6°C) front passing by each thermocouple. The derivative of this function with respect to time gave values for the tip growth rate (V_L), as shown in Figure 3a. The experimental cooling rate (Fig. 3b) was determined by considering the thermal data recorded immediately after the passing of the *liquidus* front by each thermocouple. As the solidification front advances, the values of both V_L and $\dot{T}_L$ decrease. This effect is reversely translated resulting in increasing of secondary dendritic arm spacing (see Figure 1). Additionally, it can be observed that the V_L and $\dot{T}_L$ experimental values for eutectic Sn-0.7Cu and for the ternary Sn-Bi-Cu directionally solidified solders are similar, whereas the binary Sn-Bi alloy presented a range of values much greater than those cited. This fact indicates that the addition of 0.7Cu in the eutectic Sn-40Bi tends to decrease remarkably the metal/mold thermal conductance efficiency, which could be intimately related with interfacial conditions and affinity between molten alloys and substrate. In the case of the two binary alloys inserted for comparison purposes the water-cooled bottom was also made of low carbon steel (SAE 1020). The experimental fittings were inserted to indicate possible values considering positions which were not monitored within the castings.

The secondary dendritic arm (λ_2) dependence on the growth rate (V_L) for the ternary Sn-Bi-Cu alloy is shown in Figure 4, which present the average spacing and their minimum and maximum values. The line represents empirical power law which fit the experimental scatter. It can be seen that λ_2 variation with the tip growth rate (V_L) is characterized by -2/3 exponent, which show that the increased of V_L, λ_2 values decreases. This exponent seems to represent reasonably the Sn-40wt%Bi-0.7wt%Cu solder and it has been reported for similar correlations concerning Sn-Pb [22], Pb-Sb [20], Al-Fe [19] and Sn-Cu [12] alloys. Experimental model of Sn-40wt%Bi solder [4] also solidified under unsteady-state conditions was inserted in the graph (Figure 4), having a -2/3 exponent. Figure 4 shows that for a same V_L value, lower λ_2 measurements have been achieved for ternary Sn-Bi-Cu, indicating that the addition of cooper (Cu) may be beneficial for microstructure refinement.

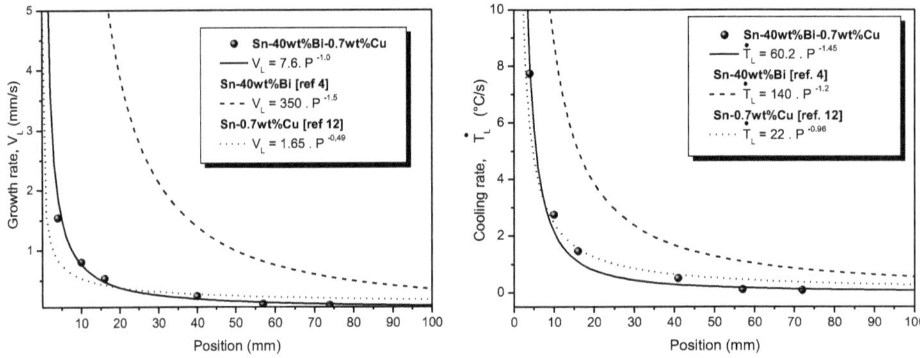

Figure 3. Experimental values of (a) growth rate and (b) cooling rate versus position along the casting length for the Sn-0.7wt%Cu, Sn-40wt%Bi and Sn-40wt%Bi-0.7wt%Cu solder alloys.

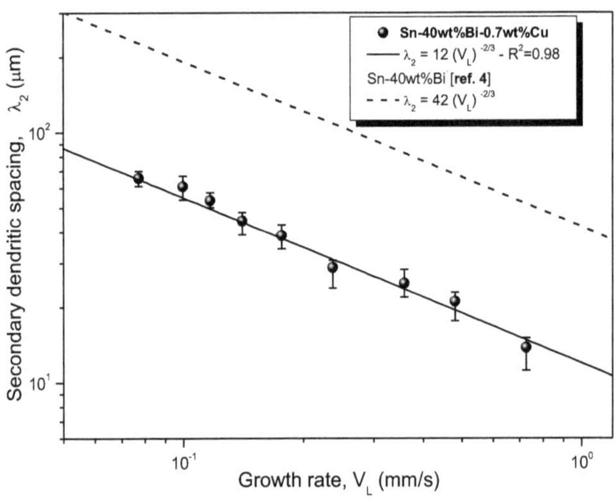

Figure 4. Secondary dendritic arm (λ_2) as a function of growth rate (V_L) for the directionally solidified Sn-40wt%Bi-0.7wt%Cu alloy casting. R^2 is the coefficient of determination.

It is known that the level of hardness is affected by phases and their size and distribution in the microstructure. Thus, Figure 5 shows a classical Hall-Petch relationship between hardness Vickers and the inverse square root of λ_2 obtained for the Sn-Bi-Cu alloy. The experimental expression $HV= 6.1 + 39.5\ \lambda_2^{-1/2}$ is able to represent the hardness behavior of the ternary Sn-40wt%Bi-0.7wt%Cu solder alloy with evolution of λ_2. It can be seen that the

HV measurements remarkably increase with decreasing secondary dendritic arm spacing. This is due to the better distribution of the reinforcing phases such as the eutectic Sn/Bi mixture associated with lower secondary dendrite arm spacing as can be seen in the microstructures inserted in Figure 5. Increase on hardness can also be attributed to smaller Bi precipitates within the dendritic matrix developed for positions near the cooled surface of the Sn-Bi-Cu alloy casting. Average hardness values about 19HV are associated with a mean λ_2=11μm while a value of 11.5HV refers to λ_2 of 61μm.

Hardness values found in this study are higher than those obtained for eutectic Sn-37%Pb solder [23], showing that Sn-40wt%Bi-0.7wt%Cu solder may be an alternative to replace the Sn-Pb alloys in applications that hardness is required.

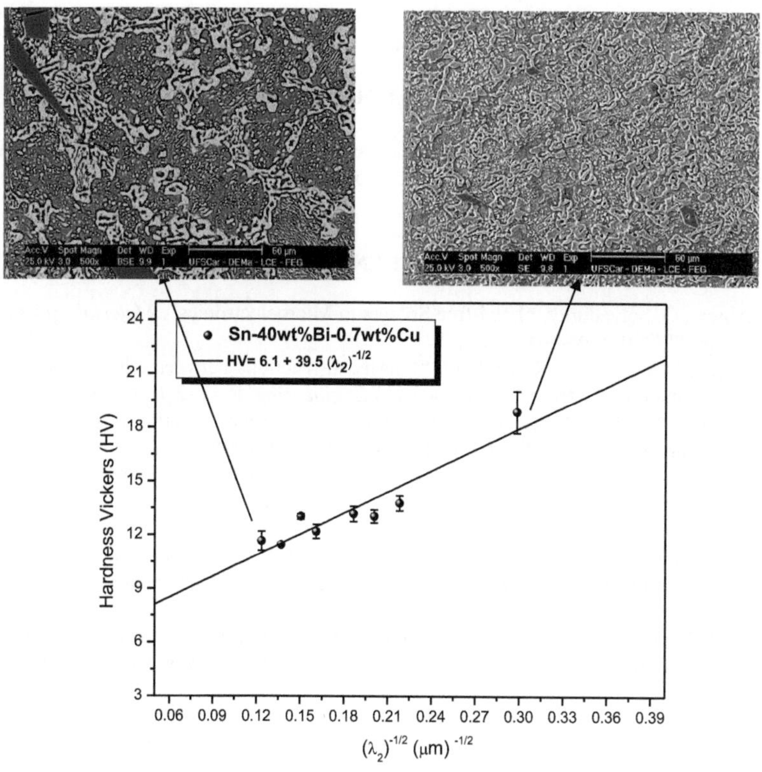

Figure 5. Hardness Vickers (HV) against (a) $\lambda_2^{-1/2}$ for Sn-40wt%Bi-0.7wt%Cu solder.

CONCLUSIONS

A microstructural dendritic matrix prevailed in the entire DS Sn-40wt%Bi-0.7wt%Cu alloy casting. The eutectic mixture located in the interdendritic regions is shown to be formed by Sn-rich and Bi-rich. Also, the dendrites were found to be decorated by Bi particles. Cu_6Sn_5 IMCs were seen non-homogeneously dispersed in both the eutectic Sn-Bi and the Sn-rich dendrites.

As the solidification front advances the experimental solidification cooling rate ($\dot{T}_L$) and growth rare (V_L) tend to shorten. The secondary dendritic arm spacing (λ_2) decreases as the growth rate V_L increases. Hardness changes remarkably along the Sn-Bi-Cu alloy casting, with higher values connected with the fineness of microstructure and of the eutectic mixture. Such refinement allows a homogeneous distribution of the eutectic and it is translated by the secondary dendritic spacing. It was found that a classical Hall-Petch correlation ($HV= 6.1 + 39.5\ \lambda_2^{-1/2}$) fit adequately the experimental scatter.

ACKNOWLEDGEMENTS

The authors acknowledge the financial support provided by FAPESP (São Paulo Research Foundation, Brazil: grants 2013/13030-5 and 2013/08259-3).

REFERENCES

[1] M. Abtew, G. Selvaduray, "Lead-free Solders in Microelectronics", *Materials Science and Engineering*, 27 (2000), 95-141.

[2] T Laurila, V Vuorinen, JK Kivilahti, "Interfacial reactions between lead-free solders and common base materials", *Materials Science and Engineering*, R 49 (2005), 1-60.

[3] N Pareck, "Parts and Packaging Program, United State America". *Nasa: Lead-free solders*, Goddard Space Flight Center, Greenbelt, Maryland, 1996.

[4] WR Osorio, LC Peixoto, LR Garcia, N Mangelinck-Noel, A Garcia, "Microstructure and mechanical properties of Sn-Bi, Sn-Ag and Sn-Zn lead-free solder alloys", *Journal of Alloys and Compounds*, 572 (2013), 97-106.

[5] Y Goh, ASMA Haseeb, MFM Sabri, "Effects of hydroquinone and gelatin on the electrodeposition of Sn-Bi low temperature Pb-free solder", *Electrochimica Acta*, 90 (2013) 265-273.

[6] F Zu, B Zhou, X Li, X Yi, Y Chen, Q Sun, "Effect of liquid-liquid structure transition on solidification of Sn-Bi alloys", *Transactions of Nonferrous Metals Society of China*, 17 (2007), 893-897.

[7] K. Nogita, J. Read, T. Nishimura, K. Sweatman, S. Suenaga and A. K. Dahle, "Microstructure control in Sn-0.7wt%Cu alloys", *Materials Transactions*, 46 (2005) 2419-2425.

[8] T Ventura, C Gourlay, K Nogita, T Nishimura, M Rappaz, AK Dahle, "The influence of 0-0.1wt% Ni on the microstructure and fluidity length of Sn-0.7Cu-xNi", *Journal of Electronic Materials*, 37 (2008), 32-39.

[9] K Nogita, "Stabilization of Cu6Sn5 by Ni in Sn-0.7Cu-0.05Ni lead-free solder alloys", *Intermetallics*, 18 (2010), 145-149.

[10] L Chen, P Septiwerdani, Z Chen, "Elastic modulus, hardness and creep performance of SnBi alloys using nanoindentation", *Materials Science & Engineering A*, 558 (2012), 253-258.

[11] AA El-Daly, AE Hammad, "Development of high strength Sn-0,7Cu solders with the addition of small amount of Ag and In", *Journal of Alloys and Compounds*, 509 (2011) 8554-8560.
[12] ITL Moura, CLM. Silva, N Cheung, P R Goulart, A Garcia, J E Spinelli, "Cellular to dendritic transition during transient solidification of a eutectic Sn-0.7wt%Cu solder alloy", *Materials Chemistry and Physics*, 132 (2012), 203-209.
[13] BL Silva, N Cheung, A Garcia, JE Spinelli, Thermal Parameters, Microstructure, and Mechanical Properties of Directionally Solidified Sn-0.7wt%Cu Solder Alloys Containing 0 ppm to 1000 ppm Ni, *Journal of Electronic Materials*, 42 (2013) 179-191.
[14] X Hu, K Li, Z Min, "Microstructure evolution and mechanical properties of Sn0.7Cu0.7Bi lead-free solders produced by directional solidification", *Journal of Alloys and Compounds*, 566 (2013) 239-245.
[15] J Xu, Q Hu, H He, F Zhang, Z Zhao, "Study of Sn-Bi-Cu Lead-free Solder". Proceedings of 10th Electronics Packaging Technology Conference, (2008) 1375-1380.
[16] L Zang, Z Yuan, H Zhao, X Zhang, "Wettability of molten Sn-Bi-Cu solder on Cu substrate", *Materials Letters*, 63 (2009) 2067-2069.
[17] H Takao, A Yamada, H Hasegawa, "Mechanical properties and solder joint reliability of low-melting Sn-Bi-Cu lead free solder alloy". R&D Review of Toyota CRDK, 39 (2004).
[18] MV Cante, JE Spinelli, IL Ferreira, N Cheung, A Garcia, "Microstructural development in Al-Ni alloys directionally solidified under unsteady-state conditions". *Metallurgical and Materials Transactions A*, 9 (2008), 1712-1726.
[19] PR Goulart, KS Cruz, JE Spinelli, IL Ferreira, N Cheung, A Garcia, "Cellular growth during transient directional solidification of hypoeutectic Al-Fe alloys", *Journal of alloys and compounds*, 470 (2009), 589-599.
[20] DM Rosa, JE Spinelli, IL Ferreira, N Cheung, A Garcia, "Cellular/dendritic transition and microstructure evolution during transient directional solidification of Pb-Sb alloys", *Metallurgical and Materials Transactions A*, 39A (2008), 2161-2174.
[21] DG McCartney, JD Hunt, "Measurements of cells and primary dendrite arm spacing in directionally solidified aluminium alloys", *Acta Metallurgica*, 29 (1981), 1851-1863.
[22] OL Rocha, CA Siqueira, CA Siqueira, A Garcia, "Cellular/dendritic transition during unsteady-state unidirectional solidification of Sn/Pb alloys", *Materials Science and Engineering A*, 347 (2003) 59-69.
[23] T Siewert, S Liu, DR Smith, JC Madeni, "Database for Solder Properties with Emphasis on New Lead-free Solder", *Colorado: National Institute of Standards and Technology & Colorado School of Mines*, 77p (2002).

Advances in the Science and Engineering of Casting Solidification
Edited by: Laurentiu Nastac, Baicheng Liu, Hasse Fredriksson, Jacques Lacaze, Chun-Pyo Hong, Adrian V. Catalina, Andreas Buhrig-Polaczek, Charles Monroe, Adrian S. Sabau, Roxana Elena Ligia Ruxanda, Alan Luo, Subhayu Sen, and Attila Diószegi
TMS (The Minerals, Metals & Materials Society), 2015

CONTRIBUTIONS ON OPTIMIZING APPROXIMATIONS IN THE STUDY OF MELTING AND SOLIDIFICATION PROCESSES THAT OCCUR IN PROCESSING BY ELECTRO-EROSION

F. L. Potra[1], T. Potra[1], V. F. Soporan[2]

[1] Technical University of Cluj-Npoca, Muncii Avenue, no. 103-105, post code 400641, city Cluj-Napoca, Province Cluj-Napoca, country Romania, email address vfsoporan@gmail.com

Keywords: Stefan Problem, Bessel Solution for Thermal Flux, Gauss Thermal Flux, Spline Thermal Flux, Cut of Triangular Fuzzy Numbers, Eigenvalues and Eigenvectors, Mean Values of Best Approximation

Abstract

We propose two optimization methods of the processes which appear in EDM (Electrical Discharge Machining). First refers to the introduction of a new function approximating the thermal flux energy in EDM machine. Classical researches approximate this energy with the Gauss' function. In the case of unconventional technology the Gauss' bell became null only for $r \to +\infty$, where r is the radius of crater produced by EDM. We introduce a cubic spline regression which descends to zero at the crater's boundary. In the second optimization we propose modifications in technologies' work regarding the displacement of the tool electrode to the piece electrode such that the material melting to be realized in optimal time and the feeding speed with dielectric liquid regarding the solidification of the expulsed material. This we realize using the FAHP algorithm based on the theory of eigenvalues and eigenvectors, which lead to mean values of best approximation. [6]
This optimization is presented in Figure 1 a) and 1 b).

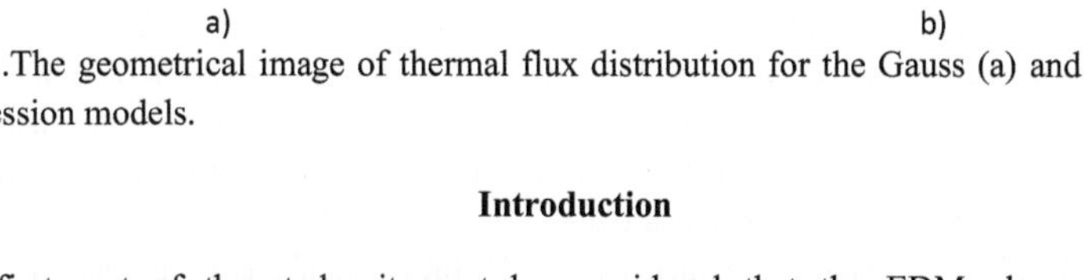

a) b)

Figure 1.The geometrical image of thermal flux distribution for the Gauss (a) and cubic Spline (b) regression models.

Introduction

In the first part of the study, it must be considered that the EDM phenomenon is an unconventional one, described by [μm] dimensions. Considering this, the boundary conditions must be established as correct as possible.

The thermal energy flux generated by the equipment used in practice is the subject of this study. But, the Gaussian flux has a significant contribution in conventional conditions, but in this case, the flux in almost 0 when approaching the proximity to the outer end of the boundary, even if the Gauss function is null in $+\infty$. This fact indicated the idea of approximating the thermal flux by a cubic spline regression, with dots on Gauss' bell and which is also null at the end of the boundary. A spline function was obtained in this way. This function is optimal approximating optimal the Gauss bell and is null at the end of the crater's radius Figure. 1 a) and 1 b).
In the second part of the study, fuzzy triangular numbers are used for studying melting and solidification of materials used in manufacturing of tool electrode in the electro erosion process. It is essential to know the behavior of the working electrode in the electro processing and form processing. This study aimed to address this on a number of six types of material, most commonly used as a result of experimental research. Thus Table 1 presents the characteristics of these materials and their explanations, followed by the algorithm FAHP (Fuzzy Analytic Hierarchy Process).[1-4]

Table 1. Essential Materials Characteristics Used in Electrode Tool Manufacturing

No.	Name of Material Characteristics	Attributed Fuzzy Number
CR_1	Copper-Graphite	$\tilde{9}_\alpha$
CR_2	Graphite	$1\tilde{1}_\alpha$
CR_3	Agietal	$\tilde{7}_\alpha$
CR_4	Tellurium-Copper	$\tilde{9}_\alpha$
CR_5	Wolfram-Copper	$1\tilde{3}_\alpha$
CR_6	Wolfram	$1\tilde{1}_\alpha$

An explanation of the contents of the characteristics shown in Table 1 is as follows:[7]

CR_1: "COPPER-GRAPHITE" is a material composed of copper and graphite. There are two ways of combining copper with graphite to get so called "copper-graphite electrode ". The first option is to make a mixture of 65% copper 35% graphite. In this case is obtained the most valuable electrode, but it has difficulties in manufacturing and involves a higher cost. A second embodiment of the electrode "copper-graphite" is carried out by coating a graphite electrode by a copper layer. Qualities are slightly lower, however, the electrode made of graphite coated with a layer of copper is less brittle and has a lower wear, and it has 10-15% higher yield than the electrode made of pure graphite.
CR_2: "GRAPHITE" - graphite electrode composed only. With respect to the electrode made only of copper (option which has not been selected in this study), it has a 50-70% higher productivity. In the EDM, the roughing shows very little wear due to the formation of a pyrographite protective film on the active surface. Although it is inferior in terms of quality, it is recommended due to lower manufacturing costs. However, the graphite electrode is not recommended in the processing of metal carbides.
CR3: "AGIETAL" is a material obtained by "Agie" firm and if has properties of similar to graphite electrode. In terms of its use, this electrode was designed to be used in semi-finishing

and finishing. This electrode has the additional ability to process parts made of steel.
CR_4: "TELLURIUM-COPPER". This electrode has the following qualities:

- the working electrode is best done pour this alloy,
- it can be processed easily by conventional methods such as turning, milling.

CR_5: "WOLFRAM-COPPER" – the tool electrode is made of two solid components in a proportion of 50-80% wolfram. The most numerous electrodes of this type are electrodes in a proportion of 75% wolfram and 25% copper. Although their manufacturing is via a high cost, it is recommended for its qualities on the implementation of EDM. Qualities for which this type of electrode is recommended are:

- it can be used in processing hardened steels and carbides;
- used for finishing, which has a high stiffness and leads to obtain fine roughness;
- the processing the deep holes of high precision;
- the processing of sharp angles, with wear 3-5 times lower than copper electrolyte.

CR_6: "WOLFRAM" although it is used to obtain filiform electrodes compared to massive electrodes, this electrode is characterized by:

- very little wear and the possibility of processing small sizes;
- processing parts with tolerance of 0,01 [mm].

Optimal Approach of Thermal Energy Flux by Cardinal Cubic Spline Model

Taking into account the definition of cardinal spline functions, given by:

$$\begin{cases} \varphi_1(x) = 1 - 9x_+^2 + 10x_+^3 - 16\left(x - \frac{1}{2}\right)_+^3 \\ \varphi_2(x) = x_+ - \frac{7}{2}x_+^2 + 3x_+^3 - 4\left(x - \frac{1}{2}\right)_+^3 \\ \varphi_3(x) = 12x_+^2 - 16x_+^3 + 32\left(x - \frac{1}{2}\right)_+^3 \\ \varphi_4(x) = -3x_+^2 + 6x_+^3 - 16\left(x - \frac{1}{2}\right)_+^3 \end{cases} \tag{1}$$

then the thermal energy flux produced by EDM process unit has the final form:

$$q_s(r) = 33 + 35\cdot 10^{-1}\cdot x_+ + \left(-297 - \frac{245}{2}\cdot 10^{-1} + 192\cdot 10^{-5} - 123\cdot 10^{-10}\right)\cdot x_+^2 + \left(330 + 105\cdot 10^{-1} - 256\cdot 10^{-5} + 246\cdot 10^{-10}\right)\cdot x_+^3 + \left(-528 - 140\cdot 10^{-1} + 512\cdot 10^{-5} - 656\cdot 10^{-10}\right)\left(x - \frac{1}{2}\right)_+^3 \tag{2}$$

where r is the radius of the crater produced by the EDM

The inhomogeneous Stefan problem it has been proposed a boundary free semiconoid leading to L* and M* integration constants, given in Table 2.

Table 2. Constants Involved in Plotting the Spline Type Crater

No.	Symbol	Name	Formula	Value/Material
1.	t_j	Unloading time $[\mu s]$	-	80.8
2.	I	Amperage $[A]$	-	6.2
3.	$q_s(r)$	Thermal flux $[W/m^2]$	$\frac{4.56 \cdot f_c \cdot U \cdot I}{\omega^2 \cdot \pi}$	$8.50278 \cdot 10^8$
4.	D_c	Crater diameter $[\mu m]$	Measured	43
5.	L^*	Integral differential equation in time variable	$1+\sqrt{1+\frac{(\mu_1)^2 \cdot t_f}{2 \cdot \alpha}}$	7860
6.	M^*	Bassel integral constant	$\frac{q_{\max}}{k \cdot u_0}$	$27 \cdot 10^4$

If the solution generally requires initial boundary conditions, Cauchy spline type and Stefan, the equation of the crater is obtained:

$$z = \frac{1}{M^*} \ln\left[\beta \cdot J_0(s \cdot \sqrt{x^2 + y^2})\right] \tag{3}$$

where J_0, is Bessel series, and:

$$s = \sqrt{L^{*2} + M^{*2}} \text{, and} \tag{4}$$

$$\beta = \frac{u_0}{G_L \cdot e^{-\alpha L^{*2} t_f}} \tag{5}$$

With these data the chart of the crater is shown in Figure 2.

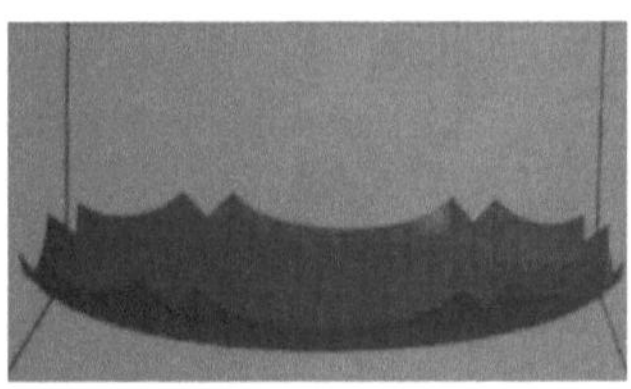

Figure 2. The geometric image of spline crater.

Conclusions:

The results obtained by this method were compared with their Gaussian functions determine if the relative experimental measurements of the EDM craters are closer to the spline case. However, the differences between the Gaussian case and the spline case are not very high, because spline was obtained using the interpolation points from Gauss Bell. These points were chosen because Gauss function has proven to be a good result for conventional processes.

Numerical Optimization Models of the EDM Process Through FAHP Algorithm

When using material characteristics proposed in Table 1, and also associate these features with fuzzy numbers, through the corresponding MOF matrices, the Flowchart hierarchy is:

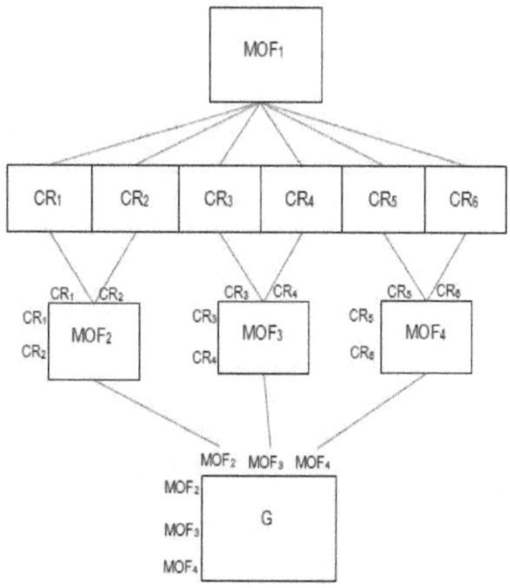

Figure 3. Flowchart hierarchy.

Based on this flowchart, MOF matrices are:

MOF₂=

	$\tilde{9}_\alpha$	$1\tilde{1}_\alpha$
$\tilde{9}_\alpha$	1	$1\tilde{1}_\alpha^{-1}$
$1\tilde{1}_\alpha$	$1\tilde{1}_\alpha$	1

MOF₃=

	$\tilde{7}_\alpha$	$\tilde{9}_\alpha$
$\tilde{7}_\alpha$	1	$\tilde{9}_\alpha^{-1}$
$\tilde{9}_\alpha$	$\tilde{9}_\alpha$	1

MOF₄=

	$1\tilde{3}_\alpha$	$1\tilde{1}_\alpha$
$1\tilde{3}_\alpha$	1	$1\tilde{3}_\alpha$
$1\tilde{1}_\alpha$	$1\tilde{3}_\alpha^{-1}$	1

Formulas for triangular fuzzy numbers assigned to the critical characteristics of EDM and, respectively, the formulas for fuzzy inverse numbers are:

$$\begin{aligned} \tilde{7}_\alpha &= (1-\beta)(5+2\alpha)+\beta(9-2\alpha) \\ \tilde{9}_\alpha &= (1-\beta)(7+2\alpha)+\beta(11-2\alpha) \\ 1\tilde{1}_\alpha &= (1-\beta)(9+2\alpha)+\beta(13-2\alpha) \\ 1\tilde{3}_\alpha &= (1-\beta)(11+2\alpha)+\beta(15-2\alpha) \end{aligned} \qquad \begin{aligned} \tilde{7}_\alpha^{-1} &= \frac{1-\beta}{9-2\alpha}+\frac{\beta}{5+2\alpha} \\ \tilde{9}_\alpha^{-1} &= \frac{1-\beta}{11-2\alpha}+\frac{\beta}{7+2\alpha} \\ 1\tilde{1}_\alpha^{-1} &= \frac{1-\beta}{13-2\alpha}+\frac{\beta}{9+2\alpha} \\ 1\tilde{3}_\alpha^{-1} &= \frac{1-\beta}{15-2\alpha}+\frac{\beta}{11+2\alpha} \end{aligned} \tag{6}$$

For chosen fuzzy numbers resulted in the following numerical values:

Table 3. Values of Defuzzyfication Fuzzy Numbers

	α	β	$\tilde{7}_\alpha$	$\tilde{9}_\alpha$	$1\tilde{1}_\alpha$	$1\tilde{3}_\alpha$
Moderate Case	$\frac{1}{2}$	$\frac{3}{4}$	7.5000	9.5000	11.5000	13.5000
Optimistic Case	$\frac{1}{4}$	$\frac{1}{2}$	7.0000	9.0000	11.0000	13.0000
Pessimistic Case	$\frac{3}{4}$	$\frac{1}{4}$	6.7500	8.7500	10.7500	12.7500

Table 4. Values of Defuzzyfication Inverse Fuzzy Numbers

	α	β	$\tilde{7}_\alpha^{-1}$	$\tilde{9}_\alpha^{-1}$	$1\tilde{1}_\alpha^{-1}$	$1\tilde{3}_\alpha^{-1}$
Moderate Case	$\frac{1}{2}$	$\frac{3}{4}$	0.1563	0.1188	0.0958	0.0804
Optimistic Case	$\frac{1}{4}$	$\frac{1}{2}$	0.1497	0.1143	0.0926	0.0780
Pessimistic Case	$\frac{3}{4}$	$\frac{1}{4}$	0.1385	0.1084	0.0890	0.0756

The first immersion of the Fuzzy number $\tilde{i}_\alpha$ in the interval means $\alpha \in [0,1]$ which has a probabilistic interpretation. If $\alpha \to 0$, $\alpha \in [0, \frac{1}{2})$, we require the projection to be achieved in a longer interval, that is we are in a pessimistic situation of achieving the aimed goals. If $\alpha \to 1$, $\alpha \in (\frac{1}{2}, 1]$, we are certain that the process is almost surely realized. If $\alpha = \frac{1}{2}$, the phenomenon is balanced. Consequently we have: $\alpha \in [0, \frac{1}{2})$, the phenomenon is pessimistic, $\alpha = \frac{1}{2}$, the phenomenon is moderate, $\alpha \in (\frac{1}{2}, 1]$, the phenomenon is optimistic. A similar study is reproduced in the relation with the convexity parameter β.

Global weights are obtained by the formula (7) and are given in the Table 5:

$$\begin{aligned} W_1 &= w_{41} \cdot w_{21} \\ W_2 &= w_{41} \cdot w_{22} \\ W_3 &= w_{41} \cdot w_{23} \\ W_4 &= w_{42} \cdot w_{31} \\ W_5 &= w_{42} \cdot w_{32} \end{aligned} \quad (7)$$

Calculation of overall weight is validated according to the flowchart:

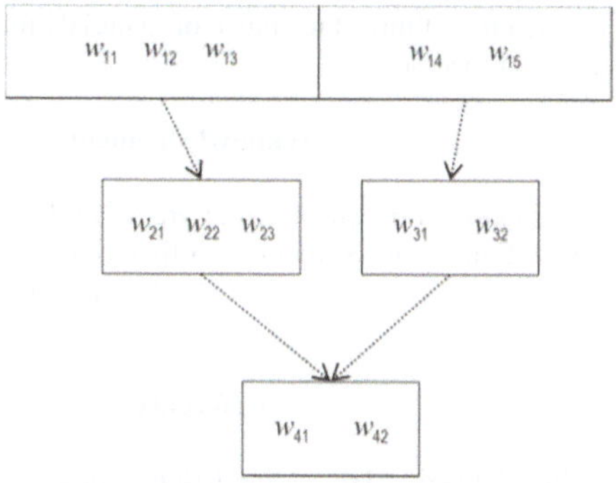

Table 5. The Values of Global Weights for All Three Cases

	W_1	W_2	W_3	W_4	W_5
Moderate Case	0.00222	0.05974	0.00964	0.85079	0.07761
Optimistic Case	0.00221	0.06004	0.00966	0.85004	0.07806
Pessimistic Case	0.00222	0.05975	0.00953	0.85107	0.07744

Histograms of global weights are shown in the following figure:

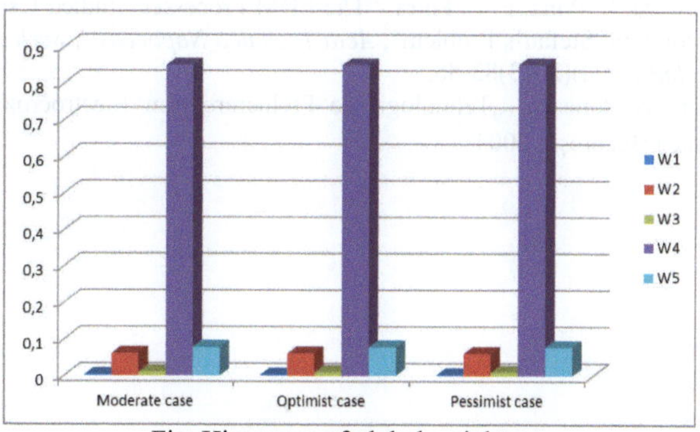

Fig. Histogram of global weights

Conclusion

Linguistic explanations based on experimental observations can not quantify the numerical importance of the contribution of different types of essential materials used in the construction of the tool electrode. FAHP approach is more than welcome. Because many of these materials are alloys, local weights can be used as such. Global weights represent a global behavior of this type of material that can be used to optimize the choice of materials for manufacturing tool electrode according to the piece to be processed.

Acknowledgement

This work was partially supported by the strategic grant POSDRU/159/1.5/S/137070 (2014) of the Ministry of National Education, Romania, co-financed by the European Social Fund – Investing in People, within the Sectorial Operational Program Human Resources Development 2007-2013.

References

1. Z. Ayan,, "A combined fuzzy AHP – simulation approach to CAD software selection", *International Journal of General System*, 39(7)(2010).
2. S.N. Joshi, S.S. Pande, "Thermo-Physical Modelling of Die-Sinking EDM Process", *Journal of Manufacturing Processes*, 12(2010), 45-56.
3. P. Kuppan, A. Rajadurai, and S. Narayanan, "Influence of EDM Process Parameters in Deep Hole Drilling of Inconel 718", *International Journal of Advance Manufacturing Technology*, 38(2007),74-84.
4. S. Mahendran, R. Devarajan, T. Nagarajan, A. Majdi, *Review of Micro-EDM, Proceedings of the International MultiConference of Engineering and Computer Science*, (2(2010), Hong-Kong, ISBN 978-988-18210-4-1, ISSN 2078-0966 (online));
5. A. Ozdagoglu, G. Ozdagoglu, "Comparison of AHP for the Multi-Criteria Decision Making Processes with Linguistic Evaluation", *Istambul Ticaret Universitesi Fen Bilimleri Dergisis Yil*, 2007.
6. G. Contiu, M. S. Popa, F. L. Potra, "The EDM Processes Studied Using the Cubic Spline of Regression type Stefan's Problem", *Acta Technica Napocensis, Series: Applied mathematics and Mechnics*, 5(I)(2013)83-88;
7. A. Vişan, N. Ionescu, „Tehnologii de Prelucrare prin Electroeroziune 1", Universitatea Politehnică Bucureşti, 2004

Advances in the Science and Engineering of Casting Solidification
Edited by: Laurentiu Nastac, Baicheng Liu, Hasse Fredriksson, Jacques Lacaze, Chun-Pyo Hong, Adrian V. Catalina, Andreas Buhrig-Polaczek, Charles Monroe, Adrian S. Sabau, Roxana Elena Ligia Ruxanda, Alan Luo, Subhayu Sen, and Attila Diószegi
TMS (The Minerals, Metals & Materials Society), 2015

INVESTIGATION OF THIN-WALLED IN718 CASTINGS BY COUNTER-GRAVITY INVESTMENT CASTING

Anping Dong[1,2], Naishun Yan[1], Jiao Zhang[1], Jun Wang[1,3], Baode Sun[1,3] Haiyan Gao[1], Da Shu[1]

[1]Shanghai Key Laboratory of Advanced High-temperature Materials and Precision Forming, Shanghai Jiao Tong University;

800 Dongchuan Rd.; Shanghai 200240, China

[2] Gas Turbine Research Institute, Shanghai Jiao Tong University;

800 Dongchuan Rd.; Shanghai 200240, China

[3] The State Key Laboratory of Metal Matrix Composites;

800 Dongchuan Rd.; Shanghai 200240, China

Keywords: Superalloys, In718, Counter-gravity, Investment casting, Pressure solidification

Abstract

Adjusted pressure casting is one kind of counter-gravity precision forming method in which the filling processes are at low counter-pressure while the solidification process are at high pressure. It aims at increasing the filling capacity for complex thin-walled castings and also provides an opportunity of introducing automatic control into the process by changing the pressure under which the metal is forced into the mold cavity. The filling capacity and solidification microstructure of complex thin-walled IN718 castings has been studied by experimental and numerical simulation. The results show that the liquid metal can flow through the mold cavity easily by imposing very low pressure. The average grain size of the 200mm×200mm×1mm thin walled IN718 casting piece varies from about 500μm to 800μm under different crystallization pressures. The average grain sizes are much smaller than those in the gravity casting process. Therefore, adjusted pressure casting is the promising technology for producing the complex thin-walled superalloy castings.

Introduction

With the rapid development of manufacturing technologies especially in the aerospace and automotive industries, more and more thin-walled casting parts need to be produced in a lightweight, holistic and precise way, with increasing attention and demand for complex thin-walled casting parts [1-4].

Filling capacity is one of very important factors in development of superalloys precision castings, especially for castings with complicated thin-walled structures [5]. Filling capacity has a significant impact on the solidification structure and performance of castings. The Laplace force and viscous force caused by the surface tension of the liquid metal would affect the flow state of the filling metal significantly, which would definitely make the filling process hard to predict. Moreover, the complex thin-walled structure increase the difficulty in controlling the distribution of filling pressure and flow path of liquid metal. Accordingly, it is essential to invent a new casting technology to produce high quality castings.

Adjusted pressure casting (APC) is an advanced technique developed based on the counter pressure casting, characterized by its higher filling capacity and better feeding capacity. It has been used in producing large scale, thin-walled and complicated nonferrous alloy castings in aviation, spaceflight, national defense and automobile industry. As for IN718 alloy, the properties and casting parameters such as the density, melting point and pouring temperature of it are quite different from Aluminum and magnesium. So that, IN718 needs greater pressure and higher temperature to ensure the cavity be completely filled. Although some people have proposed 1500~1620℃ ceramic core and shell techniques, it is still a challenge to find a suitable material for mold and rising tube for higher pouring temperature and greater filling pressure, which severely restricted the development of APC technology for superalloys.

In this paper, the APC equipment is set up and the APC process of IN718 thin-walled sheets are simulated with the commercial software ProCAST. Defects generated in the process are predicted and the effective method is put forward to solve these defects based on simulation results. To verify the simulation results, the modified process parameters are carried out by APC experiments. Finally, the microstructure of APC IN718 casts under different process parameters were studied.

IN718 APC Equipment and Casting Process

The Schematic diagram of APC equipment used in the present work is shown in Figure 1, and it includes upper and lower pressure chamber. Meanwhile, the mould was set in the upper pressure chamber and the induction heating coil was kept inside the lower pressure chamber.

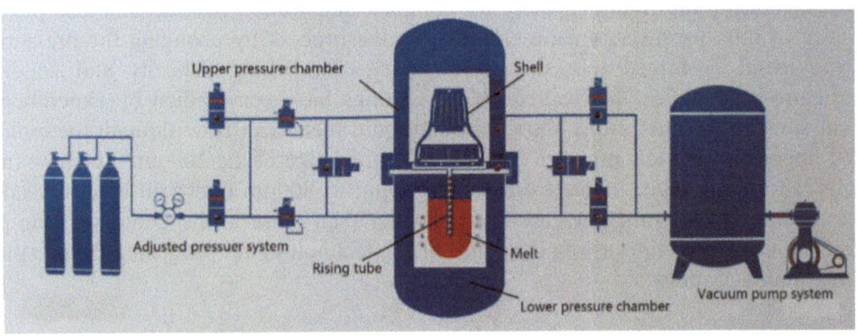

Figure 1. Schematic diagram of main part of APC equipment??

The whole counter-gravity APC process can be divided into 5 steps, i.e., vacuuming, mould filling, pressure increasing, pressure holding and pressure releasing, as shown in Figure 2. In the IN718 counter-gravity APC process, the IN718 alloy is melted by using vacuum induction melting in a refractory crucible, and the alloy is filled into the mold from the bottom by creating a vacuum in the mold cavity.

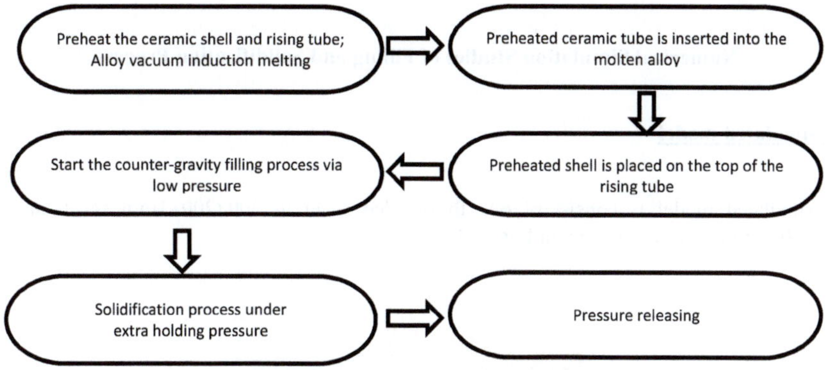

Figure 2. Flow chart of the APC process.

Before the filling process, the upper and lower chamber are evacuated to control gas of the cavity and liquid metal to reduce the porosities inside the casting. The lower chamber pressure increases to P2 value, which makes the liquid metal flow into the shell cavity along counter-gravity direction. When the filling process is finished, the mold is held under pressure several times higher than atmospheric pressure for a specific amount of time that depends on different casting structures and other casting parameters such as mold temperature prior to cast, alloy temperature, etc. During the holding pressure time interval, the ceramic rising tube is kept immersed into the crucible and the castings with in-gates are solidified. At the end of the APC process, the pressure on the upper chamber and lower chamber returned to atmospheric pressure value, which ensures the molten alloy in the mold center sprue back into the crucible. Figure 3 shows the pressure change in the upper chamber and lower chamber during the whole casting process.

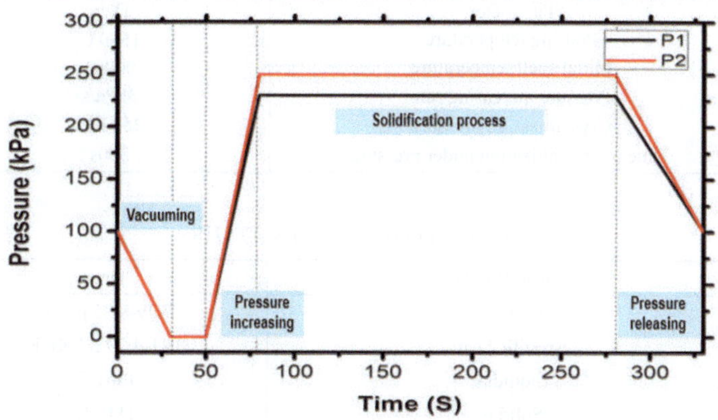

Figure 3. Pressure change in the upper chamber and lower chamber during the APC process. P1 is the upper chamber pressure and P2 is the lower chamber pressure.

Numerical Simulation Studies of Filling and Solidification Process

The simulated model

The simulated model is consist of one thin-walled sheet in 200×200×1mm sheet and two 200×200×2mm sheets, as shown in Figure 4.

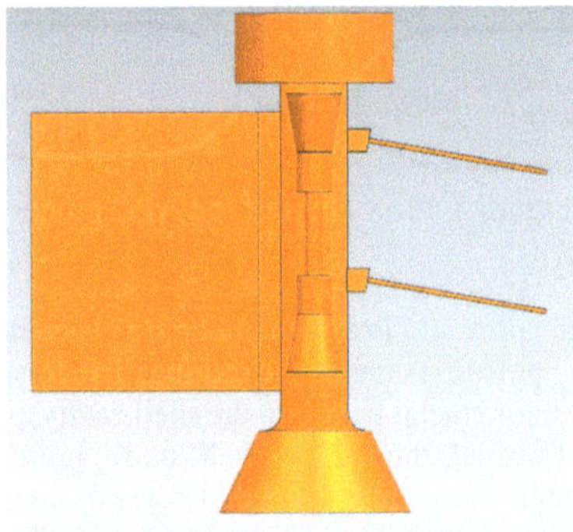

Figure 4. Schematic of filling sample

The casting parameters are used in APC process, as shown in Table I. The thermal and physical property data of IN718 are listed in Table II.

Table I. The APC parameters.

Parameter	Data
Pouring temperature	1560℃
Initial shell temperature	900°C
Pressure increasing rate	3kPa/s
Crystallization pressure	250 kPa
Time of crystallization under pressure	300s

Table II. Physical properties of IN718.

Physical Property	Data
Density	7.29-8.25 g/cm^3
Specific heat	0.41-4.29 KJ/Kg/K
Liquidus	1361°C
Solidus	1110°C
Latent heat	120 KJ/Kg
Thermal conductivity	12.28-28.47 W/m/K
Viscosity	0.0056-0.02 Pa·s

The filling and solidification simulation analysis

The numerical simulation has been carried out and the whole filling time is 18.21s. Figure 5 shows the temperature field distribution during filling process. At first, the liquid metal flows through the rising tube into the mold cavity, and then fills the sheets smoothly and steadily. At 10.9s, the liquid metal reaches the junctions between the rising tube and the casting, and starts to fill the bottom sheets at 12.31s, the temperature in front of the liquid melts is about 1460℃ at this moment. The liquid melts reaches the middle of the 1mm thick sheet at 14.38s, the temperature is about 1360℃. The liquid melts attains the top of the sheet at 18.21s and finishes the filling process completely.

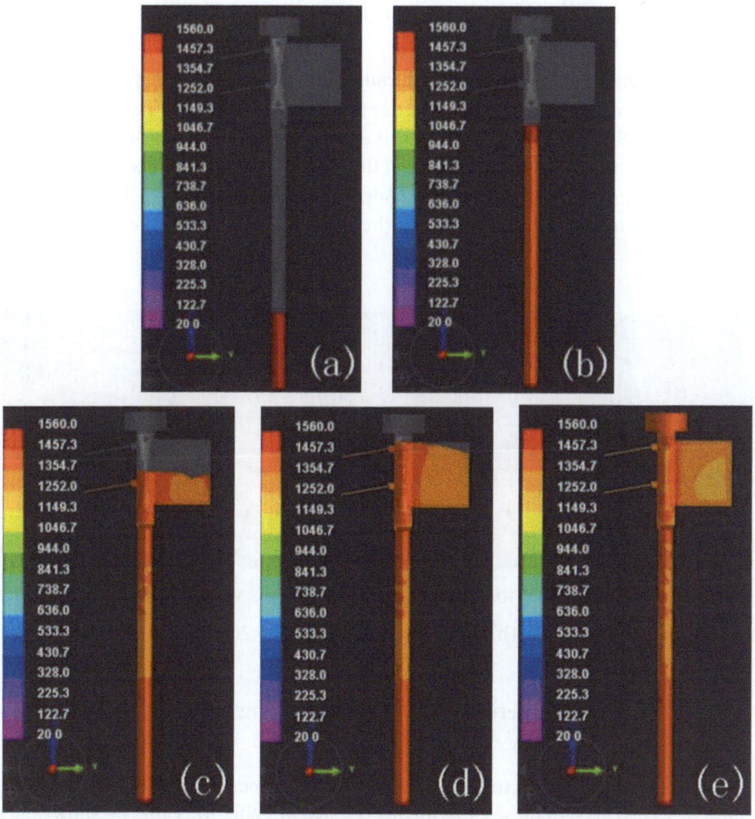

Figure 5. Temperature distribution during filling process

(a) t= 0.51s (b) t=10.90s (c) t=14.38s (d) t=16.10s (e) t=18.21s

When the liquid melt flows into the mold cavity of the casting, the metal in the front is totally in liquid. After it enters into the sheets, due to their high surface-to-volume ratios, the temperature of the melts decreases very fast. The solidification starts at the middle of 1mm thick sheet, which freeze faster than other thicker parts. Due to the higher cooling rate, the premature solidification in the middle region of sheet occurs and a critical concentration of solid forms as a slurry of crystals near the tip of the flowing stream. On the contrary, other thicker parts with relatively low cooling rate could be filled without defects formed.

Cold shuts occur when two fronts of liquid metal do not fuse properly in the mold cavity, leaving a weak spot. Both are caused by either a lack of fluidity in the molten metal or cross-sections that are too narrow. The reason leading to cold shut in this case is the relatively low temperature of liquid metal during the filling process of 1mm thick sheet. There has some methods to eliminate cold shuts defects on the sheet, such as increasing the pouring velocity, i.e., changing pressure increasing rate, and initial mold temperature [6]. The pressure increasing rate is increased from 3kPa/s to 5kPa/s and initial mold temperature is increased from 900℃ to 1000℃, and the other conditions are kept constant.

Figure 6 shows the temperature distribution during filling process of improved scheme. The whole filling time is 11.14s, which is about 7.07s shorter than original scheme。 The whole filling process of 1mm sheet samples is about 3.55s, which is 0.24s shorter than original scheme. When the melt starts to fill the bottom sheet, the temperature in the front of the liquid melts is above 1460℃, and, reaches the top of the 1mm thick sheet with a temperature of about1430℃, 100°C and 70℃ higher than original scheme respectively. Moreover, the metal in the front of the flow is still in liquid form.

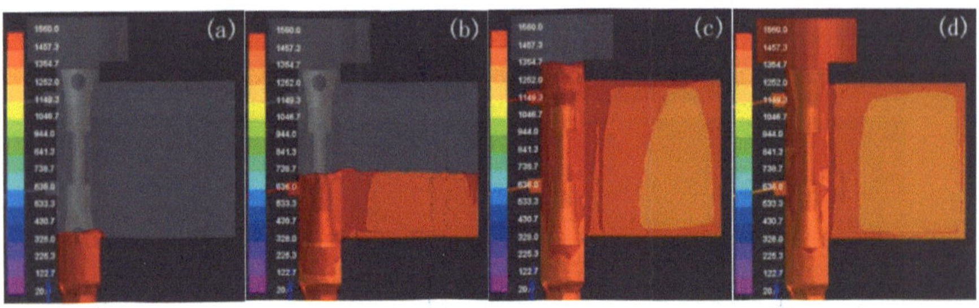

Figure 6. Temperature distribution during filling process with modified process parameters

(a) t= 7.59s (b) t=9.29s (c) t= 10.20s (d) t=11.14s

Experimental Results and Discussions

Several experiments were conducted according to the experimental procedure listed in Fig.2. The composition of IN718 used in these experiments is listed in Table III. Table IV shows the pressure parameters used in APC experiments, pouring temperature is 1560℃ and initial mold temperature is 1000℃, the pressure difference ΔP can be defined as:

$$\Delta P = P2 - P1.$$

Where $P1$ is the upper chamber pressure value, $P2$ is the lower chamber pressure value.

Figure 7 shows the 1mm thick sheet casted by using the modified parameters. It is can be seen that there is no cold shuts on the sheet, which is in good agreement with the simulation results and analysis above, indicating that increasing pressure increasing rate and initial mold temperature could minimize the cold shut of casting processed by APC.

Figure 7. Casting processed with the modified parameters

To investigate the relationship between the APC parameters and the microstructure of APC IN718 alloy castings, several experiments with different parameters were conducted. The microstructures were observed using an optical microscope and quantitatively analyzed by means of an image analyzer. Average grain size was obtained using the linear intercept method.

Table III. Chemical composition of experimental alloy

Element	Ni	Cr	Mo	Nb	Al	Ti	C	B	S	P	Fe
wt. %	52	18.4	3.05	5.3	0.55	1.05	0.05	0.0055	0.0045	0.0005	Bal.

Table IV. Experimental design of APC

Experiment No.	Pressure difference (kPa)	Crystallization pressure (kPa)	Time of holding pressure (s)
1	140	100	300
2	140	300	300
3	140	600	300

Figure 8 shows microstructure of the sheet with 1mm thickness under different crystallization pressure. It can be seen that the average grain size decreased with crystallization increased. During APC process, the crystallization pressure applied on molten metal during solidification process provide big driving force of molten metal among dendrite space. Therefore, the flow ability of molten metal is improved, which is beneficial for the increase of microstructure dense. With the driving force, molten metal flow can broke dendrites form continuous skeleton when their strength is relatively lower at the beginning of theirs formation, resulting the increase of crystal nucleus and the refinement of grain. At the meanwhile, with the increased flow ability, molten metal can flow through different parts of the sheet sample, which could promotes grain refinement.

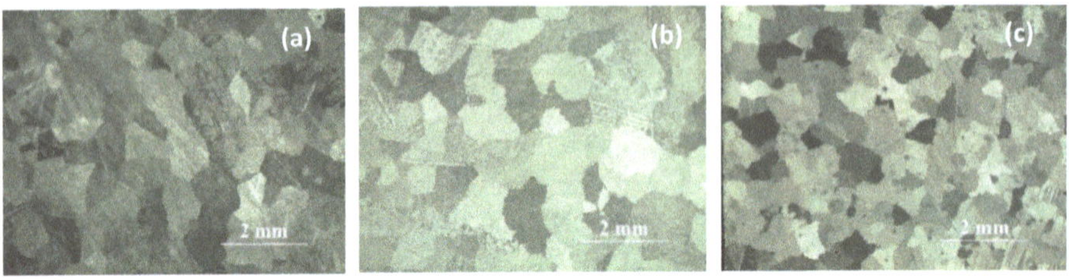

Figure 8. The relationship between the crystallization pressure and the average grain size of the IN718 sheet with 1mm in thickness. (a) 100kPa; (b) 300kPa; (c) 600kPa.

Conclusion

(1) IN718 superalloy can be casted by using the APC technology successfully.

(2) The casting defects can be eliminated when the initial shell temperature increases to 1000℃ and the pressure increasing rate is equal to 5kPa/s.

(3) The average grain size of APC IN718 alloy samples decreased with the increase of crystallization pressure.

Acknowledgement

The authors gratefully acknowledge the financial supports from National Natural Science Foundation of China (51204110) and New Teacher Project by Shanghai Jiao Tong University (12X100040087).

References

[1] Q. S. Yan, “Process and Theory of Vacuum Counter-pressure Casting Thin-wall Aluminum Alloy Castings Based on Intelligent control” (Ph.D. thesis, University of Science and Technology, 2006), 1-4.

[2] W.D. Huang, “Challenges and Countermeasures of Thermal Material Forming Technologies Raised by New Generation Aircrafts and Engines,” *Aeronautical Manufacturing Technology*. 10 (2004), 28-31.

[3] R.Z. Chen, L.B. Wang, J.H. Li. “Review and Prospect of the Development of Casting Superalloys,” *Journal of Aeronautical Materials*. 20 (2000), 55-61.

[4] X. Tang, et al, “Progress in the study of integral impeller casting technology for Superalloys,” Journal of Aeronautical Materials. 25 (2005), 57-62.

[5] Q. S. Yan et.al, “Effect of holding pressure on the microstructure of vacuum counter-pressure casting aluminum alloy,” *Journal of Alloys and Compounds*, 501(2), 352-357.

[6] Flemings, Merton C, "Solidification processing," Metallurgical Transactions, 5(10): 2121-2134.

Advances in the Science and Engineering of Casting Solidification
Edited by: Laurentiu Nastac, Baicheng Liu, Hasse Fredriksson, Jacques Lacaze, Chun-Pyo Hong, Adrian V. Catalina, Andreas Buhrig-Polaczek, Charles Monroe, Adrian S. Sabau, Roxana Elena Ligia Ruxanda, Alan Luo, Subhayu Sen, and Attila Diószegi
TMS (The Minerals, Metals & Materials Society), 2015

THERMAL TEST OF NODULAR CAST IRON COOLING STAVE

Jin Zheng[1], Haibin Zuo[1], Jianliang Zhang[2], Fengguang Li[2]

[1] State Key Laboratory of Advanced Metallurgy, University of Science and Technology Beijing, 30 Xueyuan Rd; Haidian District, Beijing, 100083, China
[2] School of Metallurgical and Ecological Engineering, University of Science and Technology Beijing, 30 Xueyuan Rd; Haidian District, Beijing, 100083, China

Keywords: Nodular cast iron, Cooling stave, Heat test, Blast furnace, Slag crust

Abstract

Nodular cast iron cooling stave is one of the cooling equipment which is widely used for the small and medium blast furnaces. The thermal resistance of cast iron cooling stave produced by lost foam process was analyzed with hot transfer theory and an experimental furnace was designed and constructed to test the using effect of the cooling stave. In condition of containing slag or not, cooling effect of the cooling stave under the condition of different temperature in furnace of 800°C, 900°C, 1,000°C and 1,100°C as well as different cooling water velocities of 0.5m/s, 1.0m/s, 1.5m/s and 2.0m/s were tested. The result indicates that the cast iron cooling stave produced by lost foam process has a good cooling ability and the temperature of the hot side of the cooling stave can be kept at a low level so that it could be used safely in blast furnace. The furnace temperature can bring a tremendous impact to cooling stave, and the slag can effectively reduce the temperature of the cooling wall. And it is not worthy to reduce the temperature of the cooling stave by increasing the water speed.

Introduction

At present, nodular cast iron cooling stave is widely used for the small and medium blast furnaces in bosh area. The area from lower stack to bosh is one of the limiting elements of the blast furnace service life. Therefore, the use of cast iron cooling stave directly influences the service life and stabilizes direct motion of the blast furnace [1]. Some researchers [2-6] have done numerical simulation to find effects of the different influence factors of the cooling stave, this paper verifies the validity of the results by thermal test. The nodular cast iron cooling stave which the test used was produced by lost foam process, the process can improve the quality of the cooling stave while reduce the cost of production. In order to test the thermal properties of the nodular cast iron cooling stave, a cooling stave thermal test furnace was designed and built, and two working conditions were simulated in this test furnace. In condition of containing slag or not, cooling effect of the cooling stave under the condition of different temperature in furnace of 800°C, 900°C, 1,000°C and 1,100°C as well as different water velocities of 0.5m/s, 1.0m/s, 1.5m/s and 2.0m/s were tested. Meanwhile, the hot transfer performances of the cooling stave under conditions were compared, make it clear that the slag plays an important role in reducing the hot face temperature of cooling stave.

Experimental Apparatus Description

This experiment was conducted under high temperature conditions to determine the cooling effect of nodular cast iron cooling stave in different conditions of gas temperature and water velocity. Specific steps are as follows: A cooling stave was installed on one side of the test furnace, two gasoil burners were used to heat face. In this experiment the temperature of the furnace was changed by adjusting the number of burner and combustion intensity (under 1,200°C). In this experiment water flow rate was changed by adjusting the pump pressure or the inlet valve, and water flow rate was measured using a flowmeter. Also, the temperature of the furnace gas and cooling stave was measured using thermocouples, and the inlet and outlet water temperature was measured using thermometers. Thermocouples are directly connected with the data acquisition system so that the experiment data can be collected in real time.

Introduction of the Thermal Test System

As shown in fig.1, the experimental setup is mainly composed of 3 parts: the thermal test furnace system, the water circulation system and information acquisition system. The test furnace system uses gasoil burner as the heater, and a layer of 200mm thick checker bricks was arranged above burner to stabilize the furnace temperature. The water loop circulation system supplies cooling water for cooling stave, and ensures that it is able to adjust flow and temperature. Then, a thermal resistance was installed in the inlet water tank to measure inlet water temperature, and electromagnetic flowmeters were installed in each water pipe to measure the inlet water flows. Outlet water temperature was also measured using thermal resistances. Each inlet pipe and outlet pipe are provided with a pressure transducer, which can record the inlet and outlet water pressure in real time.

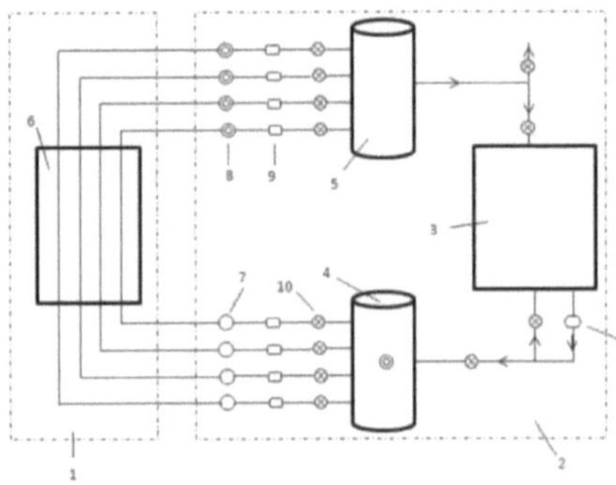

1-Thermal Test System; 2-Water Circulation System; 3-Cooling Water Tank; 4-Inlet Water Tank; 5-Outlet Water Tank; 6-Cooling Stave; 7-Electromagnetic Flowmeter; 8-Thermal Resistance; 9-Pressure Transmitter; 10-Flow Control Valve; 11-Pump

Fig.1 Thermal Test System for Cooling Stave

This data acquistion system use two sets of 32 digital channels to record data of the thermal resistances, thermocouples, electromagnetic flowmeters etc.

Experimental Cooling Stave

The size of this test cooling stave is measured at 1,680mm×800mm, and the hot face of the cooling stave was evenly provided with a dovetail groove which is 80mm wide and 100mm deep, the dovetail groove was filled with silicon-carbide refractory, so the total thickness of the cooling stave (including the dovetail groove) is 240mm. The cooling stave has four water outlets and four water entrances, and the outer diameter of the cooling water pipe is 60mm, while its thickness is 6mm. A 0.15mm thick coating layer were made on the outside of the water pipes.

The positions of the thermocouples inside the cooling stave were shown in Fig. 2.

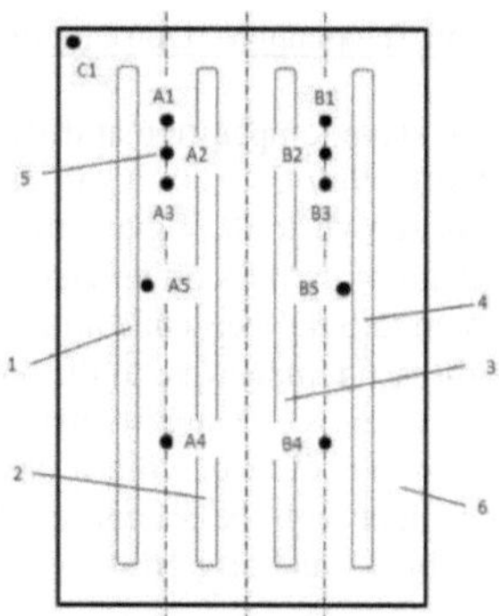

1-1# Cooling Water Pipe; 2-2# Cooling Water Pipe; 3-3# Cooling Water Pipe; 4-4# Cooling Water Pipe; 5-Thermal Couple; 6- Cooling Stave

Fig.2 Structure of the Experimental Cooling Stave and Location of the Thermal Couples

Experiment Program

The main purpose of the experiment is testing the hot transfer performance of nodular cast iron cooling stave under the conditions of different cooling water velocities and different temperatures, meanwhile its hot transfer performances with or without slag should also be tested. So there are two main parts in this experiment.

Constant Water Velocity and Different Furnace Temperature

In this part, the cooling water velocity was kept at 1.5m/s, at the same time, oil quantity of the burner was adjusted to make the furnace temperature stable for a period of time at a certain value, then the temperature data of the sensors was recorded. Then, oil quantity was adjusted and the furnace temperature got into next stage. The control value of temperature was shown in Table 1.

Table 1 Testing Program of Constant Water Velocity and Different Furnace Temperature

Water Velocity(m/s)	1.5			
Temperature(℃)	800	900	1,000	1,100

Constant Furnace Temperature and Different Water Velocity

In this part, the furnace temperature was kept at 1,000°C, at the same time the water velocity was adjusted to a large value, and then decreased the water velocity to a certain value and kept stable for a period of time according to Table 2, after the temperature data of sensors was recorded, water velocity was adjusted to next stage.

Table 2 Testing Program of Constant Furnace Temperature and Different Water Velocity

Temperature(℃)	1,000			
Water Velocity(m/s)	2.0	1.5	1.0	0.5

Results and Discussion

Influence of Furnace Temperature on Temperature Distribution of Cooling Stave

Figure 3-a shows the changes of the stave temperature at different depths under the condition of different furnace temperature of 800°C, 900°C, 1,000°C and 1,100°C when the water velocity is 1.5m/s, and has slag crust on it. As shown in the figure, maximum of the stave temperature is below 760°C, in other words, the highest permission temperature. It can be seen that the growth of temperature is slow from cold face to hot face, while a rapid increase appears close to the hot face. That means the effect of the furnace temperature to the hot face is more significant than that of the effect to the cold face. The references suggest that the temperature of the stave body increase linearly from the cold face to the hot face, while a rapid increase appears on the hot face rib [7], the present result is in agreement with the references. This may explain why the cooling stave is easy to be damaged without bricks and slag crust.

Figure 3-b shows the experiment result of the condition without slag. It can be seen that the temperature of the cooling stave still kept below 760°C. It can be concluded from Fig.3-a and Fig.3-b that the stave temperature with slag is far lower than that without slag. So the existence of slag can effectively reduce the temperature of the cooling stave.

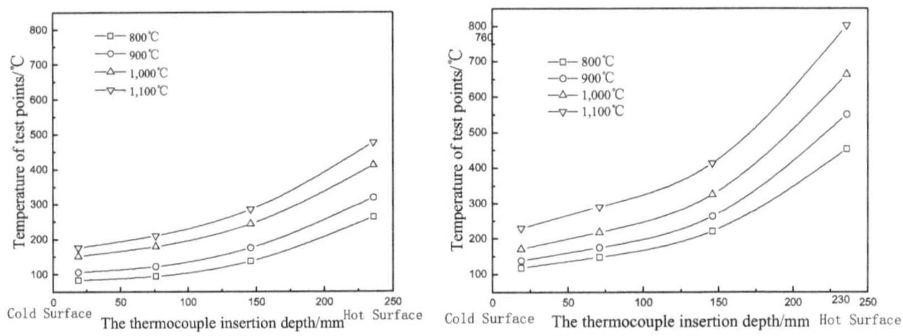

a- With Slag Crust b- Without Slag Crust

Fig.3 Influence of Furnace Temperature on Temperature Distribution of Cooling Stave

As is known to all, the thermal stress is one of the most important damage reasons of cast iron cooling stave. It can be obtained from the generalized Hooke's law of the thermal stress that the thermal stress is proportional to the stave temperature gradient under the condition of free expansion and shrinkage [8]. So, to simply calculating, the temperature difference between hot and cold face of the cooling stave was used to represent the thermal stress (see table 3). Table 3 shows that the temperature difference between the stave hot face and cold face without slag on its hot side is larger than with slag, and the increasing trend is bigger too. Conclude from the above analysis, keeping slag stable is very important to prolong the BF service life.

Table 3 Comparison of Temperature Difference between Hot and Cold Face

Furnace Temperature/°C	800	900	1,000	1,100
Temperature Difference(With Slag Crust)/°C	180	213	260	301
Temperature Difference(Without Slag Crust)/°C	335	412	492	573

Influence of Water Velocities on Temperature Distribution of Cooling Stave

Figure 4-a shows the changes of the stave temperature at different depths under the condition of different cooling water velocities of 0.5m/s, 1.0m/s, 1.5m/s and 2.0m/s when the furnace temperature is 1,000°C, and has slag crust on it. When the velocity of the cooling water increased from 0.5m/s to 2.0m/s, The temperature measured by thermocouple A1 (near cold face) decline from 161°C to 151°C, total 10°C decline; While the temperature measured by thermocouple A4 (near hot face) decline from 465°C to 445°C, total 20°C decline. The data suggests that when the cooling water velocity increases under the ordinary working conditions of the cooling stave, the stave temperature can be reduced by a certain extent, but not obvious. At the same time, the change of water velocity is more effectively to the hot face.

Figure 4-b shows the situation without slag. As can be seen from the figure, the effect is not so obvious, either.

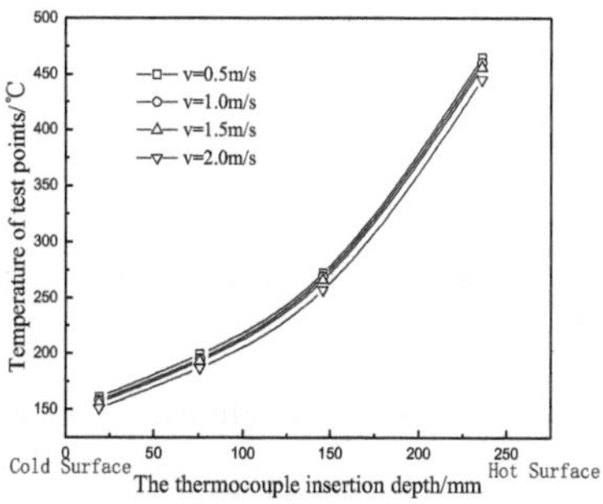

a- With Slag Crust

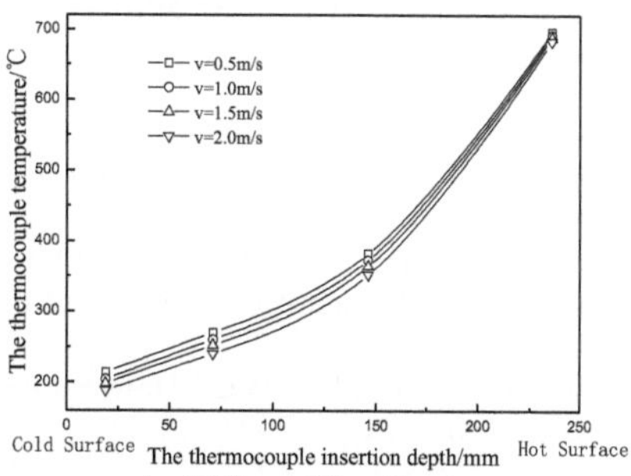

b- Without Slag Crust

Fig.4 Influence of Water Velocities on Temperature Distribution of Cooling Stave

The references suggest that when doubles the water velocity, the corresponding loss of resistance increases 3 times [9], so raising the velocity of the cooling water isn't an economical way to reduce the temperature of stave body.

Conclusion

(1) The results of the thermal test show that the tested nodular cast iron cooling stave can guarantee the temperature of stave body below the allowable using temperature in the normal operating condition with slag on its hot face.

(2) The temperature of the cooling stave rises along with the furnace temperature, and the stave temperature rising trend is sharper when the furnace temperature is higher. In addition, it's more effective on the hot face of the cooling stave. Therefore, the distribution of gas flow should be reasonably controlled during production to prolong the service life of the cooling stave.

(3) Under ordinary working conditions of the cooling stave, raising the velocity of the cooling water isn't an economical way to improve the cooling effect.

(4) The existence of slag can effectively reduce the temperature of the cooling stave body. In order to prolong the service life of the cooling stave, the thickness of slag should be reasonable controlled.

Acknowledgments

The authors would like to acknowledge the help of The School of Metallurgical and Ecological Engineering, University of Science and Technology Beijing, and Hebei Tianyu High-Tech Metallurgical Casting Co. Ltd. for providing experiment platform.

References

1. R. W. Steiger, R. E. Braun, "Utilization of computer analysis in blast furnace refractory lining and shell design", *Ironmaking Conf. Proc.*, 44 (1985), 485-504.

2. S. Chen et al., "Designing for long campaign life blast furnace(1) - the mathematical model of temperature field for blast furnace lining and cooling apparatus and new concept of long campaignship blast furnace cooler design", *J. Univ. Sci. Tech.*, 6 (3) (1999), 178-182.

3. Q. Xue et al., "Designing for long campaign life blast furnace(2) - the simulation of temperature field of lining and cooling apparatus", *J. Univ. Sci. Tech.*, 7 (1) (2000), 30-33.

4. L. J. Wu et al., "The study of cooling channel optimization in blast furnace cast steel stavebased on hot transfer analysis", *Int. J. Adv. Manufac. Tech.*, 29 (1-2) (2006), 64-69.

5. Z. Qian, Z. H. Du, L. J. Wu, "Transient hot transfer analysis of blast furnace cooling staves", *Steel Res. Int.*, 78 (1) (2007), 19-23.

6. S. Chen, Y. He, Q. Wu, "Temperature field computation of stave in blast furnace operation", *Iron Steel (Peking)* (in Chinese), 29 (1) (1994), 52-56.

7. L. Shi et al., "Thermal Test of buried pure copper tubular cooling stave", *Iron Steel (Peking)* (in Chinese), 26(03) (2007), 29-32.

8. W. T. Li, B. H. Huang, Z. B. Bi, *Theoretical analysis and application of thermal stress* (Beijing, BJ: China Electric Power Press (in Chinese), 2004), 59-60.

9. X. J. Ning, S. S. Cheng, N. Q. Xie, "Thermal state experiment and analysis of thin copper cooling stave", *Journal of University of Science and Technology Beijing* (in Chinese), 29(S2) (2007), 126-129.

AUTHOR INDEX
Advances in the Science and Engineering of Casting Solidification

SUBJECT INDEX
Advances in the Science and Engineering of Casting Solidification

MIX
Papier aus verantwortungsvollen Quellen
Paper from responsible sources
FSC® C105338

If you have any concerns about our products,
you can contact us on
ProductSafety@springernature.com

In case Publisher is established outside the EU,
the EU authorized representative is:
Springer Nature Customer Service Center GmbH
Europaplatz 3, 69115 Heidelberg, Germany

Printed by Libri Plureos GmbH
in Hamburg, Germany